CODES, GRAPHS, AND SYSTEMS

A Celebration of the Life and Career of G. David Forney, Jr. on the Occasion of his Sixtieth Birthday

THE KLUWER INTERNATIONAL SERIES IN ENGINEERING AND COMPUTER SCIENCE

COMMUNICATIONS AND INFORMATION THEORY

Consulting Editor

Robert Gallager

Other books in the series:

CODES, CURVES AND SIGNALS: ***Common Threads in Communications***, edited by Alexander Vardy; ISBN: 0-7923-8374-5

PERSPECTIVES IN SPREAD SPECTRUM, Amer A. Hassan, John E. Hershey, and Gary J. Saulnier; ISBN: 0-7923-8265-X

WIRELESS PERSONAL COMMUNICATIONS: ***Advances in Coverage and Capacity***, Jeffrey H. Reed, Theodore S. Rappaport, Brian D. Woerner; ISBN: 0-7923-9788-6

ASYMPTOTIC COMBINATORIAL CODING THEORY, Volodia Blinovsky; ISBN: 0-7923-9988-9

PERSONAL AND WIRELESS COMMUNICATIONS: ***Digital Technology and Standards,*** Kun Il Park; ISBN: 0-7923-9727-4

WIRELESS INFORMATION NETWORKS: ***Architecture, Resource Managment, and Mobile Data,*** Jack M. Holtzman; ISBN: 0-7923-9694-4

DIGITAL IMAGE COMPRESSION: ***Algorithms and Standards,*** Weidong Kou; ISBN: 0-7923-9626-X

CONTROL AND PERFORMANCE IN PACKET, CIRCUIT, AND ATM NETWORKS, XueDao Gu, Kazem Sohraby and Dhadesugoor R. Vaman; ISBN: 0-7923-9625-1

DISCRETE STOCHASTIC PROCESSES, Robert G. Gallager; ISBN: 0-7923-9583-2

WIRELESS PERSONAL COMMUNICATIONS: ***Research Developments,*** Brian D. Woerner, Theodore S. Rappaport and Jeffrey H. Reed; ISBN: 0-7923-9555-7

PLANNING AND ARCHITECTURAL DESIGN OF INTEGRATED SERVICES DIGITAL NETWORKS, A. Nejat Ince, Dag Wilhelmsen and Bülent Sankur; ISBN: 0-7923-9554-9

WIRELESS INFRARED COMMUNICATIONS, John R. Barry; ISBN: 0-7923-9476-3

COMMUNICATIONS AND CRYPTOGRAPHY: ***Two sides of One Tapestry,*** Richard E. Blahut, Daniel J. Costello, Jr., Ueli Maurer and Thomas Mittelholzer; ISBN: 0-7923-9469-0

WIRELESS AND MOBILE COMMUNICATIONS, Jack M. Holtzman and David J. Goodman; ISBN: 0-7923-9464-X

INTRODUCTION TO CONVOLUTIONAL CODES WITH APPLICATIONS, Ajay Dholakia; ISBN: 0-7923-9467-4

CODED-MODULATION TECHNIQUES FOR FADING CHANNELS, S. Hamidreza Jamali, and Tho Le-Ngoc; ISBN: 0-7923-9421-6

WIRELESS PERSONAL COMMUNICATIONS: ***Trends and Challenges,*** Theodore S. Rappaport, Brian D. Woerner, Jeffrey H. Reed; ISBN: 0-7923-9430-5

ELLIPTIC CURVE PUBLIC KEY CRYPTOSYSTEMS, Alfred Menezes; ISBN: 0-7923-9368-6

SATELLITE COMMUNICATIONS: Mobile and Fixed Services, Michael Miller, Branka Vucetic and Les Berry; ISBN: 0-7923-9333-3

WIRELESS COMMUNICATIONS: Future Directions, Jack M. Holtzman and David J. Goodman; ISBN: 0-7923-9316-3

DISCRETE-TIME MODELS FOR COMMUNICATION SYSTEMS INCLUDING ATM, Herwig Bruneel and Byung G. Kim; ISBN: 0-7923-9292-2

APPLICATIONS OF FINITE FIELDS, Alfred J. Menezes, Ian F. Blake, XuHong Gao, Ronald C. Mullin, Scott A. Vanstone, Tomik Yaghoobian; ISBN: 0-7923-9282-5

WIRELESS PERSONAL COMMUNICATIONS, Martin J. Feuerstein, Theodore S. Rappaport; ISBN: 0-7923-9280-9

SEQUENCE DETECTION FOR HIGH-DENSITY STORAGE CHANNEL, Jaekyun Moon, L. Richard Carley; ISBN: 0-7923-9264-7

DIGITAL SATELLITE COMMUNICATIONS SYSTEMS AND TECHNOLOGIES: Military and Civil Applications, A. Nejat Ince; ISBN: 0-7923-9254-X

IMAGE AND TEXT COMPRESSION, James A. Storer; ISBN: 0-7923-9243-4

VECTOR QUANTIZATION AND SIGNAL COMPRESSION, Allen Gersho, Robert M. Gray; ISBN: 0-7923-9181-0

CODES, GRAPHS, AND SYSTEMS

A Celebration of the Life and Career of G. David Forney, Jr. on the Occasion of his Sixtieth Birthday

edited by

Richard E. Blahut
Ralf Koetter

University of Illinois at Urbana-Champaign

SPRINGER SCIENCE+BUSINESS MEDIA, LLC

ISBN 978-1-4613-5292-1 ISBN 978-1-4615-0895-3 (eBook)
DOI 10.1007/978-1-4615-0895-3

Library of Congress Cataloging-in-Publication Data

A C.I.P. Catalogue record for this book is available
from the Library of Congress.

Printed on acid-free paper.

Contents

Part III Modems

Part IV Physics and Information Theory

Part V Lattices and Geometry

Part VI Behaviors and Codes on Graphs

This book is dedicated
to G. David Forney, Jr.

Contributing Authors

C. Berrou ENST-Bretagne Brest, France

R. E. Blahut University of Illinois, Urbana-Champaign.

I. F. Blake University of Toronto.

L. Brown Conexant.

H. A. Cabral University of Notre Dame.

T. Constantinescu University of Texas, Richardson.

D. J. Costello, Jr. University of Notre Dame.

M. V. Eyuboglu Airvana, Inc.

G. D. Forney, Jr. Massachusetts Institute of Technology.

M. P. C. Fossorier University of Hawaii at Manoa.

R. G. Gallager Massachusetts Institute of Technology.

J. Giles IBM T.J. Watson Research Center.

C. Gomes Cornell University.

B. Hajek University of Illinois, Urbana-Champaign.

P. A. Humblet Institute Eurecom, France.

R. Johannesson Lund University, Sweden.

J. Justesen Technical University of Denmark.

T. Kailath Stanford University.

D.-Y. Kim Airvana, Inc.

R. Koetter University of Illinois, Urbana-Champaign.

Y. Kou University of California, Davis.

S. Lin University of California, Davis.

H.-A. Loeliger ETH Zürich.

J. L. Massey Copenhagen, Denmark.

S. Mehrabanzad Airvana, Inc.

G. Nebe Universität Ulm, Germany

J. A. O'Sullivan Washington University.

K. G. Paterson Hewlett-Packard Laboratories.

H. V. Poor Princeton University.

G. Pottie University of California, Los Angeles.

E. M. Rains AT&T Shannon Laboratories

J. Rosenthal University of Notre Dame.

A. H. Sayed University of California, Los Angeles.

B. Selman Cornell University.

S. Shamai (Shitz) Israel Institute of Technology, Haifa, Israel

N. J. A.Sloane AT&T Shannon Laboratories

P. Ståhl Lund University, Sweden.

O. Y. Takeshita Ohio State University.

R. M. Tanner University of California, Santa Cruz.

V. Tarokh Massachusetts Institute of Technology.

G. Ungerboeck Broadcom Corporation.

S. Verdú Princeton University.

S. B. Wicker Cornell University.

X. Xie Cornell University.

Z. Zhao University of Illinois, Urbana-Champaign.

G. David Forney, Jr.

Foreword

G. David Forney, Jr. surprised many when, after completing his Sc.D. degree in Electrical Engineering at MIT in 1965, he joined the fledgling Codex Corporation in Watertown, MA. Dave had received considerable acclaim for his doctoral thesis, *Concatenated Codes* (which in addition to the innovation mentioned in its title had also introduced generalized minimum distance decoding and methods for erasure-and-error correction of Reed-Solomon and BCH codes) and could have had his pick of a prestigious university or industrial research laboratory.

Codex had been founded three years earlier to develop and market the "low-density parity-check (LDPC) codes" and "threshold decoders" that had been developed by Bob Gallager and Jim Massey, respectively, in previous MIT doctoral research. It says much about Dave that he accepted the financially inferior offer from Codex in order to be part of "the first organization dedicated solely to the practical application of information-theoretic research."[1] Other factors in Dave's decision were the persuasive charm of Arthur Kohlenberg, the chief technical officer at Codex and a past Editor of the *IEEE Transactions on Information Theory* (hereafter, *IT-Trans.*), and the opportunity to work closely with Gallager and Massey, who were then frequently consulting at Codex. But more than anything, it was Dave's conviction that he could make a real impact at a minuscule (twelve people) but promising start-up that led him to Codex.

Dave still recalls with pride that, in his first week at Codex, he reworked the threshold decoding equations in Codex's first product, the TD–12 error corrector, to save six flip-flops!—an important savings in those early days of discrete logical circuitry. He continued to work on burst-error-correcting convolutional codes and threshold decoders, which were at that time the coding systems most suited to practical applications and about which he and Kohlenberg wrote the paper "Convolutional coding for channels with memory" [*IT-Trans.*, 1986].

Subsequently Dave worked on coding for deep-space communications under contracts with Dale Lumb at the NASA Ames Research Center. After aggressively electing to use sequential decoding to maximize coding gain, Dave played a key role (including programming the on–site computer in assembly language to implement the Fano sequential–decoding algorithm) in realizing the convolutional coding system for the Pioneer 9 mission in 1968, which was the first coding system to operate in deep space. Two years later he designed and implemented a hardware sequential decoder with a clock speed of 15 MHz, a marvel of speed in its day.

Tragedy befell Codex in 1970 with the deaths of both Kohlenberg and James Cryer, Codex's president, from Hodgkins' disease and a heart attack, respectively. The company hovered near bankruptcy as expected defense orders for its coding systems were delayed or cancelled. Moreover, Codex' first commercial product, a 9600 bits/second modem developed by Jerry Holsinger, was not doing well in the field and Holsinger had left the company. Early in 1970, Dave took the lead in developing what became the first reliable 9600 bits/second modem, whose commercial success led to Codex's survival, and indeed to great growth and profitability in the Seventies. In 1976, the Codex design was adopted as the international 9600 bits/second modem standard.

Dave became an officer and Director of Codex in 1970, when the company was "refounded." In 1975, he became responsible for all its research, development and planning. He was deeply involved in the friendly acquisition of Codex by Motorola in 1977. He became a Vice President of Motorola and held several Motorola executive positions through 1986, when he returned to Codex (by then a division of Motorola). Subsequently he elected to become primarily technical once more, and remained so until his retirement from Motorola in 1999. Since 1996, he has been Bernard M. Gordon Adjunct Professor at MIT, where he teaches a course in data communications.

Even when deeply involved in practical and management issues, Dave kept returning to fundamental technical problems. In a contract report that he wrote in 1967 for the NASA Ames Research Center, Dave introduced "trellises" (coining this term in coding) to explain the Viterbi decoding algorithm, laying the groundwork for what has become a field in itself. Dave's paper "Convolutional codes I: Algebraic structure" [*IT-Trans.*, Nov. 1970] established much of the general theory of convolutional codes and is still mined for results; it was honored with the 1972 IEEE Browder J. Thompson Memorial Prize Award and an IT Society Golden Jubilee Paper Award in 1998. Dave's modem work inspired him to write "Maximum-likelihood sequence estimation of digital sequences in the presence of intersymbol interference" [*IT-Trans.*, May 1972], which has had an enormous impact on the entire field of statistical communication theory, and which also garnered an IT Society Golden Jubilee Paper Award in 1998. Dave's invited paper on "The Viterbi algorithm" [*Proceedings of the IEEE*, 1973] is a recognized classic for its simple and insightful treatment of this algorithm. "Minimal bases of rational vector spaces, with applications to multivariable linear systems" [*SIAM J. Control*, 1975] grew out of his fundamental work on convolutional codes, and has become a "citation classic" in the field of mathematical systems theory.

Dave's pen fell silent for about ten years (1976–1984) when he was heavily occupied with managerial responsibilities at Codex and Motorola. He awoke from this scientific hibernation with a vengeance and a rekindled interest in coding, stimulated initially by the invention of trellis-coded modulation. A good three-quarters of Dave's published papers have appeared since 1984. His invited paper "Coset codes— Part I: Introduction and geometrical classification," [*IT-Trans.*, Sept. 1988] proposed a unified lattice-based characterization of trellis codes, and was honored with the 1990 IEEE Donald G. Fink Prize Paper Award. There followed a long string of important papers that dealt with the structure of trellis codes and decoding procedures, connections with system theory, geometrical uniformity and group structure, codes on graphs, and more. This work gained impetus with the introduction of capacity-approaching codes such as turbo codes (and, ironically, the long-forgotten LDPC codes that Codex had hoped to exploit). As this part of Dave's work is well–treated in the articles within this book, no more will be said about it here.

Dave has a keen sense for the right mathematics to apply to a problem, and has often introduced new mathematical techniques to information theory. Occasionally he has committed mistakes, but those missteps have never seemed to bother him when they were pointed out by others—he has always focused much more on large truths than on small details.

Dave was born in New York City on March 6, 1940. He was awarded the B.S.E. degree *summa cum laude* from Princeton University in 1961, with a bachelor's thesis entitled "Shift-register synthesis using single-trigger flip-flops." Some believe that his manner still bears traces of his Ivy League background.

Dave has been an outstanding citizen of the IEEE Information Theory Society (hereafter, IT Society), which he served as President in 1992. His three-year (1970-1973) term as Editor of the *IT-Transactions* is legendary—he personally read and corrected every paper that appeared under his editorship. He also co-chaired the 50th-anniversary 1998 IEEE International Symposium on Information Theory. Dave has also participated actively in the U. S. National Academy of Engineering, to which he was elected in 1983.

Particularly in the last decade, Dave has been of immense assistance to young information theorists who seek his advice or help. He has been unstinting with his time and his knowledge. His comments on a draft paper often exceed its length, and he responds to e–mail messages with unbelievable alacrity. Dave also devotes considerable time to nontechnical *pro bono* activities. Perhaps closest to his heart are the Shady Hill

School and the Mount Auburn Hospital in Cambridge, on whose Boards of Trustees he has served for long terms with dedication.

Dave has two sons, Mark and William, as well as a married daughter, Jill Gates, who has presented him not only with a grandson but also with grandtriplets!

Dave's long list of scientific awards, in addition to those already mentioned, includes the 1970 IT Society Prize Paper Award, election as a Fellow of the IEEE in 1973, the 1992 IEEE Edison Medal, election as a Fellow of the American Association for the Advancement of Science in 1993, election as an Honorary Member of the Popov Society (Russia) in 1994, the 1995 IT Society Shannon Award, the 1996 Christopher Columbus International Communication Award, the 1997 Marconi International Fellowship, and election as a Fellow of the American Academy of Arts and Sciences in 1998.

It should be mentioned in closing that Dave's 1963 S.M. thesis at M.I.T. was on "The concepts of state and entropy in quantum mechanics," and that he has recently co-authored two papers in quantum information theory. Keep tuned for further developments!

James L. Massey

Notes

1. From Kohlenberg's obituary [*IT-Trans.*, March 1971].

Publications of G. David Forney, Jr.

1. G.D. Forney, Jr., "On decoding BCH codes," *IEEE Transactions on Information Theory*, vol. IT-11, pp. 549-557, Oct. 1965.

2. G.D. Forney, Jr., "Generalized minimum distance decoding," *IEEE Transactions on Information Theory*, vol. IT-12, pp. 125-131, Apr. 1966.

3. G.D. Forney, Jr., "Review of random tree codes," Appendix A of Final Report on Contract NAS2-3637, NASA CR73176, NASA Ames Res. Ctr., Calif., 1967

4. G.D. Forney, Jr., "Exponential error bounds for erasure, list, and decision feedback schemes," *IEEE Transactions on Information Theory*, vol. IT-14, pp. 206-220, Mar. 1968.

5. A. Kohlenberg and G.D. Forney, Jr., "Convolutional coding for channels with memory," *IEEE Transactions on Information Theory*, vol. IT-14, pp. 618-626, 1968.

6. G.D. Forney, Jr., "Coding and its application in space communications," IEEE Spectrum, vol. 7, pp. 47-58, 1970.

7. G.D. Forney, Jr., "Convolutional codes I: Algebraic structure," *IEEE Transactions on Information Theory*, vol. IT-16, pp. 720-738, Nov. 1970.

8. G.D. Forney, Jr., "Use of a sequential decoder to analyze convolutional code structure," *IEEE Transactions on Information Theory*, vol. IT-16, pp. 793-795, 1970.

9. G.D. Forney, Jr., "Burst-correcting codes for the classic bursty channel," *IEEE Transactions on Communications Technology*, vol. COM-19, pp. 772-781, Oct. 1971.

10. G.D. Forney, Jr., and E.K. Bower, "A high-speed sequential decoder: Prototype design and test," *IEEE Transactions on Communications Technology*, vol. COM-19, pp. 821-835, Oct. 1971.

11. G.D. Forney, Jr., "Maximum-likelihood sequence estimation of digital sequences in the presence of intersymbol interference," *IEEE Transactions on Information Theory*, vol. IT-18, pp. 363-378, May 1972.

12. G.D. Forney, Jr., "Lower bounds on error probability in the presence of large intersymbol interference," *IEEE Transactions on Communications Technology*, vol. COM-20, pp. 76-77, 1972.

13. G.D. Forney, Jr., "The Viterbi algorithm," Proceedings of the IEEE, vol. 61, pp. 268-278, 1973.

14. G.D. Forney, Jr., "Structural analysis of convolutional codes via dual codes," *IEEE Transactions on Information Theory*, vol. IT-19, pp. 512-518, July 1973.

15. G. D. Forney, Jr., "Convolutional codes II. Maximum-likelihood decoding," *Information and Control*, vol. 25, pp. 222-266, 1974.

16. G. D. Forney, Jr., "Convolutional codes III. Sequential decoding," *Information and Control*, vol 25, pp. 267-297, 1974.

17. G. D. Forney, Jr., "Minimal bases of rational vector spaces, with applications to multivariable linear systems," *SIAM Journal on Control*, vol. 13, pp. 493-520, 1975

18. G. D. Forney, Jr., R. G. Gallager, G. R. Lang, F. M. Longstaff and S. U. Qureshi, "Efficient modulation for band-limited channels," *IEEE Journal on Selected Areas in Communications*, vol. SAC-2, pp. 632-647, 1984.

19. G. D. Forney, Jr., "Coset codes Part I: Introduction and geometrical classification," *IEEE Transactions on Information Theory*, vol. 34, pp. 1123-1151, Sept. 1988.

20. G. D. Forney, Jr., "Coset codes Part II: Binary lattices and related codes," *IEEE Transactions on Information Theory*, vol. 34, pp. 1152-1187, Sept. 1988.

21. G. D. Forney, Jr., "Coset codes Part III: Ternary codes, lattices and trellis codes," *Proc. 1988 Beijing Intl. Workshop Inform. Theory*, Beijing, pp. E-6.1-5, July 1988.

22. G. D. Forney, Jr., "A bounded-distance decoding algorithm for the Leech lattice, with generalizations," *IEEE Transactions on Information Theory*, vol. 35, pp. 906-999, July 1989.

23. G. D. Forney, Jr. and L.-F. Wei, "Multidimensional constellations Part I: Introduction, figures of merit, and generalized cross constellations," *IEEE Journal on Selected Areas in Communications*, vol. 7, pp. 877-892, Aug. 1989.

24. G. D. Forney, Jr., "Multidimensional constellations Part II: Voronoi constellations," *IEEE Journal on Selected Areas in Communications*, vol. 7, pp. 941-958, Aug. 1989.

25. G. D. Forney, Jr. and A. R. Calderbank, "Coset codes for partial response channels; or, coset codes with spectral nulls," *IEEE Transactions on Information Theory*, vol. 35, pp. 925-943, 1989.

26. G. D. Forney, Jr., "Coded modulation for band-limited channels," Proc. IEEE Region 10 Colloquium, Oct. 1989. (Reprinted in IEEE Information Theory Society Newsletter, Dec. 1990.)

27. G. D. Forney, Jr., "Review of Sphere Packings, Lattices and Groups, by J. H. Conway and N. J. A. Sloane," *IEEE Transactions on Information Theory*, vol. 36, pp. 955-956, July 1990.

28. G. D. Forney, Jr., "Algebraic structure of convolutional codes, and algebraic system theory," in Mathematical System Theory (A. C. Antoulas, ed.), pp. 527-558. Berlin: Springer, 1991.

29. G. D. Forney, Jr., "Geometrically uniform codes," *IEEE Transactions on Information Theory*, vol. 37, pp. 1241-1260, Sept. 1991.

30. G. D. Forney, Jr., and M. V. Eyuboglu, "Combined equalization and coding using precoding," *IEEE Communications Magazine*, vol. 29, no. 12, pp. 25-34, Dec. 1991.

31. G. D. Forney, Jr., "Trellis shaping," *IEEE Transactions on Information Theory*, vol. 38, pp. 281-300, Mar. 1992.

32. M. V. Eyuboglu and G. D. Forney, Jr., "Trellis precoding: Combined coding, precoding and shaping for intersymbol interference channels," *IEEE Transactions on Information Theory*, vol. 38, pp. 301-314. Mar. 1992.

33. G. D. Forney, Jr., "Advances in modem technology beyond V.32/V.32bis," *Proceedings IEEE Data Transmission Conference*, London, Sept. 1992.

34. G. D. Forney, Jr., "On the duality of coding and quantizing," in *Coding and Quantization: DIMACS/IEEE Workshop* (Oct. 19-21, 1992) (R. Calderbank et al., eds.). pp. 1-14. Providence, RI: Am. Math. Soc., 1993.

35. G. D. Forney, Jr., N. J. A. Sloane and M. D. Trott, "The Nordstrom-Robinson code is the binary image of the octacode," in *Coding and Quantization: DIMACS/IEEE Workshop* (Oct. 19-21, 1992) (R. Calderbank et al., eds), pp. 19-26. Providence, RI: Am. Math. Soc., 1993.

36. G. D. Forney, Jr., "On the Hamming distance properties of group codes," *IEEE Transactions on Information Theory*, vol. 38, pp. 1797-1801, Nov. 1992.

37. M. V. Eyuboglu and G. D. Forney, Jr., "Lattice and trellis quantization with lattice- and trellis-bounded codebooks High-rate theory for memoryless sources," *IEEE Transactions on Information Theory*, vol. 39, pp. 46-59, Jan. 1993.

38. M. V. Eyuboglu, G. D. Forney, Jr., P. Dong and G. Long, "Advanced modulation techniques for V.fast," *European Transactions on Telecommunications and Related Technology.*, vol. 4, pp. 9-22, May 1993.

39. G. D. Forney, Jr., "Progress in geometrically uniform codes," in *Proceedings Sixth Joint Swedish- Russian International Workshop of Inform. Theory*, Mølle, Sweden, pp. 16-20, Aug. 1993.

40. G. D. Forney, Jr. and M. D. Trott, "The dynamics of group codes: State spaces, trellis diagrams, and canonical encoders," *IEEE Transactions on Information Theory*, vol. 39, pp. 1491-1513, Sept. 1993.

41. H.-A. Loeliger, G. D. Forney, Jr., T. Mittelholzer and M. D. Trott, "Minimality and observability of group systems," *Linear Algebra Appl.*, vol. 205-206, pp. 937-963, July 1994.

42. G. D. Forney, Jr., "Trellises old and new," in *Communications and Cryptography* (R. E. Blahut et al., eds.), pp. 115-128. Boston: Kluwer, 1994.

43. G. D. Forney, Jr., "Dimension-length profiles and trellis complexity of linear block codes," *IEEE Transactions on Information Theory*, vol. 40, pp. 1741-1752, Nov. 1994.

44. G. D. Forney, Jr., "Density-length profiles and trellis complexity of lattices," *IEEE Transactions on Information Theory*, vol. 40, pp. 1753-1772, Nov. 1994.

45. J. M. Cioffi, G. P. Dudevoir, M. V. Eyuboglu and G. D. Forney, Jr., "MMSE decision- feedback equalizers and coding Part I: Equalization results," *IEEE Transactions on Communications*, vol. 43, pp. 2582-2594, Oct. 1995.

46. J. M. Cioffi,G. P. Dudevoir, M. V. Eyuboiglu and G. D. Forney, Jr., "MMSE decision- feedback equalizers and coding Part II: Cod-

ing results," *IEEE Transactions on Communications*, vol. 43, pp. 2595-2604, Oct. 1995.

47. G. D. Forney, Jr., B. Marcus, N. T. Sindhushayana and M. D. Trott, "Multilingual dictionary: System theory, coding theory, symbolic dynamics and automata theory," in Different Aspects of Coding Theory (R. Calderbank, ed.), in *AMS Proc. Symp. Appl. Math.* Vol. 50, pp. 109-138, 1995.

48. G. D. Forney, Jr. and M. D. Trott, "Controllability, observability and duality in behavioral group systems," in *Proc. CDC '95*, pp. 3259-3264, New Orleans, Dec. 1995.

49. G. D. Forney, Jr., "Performance and complexity," Shannon Lecture in *Information Theory Society Newsletter*, pp. 3-4, 23-25, Mar. 1996.

50. G. D. Forney, Jr., "The forward-backward algorithm," in *Proc. 34th Allerton Conf. Comm. Ctrl. Comput.*, U. Ill. Urbana-Champaign, pp. 432-446, Oct. 1996.

51. J. Feigenbaum, G. D. Forney, Jr., B. H. Marcus, R. J. McEliece and A. Vardy, "Introduction to the special issue on codes and complexity," *IEEE Transactions on Information Theory*, vol. 42, pp. 1649-1659, Nov. 1996.

52. G. D. Forney, Jr., R. Johannesson and Z.-X. Wan, "Minimal and canonical rational generator matrices for convolutional codes," *IEEE Transactions on Information Theory*, vol. 42, pp. 1865-1880, Nov. 1996.

53. G. D. Forney, Jr., and A. Vardy, "Generalized minimum distance decoding of Euclidean-space codes and lattices," *IEEE Transactions on Information Theory*, vol. 42 pp. 1992-2026, Nov. 1996

54. G. D. Forney, Jr., L. Brown, M. V. Eyuboglu, and J. L. Moran III, "The V.34 high-speed modem standard," *IEEE Communications Magazine*, vol. 34, no. 12, pp. 28-33, Dec. 1996.

55. J. M. Cioffi and G. D. Forney, Jr., "Generalized decision-feedback equalization for packet transmission with ISI and Gaussian noise," in *Communications, Computation, Control and Signal Processing* (A. Paulraj et al., eds.), pp. 79-127. Boston: Kluwer, 1997.

56. G. D. Forney, Jr., "On iterative decoding and the two-way algorithm," in *Proc. Intl. Symp. Turbo Codes and Related Topics*, Brest, France, Sept. 1997.

57. G. D. Forney, Jr. and G. Ungerboeck, "Modulation and coding for linear Gaussian channels," *IEEE Transactions on Information Theory*, vol. 44, pp. 2384-2415, Oct. 1998.

58. G. D. Forney, Jr., "Transforms and groups," in *Codes, Curves and Signals: Common Threads in Communications* (A. Vardy, ed.), pp. 79-97. Boston: Kluwer, 1998.

59. A. R. Calderbank, G. D. Forney, Jr. and A. Vardy, "Classification of the 16-state tail-biting representations of the binary Golay code," in *Codes, Curves and Signals: Common Threads in Communications* (A. Vardy, ed.), pp. 127-153. Boston: Kluwer, 1998.

60. G. D. Forney, Jr., "Group codes and behaviors," in *Proc. MTNS '98* (Padova Italy). Zürich: Birkhauser, 1998.

61. A. R. Calderbank, G. D. Forney, Jr. and A. Vardy, "Minimal tail-biting trellises: The Golay code and more," *IEEE Transactions on Information Theory*, vol. 45, pp. 1435-1455, July 1999.

62. G. D. Forney, Jr., M. D. Trott and S.-Y. Chung, "Sphere-bound-achieving coset codes and multilevel coset codes," *IEEE Transactions on Information Theory*, vol. 46, pp. 820-850, May 2000.

63. B. J. Frey, R. Koetter, G. D. Forney, Jr., F. R. Kschischang, R. J. McEliece and D. A. Spielman, "Introduction to the special issue on codes on graphs and iterative algorithms," *IEEE Transactions on Information Theory*, vol. 47, pp. 493-497, Feb. 2001.

64. G. D. Forney, Jr., "Codes on graphs: Normal realizations," *IEEE Transactions on Information Theory*, vol. 47, pp. 520-548, Feb. 2001.

65. S.-Y. Chung, G. D. Forney, Jr., T. J. Richardson and R. Urbanke, "On the design of low-density parity-check codes with 0.0045 dB from the Shannon limit," *IEEE Communications Letters*, vol. 5, pp. 58-60, Feb. 2001.

66. G. D. Forney, Jr., R. Koetter, F. R. Kschischang and A. Reznik, "On the effective weights of pseudocodewords for codes defined on graphs with cycles," in *Codes, Systems and Graphical Models* (B. Marcus and J. Rosenthal, eds.), pp. 101-112. New York: Springer, 2001.

67. G. D. Forney, Jr., F. R. Kschischang, B. Marcus and S. Tuncel, "Iterative decoding of tail-biting trellises and connections with

symbolic dynamics," in *Codes, Systems and Graphical Models* (B. Marcus and J. Rosenthal, eds.), pp. 239-264. New York: Springer, 2001.

68. Y. Eldar and G. D. Forney, Jr., "On quantum detection and the square-root measurement," *IEEE Transactions on Information Theory*, vol. 47, pp. 858-872, Mar. 2001.

69. Y. Eldar and G. D. Forney, Jr., "Optimal tight frames and quantum measurement," to appear in *IEEE Transactions on Information Theory*, 2002.

70. A. Barg and G. D. Forney, Jr., "Random codes: Minimum distances and error exponents," submitted to *IEEE Transactions on Information Theory*, Sept. 2001.

71. G. D. Forney, Jr. and M. D. Trott, "The dynamics of group codes: Dual abelian group codes and systems," to appear in *IEEE Transactions on Information Theory.*

I

CONVOLUTIONAL CODES, TURBO CODES AND OTHER WORDS OF WISDOM

Chapter 1

THE MUTATIONS OF CONVOLUTIONAL CODING (AROUND THE TRELLIS)

C. Berrou
ENST-Bretagne
Brest, France

Abstract In years past, it was somewhat difficult to make up one's mind about the multiple types of error correcting coding schemes, which were categorized mainly into convolutional or block codes, with many variations in each family. Are there such great differences between all these codes? Without giving formal answers, this short paper translates a personal opinion that was initially conceived thanks to the ideas of Dave Forney.

As I was trying to find my way as a researcher at the *ENST de Bretagne*, I decided to take an interest in error correcting coding and the only serious track I had was the well-known quip: "Some codes allow us to make good communications at conferences while others make good communications in real systems". I was advised to take an interest in the codes which make good communications possible...

Belonging to that underprivileged kind of menial researcher who needs a visual representation in order to get an idea, I was looking for pictures and in the literature I found various forms of convolutional encoders, which puzzled me. It seemed to me, from comparing one with another, that genetic transformations constantly affected that kind of digital machine, whereas their principle should have been, on the whole, very easy to state. Like many others, I thought that convolutional coding had been devised by Peter Elias [1] as a sliding window Hamming code and that it would remain in its initial shape until the end of time. On the other hand, the great variety of structures displayed for this coding proved that it had not been completely treated. For example, why did Gottfried Ungerboeck [2] introduce recursivity in his coding registers while there had scarcely been any question of that before?

"Tell me how you decode and I'll be able to understand the code." When you have no particular gift for algebra, when for instance Galois fields—despite any sympathy you might have for this French revolutionary who fought a fatal duel aged 21—taste like a bitter potion, then think about the decoding side before the encoding one (that was actually the approach in the case of turbo codes [3]). Indeed, for those who are more comfortable with physics than with mathematics, decoding algorithms are more accessible than coding constructions, and help to understand them. From this standpoint, the visual and physical representation that Dave Forney gave of the decoding of convolutional codes—the trellis—was, for most of us, decisive for understanding the decoding processes (Viterbi, APP, ...) and allowed unification in the approach to convolutional codes, basically that of sequential machines with a redundant output.

What I think I know today about convolutional coding is almost entirely deduced from Figure 7 in Forney's "The Viterbi Algorithm" [4]: two simple drawings for rapid comprehension! I would say that, in the most general way:

1. A convolutional encoder is a sequential machine with ν binary memory elements (not necessarily a shift register). It can be a nonlinear device.

2. There can be any number of inputs (or stimuli), but in practice this number is limited to ν, in order to avoid parallel transitions between states.

3. The machine can have any initial state (not necessarily 0, see tail-biting or circular coding).

4. The redundant outputs of the machine can be expressed in any way, depending on the current state, possibly on inputs, or on the final state only, or on any other "photograph" of the behavior of the machine during the encoding process. The performance of the code is both linked to the connectivity of the graph and to the way of extracting redundancy from the temporal evolution of the machine.

In reality, one can see that this definition includes most error-correcting codes, binary or nonbinary: convolutional, BCH, Reed-Solomon, turbo codes, etc.! They are only distinguished by the type of graph and the way in which "photographs" are used to construct redundancy. Note that all these codes, except classical convolutional codes, are systematic

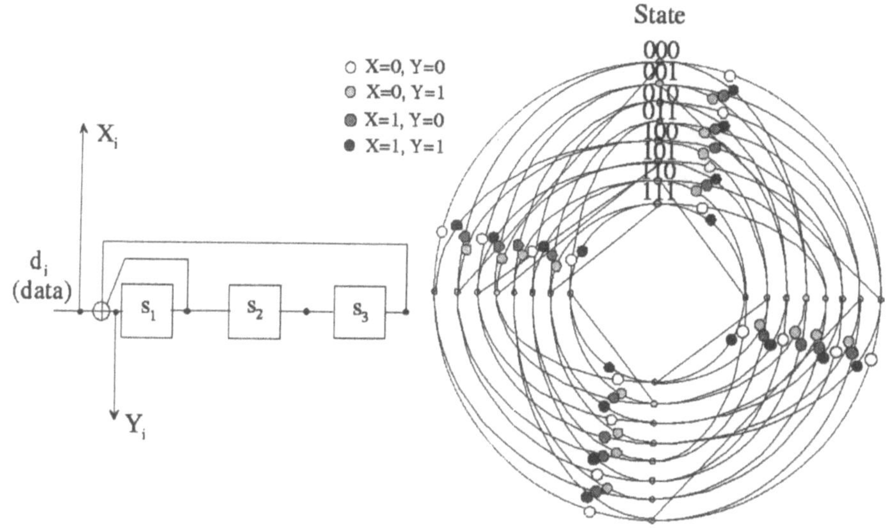

d_i	extended Hamming	Y_i	d_i	extended Hamming	Y_i
0000	0000	0000	1111	1111	1111
0001	0111	1011	1110	1000	0100
0010	1101	0111	1101	0010	1000
0011	1010	1100	1100	0101	0011
0100	1110	1110	1011	0001	0001
0101	1001	0101	1010	0110	1010
0110	0011	1001	1001	1100	0110
0111	0100	0010	1000	1011	1101

Figure 1 Constructing a (8,4,4) code with a circular (tail-biting) Recursive Systematic Convolutional code (polynomials 15,10 in octal) and rate 1/2. Each codeword is made up of 4 bits for the systematic part ($X_i = d_i$) and 4 bits for the redundant part (Y_i). It gives the same minimum distance as the perfect extended Hamming code.

and use recursive generators. For instance, Recursive Systematic Convolutional (RSC) codes and BCH codes are twins: they use the same type of sequential machine (a feedback register) but they differ in the number of states (small or large) and in the extraction of redundancy (continuously for a RSC code, at the end of the process for a BCH code). Nevertheless, in academic textbooks, they are presented in separate chapters as very different species: "block" codes and convolutional codes (a tail-biting convolutional code, as will be mentioned just below, is however a perfect block code). Even Hamming codes may be specified as sequential (convolutional-like) machines. Let us for instance consider the small circular trellis shown in Figure 1. It corresponds to the coding of a block of $k = 4$ information bits by a regular RSC encoder with eight states (polynomials 15, 10 in octal). The minimum distance of the code, given by the table, is 4, i.e., the same as that of the perfect extended Hamming code $(8, 4, 4)$. Note also that, thanks to the tail-biting principle, convolutional codes have become quasi-cyclic codes since the cyclic shifting of the systematic part leads to the same shifting for the redundancy part. What a surprising evolution for convolutional codes! Other constructions, based on the same principle, may be devised to obtain perfect or nearly perfect codes. The extended Golay code $(24, 12, 8)$ can thus be represented by a circular sixteen-state trellis [5], which is an even more surprising result than the example of Figure 1.

The examples given above are codes where the number of states is comparable to the number k of encoded information bits. When k is larger, from a few hundred to a few thousand for practical applications, obtaining perfect or nearly perfect codes is apparently possible on condition that you increase the number of states to the same level as k (assuming some scaling factor property). But the decoding complexity is no longer acceptable. A well-known trick today involves coupling at least two small component codes and decoding them with an iterative method. When the number of coupled codes, that is the dimension of the composite code, through judiciously chosen permutations, reaches four or five, the minimum distances obtained are comparable to those of random codes [6]. A daring analogy with this kind of coupling is to be found in quantum physics. Whereas the possible number of states for an electron of a isolated atom (another circular structure!) is a few units, these same atoms coupled in a crystal offer the electrons a much higher number of possible states, even a *continuum* of states (in an unlimited volume). Turbo codes and low density parity check (LDPC) codes [7] are convincing examples of this approach. In passing, note that LDPC codes can be viewed, with no great stretch of the imagination, as a very large number of small circular trellises, comparable to that of Figure 1, each one yielding one check bit.

Historically speaking, what differentiated all these codes was the mathematical approach to their design (when there was one), but when it comes down to it, the encoders are very similar, based on the temporal evolution of a finite-state machine. A more suitable terminology to distinguish these codes would refer to the types of decoding algorithms: algebraic, probabilistic, or iterative.

Shall we say, in conclusion, that the boundaries between convolutional codes and others are arbitrary? The main thing that has been elaborated until today, at various times and using different concepts, seems in fact to represent more or less original variations on a feedback register. In the very near future, most codes will perhaps meet around the same trellis (again, thanks to Dave Forney for one of the most beautiful drawings he ever made) and will enable good communications to be made.

References

[1] P. Elias, "Error-Free Coding," *IRE Transactions on Information Theory*, vol. PGIT-4, pp. 29–37, Sept. 1954.

[2] G. Ungerboeck, "Channel Coding With Multilevel/Phase Signals," *IEEE Transactions on Information Theory*, vol. IT-28, pp. 55–67, Jan. 1982.

[3] C. Berrou, A. Glavieux, and P. Thitimajshima, "Near Shannon Limit Error-Correcting Coding and Decoding: Turbo Codes," *Proceedings of the IEEE International Conference on Communications 1993*, Geneva, pp. 1064–1070, May 1993.

[4] G. D. Forney, "The Viterbi Algorithm," *Proceedings of the IEEE*, vol. 61, no. 3, pp. 268–278, Mar. 1973.

[5] A. R. Calderbank, G. D. Forney, and A. Vardy, reported in *Fundamentals of Convolutional Coding*, (R. Johannesson and K. S. Zigangirov), pp. 227–228, *IEEE Press*, 1999.

[6] C. Berrou, C. Douillard, and M. Jézéquel, "Multiple Parallel Concatenation of Circular Recursive Systematic Convolutional (CRSC) Codes," Annales des Télécommunication, tome 54, no. 3-4, pp. 166–172, Mar-April 1999.

[7] R. G. Gallager, "Low-Density Parity-Check Codes," *IRE Transactions on Information Theory*, vol. IT-8, pp. 21–28, Jan. 1962.

Chapter 2

HEAVY TAILS, PHASE TRANSITIONS, AND THE NATURE OF CUTOFF

C. Gomes
Department of Computer Science,
Cornell University,
Ithaca, New York 14853

Xi Xie
School of Electrical and Computer Engineering,
Cornell University,
Ithaca, New York 14853

S. B. Wicker
School of Electrical and Computer Engineering,
Cornell University,
Ithaca, New York 14853

B. Selman
Department of Computer Science,
Cornell University,
Ithaca, New York 14853

Abstract It has recently been shown that NP-complete problems exhibit "phase transitions"—narrow regions in which the computational complexity and character of solutions to these problems change abruptly. Algorithms operating at or near the phase transition exhibit runtimes that can be characterized by heavy tailed distributions. In this paper we provide an overview of these results, and then use them to explaine the functionality and improve the performance of sequential decoders for convolutional codes. We show that cutoff rate can be interpreted as a particular point in the onset of heavy tailed behavior, and that heuristic techniques used to avoid heavy tailed behavior can be used to support sequential decoding above the cutoff rate. Connections are also drawn to other decoding problems, including that for expander codes.

1. INTRODUCTION

Over the past ten years computer scientists have been encroaching on the domain of problems previously deemed intractable. Recent advances in complexity theory have uncovered structural tendencies in NP-complete problems that can be exploited through the use of heuristic design strategies. One of the key results in this effort has been the identification of a "phase transition" in the complexity of algorithms for solving NP-complete problems. The phase transition is a narrow region in which the computational complexity and character of solutions to these problems change abruptly as the number of constraints on the problem is increased. As a result, many problems once avoided are now under attack by carefully avoiding this phase transition.

Algorithms that operate at or near the phase transition exhibit heavy tailed behavior—running times that are characterized by probability distributions with tails that are substantially "thicker" than those associated with normal distributions. Heavy tailed behavior often manifests itself through extremely long algorithm run times that occur much more often than one would otherwise expect. An understanding of this behavior has led to the development of heuristic strategies that eliminate heavy tails, increasing the utility of these algorithms near the phase transition.

In this paper we will show that what coding theorists view as cutoff rate can be interpreted in terms of a phase transition in the complexity of sequential decoding algorithms. In particular, we show that the cutoff rate is a critical point in the approach to this transition. We use this connection to exploit recently developed strategies for coping with heavy tailed distributions, allowing for low-complexity sequential decoding above the cutoff rate.

In the next section we discuss the relevant recent work in the complexity of algorithms. This includes a discussion of heavy tailed behavior and the phase transitions exhibited by algorithms for solving NP-complete problems. The emphasis in this section is on the efficacy of restart strategies in coping with heavy tailed behavior.

In the third section, we briefly summarize Berlekamp, McEliece, and van Tilborg's demonstration that the general problem of decoding linear block codes is NP-complete. Similarly, the convolutional decoding problem can also be shown to be NP-complete. Given the fundamental equivalence between NP-complete problems in computational terms it is convenient to study the decoding task as a Boolean satisfiability problem. The satisfiability problem is the canonical example of an NP-complete computational task and has been studied extensively in recent years in order to obtain a better understanding of the practical computational properties of NP-complete problems. Therefore, in order to illustrate the general computational nature of the decoding task, we will provide a translation of the convolutional decoding problem into the Boolean satisfiability problem

In Section 4 we provide a brief review of the sequential decoding of convolutional codes, and Jacobs and Berlekamp's demonstration that sequential decoders exhibit heavy tailed behavior. We show that this heavy tailed behavior is analogous to the behavior observed in algorithms for solving NP-complete problems, and then show that this heavy tailed behavior can be eliminated through the use of a restart strategy. The restart strategy takes the form of retransmission requests triggered after a fixed number of backtracks by the decoding algorithm. We show that this eliminates heavy tailed behavior, and allows for more computationally-efficient convolutional decoding above the cutoff rate.

2. RECENT RESULTS IN THE COMPLEXITY OF ALGORITHMS

The complexity of a problem is generally characterized in terms of the rate at which the number of computations required increases with the size of the problem. The problem size is generally specified in terms of an "input length." The problem of finding the determinant of a square matrix, for example, is characterized in terms of the number of rows of the matrix. The complexity of this problem increases with the cube of the number of rows. The class of such problems for which algorithms exist that execute in time that is polynomial in the length of the input is called Class P.

The class NP is the set of problems that can be solved by a nondeterministic algorithm whose running time is bounded by a polynomial in the length of the input. A nondeterministic algorithm is an algorithm that replicates itself at each branching point in the problem, creating a copy for each potential path to be followed. These copies continue to run, replicating themselves at subsequent branching points, until one of the copies finds a solution or the space of potential solutions is exhausted. The term "NP" stands for nondeterministic polynomial." The number of copies generated by the nondeterministic algorithm may grow exponentially, but the individual copies run in polynomial time, hence the name.

Classic examples of NP problems include the problem of determining whether a given graph is Hamiltonian; i.e., whether there exists a closed path through the graph that visits each vertex exactly once. Finding such a path is quite difficult. Once a path is found, however, verification that it is indeed a Hamiltonian path is relatively simple. This leads to yet another definition of the class NP—those problems for which a given solution can be verified in polynomial time.

The class NP clearly contains the class P, but it is not known whether there are problems in the class NP that are not in P. What is known is that there is a subset of problems within class NP that can be mapped onto one another in such a manner that a solution to one can be directly translated into a solution for the others. In 1971 Cook showed that if a polynomial-time algorithm was found for a particular NP problem called the *satisfiability problem*, then that algorithm could be used to solve any other class-NP problem in polynomial time [1]. The satisfiability problem, simply put, is to determine whether the variables in a given Boolean expression have an assignment that will make the expression true. More formally, the problem is to determine if there is a truth assignment that satisfies a formula in conjunctive normal form. A formula in conjunctive normal form is a conjunction (logical "and") of clauses, while a clause is a disjunction (logical "or") of variables. A variable may occur in either positive or negative form. An example of a three clause formula in conjunctive normal form is $(A \vee \neg B \vee \neg C)\ (\neg B \vee \neg C)\ (A \vee C)$. This formula is true when, for example, $A = 1$; $B = 0$; and $C = 1$.

Shortly after Cook published his results, Karp showed that polynomial-time algorithms for several other NP problems could be translated into polynomial-time algorithms for the satisfiability problem [19]. There is thus a subset of NP problems, now referred to as NP-complete, in which polynomial-time algorithms to any one problem in the set can be translated into polynomial-time methods for all of the problems in class NP.

The class NP is also characterized as the the set of problems that can be solved by backtrack-search algorithms in time that is polynomial in the depth of the search. "Backtrack-style" search procedures are a systematic means for exploring a large space of potential solutions to find a solution or prove that no solution exists. The general strategy of backtrack search underlies popular software for solving large classes of combinatorial problems, such as integer programming (e.g., [24] and constraint-programming packages [25]). Backtrack search is also, of course, the conceptual basis for the sequential decoding of convolutional codes.

In backtrack search, a solution is constructed incrementally for a given set of variables. At any point in the execution of the algorithm, a subset of the variables have been assigned a value, and the other variables have no assignment. The algorithm repeatedly picks an unassigned variable and sets it to some value. It then checks whether the current partial assignment obviously violates any one constraint. If so, the last variable assignment is undone, and a different value is assigned. If all values for the variable have been tried, the algorithm *backtracks* to the previous variable, and attempts to reassign it. The loop repeats until a complete assignment has been found or the algorithm has exhausted all possibilities, in which case it has been demonstrated that no satisfactory complete assignment exists.

Figure 1 illustrates the application of backtrack search to the satisfiability problem. The figure shows a search tree that traces the activity of a backtrack search algorithm looking for an assignment to the variables A, B, and C that satisfies the formula: $(A \vee \neg B \vee \neg C)(B \vee \neg C)(A \vee C)$. In the search tree the variable choice heuristic picks the variables in the order A, B, then C, and always tries the value 0 before the value 1. Execution is traced along the left-most branch of the tree until a constraint is determined to be false, creating a dead-end (marked X). The algorithm then backtracks up a level in the tree and tries the right branch (assigning 1 to the variable). To backtrack from a right-most branch the algorithms goes up two levels, and so on. The run time of a backtrack search algorithm is proportional to the number of backtracks. In our experiments, we report as computational cost the number of backtracks since this notion is machine independent.

The performance of backtrack-style search methods can vary dramatically depending on the problem instance as well as on the way one selects the next variable on which to branch (the "variable selection strategy") and in what order the possible values are assigned to a variable (the "value selection strategy").

The distributions of running times of randomized backtrack search algorithms are often *heavy tailed.* Standard probability distributions, such as the normal distribution, have exponentially decreasing tails—events that are several standard deviations from the mean of the distribution ("outliers") are rare.[1] Distributions whose tails have a *power-law* decay are said to have "heavy tails." They are also called tails of the Pareto-Lévy form, and are generally modeled using the expression

$$P\{X > x\} \sim Cx^{-\alpha}, \quad x > 0 \qquad (1)$$

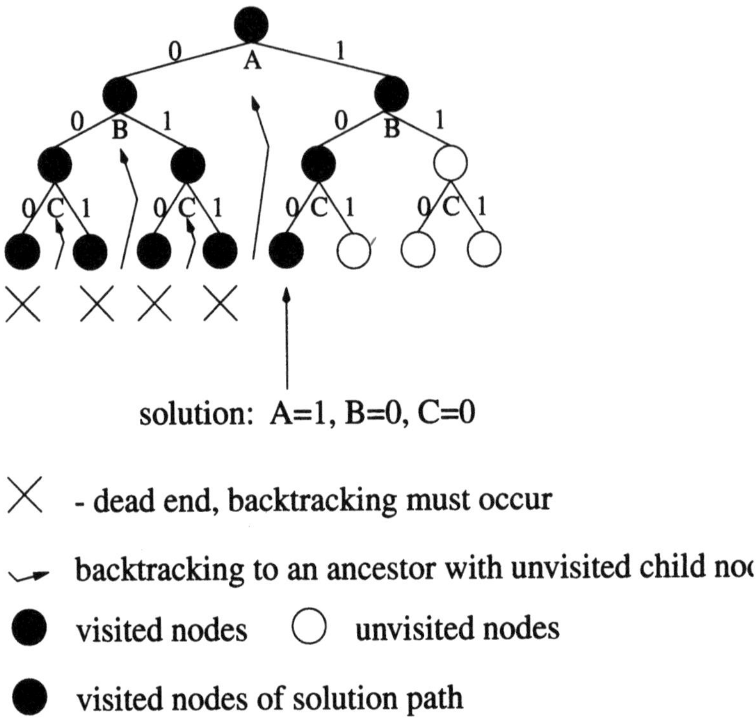

Figure 1 Backtrack search applied to the satisfiability problem.

where $0 < \alpha < 2$ and $C > 0$ are constants.

In the present context of backtrack search, heavy tails manifest themselves as follows. When solving a computationally hard problem with a randomized search method, very long runs occur much more frequently than one might intuitively expect. For example, in practice, one can observe runs that take only seconds while other runs take hours or days, even when solving the same problem instance. The exploitation of structure in many NP-complete problems is based on an understanding of the heavy tailed behavior that is often displayed by associated search algorithms. The next section provides a more detailed look at heavy tailed distributions.

2.1 HEAVY TAILED DISTRIBUTIONS

Heavy tailed distributions were first introduced by the Italian-born Swiss economist Vilfredo Pareto in the context of income distribution. They were extensively studied mathematically by Paul Lévy in the period between the world wars. Lévy worked on a class of probability distributions with heavy tails, which he called *stable* distributions. However, at the time, these distributions were largely considered probabilistic curiosities or pathological cases mainly used in counterexamples. This situation changed dramatically with Mandelbrot's work on fractals. In particular, two seminal papers of Mandelbrot [20, 21] were instrumental in establishing the use of stable distributions for modeling real-world phenomena. Recently, heavy tailed distributions have been used to model phenomena in areas as diverse as economics, statistical physics, and geophysics. More concretely, they have been applied in stock market analysis, weather forecasts, earthquake prediction, and recently, for modeling time delays on the World Wide Web (*e.g.*, [22, 26]). Heavy tailed distributions also characterize the behavior of sequential decoding algorithms when operated near, at, or above the cutoff rate, as will be shown later in the paper.

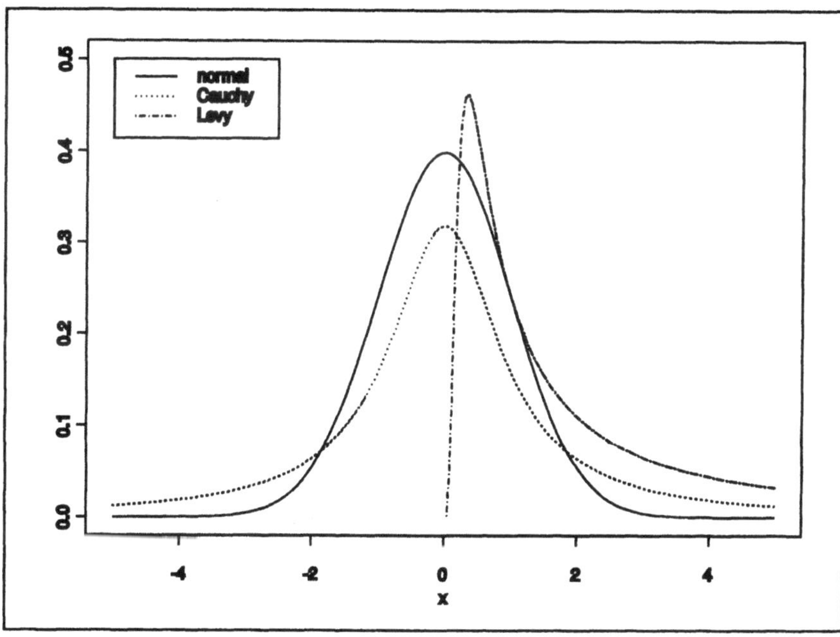

Figure 2 Standardized Normal, Cauchy, and Lévy densities. (Adapted from Nolan 1999).

Figure 2 contrasts three distributions: standard normal, Cauchy, and Lévy distributions. The key property to observe is the dramatic difference in the decay of the tails of the distributions. In Table 1, we compare the tail probabilities for the three distributions. It is clear that the tail probability for the standard normal quickly becomes negligible, whereas the other two distributions have a significant probability mass in the tail. The parameter α in Equation (1) is referred to as the *index*

c	Normal	Cauchy	Lévy
0	0.5000	0.5000	1.0000
1	0.1587	0.2500	0.6827
2	0.0228	0.1476	0.5205
3	0.001347	0.1024	0.4363
4	0.00003167	0.0780	0.3829
5	0.0000002866	0.0628	0.3453

Table 1 Comparison of tail probabilities, $\mathrm{P}\{X > c\}$, for standard Normal, Cauchy and Lévy distributions. (Adapted from Nolan 1999.)

of stability of the distribution. The lower the index, the heavier the tail. For example, the Cauchy has $\alpha = 1.0$ and the Lévy distribution has $\alpha = 0.5$.

Heavy tailed distributions have a number of surprising properties. For example, consider the moments of these distributions. For standard distributions, the moments are well-defined. In the case of heavy tailedness, due to the large probability mass in the tails, some of the integrals that define the moments do not converge. In particular, for $\alpha < 2$, moments of X of order less than α are finite while all higher order moments are infinite, i.e., $\alpha = \sup\{b > 0 : \mathrm{E}|X|^b < \infty\}$. For example, when $\alpha = 1.5$, the distribution has a finite mean but no finite variance. With $\alpha = 0.6$, the distribution has neither a finite mean nor a finite variance.

Figure 3 illustrates the "wandering sample mean" phenomenon of a randomized backtrack search algorithm, one of the cues indicating the presence of a distribution with heavy tails and infinite mean. The average number of backtracks until a solution is reached exhibits erratic behavior as the number of runs of a randomized backtrack search algorithm is increased on the same problem instance. After 200 runs, the mean is around 500 backtracks; after 600 runs, it is around 2000 backtracks; after 1000 runs, it is around 3500 backtracks. The phenomenon is due to the fact that the more runs we take into account, the more likely we are to hit some very long runs, increasing the mean number of backtracks. This particular experiment in which this data was generated consisted of a randomized backtrack search algorithm applied to the quasi-group completion problem [10]. A quasi-group or Latin square is an $n \times n$ matrix in which, given n different symbols, each symbol occurs exactly once in each row and column. The quasi-group completion problem is the problem of completing a partially filled matrix in order to obtain a Latin square. For this particular problem, in 50% of the cases the search algorithm finds a solution using only one or zero

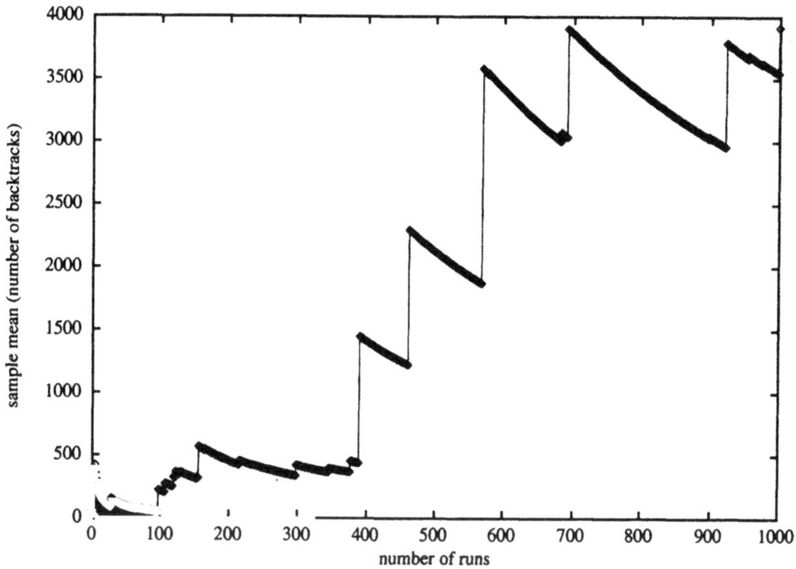

Figure 3 Wandering sample mean phenomenon of a backtrack search method

backtracks; however, approximately 1% of the runs require more than 1,000,000 backtracks to find a solution.

Many aspects of random walks involve heavy tailed distributions [5, 6]. Consider a one-dimensional random walk, where at each time step one takes a unit step to the left or right with equal probability. One can show that after starting at the origin, with probability one, the walk will eventually return to the origin. However, the *expected* time before return is infinite and, on average, the walk will reach all values on the x axis before its return. Another intriguing phenomenon involves the expected number of returns to the origin ("zero-crossings") in a given number of steps. Intuition would dictate that if in k steps one has on average l crossings, then in a walk that is m times as long, one would expect on average $m \times l$ crossings. However, it can be shown that in $m \times l$ steps, one will only observe, on average, $\sqrt{m} \times l$ crossings. This means that there can be surprisingly long periods in a walk between two crossings. In fact, when doing a series of r random walks, each terminating at the first crossing, on average, some of the walks will be of the same order as the length of all other walks combined, *no matter what the value of r is.* Such events would normally be dismissed as outliers, but when dealing with random walks, they are far from rare and are an inherent aspect of the distribution. Heavy tailed distributions can appropriately capture

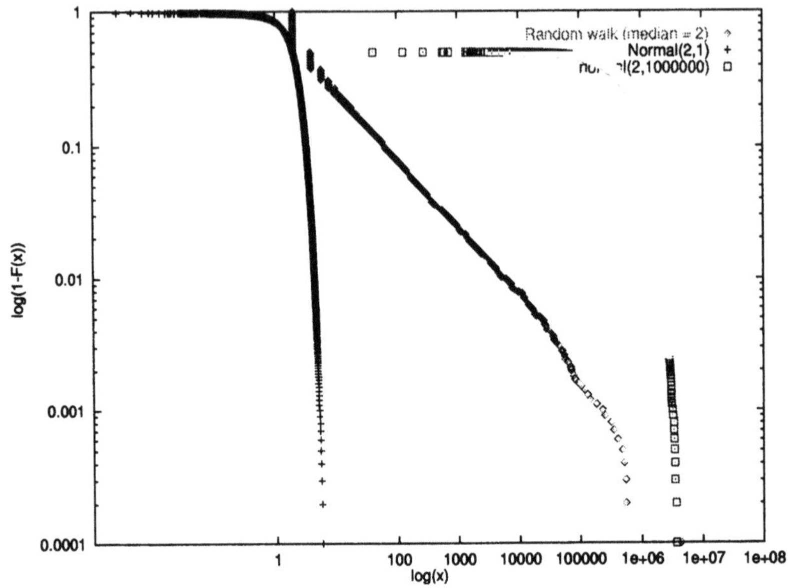

Figure 4 The heavy tailed nature of a random walk with the exponential decay of the normal distribution for comparison.

these phenomena. These distributions are therefore good models for dealing with phenomena that exhibit extreme fluctuations.

In Figure 4 we contrast the heavy tailed behavior of a random walk with the exponential decay of the tail of a normal distribution. The figure shows the data for 10,000 runs of a symmetric random walk. For each walk we recorded the number of steps before the walk returned to the origin. In the figure, we plot the complement-to-one of the cumulative distribution, i.e., $1 - F(x) = \mathrm{P}\{X > x\}$, with $F(x)$ being the cumulative distribution of $f(x)$. The probability function, $f(x)$, gives the probability that the walk returns to the origin in exactly x steps. So, $1 - F(x)$ gives us the probability that the walk returns to the origin in more than x steps. It can be seen that $f(1) = 0$ and $f(2) = 0.5$, so we have $F(x) = 0.5$, which means that with 50% chance the walk returns to the origin in at most two steps. In the figure, we give the log-log plot for $1 - F(x)$. (In the figure, the walk data is given by the diagonal straight line.) As we can see, we obtain a near straight line for the tail of the distribution. In fact $1 - F(x)$ has power-law decay, i.e., we have $1 - F(x) = \mathrm{P}\{X > x\} \sim Cx^{-\alpha}$, and thus the distribution is heavy tailed according to our definition (1) above. The slope of the line gives us an estimate of the index of stability, α, which in this case is equal to 0.5. The relatively high frequency of large outliers is clear from the figure.

For example, although 50% of the walks return in just two steps or less, 1% of the walks take more than 5,000 steps to return to the origin, and about 0.1% take over 200,000 steps. In fact, several of the walks in our sample take almost 1,000,000 steps.

Figure 4 also shows the complement-to-one of the cumulative distribution for a normal distribution. We used a mean value of 2 and give the curves for two different standard deviations ($\sigma = 1$, left-most curve, and $\sigma = 10^6$, right-most curve). The key property to observe is the sharp, faster-than-linear, decay of the normal distribution in the log-log plot, which is consistent with the exponential decay in the tail of the distribution. We included a normal distribution with $\sigma = 10^6$ to show that the drop off of the tail remains sharp even when the normal distribution has a large standard deviation. (The normal distribution is symmetrical; the figure only gives the right side.)

There is a substantial literature on heavy tailed distributions. Mandelbrot [22] provides a good introduction to these distributions with a discussion of their inherently self-similar or fractal nature. For a complete treatment of stable distributions see either Zolotarev [31], or the more modern approach of Samorodnitsky and Taqqu [26].

2.2 HEAVY TAILS IN COMBINATORIAL SEARCH

To model the heavy tailed behavior of backtrack search, we consider distributions that asymptotically have tails of the Pareto-Lévy form, i.e., $\mathrm{P}\{X > x\} \sim Cx^{-\alpha}, \quad x > 0$, with $\alpha > 0$. In order to check for the existence of heavy tails in our distributions, we proceed in two steps. First, we graphically analyze the tail behavior of the sample distributions. Second, we estimate the index of stability. As noted above, distributions of the Pareto-Lévy form have tails with power-law decay. Therefore, the log-log plot of the tail of the distribution should show an approximate linear decrease, as shown in Figure 5. The figure displays the log-log plot of the complement-to-one of the cumulative distribution (i.e., $\mathrm{P}\{X > x\} = 1 - F(x)$, where $F(x)$ is the cumulative distribution) of the solution time of backtrack search algorithms: (a) for three different instances of the quasi-group completion problem and in (b) for a timetabling problem and for a software code optimization problem [18]. For example, the bottom curve in (a) shows that approximately 90% of the runs complete in fewer than ten backtracks, and about 1% of runs have not found a solution after 10,000 backtracks. Each distribution is based on 10,000 runs of the backtrack search method.

The visual check of the linearity of the plot can be confirmed by calculating the maximum-likelihood estimates of the indices of stability (the values of α), using a variation of the Hill estimator [14] modified to take into account data truncation of extreme outliers [11]. The modified version proceeds as follows. Assume a sample of $k = n + u$ iid random variables is drawn. Let the order statistics for the smallest n $(n \leq k)$ values on the sample be $X_{n1} \leq X_{n2} \leq \ldots \leq X_{nn}$. Assume that, for $X_{n,n-r} \leq X \leq X_{nn}$, the tail distribution is of the Pareto-Lévy type. The conditional maximum-likelihood estimator for the maximal moment exponent α is:

$$\hat{\alpha}_{r,u} = \left(\frac{1}{r} \sum_{j=1}^{r-1} \ln X_{n,n-r+j} + \frac{u+1}{r} \ln X_{n,n} - \frac{u+r}{r} \ln X_{n,n-r} \right)^{-1} \tag{2}$$

and its variance is given by

$$\widehat{\mathrm{Var}}(\hat{\alpha}) = \frac{\hat{\alpha}^2 (r+1)^2}{r^2 (r-1)}.$$

For a more complete treatment, see Gomes, et al [11].

Because $\alpha < 1$ for all the distributions displayed in Figure 5, in all those cases the data is consistent with the hypothesis of infinite mean and infinite variance. For the quasi-group instances in (a) the estimates of the indices of stability α are 0.153±0.003 (top curve), 0.32±0.06 (middle curve), and 0.466±0.009 (bottom curve). For the timetabling problem in (b) α is 0.219±0.004. For the software code optimization in (b) α is 0.102±0.002.

2.3 CONSEQUENCES FOR ALGORITHM DESIGN: RESTARTS

We have shown the heavy tailed nature of the runtime distributions of backtrack search methods, and we have seen how the distributions can be modeled with Pareto-Lévy type (heavy) tails. Our estimates for α are consistent with the hypothesis of infinite variance ($\alpha < 2$), and infinite mean ($\alpha < 1$), which is consistent with the empirically observed erratic behavior of the mean and extreme variance of the cost of backtrack search methods. Given the heavy tailed phenomenon, a randomized backtrack procedure is, in a sense, most effective early on in the search, which suggests that a sequence of short runs instead of a single long run may be a more effective use of computational resources. In this section, we show how a *rapid randomized restart strategy* can take advantage of the heavy tailed behavior.

In Figure 6, we show the result of applying a strategy of fixed-length short runs ("restarts") of a randomized backtrack procedure. We con-

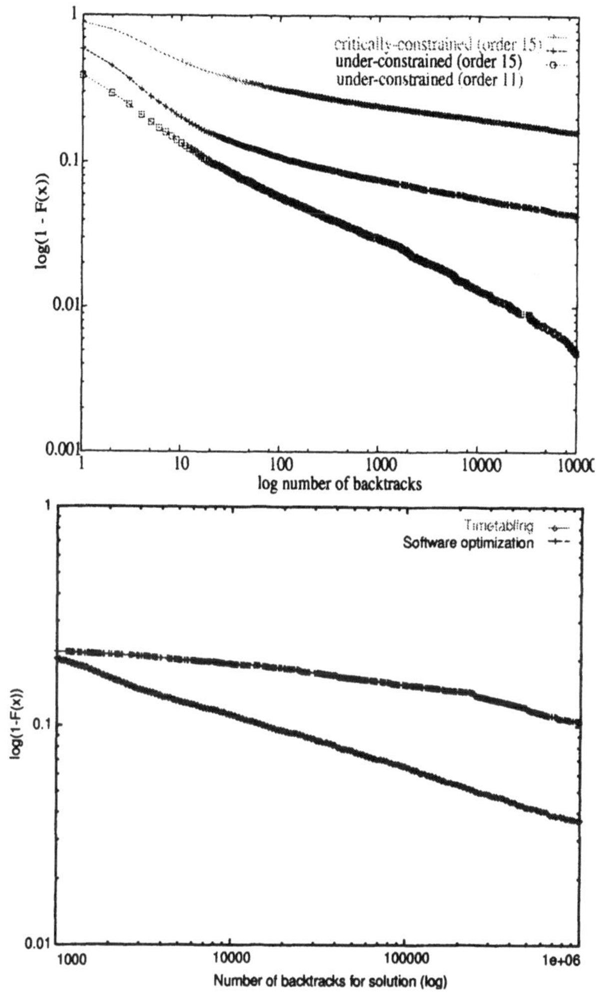

Figure 5 Heavy tailed distributions in combinatorial search: (a) for three different instances of the quasi-group completion problem and in (b) for a timetabling problem and for a software code optimization problem [18].

trast the heavy tailed behavior of a randomized backtrack search algorithm on a quasi-group completion problem with the non-heavy tailed behavior obtained by restarting the same algorithm every four backtracks. While the original algorithm fails to find a solution in 70% of the cases after 1000 backtracks, the rapid restart version finds a solution in almost all the cases (99.99%) using a total of 250 backtracks during 62 restarts. The figure shows a clear downward curve for the log-log plot of the complement-to-one of the cumulative distribution of the restart strategy, which is an indication that the heavy tailed nature

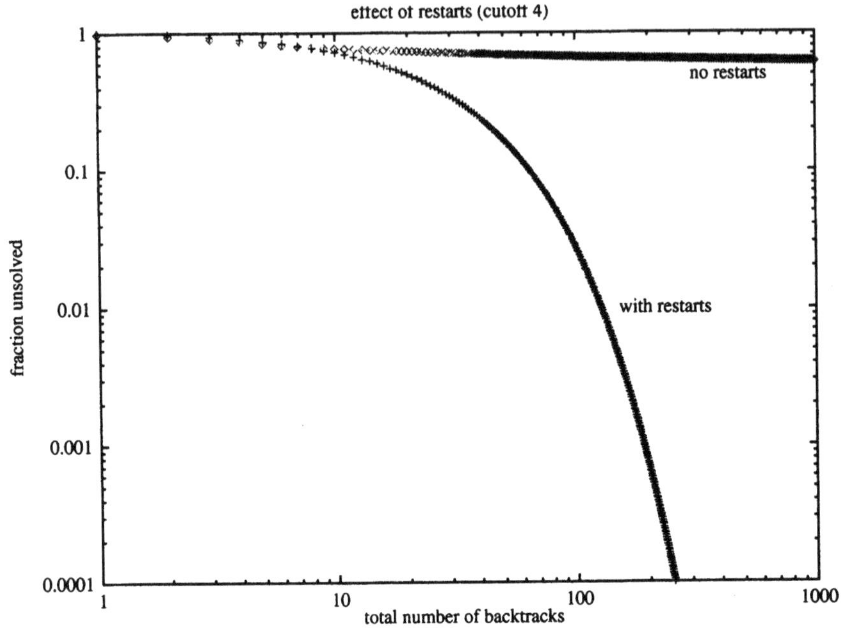

Figure 6 A restart strategy eliminates heavy tail phenomena and speeds up computational search.

of the original cost distribution has disappeared. In fact, the restart strategy follows a geometric distribution, and therefore, does not have heavy tails [29].

The rapid restart strategy with a randomized backtrack search enabled us to solve problems that previously appeared unsolvable with state-of-the-art search methods. For example, after randomizing a deterministic search procedure for a sports timetabling problem, we could solve problems with up to twenty teams using rapid restarts, while the original method could find only schedules for up to fourteen teams. Similarly, we have solved large logistics planning problems that contain up 200 activities in their solution, while previous deterministic methods handle only about thirty activities [11].

We now formalize the restart strategy of a complete randomized backtrack search method. Given a randomized backtrack search procedure, consider the number of choice points (or backtracks) performed by such a procedure. We introduce random variable B, such that, B is the number of choice points that the backtrack search procedure takes to find a solution or prove that it does not exist, $B = \{1, 2, \cdots\}$.

Now consider a rapid-randomized-restarts (RRR) strategy for running the backtrack procedure. Run the procedure up to a fixed number of choice points c (the restart threshold). If the procedure finds a solution or proves that no solution exists, then RRR has also found a solution (or proven that no solution exists) and stops; otherwise, restart the backtrack procedure from the beginning (using an independent random seed) for another c decision events, and continue. We associate with RRR the random variable S, the number of choice points that RRR takes to find a solution or prove that no solution exists.

Define a "run" as the execution of the randomized backtrack search method for up to c steps. We now define the random variable R, the number of runs executed by RRR.

Figure 7 illustrates the interrelationship of the different random variables.

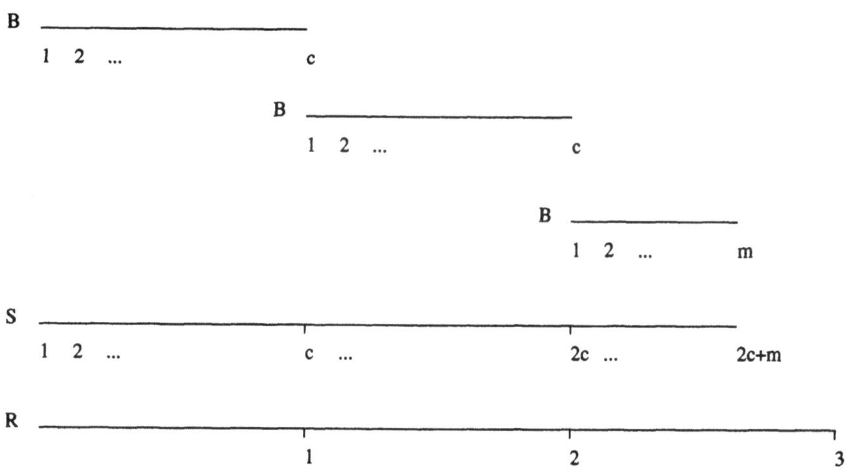

Figure 7 B — number of choice points searched by the randomized backtrack 'procedure (running with restart threshold c); S — number of choice points searched by RRR; R — number of runs executed by RRR. In this case, a solution was found in the third run ($R = 3$), with a total of $2c + m$ choice points ($S = 2c + m$). The third run of the randomized backtrack search method took $m \leq c$ choice points.

The runs executed by RRR are independent (no information is carried over between runs, and each run uses a new random seed) and therefore can be seen as a sequence of Bernoulli trials in which the success of a trial corresponds to finding a solution (or proving that one does not exist) during a run. The probability of success is given by $P[B \leq c]$. Therefore, R follows a *geometric distribution* with parameter $p = P[B \leq c]$.

The probability of the tail of S, $P[S > s]$, corresponds to the probability of not finding the solution in the first $\lfloor s/c \rfloor$ runs of RRR, and finding it with more than (s mod c) choice points in the next run. We obtain the following expression:

$$P[S > s] = P[B > c]^{\lfloor s/c \rfloor} \; P[B > s \; mod \; c]. \tag{3}$$

The distribution of S is not heavy tailed because its tail exhibits exponential decay. The tail of S is limited from above by the tail of R.

$$P[S > s] \leq P[B > c]^{\lfloor s/c \rfloor} = P[R > \lfloor s/c \rfloor]. \tag{4}$$

Since R follows a geometric distribution, it has finite moments, and therefore so does S. The full distribution of S is given by the following expression:

$$P[S = s] = \begin{cases} P[B > c]^{\lfloor s/c \rfloor} \; P[B = s \bmod c] & s \bmod c \neq 0 \\ \\ P[B > c]^{\lfloor s/c \rfloor - 1} \; P[s \; = \; c] & \text{otherwise} \end{cases} \tag{5}$$

Note that the second branch of Equation (5) corresponds to the case in which the total number of choice points executed by strategy S is a multiple of c. This situation occurs when the solution is found when the restart threshold c is reached.

Based on the distribution of B, we can determine a restart threshold, c, that minimizes the expected runtime of S. Alternatively, one can determine a restart threshold that minimizes both the expected runtime and variance of the RRR strategy, using the tools from portfolio analysis (e.g., [15, 9]). In our experiments, we determined the restart threshold for the restart strategy (RRR) based on the empirical distribution of B, which was computed, when possible, by performing 10,000 runs of the backtrack search methods with a very high restart threshold.

2.4 PHASE TRANSITIONS

We conclude this section with a description of a phenomenon that is tied to heavy tailed behavior in a manner that is not yet completely understood; specifically, the tendency for NP-complete problems to show "phase transitions" in complexity. To do so we turn to a stricter version of the satisfiability problem referred to as K-SAT. In K-SAT the Boolean expression consists of N variables arranged (with repetition) into M clauses. Each clause is the logical OR of K variables, with the overall expression in conjunctive normal form. The problem can be solved by testing each of the 2^N possible assignments for the variables, though clearly the complexity of this task increases exponentially with the number of variables. More thoughtful approaches, however, often lead to simpler solutions. 1-SAT problems, for example, can always be solved. Simply assign the first $(N-1)$ values at random; the final value can always be chosen to make the overall expression true. The 2-SAT problem is much more interesting. This problem can also be solved, in the sense of determining whether or not a valid assignment exists, in polynomial time. There is not, however, always a valid assignment. When the ratio of clauses to variables (γ) is less than one, there is almost always a valid assignment. But when γ exceeds one, there is almost never such an assignment.

For values of $K \geq 3$, the existence of a valid assignment also shows a sudden transition at a particular value of γ. For the 3-SAT problem, for example, the transition occurs at $\gamma = 4$. As K increases, the critical value of γ seems to increase as well. It has been shown that such a threshold exists for any value of K, though determining the actual value may be difficult [23]. Unlike the 2-SAT problem, the K-SAT problem for $K \geq 3$ is NP-complete. The complexity of the problem of determining whether or not a solution exists shows a phase transition at the critical value of γ. The computational requirements are mild on either side of the critical value of γ, but increase dramatically as the critical value is approached. Figure 8(a) shows the increase in complexity around the phase transition for 3-SAT, while Figure 8(b) shows the abrupt transition in the probability that a solution exists. When applied to K-SAT problems, backtrack search algorithms seem to exhibit heavy tailed behavior when they are operating in the vicinity of the problem's phase transition.

This result suggests a connection that may provide a great deal of insight into the function of backtrack decoders for error control codes. There is an important distinction to be made, however, between problems like K-SAT and error control decoding. K-SAT is formulated as a decision problem. In short, solutions to the problem provide either a "yes" or "no" result. An error control decoder, on the other hand, provides as output a codeword that, to a predetermined extent, satisfies a given set of constraints. It is, in short, a solution to an optimization problem. In the next section we will show how decoding can be formulated as a decision problem.

When considering the phase transition phenomenon it is also important that the algorithms be *complete.* Within the context of a decision problem, a complete algorithm will either demonstrate that a solution exists by providing one, or it will prove that no solution exists. At the end of the paper we will suggest a connection between phase transitions and channel capacity, but it will be important to remember that the two cannot be firmly connected without the underlying use of a "complete" decoding algorithm.

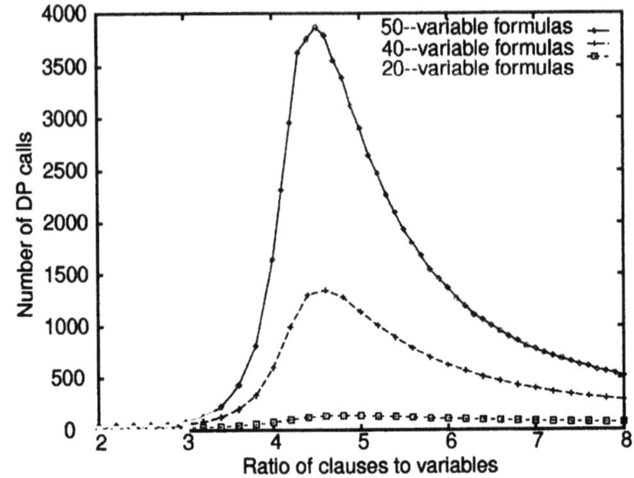

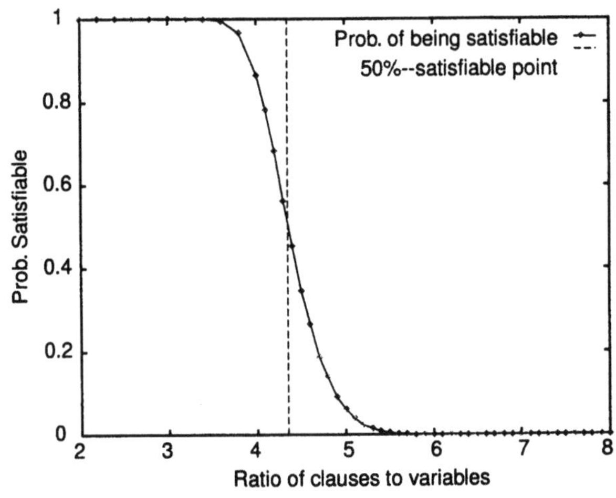

Figure 8 (a) Complexity and (b) phase transition behavior for 3-SAT.

3. CODES AND COMPLEXITY

We will now consider the sequential decoding of convolutional codes in light of the above results. We first revisit Berlekamp, McEliece, and van Tilborg's demonstration that the decoding of linear block codes is NP-complete. We then show that the decoding of convolutional codes can be encoded as a SAT problem to faciliate study of its computational properties.

3.1 DECODING OF LINEAR BLOCK CODES

The hard-decision decoding problem for linear block codes can be characterized as follows. Consider a code $\mathcal{C}$ defined by a partiy-check matrix $\mathbf{H}$. A codeword $\mathbf{c}$, where by definition $\mathbf{cH^T} = \mathbf{0}$, is transmitted across a noisy channel. The received word is $\mathbf{y} = \mathbf{c} + \mathbf{n}$, where $\mathbf{n}$ is the noise introduced by the channel. The receiver computes a syndrome $\mathbf{S} = \mathbf{yH^T} = (\mathbf{c} + \mathbf{n})\mathbf{H^T} = \mathbf{nH^T}$. It is generally assumed that the minimum-weight noise vector that satisfies $\mathbf{S} = \mathbf{nH^T}$ is the vector most likely to have corrupted the received word.

Berlekamp et al. put the decoding problem in the following standard form, creating an existence problem that they labeled "COSET WEIGHTS" [3].

Given a binary matrix $\mathbf{H}$*, a binary vector* $\mathbf{y}$*, and a nonnegative integer* w*, does there exist a vector* $\mathbf{x}$ *of Hamming weight* $\leq w$ *such that* $\mathbf{xH}^T = \mathbf{y}$*?*

They showed that this problem was NP-complete by using a reduction from a combinatorial decision problem called "Three-Dimensional Matching." Given the existence of this mapping to an NP-complete problem, there must also exist a mapping to the K-SAT problem. This connection is quite obvious for some specific decoding problems (consider, for example, the hard decision decoding of expander codes). In other cases, the connection is more opaque. We note in general that a linear code is defined by a parity check matrix $\mathbf{H}$. The matrix $\mathbf{H}$ imposes a series of constraints on the values of the codeword symbols that we can express in Boolean form. A similar approach can be used to express the decoding of convolutional codes as an instance of the satisfiability problem.

3.2 DECODING OF CONVOLUTIONAL CODES AS A SAT PROBLEM

In order to represent the decoding of convolutional codes as a satisfiability problem, we must first re-express decoding as a decision problem, i.e., a problem for which there is a "yes" or "no" answer, as opposed to our usual perspective of decoding as an act of optimization. To make it more concrete we consider the rate-1/2 linear convolutional encoder depicted in Figure 9, with binary input datastream of length k, i.e., $x = (x_0, x_1, \ldots, x_{k-1})$. Furthermore we assume we have a binary symmetric channel with crossover probability $p < 0.5$. In this case the metric for the maximum-likelihood hard-decision decoder corresponds to the Hamming distance between the received word and the decoded word.

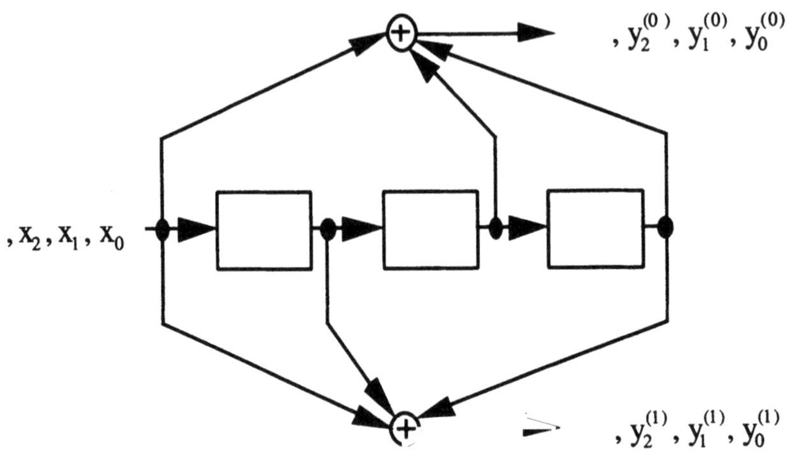

Figure 9 A Feedforward Convolutional Encoder.

Consider the binary variables

- x_i - input bit i , $i = 0, 1, \ldots, (k-1)$; we also define: $x_i = 0$, $i = -3,\ -2,\ -1,\ k,\ (k+1),\ (k+2)$;
- y_i^j - jth output bit corresponding to x_i, $i = 0, 1, \ldots, (k+2)$; $j = 0,\ 1$;
- r_i^j - received bit corresponding to encoded bit y_i^j, $i = 0, 1, \ldots,$ $(k+2)$; $j = 0, 1$;

The following equations define the rate-1/2 linear convolutional encoder in Figure 9:

$$y_i^0 = x_i \oplus x_{i-2} \oplus x_{i-3} \tag{6}$$

$$y_i^1 = x_i \oplus x_{i-1} \oplus x_{i-3} \tag{7}$$

We define the threshold T for the Hamming distance as:

$$\sum_{j=0}^{1}\sum_{i=0}^{k+2} |r_i^j - y_i^j| \leq T = 0, 1, \ldots, (2 \times (k+3)). \tag{8}$$

Note that the r_i^j's are the only known variables. We can now state the problem of decoding a k-bit message encoded with a rate-1/2 linear convolutional encoder as a problem of existence:

Given a received word $r = (r_0^0, r_1^0, \ldots, r_{k+2}^0;\ r_0^1,\ r_1^1,\ \ldots,\ r_{k+2}^1)$ and a threshold T, is there an assignment for $x = (x_0, x_1, \ldots, x_{k-1})$ and $y = (y_0^0, y_1^0, \ldots, y_{k+2}^0;\ y_0^1, y_1^1, \ldots, y_{k+2}^1)$ that satisfies Equations (6), (7), and (8)?

We now discuss how to encode this problem as a satisfiability problem. We can naturally map our binary variables into Boolean variables. For simplicity, we use the same variable names.

A - Clauses encoding the constraints that characterize the convolutional encoder

For simplicity, we only give examples of the different types of clauses for representing the Equations (6) and (7).

1. $y_0^0 \equiv x_0$ - this logical equivalence can be represented with two disjunctive clauses:

   $y_0^0 \Rightarrow x_0$ corresponding to the disjunction: $(\neg y_0^0 \vee x_0)$;
   $x_0 \Rightarrow y_0^0$ corresponding to the disjunction: $(\neg x_0 \vee y_0^0)$;

2. $y_1^1 \equiv (x_1 \oplus x_0)$ - in this case, because we have three variables, we obtain four disjunctive clauses:

   $\neg x_1 \vee \neg x_0 \vee \neg y_1^1;\ x_1 \vee \neg x_0 \vee y_1^1;\ \neg x_1 \vee x_0 \vee y_1^1;\ x_1 \vee x_0 \vee \neg y_1^1;$

3. $y_i^0 \equiv (x_i \oplus x_{i-2} \oplus x_{i-3})$ - In this case, because we have four variables, we obtain eight disjunctive clauses:

   $\neg x_i \vee \neg x_{i-2} \vee \neg x_{i-3} \vee y_i^0;\ \neg x_i \vee \neg x_{i-2} \vee x_{i-3} \vee \neg y_i^0;\ \neg x_i \vee x_{i-2} \vee \neg x_{i-3} \vee \neg y_i^0;\ \neg x_i \vee x_{i-2} \vee x_i \vee y_i^0;$
   $x_i \vee \neg x_{i-2} \vee \neg x_{i-3} \vee \neg y_i^0;\ x_i \vee x_{i-2} \vee \neg x_{i-3} \vee y_i^0;\ x_i \vee \neg x_{i-2} \vee x_{i-3} \vee y_i^0;$
   $x_i \vee x_{i-2} \vee x_{i-3} \vee \neg; y_i^0;$

B - Clauses encoding the threshold for the Hamming distance

Below are examples of the different types of clauses for representing Equation (8). We consider an encoding for a threshold $T = 2$. We introduce the auxiliary variables: n_i^j - noise affecting bit y_i^j, $i = 0, 1, \ldots, (k+2)$; $j = 0, 1$; Note that the n_i^j's correspond to the Hamming distance, i.e.,:

$$|r_i^j - y_i^j| = n_i^j \quad i = 0, 1, \ldots, (k+2);\ j = 0, 1; \tag{9}$$

1. Equation 9 can be rewritten in logical form as:

   $$(y_i^j \oplus n_i^j) \equiv r_i^j \quad i = 0, 1, \ldots, (k+2);\ j = 0, 1; \tag{10}$$

   This set of equations can be represented by $(4 \times 2 \times (k+3))$ clauses.

Equation (8) becomes:

$$\sum_{j=0}^{1}\sum_{i=0}^{k+2} n_i^j \leq T;\ T = 0, 1, \ldots, (2 \times (k+3)). \tag{11}$$

In order to encode Equation (11), we introduce the auxiliary variables Z_0, Z_1, and Z_2 for the cases in which there is no noise (Hamming distance is 0), there is one noisy bit (Hamming distance is 1), and there are two noisy bits (Hamming distance is 2), respectively.

We can now encode Equation (11) using the following clauses:

2 The Hamming distance is no more than 2, i.e., $(Z_0 \vee Z_1 \vee Z_2)$.

3 Exactly one of the cases in 1, Z_0, Z_1, or Z_2, holds, i.e.,:

   $Z_0 \Rightarrow \neg Z_1$, corresponding to the disjunction: $(\neg Z_0 \vee \neg Z_1)$;
   $Z_0 \Rightarrow \neg Z_2$, corresponding to the disjunction: $(\neg Z_0 \vee \neg Z_2)$;
   $Z_1 \Rightarrow \neg Z_0$, corresponding to the disjunction: $(\neg Z_1 \vee \neg Z_2)$;
   ...
   (6 clauses).

4 In case Z_0, there is no noise, i.e., $Z_0 \Rightarrow \neg n_i^j$, which corresponds to the disjunctions: $(\neg Z_0 \vee \neg n_i^j)$ $i = 0, 1, \ldots, (k+2)$; $j = 0, 1$ $(2 \times (k+3)$ clauses);

5 In case Z_1, there is one noisy bit, i.e., $Z_1 \Rightarrow (n_1^0 \vee n_2^0 \ldots \vee n_{k+2}^0 \vee n_0^1 \ldots \vee n_{k+2}^1)$, which corresponds to the disjunction: $(\neg Z_1 \vee n_1^0 \vee n_2^0 \ldots \vee n_{k+2}^0 \vee n_0^1 \ldots \vee n_{k+2}^1)$;

6 Furthermore, we have to ensure that exactly only one of the cases in 4, $n_1^0, n_2^0, \ldots, n_{k+2}^0, n_0^1, \ldots, n_{k+2}^1$, holds, i.e.,:

   $(Z_1 \wedge n_1^0) \Rightarrow \neg n_2^0$, which corresponds to the disjunction: $(\neg Z_1 \vee \neg n_1^0 \vee \neg n_2^0)$;
   ...
   $(Z_1 \wedge n_1^0) \Rightarrow \neg n_0^1$, which corresponds to the disjunction: $(\neg Z_1 \vee \neg n_1^0 \vee \neg n_0^1)$;
   ...
   $(Z_1 \wedge n_2^0) \Rightarrow \neg n_1^0$, which corresponds to the disjunction: $(\neg Z_1 \vee \neg n_1^0 \vee \neg n_1^0)$;
   ...
   (total number of clauses: $\binom{2\times(k+3)}{2}$).

7 To capture case Z_2, we introduce the auxiliary variables:

m_j - pair of noisy bits, $j = 1, 2, \ldots, \binom{2\times(k+3)}{2}$;
$m_1 \equiv (n_0^0 \wedge n_1^0)$ (4 clauses);
...
$m_{\binom{2\times(k+3)}{2}} \equiv (n_{k+1}^1 \wedge n_{k+2}^1)$ (4 clauses);
(total number of clauses: $4 \times \binom{2\times(k+3)}{2}$).

8 We can now state: $Z_2 \Rightarrow (m_1 \vee m_2 \ldots, m_{\binom{2\times(k+3)}{2}})$, which corresponds to the disjunction: $(\neg Z_2 \vee m_1 \vee m_2, \ldots, m_{\binom{2\times(k+3)}{2}})$;

9 Furthermore we have to ensure that exactly only one of cases $m_1, m_2, \ldots, m_{\binom{2\times(k+3)}{2}}$, holds, analogously to above:

$(Z_2 \wedge m_1) \Rightarrow \neg m_2$ which corresponds to the disjunction: $(\neg Z_1 \vee \neg m_1 \vee \neg m_2)$;
...
(total number of clauses $\binom{\binom{2\times(k+3)}{2}}{2}$).

The SAT encoding for the problem of decoding a k-bit message encoded with a rate-1/2 linear convolutional decoder, with total Hamming distance threshold $T \leq 2$, has $(7k + 18 + \binom{2\times(k+3)}{2})$ variables and $(30 + 10k + (5 \times \binom{2\times(k+3)}{2})) + \binom{\binom{2\times(k+3)}{2}}{2})$ clauses. For the case depicted in Figure 9, where $k = 4$, the number of variables is 137 and the number of clauses is 4681. Whenever the clauses have three or more variables, the decoding problem is NP-complete.

4. THE COMPLEXITY OF SEQUENTIAL DECODING AND DECODING BEYOND CUTOFF

Convolutional encoders are finite state machines, as shown in Figure 9, and their operation is readily represented through the use of trees. A small portion of the tree for the encoder in Figure 9 is shown in Figure 10. Each information bit corresponds to a decision point in the tree, with the value of the bit determining the direction of the branching in the tree. A complete sequence of information bits corresponds to a unique path from the root of the tree to a leaf. The codeword at the output of the encoder is the sequence of branch labels associated with the resulting path.

If a convolutional codeword **c** is transmitted across a noise-free channel, then the decoding problem is simply one of matching the appropriate portion of the received word to the labels of the branches at each decision point. It is possible in this manner to move from the root to the

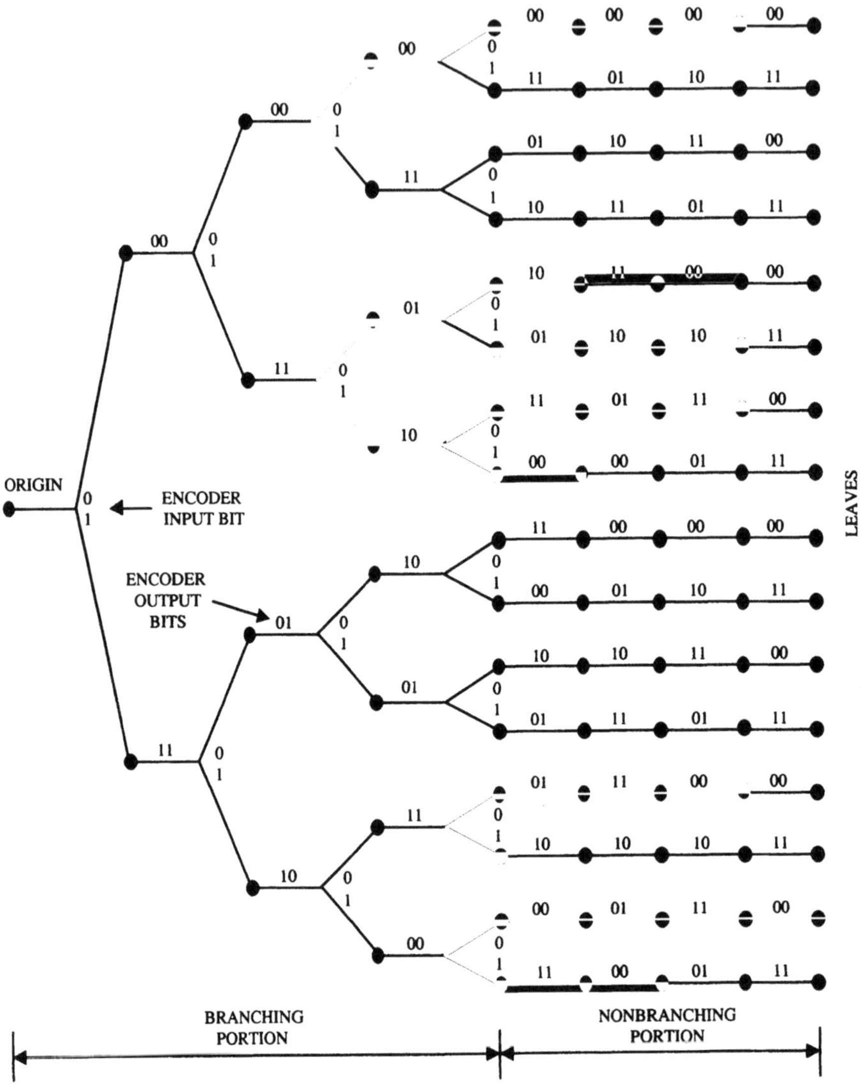

Figure 10 Tree Representation for a Convolutional Encoder.

appropriate leaf in polynomial time. If the channel is noisy, the decoding problem is more difficult, requiring a true backtrack tree search. A backtrack heuristic is defined to control the tree search.

4.1 THE FANO ALGORITHM AND BACKTRACK SEARCH

In the specific case of the Fano algorithm, the backtrack heuristic takes the form of what we now call the Fano metric [30].

$$M\left(r_i^{(j)}|y_i^{(j)}\right) = \log_2\left[\frac{P\left(r_i^{(j)}|y_i^{(j)}\right)}{P\left(r_i^{(j)}\right)}\right] - R, \tag{12}$$

where R is the rate of the code. Given a binary symmetric channel with crossover probability p, the Fano metric reduces to

$$M\left(r_i^{(j)}|y_i^{(j)}\right) = \begin{cases} \log_2 2(1-p) - R & r_i^{(j)} = y_i^{(j)} \\ \log_2 2p - R & r_i^{(j)} \neq y_i^{(j)}. \end{cases} \tag{13}$$

The Fano metric introduces a length-bias term that allows for a fair comparison of paths of varying length (longer paths will, on average have accumulated large (worse) path metrics).

$$M^k(r|y) = \sum_{i=0}^{k-1}\sum_{j=0}^{n-1} P\left(r_i^{(j)}|y_i^{(j)}\right) + \underbrace{\left[\sum_{i=0}^{k-1}\sum_{j=0}^{n-1}\log_2\frac{1}{P\left(r_i^{(j)}\right)} - nkR\right]}_{\textit{length bias term}}. \tag{14}$$

The decoder's movement through the tree is dictated by the partial path metrics for the paths ending at the various tree nodes and a threshold which is varied during decoding. A flow chart for the process is shown in Figure 11. At the beginning of the decoding operation the position of the decoder is placed at the root of the tree and the decoder threshold is set to zero. The decoder then proceeds to compute the partial path metric for a forward node. Generally there are two or more forward nodes from which to choose at each node in the branching portion of the tree (one for each possible input symbol value). The first forward node tested at any position is usually the node whose output block is closest in Hamming distance to the corresponding received block. If this node proves to be a bad choice, the other forward nodes may later be tested in order of their corresponding input values (lexicographic order), without regard to the Hamming distance between the received block and the branch labels. This simplifies the implementation of the decoder.

If the partial path metric for the forward node tested exceeds the threshold, then the decoder moves its position to the forward node (this movement is represented by P+ in the flow chart). If this new position is a leaf, decoding is complete. Otherwise, if this is the decoder's first visit to the current node, the threshold is "tightened". This tightening involves raising the threshold by integer multiples of the threshold increment Δ to the highest possible value less than or equal to the partial path metric associated with the decoder's current position. The decoder then continues by testing a forward node referenced to the new position.

If the next forward node to be tested has a partial path metric that does not equal or exceed the current threshold, then the decoder examines the preceding, or "backward" node (note that, except for the root, there is exactly one node preceding a given node in the tree). If the partial path metric at the backward node does not exceed the current threshold, then the threshold is reduced by the threshold increment and the decoder tests the next forward node. This event occurs when the threshold is too high for the number of errors corrupting the received word. The lowering of the threshold also ensures that no node is ever tested twice at the same threshold value. There is thus no chance of an infinite loop occurring during the execution of the algorithm—eventually the threshold is lowered to a point that allows the decoder to move all of the way through the tree to a leaf.

If the partial path metric of the backward node exceeds the threshold, then the decoder backs up (denoted by P-). If all nodes forward from that point have been tested, then the decoder considers backing up once again. Otherwise, the decoder tests one of the remaining forward nodes. The node preceding the origin is assumed to have a metric of $-\infty$. This forces the algorithm to lower the threshold and return to the testing of the forward paths.

4.2 HEAVY TAILED BEHAVIOR IN SEQUENTIAL DECODING

Suppose that the correct path through the tree for a particular decoding operation is that shown in Figure 12 as a bold line. At each branch (choice) point an incorrect subtree is defined as all portions of the tree extending beyond the incorrect branch. Given this definition, we can then say that a given branch has been decoded once the decoder has performed its last computation in the incorrect subtree with respect to that branch.

The standard random coding argument proceeds by examining the question of how much time the decoder spends in an incorrect subtree for the ensemble of all convolutional codes. Let C_j be the number of computations performed by the decoder in the jth incorrect subtree.

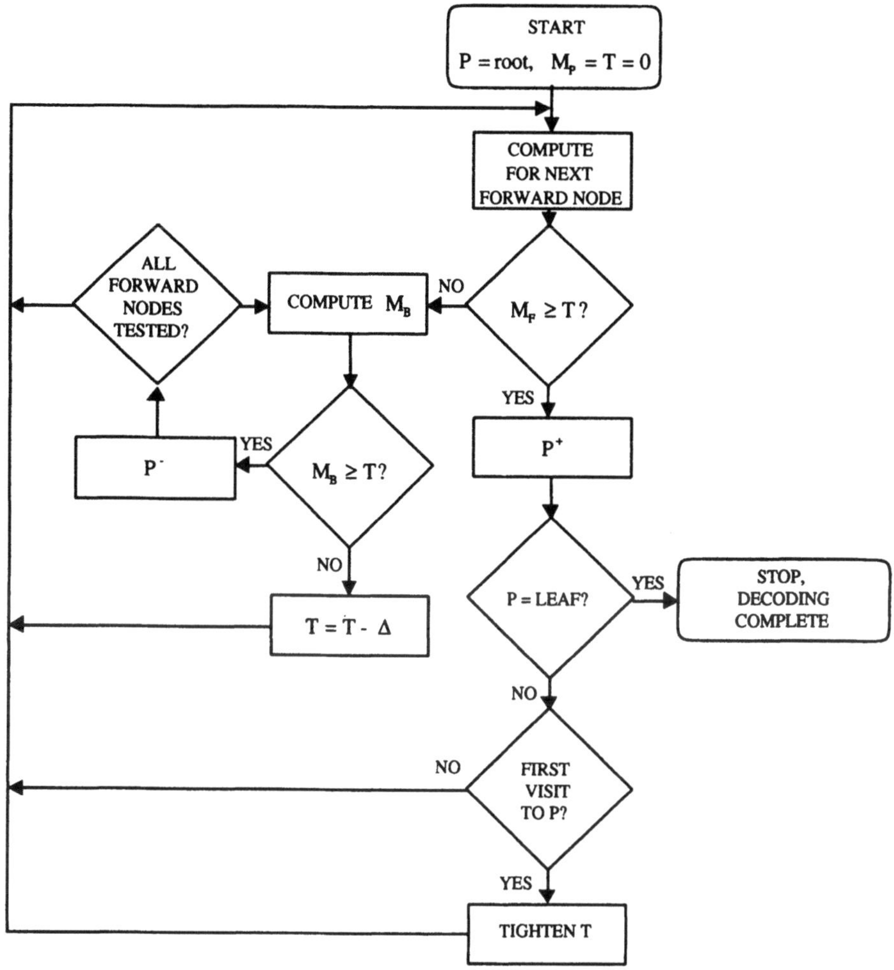

Figure 11 Flow Chart for Fano Sequential Decoding Algorithm.

The average distribution on C_j can be shown to have a Pareto distribution [30].

$$P\{C_j \geq \eta\} \approx A\eta^{-\alpha}, \quad 0 < \alpha < \infty, \quad 0 \leq j \leq L-1. \tag{15}$$

Jacobs and Berlekamp [16] formally proved this result using two arguments. The first is that with sufficiently high probability the channel noise will be extremely severe for a certain initial interval, the burst length N. The second one is that in order to find the correct path through such a noise burst the sequential decoder is expected to perform more than L computations. This fact results from the essential feature of sequential decoders in which branches are examined sequen-

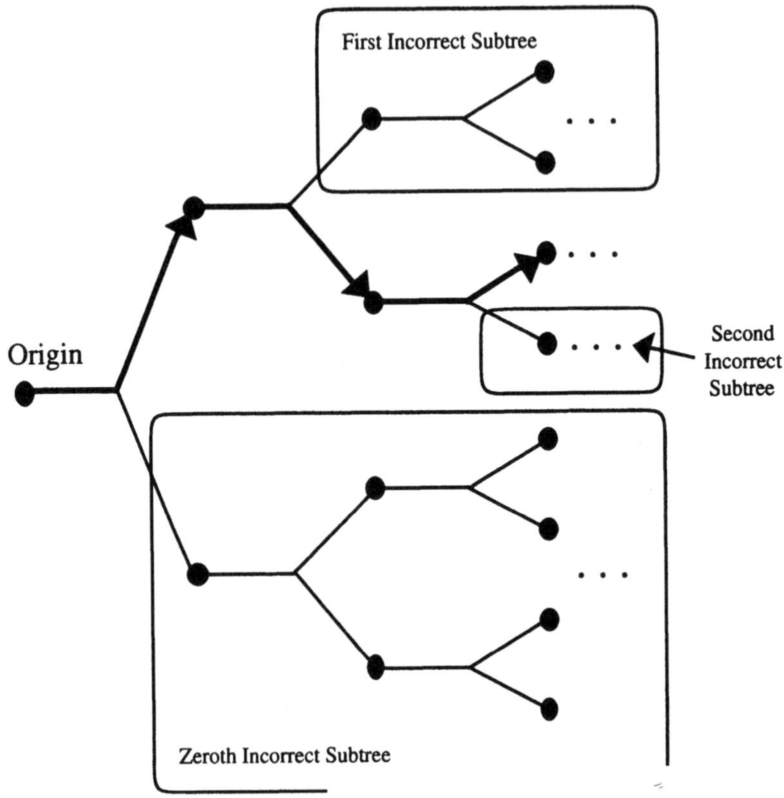

Figure 12 Incorrect Subtrees Associated with a Correct Path.

tially, in the order the bits are received i.e., at each node the decoder's choice does not depend on bits corresponding to branches deeper in the tree. The Pareto distribution of computation results from the combination of two opposite effects: (1) the number of computations necessary to find the correct path through the burst increases exponentially in N; (2) the probability of severe noise lasting for a burst length N decreases exponentially in N. (See also [8, 17, 27])

The *cutoff rate* is defined to be the code rate at which the Pareto parameter α equals 1. The cutoff rate can be found for specific channels through the use of the Gallager function. Consider a discrete memoryless channel whose transition probabilities are defined by the conditional probabilities $P(j|k)$. The Gallager function for this channel is defined as

$$E_0(\alpha) = \alpha - \log_2 \left\{ \frac{1}{2} \sum_j \left[P(j|0)^{1/1+\alpha} + P(j|1)^{1/1+\alpha} \right]^{1+\alpha} \right\}. \quad (16)$$

For a binary symmetric channel (BSC) with crossover probability ϵ, Equation (16) reduces to $E_0(\alpha) = \alpha - \log_2 \left[\varepsilon^{1/1+\alpha} + (1-\varepsilon)^{1/1+\alpha}\right]^{1+\alpha}$.

The rate R of the code in use is then related to the Pareto exponent by

$$R = \frac{E_0(\alpha)}{\alpha}, \quad 0 < R < C. \tag{17}$$

The upper limit C in this expression is the channel capacity. For fixed channel conditions, the Pareto exponent increases as the code rate decreases. An increase in the Pareto exponent causes the average number of computations in a given incorrect subtree to be reduced. Setting $\alpha = 1$ and substituting into Equation (17), we obtain an expression for the maximum code rate at which the mean of the decoding process is finite.

$$\alpha > 1 \Rightarrow R = \frac{E_0(\alpha)}{\alpha} < E_0(1) \triangleq R_0. \tag{18}$$

As mentioned earlier, when $1 < \alpha < 2$ the Pareto distribution has finite mean but infinite variance, and when $\alpha \leq 1$ it has infinite mean and infinite variance. These results represent a strong limitation for efficient sequential decoding at values of $\alpha < 2$, since, even in the case $1 < \alpha < 2$, for which the average is bounded, the variance of the distribution will be infinite and therefore the decoder will require a significantly greater than average number of computations more often than can be allowed in most applications. For this reason, it was found that, even when operating below the cutoff rate, it was often necessary to terminate sequential decoding before the search was complete, and to declare an erasure.

The parameter R_0 has come to be called the (*computational*) *cutoff rate*. It was often treated as a general parameter for a communication channel, though its meaning outside of the context of backtrack decoding algorithms was never clear. The work of Hagenauer and others on iterated versions of the CCSDS standard in the early 1990's undermined the notion of cutoff rate as a general limit (see, for example, [12]), while more recent results with turbo codes made it clear that cutoff rate was more an obstacle for a particular class of decoders than an absolute limitation [13]. Cutoff rate remains a serious problem, however, for all practical applications of backtrack decoding algorithms. In the next section we will characterize cutoff rate in greater detail and show how its impact can be eliminated.

4.3 DECODING BEYOND THE CUTOFF RATE

Figure 13 shows a pair of performance curves for a sequential/Fano decoder at two signal-to-noise ratios corresponding to operation below and above the cutoff rate. Note that the curve for the decoder operating at 9dB, well below the cutoff rate, falls off rapidly. The decoder operating at 3dB, just above cutoff, has a linear performance curve similar to that exhibited by the middle curve in Figure 4. The value of α for this second curve was found through the means described earlier to be 0.9, and therefore it is indicative that the underlying distribution has infinite mean.

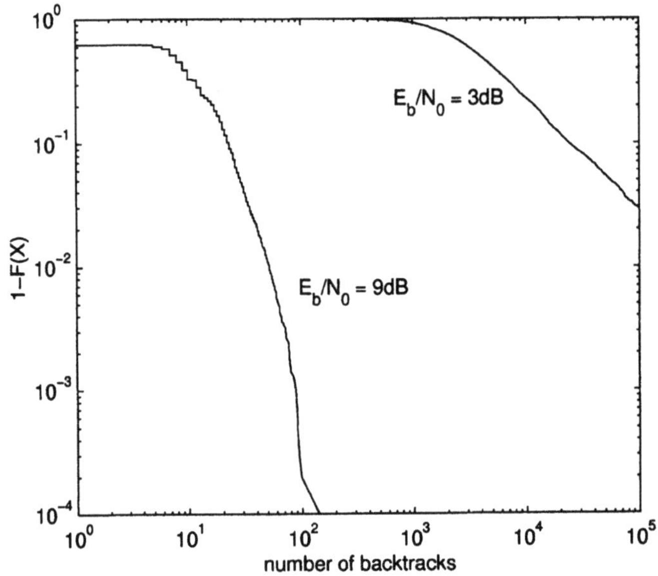

Figure 13 Decoder Performance Above and Below the Cutoff Rate.

In Figure 14 the average number of backtracks per decoding operation at a rate above cutoff is plotted as a function of the number of runs used in computing the average. The instability of the mean is evident, again consistent with the hypothesis of infinite mean (compare also with the plot in Figure 3).

We will now consider the impact of restarts. In the generalized backtrack search algorithms described above, a randomized search algorithm was applied to a data set. If the number of backtracks in the search passed a predetermined threshold, then the search was restarted on the same data set. This technique was effective in part because branch de-

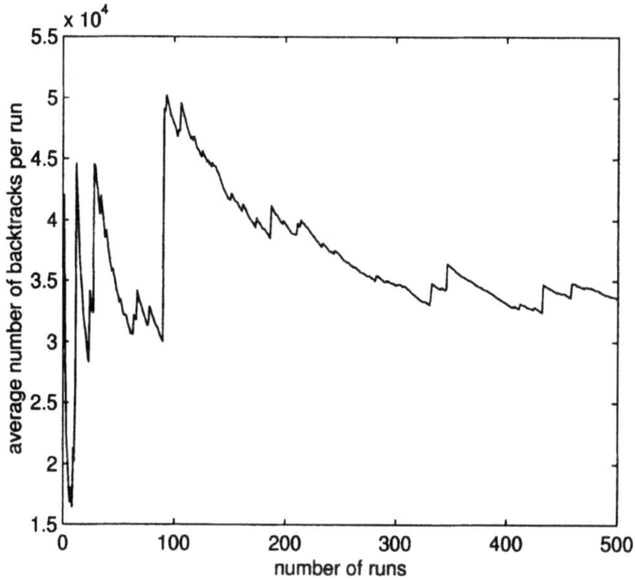

Figure 14 Erratic Behavior of the Mean Number of Backtracks Per Decoding Run Above Cutoff Rate.

cisions in the randomized algorithm were not entirely deterministic, and a restart would not result in the same path being followed. In the sequential decoding algorithm described above, there is no randomization at the branch decision points. Various attempts to introduce randomization were not successful in improving decoder performance. On the other hand, performance was greatly improved when the probabilistic nature of the channel itself was exploited. It was found that retransmission requests for channel decoders have the same impact as restarts in randomized backtrack algorithms. It was found that well-timed retransmission requests would eliminate heavy tailed behavior above the cutoff rate, as supported by Figure 15. This set of curves shows the complexity performance for a sequential decoder operating above the cutoff rate both with and without retransmission requests. The retransmission requests are triggered whenever the number of backtracks performed by the decoder exceeds a fixed threshold.

As shown in the previous section, an optimal threshold setting can be readily computed. The sensitivity of performance to the setting of this threshold is shown in Figure 16.

These results clearly show that the cutoff rate exhibited by sequential decoders is a critical point in the onset of heavy tailed behavior as gen-

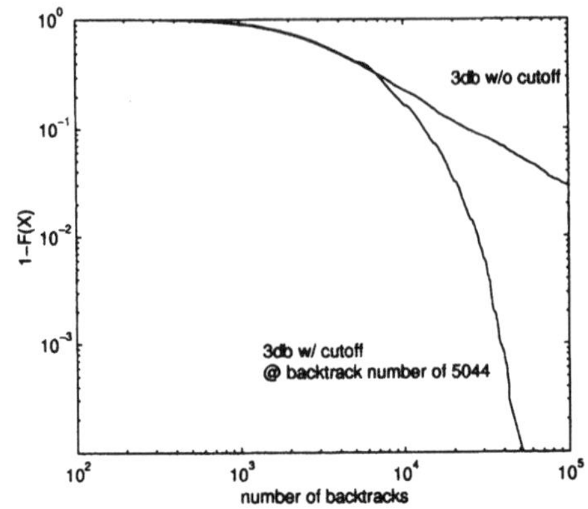

Figure 15 Distribution of the Number of Backtracks Below the Cutoff Rate. Retransmissions Eliminate Heavy Tails

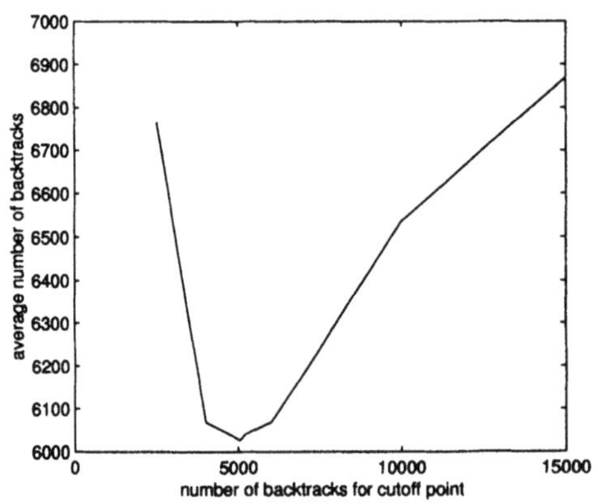

Figure 16 Sensitivity of Average Number of Backtracks to Retransmission Threshold

erally exhibited by solutions to NP-hard problems. As such, the heavy tailed behavior exhibited by sequential decoders near the cutoff rate can be eliminated through the adoption of a restart strategy.

We note that in 1983 Drukarav and Costello explored the use of retransmission requests in improving the performance of sequential decoding. They showed that the performance of sequential decoders at low signal-to-noise ratios could be greatly improved by using the elapsed decoding time and the rate of increase in the path metric as triggers for retransmission requests. Since both the elapsed time and slope of the metric can be related to the number of backtracks, the above work supports their intuition.

4.4 PHASE TRANSITIONS AND CAPACITY

If we are to relate the concepts of capacity and phase transitions in NP-hard problems, it is critical that the underlying algorithms be complete. As noted earlier, a complete algorithm provides a definitive solution to a decision problem by either providing a solution or proving that no solution exists. Most decoding algorithms are designed to find a codeword that provides the best fit to a given set of constraints. In other words, decoders perform optimization, and there is always a solution to an optimization problem (though it may not be a particularly good one). If we ignore this distinction and forge ahead with a complexity analysis of the Fano algorithm, we do get a set of curves that provide initial evidence for the existence of a phase transition.

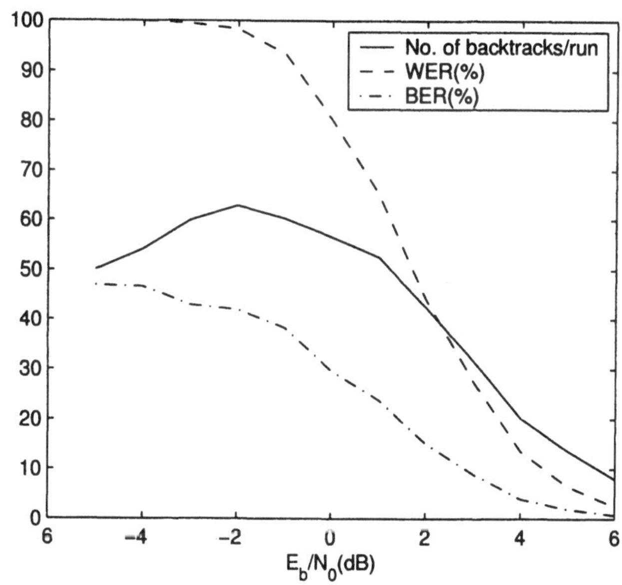

Figure 17 Phase Transition in Sequential Decoding at Capacity

The curves in Figure 17 show that as capacity is approached, the complexity of the algorithm, as indicated by the number of backtracks, peaks in the vicinity where the word error rate changes from roughly 0% to 100%, and then falls back. This indicates a connection between capacity and the more general notion of the phase transition. A conclusive link, however, must await research that focuses on decoding algorithms that are also complete decision algorithms.

5. CONCLUSION

In this paper we discussed the complexity of randomized backtrack search algorithms, focusing on their heavy tailed behavior. Heuristic strategies were shown to eliminate this behavior, significantly increasing the range of applicability of backtrack search. It was shown that this heavy tailed behavior is tied to the operation of the algorithm near a phase transition in complexity. At this transition the difficulty in showing whether a solution exists peaks, while the probability that a solution exists drops abruptly. These results were then applied to the specific case of the sequential decoding of convolutional codes. It was shown that the heavy tailed behavior of sequential decoders near and at the cutoff rate mimics the heavy tailed behavior of the more general class of randomized backtrack search algorithms. A heuristic retransmission strategy was then adopted to eliminate this heavy tailed behavior, allowing for complexity-efficient sequential decoding beyond the cutoff rate.

It follows that cutoff rate is not a barrier to the use of a decoding algorithm; it is a point along a continuous approach to a complexity phase transition. Cutoff is simply heavy tailed behavior in which the mean is no longer finite. This behavior is exhibited by a large class of algorithms, algorithms that can be effectively used well beyond this point through the use of appropriate heuristic strategies. The phase transition itself may be related to capacity—the point at which the probability that a solution exists drops to near zero. Capacity is the true barrier to effective operation of the decoding algorithm. We noted, however, that the notion of the phase transition assumes the use of a "complete" algorithm for solving a decision problem. In future work, we plan to use a complete algorithm for studying the problem of decoding convolutional codes, namely a randomized complete backtrack search applied to the satisfiability formulation of the problem. An important issue to further investigate is the relationship between the complexity curve and phase transition at capacity.

The results in this paper are readily extended to soft-decision backtrack decoding. The quantization of the received symbol is accounted for in the variable selection strategy—the process by which branch selections are made. The form of the search tree, of course, remains fixed. Since soft decision "solutions" necessarily entail a solution to the associated hard decision problem, the complexity of the former problem is at least as great as that of the latter.

We also note that the results of this paper have obvious applications in the more general area of decoding on graphs. The "bit flipping" decoding algorithm for expander codes, for example, is clearly a simple approach to solving the satisfiability problem. Recent work by one of the authors has indicated the existence of a relationship between turbo decoding and backtrack search, though the nature of the search area has not yet been fully characterized.

In short, recent results in complexity theory may lead to better decoding algorithms, and will certainly lead to a better understanding of the limitations posed by the cutoff rate and channel capacity.

6. ACKNOWLEDGMENTS

This paper is dedicated to G. David Forney on his 60th birthday, with thanks for his innumerable contributions to the field of error control and his never-failing support of young research scientists and engineers.

The authors would like to thank the National Science Foundation and DARPA for their generous support.

Notes

1. The tail of the standard normal distribution (mean 0 and standard deviation 1) decays as $\mathrm{P}\{X > x\} \sim \frac{1}{x\sqrt{2\pi}} e^{-x^2/2}$ [5]. We write $h(x) \sim g(x)$ to mean $\lim_{x \to \infty} h(x)/g(x) = 1$.

References

[1] A. V. Aho, J. E. Hopcroft, and J. D. Ullman, *The Analysis and Design of Computer Algorithms*, Reading, MA: Addison-Wesley, 1974.

[2] A. Barg, "Complexity Issues in Coding Theory," Chapter 24, *Hanbook of Coding Theory* (ed. V. S. Pless and W. C. Huffman), Amsterdam: North-Holland, 1998.

[3] E. R. Berlekamp, R. J. McEliece, and H. C. A. van Tilborg, "On the Inherent Intractability of Certain Decoding Problems," *IEEE Transactions on Information Theory*," vol. IT-24, no. 3, pp. 384–386, May 1978.

[4] A. Drukarev and D. J. Costello Jr., "Hybrid ARQ Error Control Using Sequential Decoding," *IEEE Transactions on Information Theory*, vol. IT-29, pp. 521–535, July 1983.

[5] W. Feller. *An Introduction to Probability Theory and its Applications Vol. I.* John Wiley & Sons, New York, 1968.

[6] W. Feller. *An Introduction to Probability Theory and its Applications Vol. II.* John Wiley & Sons, New York, 1971.

[7] R. G. Gallager, *Low Density Parity-Check Codes*, Cambridge, MA: MIT Press, 1963.

[8] G. D. Forney Jr., "Convolutional Codes. III. Sequential Decoding," *Information and Control*, vol. 25, no. 3, pp. 267–297, July 1974.

[9] C. P. Gomes and B. Selman, Algorithm Portfolio Design: Theory vs. Pratice, in *Proceedings of the Thirteenth Conference On Uncertainty in Artificial Intelligence (UAI-97)*, Linz, Austria., Morgan Kaufman, 1997.

[10] C. P. Gomes and B. Selman, Problem Structure in the Presence of Perturbations, in *Proceedings of the Fourteenth National Conference on Artificial Intelligence (AAAI-97)*, New Providence, RI, AAAI Press, 1997.

[11] C. P. Gomes, B. Selman, N. Crato, and H. Kautz, "Heavy-Tailed Phenomena in Satisfiability and Constraint Satisfaction Problems," *Journal of Automated Reasoning*, vol. 24, no. 1–2, pp. 67–100, 2000.

[12] J. Hagenauer, E. Offer, and L. Papke, "Matching Viterbi Decoders and Reed-Solomon Decoders in Concatenated Systems," *Reed-Solomon Codes and Their Applications* (S. B. Wicker and V. K. Bhargava, eds.), Piscataway: IEEE Press, pp. 242–271, 1994.

[13] C. Heegard and S.B. Wicker, *Turbo Coding*, Boston: Kluwer Academic Publishers, 1999.

[14] B. Hill, "A Simple General Approach to Inference About the Tail of a Distribution," *Annals of Statistics*, vol. 3 pp. 1163–1174, 1975.

[15] B. Huberman, R. Lukose, and T. Hogg, "An Economics Approach to Hard Computational Problems," *Science*, vol. 275, pp. 51–54, 1993.

[16] I. M. Jacobs and E. R. Berlekamp, "A Lower Bound to the Distribution of Computation for Sequential Decoding," *IEEE Transactions on Information Theory*, vol. IT-13, no. 2, pp. 167–174, Apr. 1967.

[17] F. Jelinek, "An Upper Bound on Moments of Sequential Decoding Effort," *IEEE Transactions on Information Theory*, vol. IT-15, no. 1, pp. 140–149, Jan. 1969.

[18] D. Johnson and M. Trick, Cliques, Coloring, and Satisfiability: Second Dimacs Implementation Challenge, in *Dimacs Series in Discrete Mathematics and Theoretical Computer Science*, vol. 36, 1996.

[19] R. M. Karp, "Reducibility Among Combinatorial Problems," in *Complexity of Computer Computations*, (ed. R. Miller and J. Thatcher), New York: Plenum, pp. 85–103, 1972.

[20] B. Mandelbrot, "The Pareto-Lévy Law and the Distribution of Income," *International Economic Review*, vol. 1, pp. 79–106, 1960.

[21] B. Mandelbrot, "The Variation of Certain Speculative Prices," *Journal of Business*, vol. 36, pp. 394–419, 1963.

[22] B. Mandelbrot, *The Fractal Geometry of Nature*, Freeman: New York, 1983.

[23] R. Monasson, R. Zecchina, S. Kirkpatrick, B. Selman, and L. Troyansy, "Determining Computational Complexity from Characteristic 'Phase Transitions'," *Nature* 400(8), pp.133-137, 1999.

[24] G. Nemhauser and L. Wolsey, *Integer and Combinatorial Optimization*, John Wiley, 1988.

[25] J. F. Puget and M. Leconte, "Beyond the Black Box: Constraints as Objects," in *Proceedings of International Logic Programming Symposium 1995*, MIT Press, 1995.

[26] G. Samorodnitsky and M. Taqqu, *Stable Non-Gaussian Random Processes: Stochastic Models with Infinite Variance*, Chapman and Hall, 1994.

[27] J. Savage, "Sequential Decoding - The Computation Problem," *Bell System Technical Journal*, vol. 45, pp. 149–175, Jan. 1966.

[28] M. Sipser and D. A. Spielman, "Expander Codes," *IEEE Transactions on Information Theory*, vol. 42, pp. 1710–1722, Nov. 1996.

[29] M. Veloso, "Learning by Analogical Reasoning in General Problem Solving," Ph.D. Thesis, CMU, CS Techn. Report CMU-CS-92-174, 1992.

[30] S. B. Wicker, *Error Control Systems for Digital Communication and Storage*, Englewood Cliffs: Prentice Hall 1995.

[31] V. Zolotarev, One-Dimensional Stable Distributions, vol. 65 of "Translations of Mathematical Monographs," American Mathematical Society. Translation from the original 1983 Russian Ed, 1986.

Chapter 3

CONCATENATED CODES WITH CONVOLUTIONAL INNER CODES

J. Justesen
Technical University of Denmark
DK-2800 Kgs. Lyngby
Denmark

Abstract Concatenated codes have been highly successful, particularly with inner convolutional codes. We discuss how the parameters of convolutional codes determine their performance as inner codes. In particular we consider bounds on minimum distances and generalized minimum distance decoding. Finally we consider the alternative of using tail-biting codes as inner codes.

1. INTRODUCTION

The possibility of using convolutional codes as inner codes in concatenated constructions was discussed in [1]. The particular choice of codes for space communications introduced an approach to error-correction that has been highly successful in applications. On the other hand, the analysis of such codes is not easy, and their properties are discussed only in a few papers. The major obstacle may be that the theory of convolutional codes has not developed as quickly as the theory of block codes. Most of the analysis tends to concentrate on encoders as systems, while the details of error-correction and decodable errors are not well understood. In this paper we shall explore some relations between parameters of concatenated codes and convolutional inner codes, and we shall analyze what properties of convolutional codes are important for their performance in this context.

2. THE CONCATENATION AND INVERSE CONCATENATION CONSTRUCTIONS

If the constituent codes of a concatenated system are an (N, K) Reed-Solomon code over $GF(2^k)$ and a binary linear (n, k, d) code, it is now a common approach to see the structure as a single linear binary (Nn, Kk) block code. This point of view was not adopted from the beginning, or the terminology for these codes might have been different. The minimum distance of the concatenated code is lower bounded by

$$D \geq d(N - K + 1) \tag{1}$$

which gives us the lower bound often referred to as the Zyablov bound [2]. In Figure 1, the minimum distance of the inner block code is represented

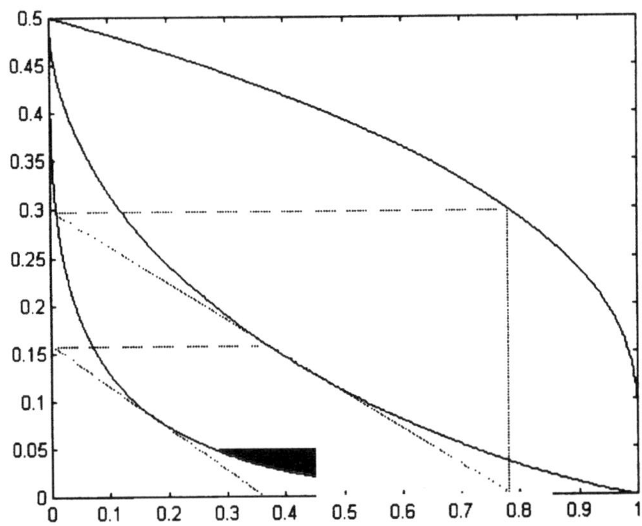

Figure 1 Concatenation and inverse concatenation constructions.

by the Varshamov-Gilbert bound, and the Zyablov bound is obtained by what may be called the "concatenation construction": A point on the $d(r)$ curve is projected on the axes, and the two points are joined by a straight line, which becomes a tangent to the lower bound. If the minimum distance is actually on this bound, there is a loss in rate and relative distance compared to the best block codes.

Forney [3], in a report which was widely circulated prior to actual publication, suggested an approach to the analysis of convolutional codes which was one of the major advances in the understanding of these codes.

A (ν, κ) memory M convolutional code may be terminated after $j\kappa$ input symbols by applying M additional zero input blocks. The resulting family of $(\nu(M+j), \kappa j)$ block codes provide much information about the convolutional code, and in particular the smallest minimum distance in the family is a bound on the free distance of the convolutional code. This translation of block code bounds to convolutional code bounds was referred to by Forney as the "inverse concatenation construction", since it happens to follow the steps described above in the reverse order. Figure 1 shows how the Costello lower bound on free distance is obtained from the Varshamov-Gilbert bound.

$$d_f/(M\nu) \;\geq\; r/(-\log(2^{l-r}-1)) \qquad (2)$$

As a conclusion of the analysis it was accurately stated [3] that *a properly terminated convolutional code is a good block code.* While this point of view provides useful bounds, the resulting block codes are interesting only in special cases.

One might be tempted by the apparently superior distance properties of convolutional codes to make a more optimistic conclusion: A *convolutional code has the same performance as a longer block code with lower rate.* In fact, since the two constructions happen to be related as discussed above, the choice of a convolutional code as the inner code might make up for the loss in performance of the concatenated code. If a convolutional code with free distance d_f is selected as the inner code, and the constraint length is equal to k, the length of the Reed-Solomon code symbols, an isolated nonzero symbol would be encoded as an inner codeword of weight at least d_f. By choosing the rate of the overall code to agree with the point where the tangent touches the Varshamov-Gilbert bound, we might get a relative weight of the encoded sequence on this bound. Clearly the argument does not provide a bound on the minimum distance in general, and the statement about convolutional codes is too optimistic. Nevertheless it has been proved that concatenated codes with suitably randomized mappings of the outer code symbols satisfy the Varshamov-Gilbert bound [5], but so far we have no method of exploiting this added performance in practical decoding algorithms.

3. PRODUCT BOUNDS ON MINIMUM DISTANCE

Since all practical decoding methods rely on decoding the two constituent codes separately and within their error-correcting capability, we are led to consider product bounds on minimum distance such as (1) for inner block codes. We would like a similar lower bound for inner convolutional codes with d replaced by d_f. This bound may be seen as a test of the relevant measures of distance for a convolutional code, and while it may be true that the free distance is the single most important parameter, it is not sufficient for this purpose.

A derivation of a product bound was given in [4], but we shall provide a slightly different interpretation. Inner codes are most often used with several interleaved outer codes. We shall assume that I codes are interleaved symbolwise, and we refer to a column of a codeword as consisting of one symbol from each of the I Reed-Solomon words. There are three cases that need to be considered as illustrated in Figure 2. If a column

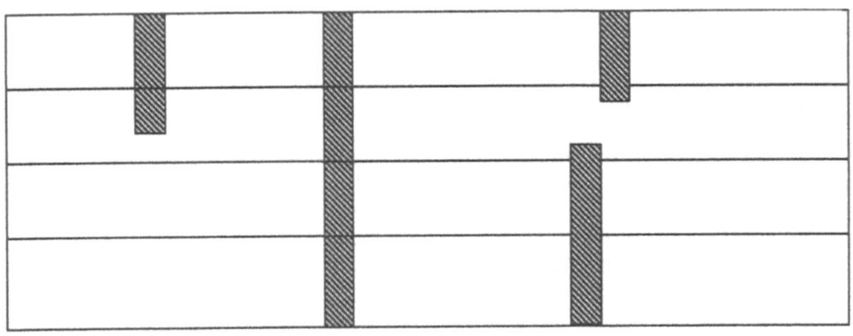

Figure 2 Product bound for concatenated code with convolutional inner code.

includes at least one nonzero symbol, but also enough zero symbols to allow the convolutional encoder to return to the zero state, the weight of the column is at least d_f. If a column is part of a long nonzero codeword in the convolutional code, the weight is at least Isb, where s is the number of blocks per symbol and b is the slope of the extended row distance. The final situation is associated with the first and last nonzero symbol that contributes to a codeword in the convolutional code. This input may be much shorter than the symbol length, but this reduced length may be balanced by the extra weight of the encoded sequence.

If the extended row distance of the inner code is lower bounded by

$$d_j^r \geq \min\{d_f, a + bj\} \quad (3)$$

the positive offset, a, is a characteristic property of a convolutional code. By choosing the parameter s right, this offset will provide the required extra weight. It may be noted that although this offset is frequently shown in graphs of distances, the analytical expression is rarely mentioned. For a typical long code it may be found as

$$a/(M\nu) = (1 + \log(1 - \delta))/(\log(1 - \delta) - \log(\delta)) \quad (4)$$

where $\delta(r)$ is the value of the Varshamov-Gilbert bound. Figure 3 shows a graph of this function.

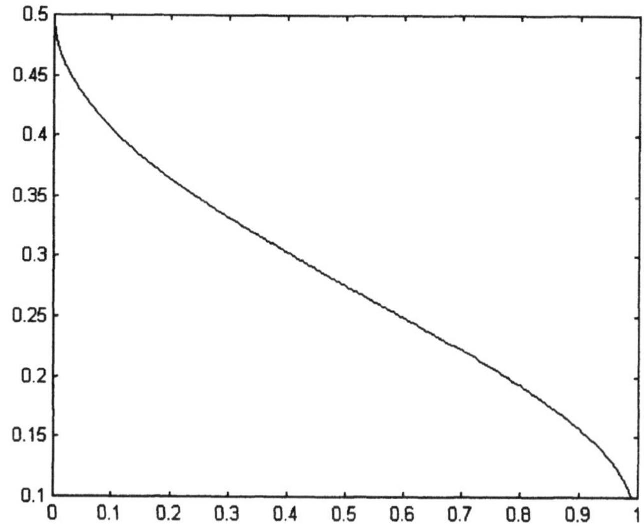

Figure 3 Weight offset for convolutional code.

We may balance the three cases to get a product bound of the desired form by choosing the parameters to satisfy (approximately) the conditions

$$d_f = Isb \qquad\qquad a = 2g(s-1) \tag{5}$$

As noted in [5], the last condition in (4) is not readily satisfied for inner codes of low rates, but we shall not discuss this case here.

Example:. For an inner code of rate $\frac{1}{2}$, (4) gives $a/(M\nu) = 0.27$, and the weight per symbol should be $bs = 0.11s\nu$. It then follows from (5) that $s \approx M$, and $I \approx 4$.

In actual system specifications, the parameters are chosen as a trade-off between performance and complexity, but the values commonly used show a good agreement with the above conditions.

4. DECODING UP TO THE PRODUCT BOUND BY GENERALIZED MINIMUM DISTANCE DECODING

Decoding the inner and outer codes separately may allow many error patterns of weight greater than half the minimum distance to be decoded, but as a worst case situation only patterns of weight $\frac{1}{4}$ of the minimum distance are decoded. Generalized minimum distance decoding may be used in these cases. Although this concept was developed in [1], the application to concatenated codes would have to follow the introduction of the product bound. With an inner block code of minimum distance d, the weight α may be assigned as a function of the number of errors corrected, t

$$\alpha = \begin{cases} 1 - 2t/d, & t < d/2 \\ 0 & \text{otherwise} \end{cases} \tag{6}$$

A slightly more general assignment of weights is possible if a list of the two closest codewords is produced by the inner decoder. If they have distances $t_1 < t_2$ from the received word, we may take

$$\alpha = \min\{(t_2 - t_1)/d, 1\} \tag{7}$$

Clearly if the distance between the two words is d, (7) reduces to (6). While GMD decoding usually does not improve the error probability much, it does serve to establish a link between the decoding algorithm and the distance bound. It is known that even if the inner code is decoded significantly beyond half the minimum distance (by a table-based decoder), the list of possible codewords is very small. However, efficient decoding of the Reed-Solomon code using these multiple inputs has not been demonstrated.

For an inner convolutional code this weight assignment is much more complicated. There have been attempts to increase the error-correcting capability by erasing unreliable segments, but the choice of the segments has not been based on the product bound on minimum distance. For a segment of the received sequence of a particular length, the reliability may be expressed by (6) with d replaced by the extended row distance (3). However, since overlapping segments of various lengths have to be considered, the weight must be calculated for each symbol as discussed in [5]. If the weights are calculated from the approximation (3), there may occur a problem illustrated in Figure 4. In a worst case situation, the number of errors corrected in a short segment is close to half the free distance or $Isb/2$, and the number of such events could be $N - K$. Thus each segment of length I symbols can have a weight close to zero and should be erased. If only $I - 1$ symbols are erased, a single error may be left, and the outer code cannot be decoded. If $I + 1$ symbols are erased, there may be too many erasures in the outer codes. However, the number of errors corrected does not provide us with information about which additional symbol should be erased. If we use list-of-two decoding of the inner code, however, we can select the closest alternative path, and erase the segment where the two paths disagree. It is not possible for

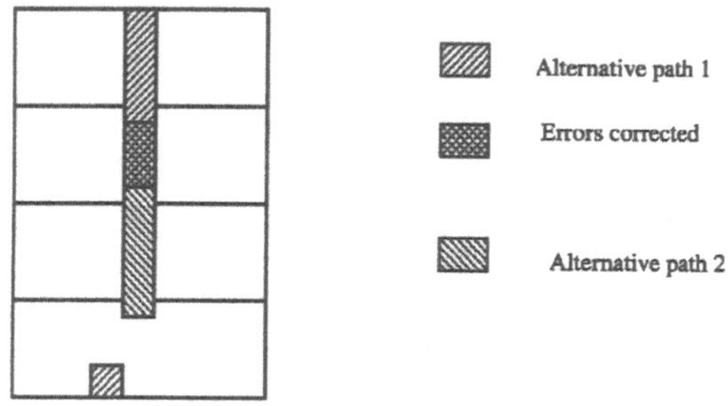

Figure 4 An error event that may require list decoding of the inner code.

both alternative paths to have distance equal to the row distance and to agree in the positions where errors have been corrected. A simple calculation shows that the errors corrected cannot be consistent with two different alternative paths.

$$2d^r_{(I-2)sb} - Isb \approx Isb \quad < \quad (I+1)sb \leq d^r_{(Is-s+2)b} \tag{8}$$

In this way a concatenated code with an inner convolutional code can always be decoded up to half the minimum distance, but the algorithm requires list decoding of the inner code. A more accurate expression for the row distance may eliminate this problem by allowing a sufficiently long segment to be erased, however list decoding still has the advantage of reducing the length of the erased segments. For decoding with high performance it is important to decode beyond half the minimum distance. In most cases, however we do not know of results relating the number of alternative paths to the number of errors corrected.

5. TAIL-BITING INNER CODES

A discussion on the internet about the results published in [6], called the attention of several researchers to the properties of short block codes related to convolutional codes by the process known as tail-biting. Even though this was not a new concept, and trellis decoding of block codes had been studied for some time, this discussion established a closer connection between block codes and convolutional codes. The results in [7] indicate that in many cases such block codes have excellent distances.

If we consider a particular noncatastrophic convolutional code of rate r, and vary the length of the tail-biting code, n_t, the minimum distance will clearly equal the free distance when n_t is sufficiently large. It follows from the inverse concatenation construction, that for average codes, termination of the convolutional code at the right length gives a block code on the Varshamov-Gilbert bound, $\delta(r')$, where r' is the rate of the terminated code. However, the tail-biting code may have codewords which do not pass through the zero state, but correspond to a cycle through the encoder states. Since the fractional weight of such a cycle is at least equal to the slope, b, of the row distance, the typical tail-biting code is on the Varshamov-Gilbert bound, $\delta(r)$. Much like the discussion in Section III, we get a balanced construction by taking

$$n_t = d_f/\delta(r) \tag{9}$$

For such a code we may conclude that *a tail-biting convolutional code of the right length is a good block code.* Thus, in such a code there are some minimum weight codewords which the code inherits from the convolutional code, but also additional words that do not pass through the zero state. In this context we are not concerned about the pseudo codewords that may occur in the decoding algorithm.

Figure 5 shows some nonzero codewords of a tail-biting code used

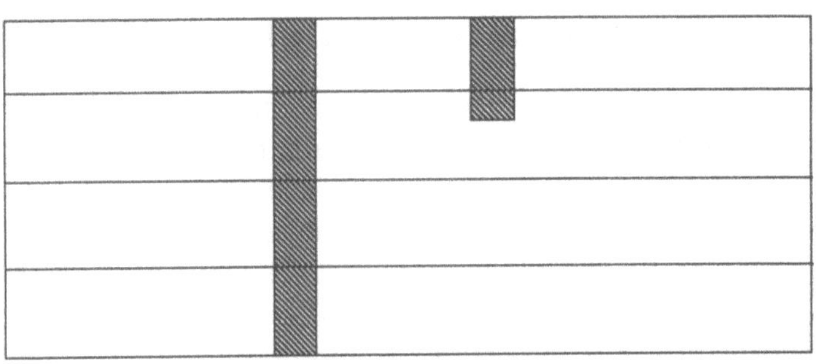

Figure 5 Tail-biting code as inner code.

as the inner code in a concatenated code. In this case the codewords which are specific to the tail-biting code replace the segments of long codewords we considered earlier.

Experiments indicate that the performances of tail-biting codes of suitable lengths are close to the performance of convolutional codes. One of the problems in making this comparison is the difference in the way error probabilities are measured, but for the present purpose the symbol error probability is the relevant quantity. For high signal-to-noise ratios, the minimum weight codewords make the most significant contribution, and clearly the performance of the tail-biting code converges to that of the convolutional code as n_t increases. Since a tail-biting code with length given by (9) has additional minimum weight codewords, it has a higher error probability. However, the difference is partially compensated by a more favorable distribution of the errors among the outer codes.

The use of tail-biting codes as inner codes may have some interesting advantages, one of which is that we avoid the problem with generalized minimum distance decoding discussed in the previous section. If a block has low weight, it may be erased, and the interaction with neighboring blocks is eliminated. Tail-biting codes may allow more efficient decoding of the inner code and more effective iteration between decoding of inner and outer codes.

6. ACKNOWLEDGEMENT

The author would like to acknowledge the inspiration from several papers by David Forney as well as from informal communications.

References

[1] Forney, Jr., G. D., *Concatenated Codes*, MIT Press, Cambridge, MA, 1966.

[2] Zyablov, V. V., "On Estimation of Complexity of Construction of Binary Linear Concatenated Codes," *Problemy Peredachi Informatsii*, vol. 7, pp. 5–13, 1971.

[3] Forney, Jr., G. D., "Convolutional Codes II: Maximum Likelihood Decoding," *Information and Control*, vol. 25, pp. 222–266, 1974.

[4] Justesen, J., C. Thommesen, and V. V. Zyablov, "Concatenated Codes With Convolutional Inner Codes," *IEEE Transactions on Information Theory*, pp. 1217–1225, 1988.

[5] Zyablov, V. V., J. Justesen, U. Dettmar, and U. Sorger, "Decoding Concatenated Codes With Convolutional Inner Codes," *Problems of Information Transmission*, pp. 3–7, 1994.

[6] Calderbank, A. R., G. D. Forney, Jr., and A. Vardy, "Minimal Tail-Biting Trellises: The Golay Code and More," *IEEE Transactions on Information Theory*, pp. 1435–1455, 1999.

[7] Stahl, P., J. B. Anderson, and R. Johannesson, "Optimal and Near-Optimal Encoders for Short and Moderate Length Tail-Biting Trellises," *IEEE Transactions on Information Theory*, pp. 2562–2571, 1999.

Chapter 4

SOME THOUGHTS ON THE EQUIVALENCE OF SYSTEMATIC AND NONSYSTEMATIC CONVOLUTIONAL ENCODERS*

D. J. Costello, Jr.
Department of Electrical Engineering
University of Notre Dame

H. A. Cabral
Department of Electrical Engineering
University of Notre Dame

O. Y. Takeshita
Department of Electrical Engineering
Ohio State University

Abstract In this paper, we re-examine the notion of equivalent systematic and nonsystematic convolutional encoders within the context of turbo coding. In particular, we argue that using equivalent systematic and nonsystematic constituent encoders in a parallel concatenated system does not yield the same performance. In fact, the performance can differ greatly, depending on the effective free distance and the encoder mapping. Since nonsystematic encoders typically have larger effective free distances than their systematic counterparts, they are better in the "error floor", or low BER, region of the performance curve. On the other hand, since systematic encoders typically have better encoder mappings than nonsystematic encoders, they are better in the "waterfall", or high BER, region of the performance curve.

*This work was supported by NSF grant NCR95-22939 and NASA grant NAG5-8355.

1. INTRODUCTION

In Forney's seminal 1970 paper, "Convolutional codes I: algebraic structure" [1], he introduced the notion of equivalent convolutional encoders, i.e., encoders whose set of output sequences is identical over the entire space of possible input sequences. Equivalent encoders are thus considered to generate the same code, i.e., the same set of codewords. Additional work in this area by Forney and others has focused on finding the encoder with the smallest state complexity, a so-called minimal encoder, that generates a particular code. (See the new book by Johannesson and Zigangirov [2] for a complete summary of this work.)

As an example of this notion of equivalence, it is well known that any nonsystematic feedforward encoder is equivalent to a minimal systematic feedback encoder [1, 3]. In other words, the weight spectrum of the code generated by the two encoders is identical. (The code generated by a systematic feedback encoder has been referred to in the recent literature on turbo coding as a systematic recursive convolutional code (SRCC) [4].)

Example 1 Consider the $(2, 1, 3)$ nonsystematic feedforward encoder[1] given by

$$G_{\text{ns}}(D) = [\ 1 + D + D^3 \quad 1 + D^2 + D^3\ ]. \tag{1}$$

The equivalent systematic feedback encoder is given by

$$G_{\text{s}}(D) = [\ 1 \quad (1 + D^2 + D^3)/(1 + D + D^3)\ ]. \tag{2}$$

The input sequence $x_{\text{s}}(D) = 1 + D^7$ to the systematic encoder generates the codeword

$$y(D) = x_{\text{s}}(D) G_{\text{s}}(D) = [\ 1 + D^7 \quad 1 + D + D^4 + D^5 + D^6 + D^7\ ], \tag{3}$$

whereas for the nonsystematic encoder the input sequence $x_{\text{ns}}(D) = 1 + D + D^2 + D^4$ generates the same codeword, i.e.,

$$y(D) = x_{\text{ns}}(D) G_{\text{ns}}(D) = [1 + D^7 \quad 1 + D + D^4 + D^5 + D^6 + D^7\ ]. \tag{4}$$

Note that, although the codeword is the same in both cases, the input sequence has weight two in the systematic case and weight four in the nonsystematic case.

This example illustrates an important property of equivalent encoders, viz., the input sequences that generate a given codeword are different and may have different weights. In other words, although the two codes are identical, the encoder mapping between input sequences and codewords is different.

Nonsystematic feedback encoders, which can be used to generate nonsystematic recursive convolutional codes (NSRCC's), have not received much attention in the literature because it can be shown that they are equivalent to systematic feedback encoders. For example, consider a rate $R = k/n = 1/2$ NSRCC generated by

$$G_{\rm ns}(D) = [\ n_1(D)/d_1(D) \quad n_2(D)/d_2(D)\ ]. \tag{5}$$

Multiplying $G_{\rm ns}(D)$ by $\mathrm{LCM}[d_1(D), d_2(D)]$ results in an equivalent nonsystematic feedforward encoder, which in turn has an equivalent systemetic feedback encoder.

A general encoder for the class of rate 1/2 NSRCC's described above, realized using two shift registers, is shown in Figure 1. If we let $m_i =$

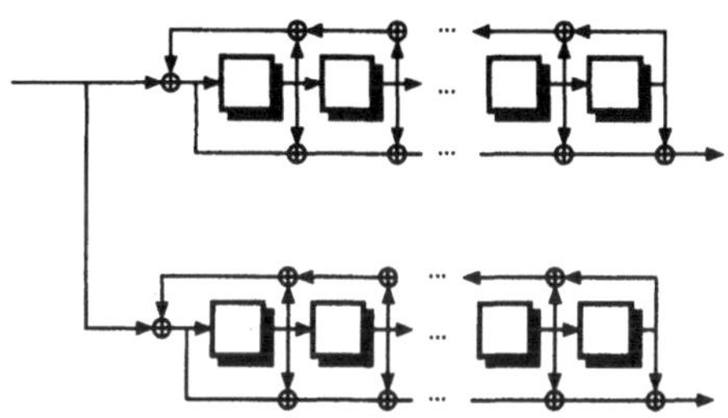

Figure 1 A general encoder for a rate 1/2 NSRCC.

$\max\{\deg[n_i(D)], \deg[d_i(D)]\}$ and assume that $n_i(D)$ and $d_i(D)$ have no common factors, $i = 1, 2$, then the encoder of Figure 1 has $2^{m_1+m_2}$ states. An interesting subclass of rate 1/2 NSRCC's results when we require that $d_1(D) = d_2(D) = d(D)$, i.e.,

$$G_{\rm ns}(D) = [\ n_1(D)/d(D) \quad n_2(D)/d(D)\ ]. \tag{6}$$

In this case the NSRCC can be realized with a single shift register, as shown in Figure 2, and the encoder has only 2^m states, where $m = \max\{\deg[n_1(D)], \deg[n_2(D)], \deg[d(D)]\}$ and $n_1(D)$, $n_2(D)$, and $d(D)$ have no common factors. Because of its simpler description, this is the form of NSRCC we will assume throughout the paper. (Note that the NSRCC becomes systematic if either $n_1(D) = d(D)$ or $n_2(D) = d(D)$.)

Example 2 Consider the (2,1,3) nonsystematic feedback encoder given by

$$G_{\rm ns}(D) = [\,(1+D+D^3)/(1+D^2+D^3) \quad (1+D+D^2+D^3)/(1+D^2+D^3)\ ]. \tag{7}$$

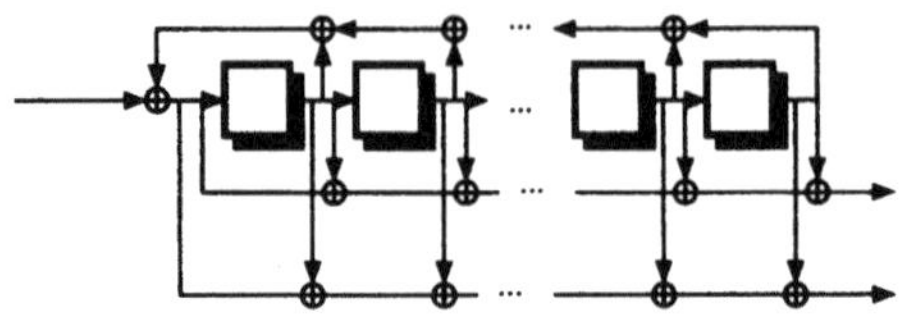

Figure 2 A rate 1/2 NSRCC realized with one shift register.

The equivalent systematic feedback encoder is given by

$$G_s(D) = [\ 1 \quad (1 + D + D^2 + D^3)/(1 + D + D^3)\ ]. \tag{8}$$

The input sequence $x_s(D) = 1 + D^7$ to the systematic encoder generates the codeword

$$y(D){=}x_s(D)G_s(D){=}[\ 1 + D^7 \quad 1 + D^2 + D^3 + D^4 + D^6 + D^7\ ], \tag{9}$$

whereas for the nonsystematic encoder the input sequence $x_{ns}(D) = 1 + D + D^4 + D^5 + D^6 + D^7$ generates the same codeword, i.e.,

$$y(D){=}x_{ns}(D)G_{ns}(D){=}[\ 1 + D^7 \quad 1 + D^2 + D^3 + D^4 + D^6 + D^7\ ]. \tag{10}$$

Again note that the input sequences that generate the same codeword have different weights.

With standard decoding methods for convolutional codes such as the maximum likelihood (ML) or maximum posterior (MAP) algorithms, the performance of a code generated by equivalent encoders is expected to be essentially identical. In fact, if we plot the codeword bit error rate (CBER), the performance with ML decoding is exactly the same. However, because of the different encoder mapping between information sequences and codewords, if we plot the information bit error rate (IBER)[2], the performance with either ML or MAP decoding will differ slightly. This is illustrated in Figure 3 for the two equivalent eight-state convolutional encoders of Example 1. However, the difference in performance is typically so small that it is commonly ignored.

2. TURBO CODES WITH EQUIVALENT CONSTITUENT ENCODERS

The recently discovered parallel concatenated convolutional codes, or turbo codes [4], have focused the attention of the research community on the properties of SRCC's, since feedback in the encoder is necessary to realize the astounding "interleaver gain" property of turbo codes. In

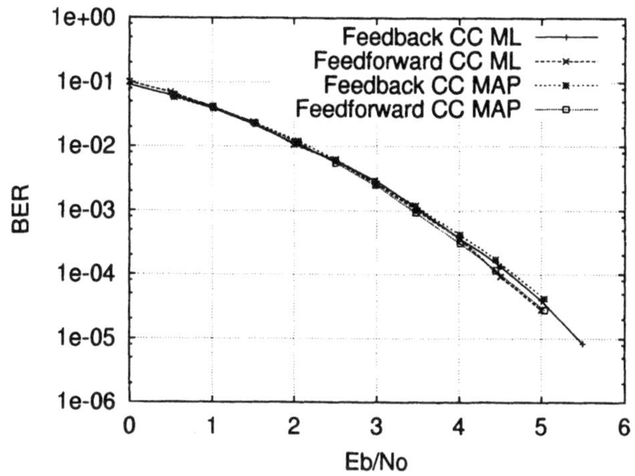

Figure 3 The performance of two equivalent convolutional encoders for ML and MAP decoding.

other words, the performance of a turbo code degrades drastically when a nonsystematic feedforward encoder is used as a constituent code instead of its equivalent systematic feedback encoder.

Example 3 Consider the two turbo codes illustrated in Figure 4, where a $(2,1,2)$ nonsystematic feedforward convolutional encoder with generator matrix

$$G_{\text{ns}}(D) = [\ 1 + D + D^2 \quad 1 + D^2\ ] \tag{11}$$

and its systematic feedback equivalent with generator matrix

$$G_{\text{s}}(D) = [\ 1 \quad (1 + D^2)/(1 + D + D^2)\ ] \tag{12}$$

are used as constituent codes in a rate $1/3$ turbo code.

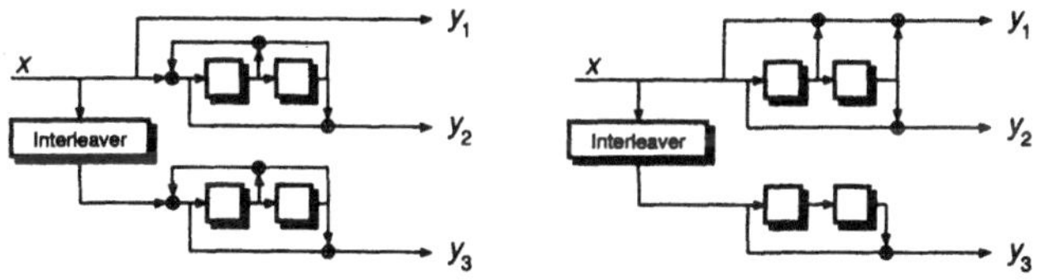

Figure 4 Two rate $1/3$ turbo codes with equivalent $(2,1,2)$ constituent encoders.

Simulation results using iterative a posteriori probability (APP) decoding show that, for an information block length of $N = 1024$ and a signal-to-noise ratio (SNR) of $E_b/N_0 = 1.76$dB, the BER of this turbo code with the systematic feedback encoder is 1.0×10^{-5}. On the other hand, the BER with the nonsystematic feedforward encoder is 7.2×10^{-3}, almost three orders of magnitude worse. Consequently, nonsystematic encoders have not previously been considered for use in turbo codes. ∎

However, a recently proposed modified iterative decoding algorithm [5] has been shown to be able to effectively decode turbo codes whose constituent codes are NSRCC's rather than SRCC's. In other words, by replacing the nonsystematic *feedforward* constituent encoders in Figure 4 with nonsystematic *feedback* encoders, we remove the usual restriction that the constituent encoders must be systematic. This opens up a large class of additional constituent encoders that can be employed in turbo codes, viz., NSRCC's, thus creating the possibility of improved performance. An example of a rate 1/3 turbo code with the NSRCC constituent encoder given by Equation (7) is shown in Figure 5. Note that it is only the top encoder that differs from the usual turbo code, since there is no systematic output sequence.

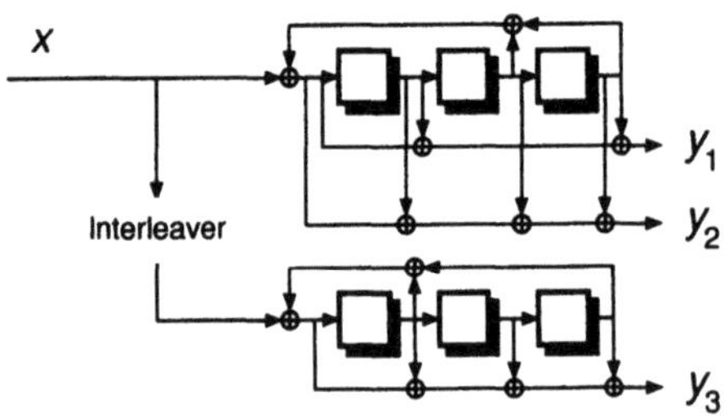

Figure 5 A rate 1/3 turbo code with an NSRCC constituent encoder.

As noted above, an NSRCC always has an equivalent SRCC, and thus it is reasonable to ask if there can be any advantage to using NSRCC's instead of SRCC's as constituent encoders in a turbo code. In other words, would we expect there to be any difference in the performance of the rate 1/3 turbo code shown in Figure 5 (with an NSRCC constituent encoder) and the one shown in Figure 6, with the equivalent SRCC constituent encoder given by Equation (8)?

In Example 3, we saw that equivalent encoders, one feedback and one feedforward, resulted in drastically different BERs when employed as constituent encoders in a turbo code. Hence, even though two encoders may be equivalent, their performance when used as constituent encoders in a turbo code may be quite different. Thus it does make sense to consider the possible use of NSRCC's in turbo codes. In this paper

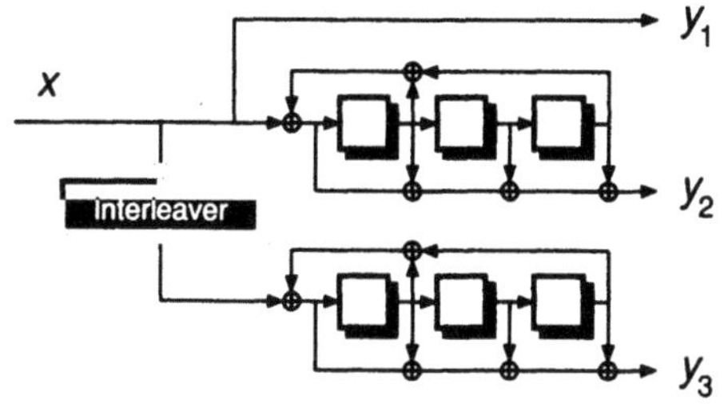

Figure 6 A rate 1/3 turbo code with an SRCC constituent encoder.

we examine some of the properties of NSRCC's that may make them attractive alternatives to SRCC's for use in turbo codes.

3. ITERATIVE DECODING OF TURBO CODES WITH NSRCC CONSTITUENT ENCODERS

A block diagram of the recently proposed modified iterative algorithm [5] for decoding turbo codes with NSRCC constituent encoders is shown in Figure 7, along with the standard iterative decoder used with SRCC constituent encoders. Each APP decoder employs the usual BCJR algorithm [6].

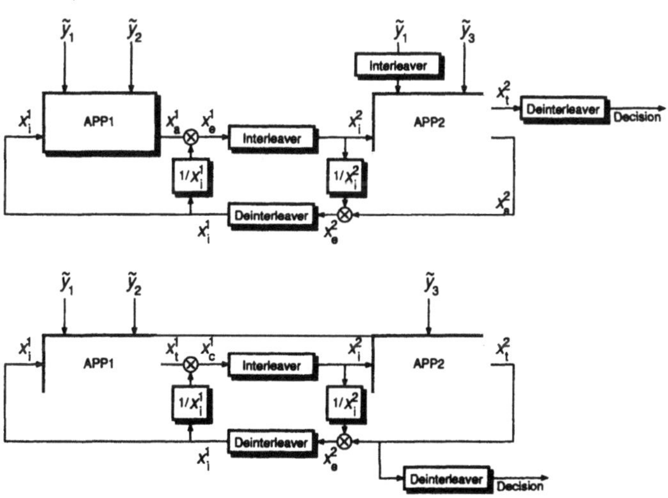

Figure 7 Decoder block diagrams for systematic (upper diagram) and nonsystematic (lower diagram) turbo codes.

The output probabilities of each APP decoder can be considered as the product of three terms [7]: the channel values, the a priori probabilities, and the extrinsic probabilities, which are passed to the next decoder as a priori probabilities. In Figure 7, we have adopted the following notation (the superscript $j = 1$ or 2 denotes values associated with APP decoder 1 or 2, respectively):

- x_t^j represents the total APP value (channel, a priori, and extrinsic),
- x_a^j represents only the a priori and extrinsic values,
- x_e^j represents only the extrinsic value,
- x_i^j represents the input values to each decoder,
- x_c^j represents the total APP value with the a priori value removed.

We see that decoder 1 is identical in the two cases. In the NSRCC case, however, decoder 1 passes the total APP value for each information bit, with only the a priori value removed, to decoder 2, whereas in the SRCC case, both the a priori and channel values are removed. This is because in the NSRCC case, there are no channel values for the information bits entering decoder 2. Thus soft values for the information bits must be passed to decoder 2 from decoder 1. These soft values are contained in the x_c^1 sequence. (Note that in the NSRCC case these soft values for the information bits are not received channel values.) The only difference in decoder 2 is that there is only one channel input in the NSRCC case. Also, the extrinsic values that are passed from the output of decoder 2 back to decoder 1 as a priori inputs are extracted from x_t^2 in the NSRCC case, whereas in the SRCC case they are extracted from x_a^2.

Finally, it is worth noting that, when either y_1 or y_2 is a systematic bit, the modified iterative algorithm for decoding turbo codes based on NSRCC's shown in Figure 7 will also decode turbo codes based on SRCC's. The only difference from the standard decoder is that, instead of entering the soft received channel values for the systematic bits directly into decoder 2, they are passed through to the a priori input of decoder 2 along with the extrinsic values from decoder 1. Thus, in the case of SRCC's, the two decoders shown in Figure 7 yield identical results. However, the implementation differs in the two cases, because nothing in the modified decoder assumes the transmission of systematic bits, and hence it is a completely general and transparent decoder.

4. EFFECTIVE FREE DISTANCE OF NSRCC'S

As noted in Example 2, every nonsystematic feedback encoder has an equivalent systematic feedback encoder. So as normal convolutional codes with either ML or MAP decoding, their BER performance is essentially identical. However, equivalent encoders do not necessarily give equivalent performance in a turbo coding arrangement, as shown in Example 3. In particular, the encoder mapping between information sequences and code sequences, which has only a minor effect on performance for normal convolutional codes, plays a critical role in the performance of a turbo code.

This is due to the action of the interleaver which, after encoding by the first constituent code, scrambles the information sequence before encoding by the second constituent code. The weight of the information sequence is preserved by the interleaver, but the ordering of input bits to the two encoders is different. Since a weight 1 input sequence cannot terminate a feedback encoder, it must generate an infinite weight output sequence, and hence weight 1 inputs can be ignored in computing the low weight terms in the distance spectrum of turbo codes. Weight 2 inputs, however, can terminate a feedback encoder, and thus may lead to a low weight output sequence. If a weight 2 input sequence that terminates the first encoder, after interleaving, also terminates the second encoder, the overall codeword can have low weight. Thus codewords corresponding to weight 2 input sequences are important in determining the distance spectrum of turbo codes. For a typical random interleaver, relative to weight 2 input sequences, higher weight input sequences that terminate the first encoder are much less likely to also terminate the second encoder and generate a low weight codeword [8]. Thus, compared to weight 2 input sequences, higher weight input sequences contribute relatively few low weight codewords to the distance spectrum of turbo codes, and the codewords corresponding to weight 2 input sequences are said to determine the effective free distance d_{eff} of a turbo code. For this reason, d_{eff} is considered to be the single most important distance parameter determining turbo code performance [9].[3]

Definition 1 The effective free distance [10] d_{eff} of a turbo code is the minimum weight output sequence corresponding to an input sequence of weight 2.

The output of a systematic (n, k, m) convolutional encoder consists of k information sequences and $r = n - k$ parity sequences, where each of the last r columns, i.e., the nonsystematic part, of the generator matrix generates a parity sequence. (In the case of a nonsystematic encoder, every one of the n columns of the generator matrix generates a "parity" sequence.) Now let $g_i(D)$ denote the column of the generator matrix $G(D)$ that generates the ith parity sequence ($1 \leq i \leq r$ for systematic encoders and $1 \leq i \leq n$ for nonsystematic encoders) and call $g_i(D)$ the i-th parity generator.

Definition 2 ([11]) The weight-2 input minimum distance $d_2(i)$ of the i-th parity generator $g_i(D)$ of an (n, k, m) convolutional encoder is the minimum weight parity output sequence corresponding to an input sequence of weight 2.

The following theorem follows directly from the above definition.

Theorem 1 *The weight-2 input minimum distance d_2 of an (n,k,m) convolutional encoder is given by*

$$d_2 = 2 + \sum_{i=1}^{r} d_2(i) \qquad \textit{(systematic case)} \qquad (13)$$

$$d_2 = \sum_{i=1}^{n} d_2(i) \qquad \textit{(nonsystematic case).} \qquad (14)$$

Tables of d_2-optimal code generators for systematic encoders with rates $R = k/n = 1/2$, $2/3$, $1/3$, $3/4$, $2/4$, and $1/4$ and memory orders as high as $m = 6$ are given in [11]. (We note from Equations (13) and (14) that the tables in [11] for rate $1/3$ and $1/4$ systematic encoders also serve as d_2-optimal nonsystematic encoders for rates $1/2$ and $1/3$, respectively.) Since, except for trivial parity generators of the form $g_i(D) = D^j/(1 + D^m)$, $0 \leq j \leq m$, $d_2(i) \geq 2$, it is clear from Equations (13) and (14) that nonsystematic encoders can achieve larger values of d_2 than systematic encoders of the same complexity.

For the rate $1/3$ turbo codes considered in this paper, the first constituent encoder is either a systematic or nonsystematic rate $1/2$ feedback convolutional encoder whose weight-2 input minimum distance is denoted either d_2^{s} or d_2^{ns}, respectively, and the second constituent encoder is a feedback parity generator whose weight-2 input minimum distance is denoted $d_2^{(2)}$. Then we can write the effective free distance of the corresponding rate $1/3$ turbo code as

$$d_{\mathrm{eff}}^{\mathrm{s}} = d_2^{\mathrm{s}} + d_2^{(2)} \qquad \text{(systematic case)} \qquad (15)$$

$$d_{\mathrm{eff}}^{\mathrm{ns}} = d_2^{\mathrm{ns}} + d_2^{(2)} \qquad \text{(nonsystematic case).} \qquad (16)$$

Since larger values of d_2 can be achieved with nonsystematic encoders than with systematic encoders of the same complexity, i.e., $d_2^{\mathrm{ns}} \geq d_2^{\mathrm{s}}$, Equations (15) and (16) imply that turbo codes employing nonsystematic rather than systematic constituent encoders can achieve larger values of d_{eff}. In Table 1, we list the largest possible effective free distances of rate 1/3 turbo codes, for both systematic and nonsystematic constituent encoders, constructed using d_2-optimal code generators obtained from the tables in [11]. We see that, except for the trivial $m = 1$ case, turbo codes with nonsystematic constituent encoders achieve the largest values of d_{eff}.

m	$d_{\mathrm{eff}}^{\mathrm{s}}$	$d_{\mathrm{eff}}^{\mathrm{ns}}$
1	4	3
2	10	12
3	14	18
4	22	30
5	38	54
6	70	102

Table 1 Optimal effective free distances of rate 1/3 turbo codes.

Although it is normal in practice to use a common parity generator for both constituent encoders in a turbo code (i.e., to use a symmetric design), it is also possible to use an asymmetric design in which the two constituent encoders do not have a parity generator in common [12]. (It should be noted that Equations (15) and (16) for computing d_{eff} hold regardless of whether the turbo code is symmetric or asymmetric.) In fact, some of the nonsystematic turbo codes[1] listed in Table 1 use asymmetric designs. For example, the eight-state ($m = 3$) nonsystematic rate 1/3 turbo code shown in Figure 5, which has optimal $d_{\mathrm{eff}} = 18$, is asymmetric since a different feedback polynomial is used in the two constituent encoders.

The fact that nonsystematic turbo codes have larger values of d_{eff} than systematic turbo codes of the same complexity makes it reasonable to expect that they will exhibit better performance. To test this hypothesis, we simulated the two "equivalent" eight-state turbo codes, one nonsystematic and the other systematic, shown in Figures 5 and 6, respectively. (The first constituent encoder in the systematic code is equivalent to the first constituent encoder in the nonsystematic code.) The nonsystematic code has $d_{\mathrm{eff}} = 18$, while the systematic code has

$d_{\text{eff}} = 14$. The results of the simulation for an information block size of $N = 1024$ are shown in Figure 8.

We note that the nonsystematic turbo code has a lower "error floor" than the systematic turbo code, i.e., it performs significantly better at high SNR's. However, its performance is slightly worse than the systematic turbo code in the "waterfall" region of the BER curve, i.e., at low SNR's. This example illustrates the well-known fact that minimum distance, in this case d_{eff}, is known to be the primary influence on performance at high SNR's, i.e., in the "error floor" region of turbo codes, but at lower SNR's other factors must also be considered. This is because at high SNR's the iterative decoder is essentially maximum likelihood, but at low SNR's the entire low weight portion of the distance spectrum must be considered, as well as the convergence properties of the iterative decoding algorithm.

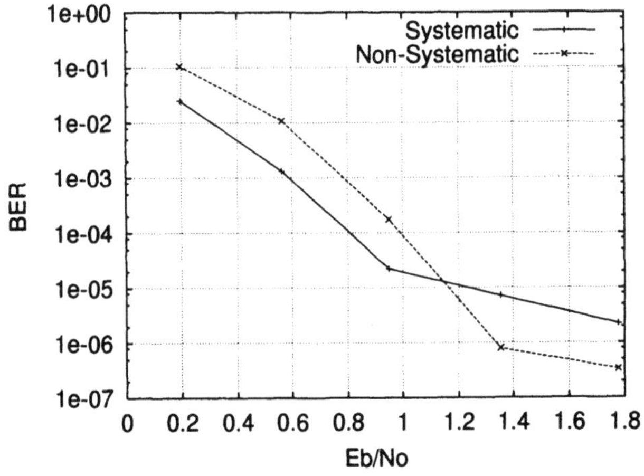

Figure 8 The BER performance of two "equivalent" 8-state nonsystematic and systematic turbo codes.

Is it possible to eliminate the performance loss of nonsystematic turbo codes in the "waterfall" region of the BER curve suggested by the above example while maintaining their superior performance performance in the "error floor" region? In the next section, we examine a possible explanation for the subpar performance of nonsystematic turbo codes in the "waterfall" region based on the distance spectrum of the code.

5. NONSYSTEMATIC TURBO CODES WITH CATASTROPHIC ENCODERS

An anomaly noted in [11] occurs when convolutional encoders have repeated parity generators. For example, the systematic feedback $(3,1,2)$ encoder with the best value of d_2, viz., $d_2 = 10$, is given by the generator matrix

$$G_s(D) = [1 \;\; (1+D^2)/(1+D+D^2) \;\; (1+D^2)/(1+D+D^2) \;]. \quad (17)$$

However, it is stated in [11] that using such encoders as constituent encoders in a turbo code "yields poor results", and whenever the encoder with the best value of d_2 contains repeated generators, as above, the tables in [11] also list the encoder without repeated generators having the best value of d_2. A natural question arises. Why should the encoder of Equation (17), with the best value of d_2 of any systematic feedback $(3,1,2)$ encoder, perform poorly when used as a constituent encoder in a turbo code? (In this case, because the constituent encoder has rate 1/3, the associated turbo code would have rate 1/5.)

Insight into the answer to this question can be obtained by viewing the nonsystematic portion of the above encoder as a $(2,1,2)$ nonsystematic encoder with generator matrix

$$G_{ns}(D) = [\; (1+D^2)/(1+D+D^2) \;\; (1+D^2)/(1+D+D^2) \;]. \quad (18)$$

This encoder is catastrophic. However, in a nonsystematic turbo code, it is not immediately obvious that the constituent encoder should not be catastrophic, since the randomization introduced by the interleaver implies that the overall encoder would still be noncatastrophic. In order to see if catastrophic encoders with large values of d_2 can be effectively used as constituent encoders in a turbo code, we simulated a rate 1/3, four-state turbo code under two conditions:

1. with the nonsystematic catastrophic encoder $G_{ns}(D) = [\; g(D) \;\; g(D) \;]$ as the first constituent encoder and the parity generator $g(D)$ as the second constituent encoder, and

2. with the systematic (and hence noncatastrophic) encoder $G_s(D) = [\; 1 \;\; g(D) \;]$ as the first constituent encoder and the parity generator $g(D)$ as the second constituent encoder,

where $g(D) = (1+D^2)/(1+D+D^2)$. Since $G_{ns}(D)$ has $d_2 = 8$, the effective free distance of the nonsystematic turbo code is $d_{\text{eff}} = 12$, whereas $G_s(D)$ has $d_2 = 6$ and the effective free distance of the systematic turbo code is only $d_{\text{eff}} = 10$. The results are shown in Figure 9 for an information block size of $N = 1024$. We see clearly that the nonsystematic turbo code performs much worse than the systematic turbo code in both the "waterfall" and "error floor" regions, even though it has a larger d_{eff}. Thus, in the design of nonsystematic turbo codes, the use of catastrophic constituent encoders should be avoided. Also, since systematic encoders with repeated generators become catastrophic when the systematic part is removed, this may explain why such encoders should be avoided in the construction of systematic turbo codes, as noted in [11].

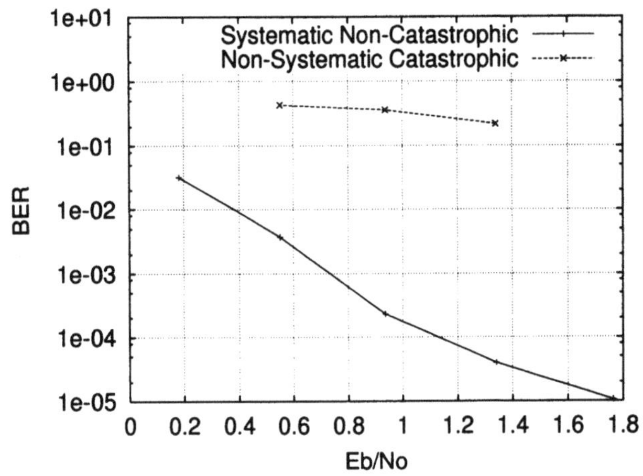

Figure 9 The BER performance of two 4-state turbo codes with noncatastrophic and catastrophic constituent encoders.

The probable explanation for the poor performance of nonsystematic turbo codes with catastrophic constituent encoders is that the iterative decoding algorithm is ineffective in these cases. In other words, even though the turbo code is noncatastrophic and may have excellent distance properties, which would allow it to be effectively decoded with a maximum likelihood decoder, the constituent decoders cannot generate reliable extrinsic information to allow iterative decoding to converge. In the case of a rate 1/3 nonsystematic turbo code whose first constituent encoder is catastrophic, the first constituent decoder is trying to decode the output of a rate 1/2 catastrophic encoder. For the low SNR's at which turbo codes operate, this leads to unreliable extrinsic values being passed to the second constituent decoder and thus the algorithm fails to converge.

For example, consider the above four-state nonsystematic turbo code. The catastrophic constituent encoder has a weight two codeword corresponding to the infinite weight input sequence $x(D) = (1+D+D^2)/(1+D^2)$. Since this codeword and the all-zero codeword differ in only two positions, even moderate noise conditions can cause these two codewords to have almost the same APP value at the output of the corresponding decoder. Thus the associated log-likelihood ratios, which are passed to the other decoder as extrinsic information, will be close to zero for all the information bits in which the two codewords differ, i.e., all the extrinsic values will be close to zero. In other words, the extrinsic values passed to the other decoder contain no useful information, and the iterative decoding process will not converge.

This explanation suggests that, in order for a turbo decoder to properly converge, it is important that the low weight codewords in a constituent code be generated by even lower weight information sequences. This is always true for systematic codes, but it is not always true for nonsystematic codes. In particular, catastrophic encoders have low weight codewords generated by infinite weight information sequences. Thus it is tempting to conjecture that the encoder mapping between information sequences and low weight codewords for the constituent encoders in a turbo code determines the convergence characteristics of the decoder[5] and that this may explain why systematic turbo codes typically outperform nonsystematic turbo codes in the "waterfall" region of the BER curve. The following example supports this conjecture.

Example 4 In this example, we examine the encoder mapping between information sequences and low weight codewords for the equivalent rate 1/2 encoders, one nonsystematic and one systematic, whose generator matrices are given in Equations (7) and (8), respectively. When used as constituent encoders in a turbo code, their relative performance was shown in Figure 8. From Equations (7) and (8), we see that the information sequence

$$x_{\mathrm{ns}}(D) = (1 + D^2 + D^3)u(D) \tag{19}$$

to the nonsystematic encoder produces the codeword

$$y(D) = [\ (1 + D + D^3)u(D) \quad (1 + D + D^2 + D^3)u(D)\ ], \tag{20}$$

and that the same codeword is produced by the information sequence

$$x_{\mathrm{s}}(D) = (1 + D + D^3)u(D) \tag{21}$$

to the equivalent systematic encoder. Thus we can examine the encoder mapping between information sequences and low weight codewords for these two encoders by considering polynomials $u(D)$ of small degree. (Clearly, if the degree of $u(D)$ is large, the codeword $y(D)$ will have high weight.) Table 2 shows the mapping between information weight and codeword weight for these two encoders for all codewords $y(D)$ of weight 9 or less.

$w_H[y(D)]$	$w_H[x_s(D)]$	$w_H[x_{ns}(D)]$
6	4	4
7	3	3
7	3	3
7	*3*	*7*
8	*2*	*6*
8	4	4
8	4	4
8	*4*	*8*
8	6	6
9	*3*	*7*
9	*3*	*7*
9	5	5
9	5	5
9	5	5
9	5	5
9	5	5
9	5	5
9	*5*	*9*
9	*5*	*9*
9	*5*	*9*

Table 2 Mapping between information weight and codeword weight for two equivalent 8-state encoders.

In this case we note that, for all reversible[6] polynomials $u(D)$, the mapping between information weight and codeword weight is the same for the two encoders. Thus, only polynomials $u(D)$ that are not reversible can result in different mappings. For example, the polynomial $u(D) = 1 + D + D^3$, which is not reversible, produces a codeword of weight 7. In this case, the nonsystematic encoder has information weight 7, whereas the systematic encoder has information weight only 3. Similarly, all table entries with different mappings (viz., the italicized entries) have a lower information weight associated with the systematic encoder than with the nonsystematic encoder. Hence we would expect the extrinsic values on the information bits passed to the other decoder to have fewer unreliable bits in the systematic case, thus leading to better decoder convergence for the systematic turbo code. Our conjecture is that this is at least part of the reason for the better "waterfall" region performance of the systematic turbo code shown in Figure 8.

Clearly the above example does not provide "proof" of our conjecture, and further investigation is needed before any conclusions can be drawn. In particular, it would be interesting to examine whether nonsystematic encoders that satisfy the expansion property of systematic encoders, i.e., every information sequence produces a codeword of higher weight, result in good "waterfall" region performance when used as constituent encoders in a turbo code.

6. CONCLUSIONS

In this paper, we have re-examined the original work of Forney [1] on the equivalence of systematic and nonsystematic convolutional encoders from the perspective of turbo coding and have reached the following conclusions:

- Equivalent convolutional encoders, when used as constituent encoders in a turbo code, do not produce similar results. Rather, the different encoder mapping between input and output sequences of equivalent encoders can have a substantial impact on the performance of the associated turbo codes.

- Turbo codes with nonsystematic constituent encoders can be effectively decoded using a modification of the usual iterative APP decoding algorithm.

- Nonsystematic turbo codes have larger effective free distances than systematic turbo codes of the same complexity. Thus they will have better performance in the "error floor" region of the BER curve.

- Catastrophic constituent encoders can produce noncatastrophic turbo codes with maximum effective free distance. However, iterative decoding cannot effectively decode such codes because of the decoder's inability to converge.

- Systematic turbo codes generally have better "waterfall" region performance than nonsystematic turbo codes. This may be explained by the fact that systematic constituent encoders, unlike most of their nonsystematic counterparts, always map low weight information sequences into low weight codewords.

Notes

1. Throughout the paper, the notation (n, k, m) will be used to denote a convolutional encoder with k input sequences, n output sequences, and memory order m.

2. Throughout the remainder of the paper, we will use BER to refer to the information bit error rate.

3. It is important to recognize that the relative importance of different weight input sequences on the distance spectrum of turbo codes depends critically on the interleaver design. Weight 2 input sequences have their greatest relative influence for random-like interleavers.

4. Henceforth, we use the term "nonsystematic turbo code" to refer to a turbo code constructed with nonsystematic constituent encoders. Similarly, a "systematic turbo code" refers to a turbo code constructed with systematic constituent encoders.

5. In a private communication, Urbanke indicated that this has been shown to be true in the case of a binary erasure channel.

6. A polynomial $u(D)$ of degree m is said to be reversible if $u(D) = D^m u(D^{-1})$.

References

[1] G. D. Forney, Jr., "Convolutional Codes I: Algebraic Structure," *IEEE Transactions on Information Theory*, vol. IT-16, pp. 720–738, Nov. 1970.

[2] R. Johannesson and K. Sh. Zigangirov, *Fundamentals of Convolutional Coding*, New York: IEEE Press, 1999.

[3] D. J. Costello, Jr., *Construction of Convolutional Codes for Sequential Decoding*, Ph.D. Dissertation, Department of Electrical Engineering, University of Notre Dame, Notre Dame, IN, Aug. 1969.

[4] C. Berrou, A. Glavieux, and P. Thitimajshima, "Near Shannon Limit Error-Correcting Coding and Decoding: Turbo-Codes," in *Proceedings of the IEEE International Conference on Communication*, Geneva, Switzerland, pp. 1064–1070, May 1993.

[5] O. M. Collins, O. Y. Takeshita, and D. J. Costello, Jr., "Iterative Decoding of Nonsystematic Turbo-Codes," in *Proceedings of the IEEE International Symposium on Information Theory*, Sorrento, Italy, p. 172, June 2000.

[6] L. R. Bahl, J. Cocke, F. Jelinek, and J. Raviv, "Optimal Decoding of Linear Codes for Minimizing Symbol Error Rate," *IEEE Transactions on Information Theory*, vol. 20, pp. 284–287, Mar. 1974.

[7] J. Hagenauer, E. Offer, and L. Papke, "Iterative Decoding of Binary Block and Convolutional Codes," *IEEE Transactions on Information Theory*, vol. 42, pp. 429–445, Mar. 1996.

[8] S. Benedetto and G. Montorsi, "Unveiling Turbo Codes: Some Results on Parallel Concatenated Coding Schemes," *IEEE Transactions on Information Theory*, vol. 42, pp. 409–428, Mar. 1996.

[9] L. C. Perez, J. Seghers, and D. J. Costello, Jr., "A Distance Spectrum Interpretation of Turbo Codes," *IEEE Transactions on Information Theory*, vol. 42, pp. 1698–1709, Nov. 1996.

[10] C. Berrou and A. Glavieux, "Turbo Codes: General Principles and Applications," in *Proceedings of the 6th Tirrenia International Workshop on Digital Communication*, Tirrenia, Italy, pp. 215–226, Sep. 1993.

[11] D. Divsalar and R. J. McEliece, "Effective Free Distance of Turbo Codes," *Electronics Letters*, vol. 32, pp. 445–446, Feb. 1996.

[12] O. Y. Takeshita, O. M. Collins, P. C. Massey, and D. J. Costello, Jr, "A Note on Asymmetric Turbo-Codes," *IEEE Communications Letters*, vol. 3, pp. 69–71, Mar. 1999.

Chapter 5

SYSTEMATIC BITS ARE BETTER AND NO BUTS ABOUT IT

R. Johannesson
Department of Information Technology, Information Theory Group
Lund University. P.O. Box 118, SE-221 00 LUND, Sweden

J. L. Massey
Trondhjemsgade 3, 2t.h.
DK-2100 Copenhagen, Denmark

P. Ståhl
Department of Information Technology, Information Theory Group
Lund University. P.O. Box 118, SE-221 00 LUND, Sweden

Abstract An explanation is given for the paradoxical fact that, at low signal-to-noise ratios, the systematic feedback encoder results in fewer decoding bit errors than does a nonsystematic feedforward encoder for the same convolutional code, which is the opposite of the situation at high signal-to-noise ratios. The analysis identifies a new convolutional code property, the *weight density of distance-d detours* for large d. For a given distance-d weight density, the decoding bit error probability depends on the number of taps in the realization of the encoder inverse. Among all encoders for a given convolutional code, the systematic one has the inverse with fewest taps and, hence, gives the smallest bit error probability at low signal-to-noise ratios where decoding error bursts typically result from the decoder following a detour at large distance d from the transmitted codeword.

*This research was supported in part by the Foundation for Strategic Research—Personal Computing and Communication under Grant PCC-9706-09.

1. INTRODUCTION

Forney has long maintained that there is nothing of particular practical interest about systematic encoders for convolutional codes. While ready to concede that the use of a systematic encoder might reduce the decoding bit error rate by a factor of two or so, his enthusiasm for such a gain was reflected in a muttered "big deal!" [1]. But even "little deals" are important in the iterated decoding algorithms that turbo coding [2] has fostered, and "any deal" becomes interesting when a theory can be found to explain it. This paper proposes a theoretical explanation for Forney's "big deal".

For simplicity, we consider only binary convolutional codes. It is well-known that the errors in the output from a Viterbi ecoder are grouped into error bursts. Forney [3] (see also [4]) called the most probable length of such a decoding error burst the *critical length.* In Figure 1 we show the (normalized by the memory) critical length, ℓ_{crit}, as a function of the code rate R for the binary symmetrical channel (BSC) with crossover probability $\epsilon = 0.045$ corresponding to $R_0 = 1/2$, where R_0 is the cutoff rate. We notice that, for rates below R_0, the length of a typical decoding error burst is very short but ℓ_{crit} tends to infinity as R approaches channel capacity. Thus, at rates close to channel capacity, typical decoding error bursts are very long and one cannot draw conclusions about the decoding error burst probability (first-event error probability) from the initial terms in the path weight enumerator. It follows that, at low signal-to-noise ratios, the codewords with weights substantially larger than the free distance are those that determine decoder performance.

It is also well-known that the free distance of a convolutional code is the principle determiner of the decoding error burst probability for high signal-to-noise ratios when maximum-likelihood decoding or nearly so is used. Since the free distance is a code parameter, the decoding error burst probability will be the same whether the convolutional code is encoded by a nonsystematic feedforward encoder or by its systematic feedback encoder. Bit error probability, however, depends on the encoder used. Typically, at high signal-to-noise ratios where most decoding error bursts are made to codewords at the free distance from the transmitted codeword, the systematic feedback encoder results in more bit errors than a nonsystematic feedforward encoder. The purpose of this paper is to explain the paradoxical fact, often observed in practice, that at low signal-to-noise ratios the systematic feedback encoder results in fewer bit errors than does a nonsystematic feedforward encoder.

Our analysis is based on consideration of what we call the *weight density of distance-d detours*, p_d, and define as the fraction of ones in the

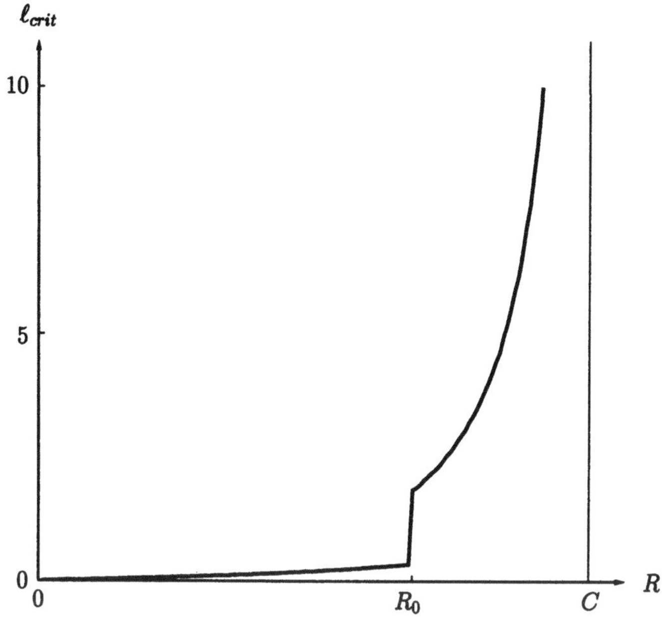

Figure 1 The critical length ℓ_{crit} for the BSC.

detours of weight d in the binary convolutional code, i.e., the detours at distance d from the all-zero codeword. We give computations showing that, for large values of d, the code parameter p_d is essentially independent of the memory of the convolutional code. Equivalently, for large values of d, the distance-d detours have a typical length that is essentially independent of the memory. We use the code parameter p_d in a model of the internal codeword structure, together with the structure of the encoder inverse, to estimate the number of information bit errors that result from each one in a decoding error burst that is a detour of weight d. The weights of the detours are code parameters, not encoder parameters, and hence these estimates reveal which type of convolutional encoder gives the best bit error probability performance. Because at low signal-to-noise ratios, i.e., at code rates close to channel capacity, decoding error bursts are typically very long, the codewords with weights substantially larger than the free distance of the code are those that determine decoding bit error probability.

2. WEIGHT DENSITY OF DISTANCE-d DETOURS

Consider all detours at distance d from the all-zero codeword in a fixed $R = b/c$ binary convolutional code. We are interested in the weight density of distance-d detours p_d, i.e., in the fraction of code symbols that are ones in these detours. This fraction is given by

$$p_d = \frac{d}{cE[L_d]}, \tag{1}$$

where $E[L_d]$ denotes the arithmetic average of the length (in c-tuples) of the detours of weight d and is given by

$$E[L_d] = \sum_{L_d=\lceil \frac{d}{c} \rceil}^{\infty} L_d P_{L_d}, \tag{2}$$

where P_{L_d} is the fraction of detours of weight d that have length L_d (in c-tuples). We now make use of (1) and (2) to study how p_d depends on d. Figure 2, in which p_d is plotted versus d for specific rate $R = 1/2$ codes with quick-look-in (QLI) encoders of various memories [4, 5], shows the typical behavior. For small d, the number of detours of weight d, n_d, is also small and the value of p_d fluctuates widely. For larger d, however, the number of detours of weight d increases rapidly and the value of p_d stabilizes. Evidently, p_d tends towards an asymptotic value, which we denote by p_∞, as d increases. This asymptote is slightly memory dependent—the larger the memory m, the smaller is p_∞.

In Table I we compare the value of p_{24} calculated via (1) and (2) for $R = 1/2$ convolutional codes encoded by systematic feedforward encoders, by quick-look-in (QLI) encoders, by easy-look-in-transparent (ELIT) encoders [6], and by nonsystematic feedforward (optimized for bit error rate) encoders. For all these encoders, p_{24} decreases somewhat as the code memory m increases—the larger m, the smaller the decrease in p_{24}. Evidently, p_{24} approaches an asymptotic value as m increases. The differences in p_{24} for a fixed memory m among the codes with different encoder types are not significant, particularly not for the larger values of m that are the most relevant in practice.

Figure 3 compares the values of p_d for some convolutional codes of different rates, viz., $R = 2/3$, $R = 1/2$, $R = 1/3$, and $R = 1/4$. One sees clearly that the lower the rate, the larger the value of p_d. As d increases, p_d decreases towards its asymptotic value.

From Table I and Figure 3 we conclude that, for large memory m, the code parameter p_d will be close to an asymptotic value that depends only on the code rate, not on the specific code!

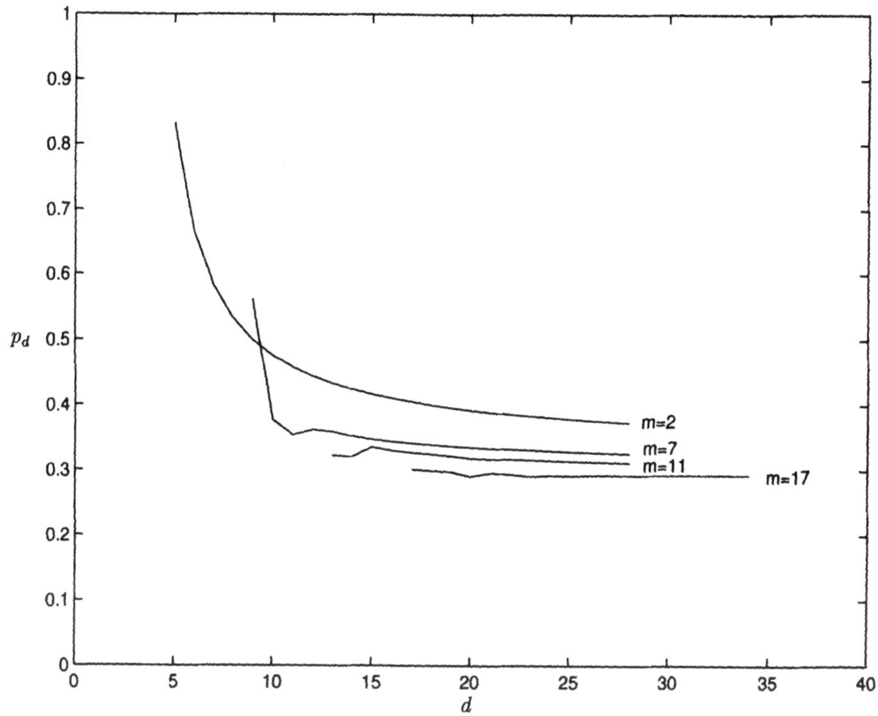

Figure 2 The weight density of distance-d detours p_d as a function of d for $R = 1/2$ QLI codes.

Table I
The weight density of distance-24 detours p_{24} for convolutional codes with feedforward systematic, quick-look-in (QLI), easy-look-in-transparent (ELIT) and nonsystematic (optimized for bit error rate) rate $R = 1/2$ encoders.

	systematic ff		QLI		ELIT		nonsystematic ff	
m	G	p_{24}	G	p_{24}	G	p_{24}	G	p_{24}
2	(4,7)	0.473	(5,7)	0.381			(5,7)	0.381
3	(40,64)	0.425	(54,74)	0.374	(54,64)	0.367	(64,74)	0.374
4	(40,66)	0.373	(46,66)	0.358	(52,76)	0.348	(46,72)	0.355
5	(40,67)	0.365	(55,75)	0.350	(51,73)	0.345	(53,75)	0.346
8	(400,727)	0.325	(551,751)	0.322	(513,735)	0.331	(561,753)	0.349
11	(4000,6457)	0.304	(5503,7503)	0.315	(5643,6352)	0.316	(5261,7173)	0.351
13			(46716,66716)	0.303	(51162,73226)	0.307	(56522,76746)	0.343
17			(454643,654643)	0.293				

3. THE ASYMPTOTE p_∞

To determine the asymptote p_∞, we considered randomly chosen rate $R = b/c$, time-varying binary convolutional codes. For such codes, the probability of a given bit in a codeword generator being a one is $\frac{1}{2}$. The

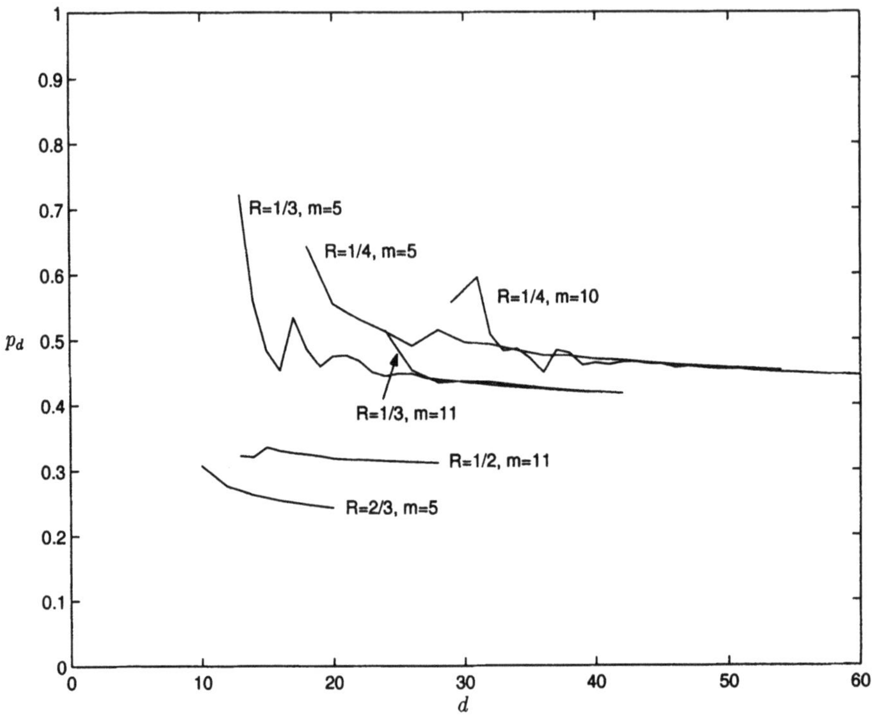

Figure 3 The weight density of distance-d detours p_d as a function of d for $R = 2/3$, $R = 1/2$, $R = 1/3$, and $R = 1/4$ codes.

fraction of detours of weight d that have length L_d is given by

$$P_{L_d} = \frac{\binom{cL_d}{d} 2^{-cL_d} N_{L_d}}{\sum_{l=l_{\min}}^{\infty} \binom{cl}{d} 2^{-cl} N_l} \tag{3}$$

where $l_{\min} = \max\{m+1, \lceil \frac{d}{c} \rceil\}$ and N_l is the number of paths in the trellis that diverge from the zero state at the root and return to the zero state for the first time at depth l. Figure 4 shows p_d as a function of d for rate $R = 1/2$ and various memories. These values were obtained by using (1)-(3). The asymptotic value depends slightly on the encoder memory. For small m, p_d is slightly greater than for larger m. However for memory $m \geq 10$, the asymptotic value is approximately the same and we denote this value p_∞. Figure 5 illustrates the memory dependence in the asymptotic value of p_d. Here N_l/N_l^* is plotted versus l, where N_l^* is the number of paths in the trellis that diverge from the zero state

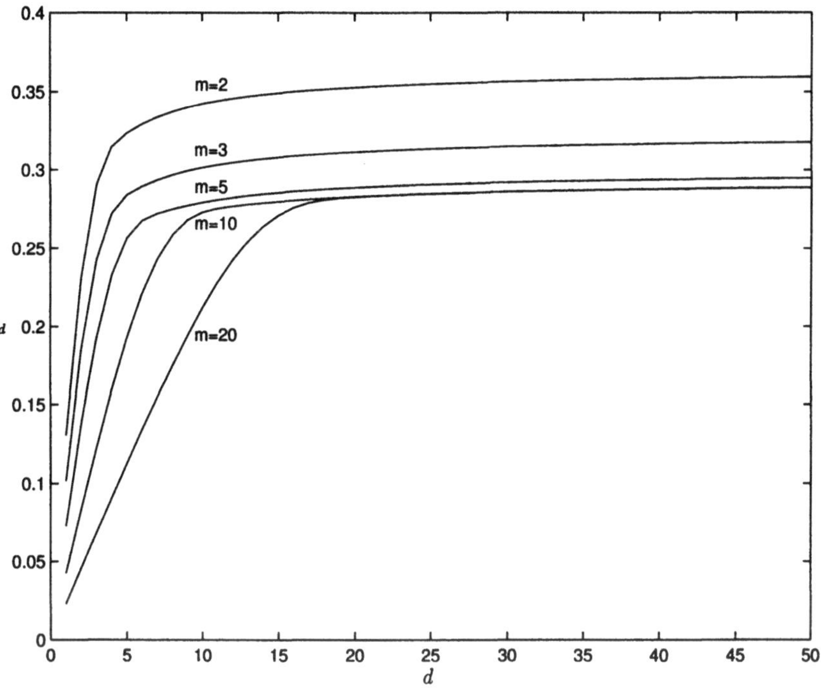

Figure 4 The weight density of distance-d detours p_d as a function of d for randomly chosen rate $R = 1/2$ time-varying convolutional codes of various memories.

at the root and return to the zero state at depth l, not necessarily for the first time. The parameter N_l^* is $2^b - 1$ for length $l = m + 1$ and $(2^b-1)^2 2^{b(l-(m+2))}$ for lengths $l \geq m+2$. Figure 5 shows that, as memory increases, N_l/N_l^* tends to 1 and that $N_l \approx N_l^*$ for $m \geq 10$. Then

$$P_{L_d} \rightarrow \frac{\binom{cL_d}{d} 2^{-cL_d}(2^b-1)^2 2^{b(L_d-(m+2))}}{\sum_{l=l_{\min}}^{\infty} \binom{cl}{d} 2^{-cl}(2^b-1)^2 2^{b(l-(m+2))}} = \frac{\binom{cL_d}{d} 2^{-cL_d} 2^{bL_d}}{\sum_{l=l_{\min}}^{\infty} \binom{cl}{d} 2^{-cl} 2^{bl}}, \tag{4}$$

which depends on the memory m via $l_{\min}$ but becomes essentially independent of m for large d. The behavior illustrated in Figure 4 and Figure 5 holds for all rates $R = b/c$. How fast the asymptotic value of p_d converges to p_∞ depends primarily on the value of b—the larger b, the smaller the memory required for convergence. For rate $R = 2/c$, convergence occurs at about $m = 8$.

Figure 6 shows the asymptotic values p_∞ as a function of the code rate R for large memory ($m \geq 10$)—the smaller the rate, the larger p_∞.

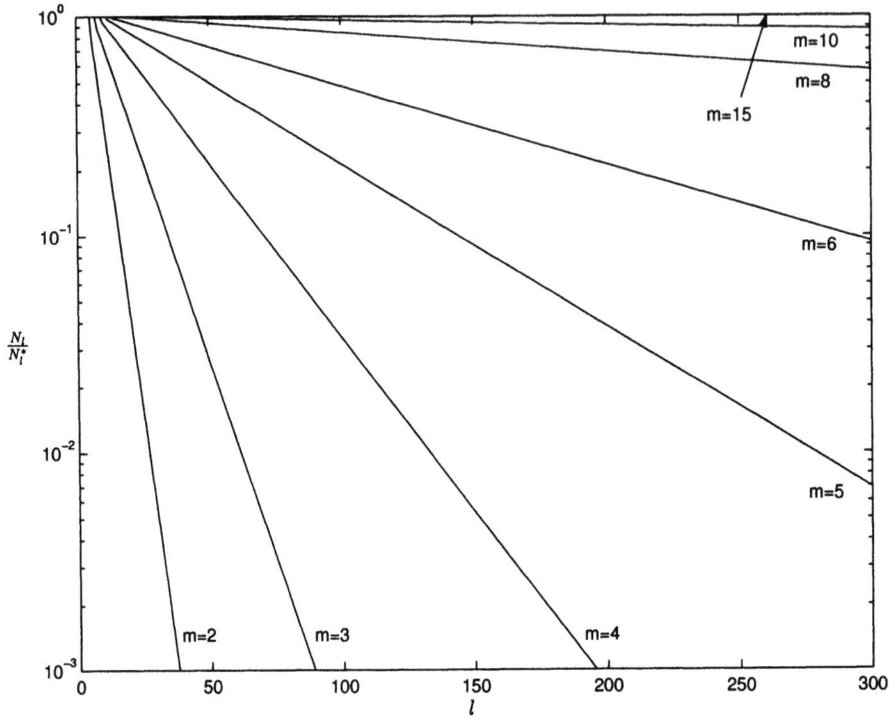

Figure 5 The fraction N_l/N_l^* of detours of length l as a function of l.

The figure shows that $p_\infty \approx 0.29$ for $R = 1/2$, $p_\infty \approx 0.37$ for $R = 1/3$, and $p_\infty \approx 0.40$ for $R = 1/4$. Rate $R = 0$ corresponds to an infinite number of codeword bits on each branch. We can think of p_d as the fraction of ones on each branch. Since the probability of a one is $\frac{1}{2}$, on the average half of the symbols on each branch are ones, which gives $p_\infty = 0.5$ at rate $R = 0$. At rate $R = 1$, there is only one binary digit on each branch in the trellis so that combining (4) and (2) gives

$$E[L_d] = \sum_{L_d=l_{\min}}^{\infty} L_d \frac{\binom{L_d}{d}}{\sum_{l=l_{\min}}^{\infty} \binom{l}{d}}, \tag{5}$$

which diverges. It follows that $p_\infty = 0$ for $R = 1$.

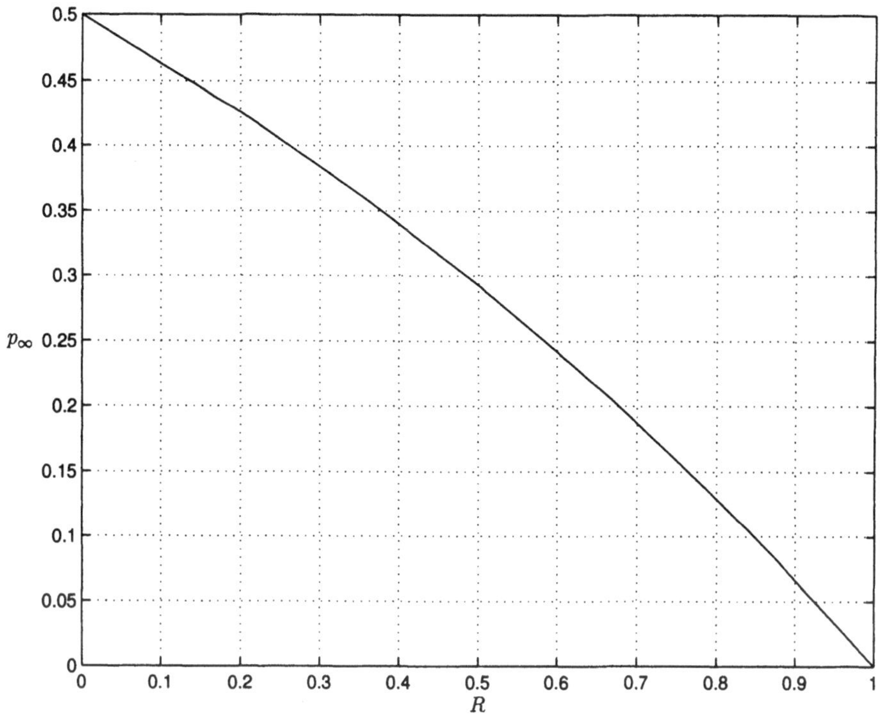

Figure 6 The asymptotic value p_∞ as a function of the code rate R for randomly chosen time-varying convolutional codes.

4. BIT ERROR PROBABILITIES VIA ENCODER INVERSES

We now compare the bit error probability for systematic and nonsystematic encoders. Let q_d denote the arithmetic average for a particular encoder of the number of decoding bit errors per codeword taken over all detours of weight d. Somewhat surprisingly, we found that q_d is essentially an affine function of d with a slope that depends on the encoder type. The different slopes can be explained using an argument involving encoder inverses.

If $G(D)$ is a $b \times c$ generator matrix for a rate $R = b/c$ convolutional code, then the inverse map is represented by a $c \times b$ right inverse matrix $G^{-1}(D)$ of $G(D)$, that is,

$$G(D)G^{-1}(D) = I_b. \tag{6}$$

The matrix $\widetilde{G^{-1}}(D)$ is called a *right pseudo inverse* of $G(D)$ if

$$G(D)\widetilde{G^{-1}}(D) = D^s I_b \tag{7}$$

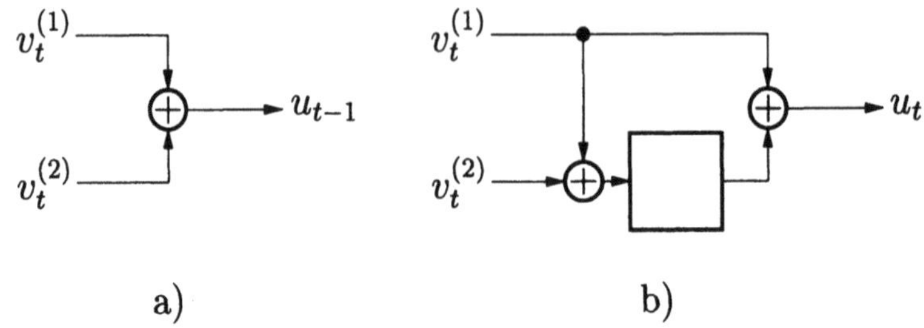

Figure 7 a) Encoder pseudo inverse for QLI encoders. b) Encoder inverse for ELIT encoders.

for some integer $s \geq 0$. The decoding bit error probability depends on the number of taps in the realization of the encoder inverse, which number can be smaller for a pseudo inverse than for an inverse. We begin by considering rate $R = 1/2$ encoders and their encoder inverses.

Systematic encoders have the simplest encoder inverses. A systematic generator matrix can be written as

$$G(D) = \begin{pmatrix} 1 & g(D) \end{pmatrix}, \tag{8}$$

where $g(D)$ is a realizable rational function. The information sequence appears unchanged in the codeword. The following simple matrix is a right inverse

$$G^{-1}(D) = \begin{pmatrix} 1 \\ 0 \end{pmatrix} \tag{9}$$

and uses only one tap. The QLI encoders have encoder pseudo inverses with two taps. The generator matrix of a QLI encoder has the form

$$G(D) = \begin{pmatrix} g(D) & g(D) + D \end{pmatrix} \tag{10}$$

so that the matrix

$$\widetilde{G^{-1}}(D) = \begin{pmatrix} 1 \\ 1 \end{pmatrix} \tag{11}$$

is a right pseudo inverse. Adding the two encoder bitstreams modulo 2 gives the information sequence delayed by one time instant. ELIT encoders have right inverses with three taps. Figure 7 shows the right pseudo inverse and right inverse for QLI and ELIT encoders, respectively.

We now model the appearance of ones within a detour of weight d by a binary memoryless source which outputs a one with probability

p_d. We can then calculate the average number of decoded bit errors per detour 1 for detours of weight d as

$$q_d = \frac{E[\#\,\text{decoding bit errors per branch}]}{E[\text{weight of a branch}]}. \qquad (12)$$

The denominator E[weight of a branch] is independent of the type of encoder inverse and equals $2p_d$ since there are two bits on each branch for a rate $R = 1/2$ code. The numerator $E[\#$ decoding bit errors per branch$]$, however, depends on the number of taps in the encoder inverse. For systematic encoders where the information sequence equals one codeword bitstream, $E[\#$ decoding bit errors per branch$] = p_d$. The average number of bit errors per detour one is then $q_d = p_d/(2p_d) = 1/2$. This seems reasonable since one would expect that half of all codeword ones occur in the systematic bitstream and, thus, also create information bit errors.

For QLI encoders whose encoder pseudo inverses have two taps, we have

$$\begin{aligned} E[\#\,\text{decoding bit errors per branch}] &= P(v_t^{(1)} \oplus v_t^{(2)} = 1) \\ &= 2p_d(1-p_d), \end{aligned} \qquad (13)$$

which gives $q_d = 2p_d(1-p_d)/(2p_d) = 1 - p_d$.

For ELIT encoders whose encoder inverses have three taps, we have

$$\begin{aligned} E[\#\,\text{decoding bit errors per branch}] &= P(v_t^{(1)} \oplus v_{t-1}^{(1)} \oplus v_{t-1}^{(2)} = 1) \\ &= 3p_d(1-p_d)^2 + {p_d}^3, \end{aligned} \qquad (14)$$

which gives $q_d = (3 - 6p_d + 4{p_d}^2)/2$.

For the general case of an encoder inverse with n taps, we have [7]

$$\begin{aligned} E[\#\,\text{decoding bit errors per branch}] &= P(v_{t_1}^{(i_{t_1})} \oplus \cdots \oplus v_{t_n}^{(i_{t_n})} = 1) \\ &= \frac{1}{2}(1 - (1-2p_d)^n) \to \frac{1}{2}, \text{ when } n \to \infty. \end{aligned} \qquad (15)$$

In Table II we list the values of q_∞^{calc} for rate $R = 1/2$ encoder inverses with different number of taps, where we have inserted the asymptotic value $p_\infty = 0.293$ in the above formulas. Table II also gives the measured slopes q_∞^{meas} of the straight lines obtained from plots of the average number of decoding bit errors per detour of weight d as a function of d for memory $m = 11$ codes with different types of encoders. The measured and calculated values are in excellent agreement.

Using the binary memoryless source to model the error patterns that occur when the decoder chooses an erroneous codeword is a reasonable approximation for large memories. However, if we compare the measured and calculated values of q_d for small memories, they can differ

Table II
The encoder parameter q_∞, for rate $R = 1/2$ encoders as a function of n, the number of taps in the encoder inverse. The values of q_∞^{calc} were obtained using the equations in the text with the asymptotic value $p_\infty = 0.293$ for rate $R = 1/2$. The values q_∞^{meas} were obtained using encoders with memory $m = 11$.

n	q_∞^{calc}	q_∞^{meas}
1	0.500	0.50
2	0.707	0.71
3	0.793	0.78
4	0.828	0.83
5	0.843	
6	0.849	
7	0.851	
8	0.853	
$\vdots$	$\vdots$	
∞	0.853	

substantially. For example, the QLI code of memory 2 with generator matrix $G(D) = (1 + D + D^2 \quad 1 + D^2)$ has a measured value $q^{\text{meas}}_{\infty,m=2} = 1$ while from Figure 4 we obtain $p_{\infty,m=2} \approx 0.36$ and, hence, (for a randomly chosen time-varying memory 2 convolutional code with a two tap encoder inverse) $q^{\text{calc}}_{\infty,m=2} = 1 - p_{\infty,m=2} \approx 0.64$. To explain this large difference we consider a section of the code trellis. When we enter and leave state zero, the corresponding code symbols are 11 and this symbol pair does not occur on any other branch in the trellis. Hence, the weight contribution for the remaining path comes from symbol pairs 10 and 01. Since the encoder inverse only has two taps, every symbol pair 10 or 01 will also increase the number of bit errors by one, which gives $q^{\text{meas}}_{\infty,m=2} = 1$. A generator for which the symbol pair 11 is not confined to branches that enter or leave the zero state will give a lower value of $q^{\text{meas}}_{\infty,m=2}$.

We now study q_d for rate $R = b/c$ encoders. If the encoder inverse has n taps, the average number of decoded bit errors per detour 1 for detours of weight d can be expressed as

$$q_d = \frac{\sum_{i=1}^{b} \frac{1}{2}\left(1 - (1 - 2p_d)^{n_i}\right)}{c p_d}, \tag{16}$$

where n_i is the number of taps for the ith output of the encoder inverse and $\sum_{i=1}^{b} n_i = n$. The relation between the values of q_d for systematic encoders and nonsystematic encoders depends on the rate R. As an example, we consider the ratio between q_d for systematic encoders ($n = b$)

and for nonsystematic encoders with infinitely many taps in the encoder inverse. Inserting $p_d = p_\infty$, we find,

R	1/8	1/2	7/8
$q_\infty(n=b)/q_\infty(n=\infty)$	0.125/0.138=0.906	0.500/0.853=0.586	0.875/5.271=0.166

The higher the rate, the lower the ratio. It follows that the higher the rate, the larger the gain obtained by using a systematic encoder rather than a nonsystematic encoder.

5. ENVOI

Private communication to G. D. Forney, Jr.: Turning 60 is no "big deal!" Happy birthday!

References

[1] Private communication between G. D. Forney, Jr. and R. Johannesson, Bologna, April 25, 1999.

[2] C. Berrou, A. Glavieux, and P. Thitimajshima, "Near Shannon Limit Error-Correcting Coding and Decoding: Turbo Codes (1)," *Proceedings of the IEEE International Conference on Communications 1993*, pp. 1064-1070, Geneva, Switzerland.

[3] G. D. Forney, Jr., "Convolutional Codes II: Maximum-Likelihood Decoding and Convolutional Codes," *Information and Control*, vol. 25, pp. 222-226, 1974.

[4] R. Johannesson and K. Sh. Zigangirov, *Fundamentals of Convolutional Coding*, IEEE Press, Piscataway, NJ, 1999.

[5] J. L. Massey and D. J. Costello Jr., "Nonsystematic Convolutional Codes for Sequential Decoding in Space Applications," *IEEE Transactions on Communications Technology*, COM-19, pp. 806-813, 1971.

[6] R. Johannesson and E. Paaske, "Further Results on Binary Convolutional Codes with an Optimum Distance Profile," *IEEE Transactions on Information Theory*, IT-24, pp. 264-268, March 1978.

[7] J. L. Massey, *Threshold Decoding.* MIT Press, Cambridge, MA, 1963.

II

DETECTION AND EQUALIZATION

Chapter 6

SEQUENCE DETECTION: BACKWARD AND FORWARD IN TIME

H. V. Poor
Department of Electrical Engineering
Princeton University
Princeton, NJ 08544 USA

Abstract The classical theory of signal detection deals principally with the problem of choosing between two statistical hypothess about a signal (or two) embedded in noise. While the extension of this theory to multiple hypotheses is obvious, the raw complexity of such an extension generally grows at least linearly with the number of hypotheses to be tested. Data transmission gives rise to several problems in which a very large number of hypotheses must be tested in order to make simultaneous decisions about an entire sequence of data symbols. These problems include decoding of convolutional codes, channel equalization, and multiuser detection. In each of these problems, the complexity of optimal detection can be greatly reduced through the use of dynamic programming, which efficiently finds optimal (e.g., maximum-likelihood or minimum-error-probability) sequence decisions. This chapter reviews this area, including the early contributions of Forney and others, and continuing with more recent work on multiuser detection and turbo decoding.

*This work was prepared under the support of the National Science Foundation under Grant CCR-99-80590.

1. INTRODUCTION

The classical theory of signal detection deals with the problem of deciding between the presence and absence of a signal in a set of measurements corrupted by random noise. The basic issue in such problems is the determination of the likelihood ratio between the two statistical hypotheses described by "signal present" and "signal absent" conditions [9]. After the likelihood ratio has been extracted (typically by some kind of filtering), an optimal decision is made by comparison of the result with an appropriate threshold [9, 15]. Likelihood ratios for such binary detection problems can be used straightforwardly to write likelihood functions for deciding among multiple possible signals. This is accomplished by constructing pairwise likelihoods between each of the implied multiple hypotheses and a "catalyst" hypothesis, such as that of noise only. Thus, in principle, problems with multiple possible signals can be solved straightforwardly as generalizations of the single-signal-in-noise case. Typically, the complexity of such solutions grows linearly with the number of signals to be decided among.

In digital communications, there are two basic types of problems in which such multisignal detection arises: the first of these is that of demodulating a single symbol from an M-ary signaling alphabet. In such situations, the number of possible different signals, M, is typically small (e.g., eight or sixteen) and the above-noted linear complexity is manageable. The second basic situation in which multisignal detection problems arise is that in which signals from multiple symbol transmissions are received in a manner such that there is statistical dependence among them. Applications in which this situation can arise include coded communications, transmission through dispersive channels, and multiaccess communications. In such situations, optimal data detection can involve the joint, or simultaneous, detection of an entire group or sequence of symbols (e.g., an entire frame), and so such problems are known as *sequence detection* problems.

If the number of M-ary symbols to be jointly detected in such a problem is N, then the number of possible hypotheses (assuming no constraints) is M^N, a number that can be prohibitively large for practical detection systems if linear complexity is required. For example, even though the signaling alphabet may typically be binary ($M = 2$), the number of symbols per frame could be on the order of hundreds. So, in order for optimal detection to be practical in such situations, some form of complexity reduction is necessary. Fortunately, many applications in which such situations arise admit dynamical models for symbol dependence that allow dynamic programming to be used to provide significant reduction of the complexity of optimal detection.

This basic principle has resulted in a number of key communication processing algorithms over the past few decades, including the 1960's *Viterbi algorithm* and the 1970's *BCJR algorithm* for detecting convolutionally encoded data transmitted over memoryless channels [2, 25], Forney's 1972 and Ungerboeck's 1974 *maximum-likelihood sequence detectors* (MLSDs) for equalizing linearly dispersive channels [7, 20] (similar algorithms were also developed independently by Kobayashi [11] and Omura [14]), and Verdú's 1984 maximum-likelihood and minimum-error-probability *multiuser detectors* (MUDs) for demodulating nonorthogonally multiplexed data [21]. Interest in these algorithms continues to be strong due to their key roles in modern applications such as turbo decoding [13] and third-generation code-division multiple-access (CDMA) communications [16].

The purpose of this chapter is to provide a brief review of these developments in a unified framework. Given the influence of Dave Forney's MLSD as a centerpiece for this field, it is fitting that such a review be included in this volume honoring his sixtieth birthday.

This chapter is organized as follows. We first give, in Section 2, a basic model for a received communication signal within which all of the results of interest can be discussed. Then, in Section 3, we consider the problem of maximum-likelihood (ML) sequence detection, in which the symbols in a sequence are detected on the basis of the maximum-likelihood principle without benefit of a prior model for the symbol statistics. This criterion gives rise to the Viterbi algorithm, the Forney and Ungerboeck MLSD's, and maximum-likelihood MUD. Such algorithms are typically viewed as *forward* dynamic programs. In Section 4 we consider the problem of Bayesian sequence detection, in which a prior probability distribution can be assumed for the symbols in the sequence. In this situation, the sequence detector's role is to compute a posterior probability distribution on the symbols in the sequence, conditioned on the values of the observations. This computation facilitates data detection based on the maximum-posterior probability (MPP) criterion (also known as the maximum *a posteriori* probability (MAP) criterion). The BCJR algorithm provides this computation in the case of convolutionally encoded data, while the minimum-error-probability MUD provides this computation in the case of nonorthogonally multiplexed data. In either case, the problem has hidden-Markov model (HMM) structure from which efficient backward-forward dynamic programming solutions can be derived. Finally, Section 5 includes some brief concluding remarks.

2. A SIMPLE MODEL FOR A RECEIVED COMMUNICATION SIGNAL

Throughout this chapter, we will restrict attention to the situation in which the only stochastic channel impairment is additive white Gaussian noise (AWGN), in which case can we can consider a received signal model of the general form

$$y(t) = m_{\mathbf{b}}(t) + v(t), \quad -\infty < t < \infty, \quad \mathbf{b} \in \mathcal{B}, \tag{1}$$

where $m_{\mathbf{b}}(t)$ and $v(t)$ represent the useful signal and AWGN, respectively, and where $\mathcal{B}$ is a set of N-tuples with elements from the set $\{-1, +1\}$; i.e.,

$$\mathcal{B} \subset \{-1, +1\}^N. \tag{2}$$

The signal $m_{\mathbf{b}}(\cdot)$ denotes the signal that will be received when the specific sequence of symbols $\mathbf{b}$ is transmitted, and the noise is assumed to be independent of the symbol vector $\mathbf{b}$. We shall assume generally that $m_{\mathbf{b}}(\cdot)$ is known for each $\mathbf{b} \in \mathcal{B}$, which corresponds to the situation in which any parameters of the transmission channel (multipath delays and gains, carrier amplitude, frequency and phase, etc.) are known. As a mild technical condition, we will assume that

$$\int_{-\infty}^{\infty} [m_{\mathbf{b}}(t)]^2 dt \ < \ \infty, \ \forall \ \mathbf{b} \in \mathcal{B}. \tag{3}$$

For simplicity, we limit attention to real-valued signals; the extension to the complex-valued case is straightforward. And, finally, we note that the extension to more general finite alphabets is also straightforward.

Under the restriction (3) the likelihood function (i.e., the likelihood ratio with respect to a noise-only catalytic hypothesis) for the model (1) can be written via the Cameron-Martin formula [4] as

$$\ell(\{y(t); -\infty < t < \infty\} | \mathbf{b}) = C \exp\{\Omega(\mathbf{b})/2\sigma^2\} \tag{4}$$

where C is a constant, σ^2 is the noise intensity, and

$$\Omega(\mathbf{b}) = 2 \int_{-\infty}^{\infty} m_{\mathbf{b}}(t) y(t) dt - \int_{-\infty}^{\infty} [m_{\mathbf{b}}(t)]^2 dt. \tag{5}$$

Without further structure either on the set $\mathcal{B}$ of symbol sequences or on the set $\{m_{\mathbf{b}}(\cdot)\}$ of corresponding received waveforms, not much of practical interest can be said about demodulation in this model. For example, to make unconstrained ($\mathcal{B} = \{-1, +1\}^N$) maximum-likelihood symbol decisions, we need to solve the optimization problem

$$\max_{\mathbf{b} \in \{-1,+1\}^N} \Omega(\mathbf{b}). \tag{6}$$

The basic complexity of this problem without further structure is $\mathcal{O}(2^N)$, which is the complexity of searching exhaustively through all possible values of **b** for the maximum in (6). As noted in the introduction, this is a prohibitive degree of complexity for most problems of interest. However, this complexity can be mitigated significantly for certain signaling structures arising in practical communication systems.

To examine this issue, we will restrict attention in the remainder of this chapter to the case of *linear modulation*, in which the signals $m_{\mathbf{b}}(t)$ are of the form

$$m_{\mathbf{b}}(t) = \sum_{n=1}^{N} b_n s_n(t), \tag{7}$$

where b_n denotes the nth symbol in **b**, and $s_n(t)$ is a known waveform, depending of course on n. Although complexity reduction can result in cases of nonlinear modulation, the linear case can be used to illustrate all of the algorithms mentioned in the introduction. Also, as we will show below, this linear case is sufficient to consider many channels of interest, including coded channels, intersymbol-interference channels, and multiple-access channels.

In this case of linear modulation, the objective $\Omega(\mathbf{b})$ from (5) becomes

$$\Omega(\mathbf{b}) \;=\; 2\sum_{n=1}^{N} b_n y_n \;-\; \sum_{n=1}^{N}\sum_{m=1}^{N} b_n b_m H_{n,m}, \tag{8}$$

with

$$y_n = \int_{-\infty}^{\infty} s_n(t) y(t)\, dt\ , \quad n \;=\; 1, 2, \ldots, N \tag{9}$$

and

$$H_{n,m} = \int_{-\infty}^{\infty} s_n(t) s_m(t)\, dt, \quad m, n \;=\; 1, 2, \ldots, N. \tag{10}$$

If we let **b** and **y** denote the (column) vectors with components $\{b_n\}$ and $\{y_n\}$, respectively, and **H** denote the matrix with elements $\{H_{n,m}\}$, then we can write (8) compactly as

$$\Omega(\mathbf{b}) \;=\; 2\mathbf{b}'\mathbf{y} \;-\; \mathbf{b}'\mathbf{H}\mathbf{b}. \tag{11}$$

We see from (4) and (11) that **y** is an N-dimensional sufficient statistic for detecting the N-dimensional symbol vector **b**. This already can provide a substantial reduction in problem complexity since, in its general form (5) the least dimensional sufficient statistic is the collection

$$\int_{-\infty}^{\infty} m_{\mathbf{b}}(t) y(t) dt, \quad \mathbf{b} \in \mathcal{B}, \tag{12}$$

which has $|\mathcal{B}|$ elements. (Typically $|\mathcal{B}| \gg N$.)

Unfortunately, this reduction in "hardware" complexity in going from the general case to the linear modulation case, is not generally accompanied by a corresponding decrease in "software" complexity. For example, the unconstrained maximum-likelihood detection problem (6) becomes

$$\max_{\mathbf{b}\in\{-1,+1\}^N} \left[2\mathbf{b}'\mathbf{y} - \mathbf{b}'\mathbf{H}\mathbf{b}\right]. \tag{13}$$

The problem (13) is an integer quadratic program, an NP-complete problem that in general offers little improvement in complexity over the exhaustive search needed for the more general problem (6) (see [22]). However, as we will see in the following sections, we will consider sequence detection for specific signaling situations that introduce restrictions either on the set $\mathcal{B}$ or on the matrix $\mathbf{H}$ that in turn give rise to much lower complexity algorithms for data detection.

3. FORWARD IN TIME: MAXIMUM-LIKELIHOOD SEQUENCE DETECTION

In this section, we will consider the problem of maximum-likelihood sequence detection, which for the linear-modulation model (1) and (7), is given by

$$\hat{\mathbf{b}} = \text{argmax}_{\mathbf{b}\in\mathcal{B}} \left[2\mathbf{b}'\mathbf{y} - \mathbf{b}'\mathbf{H}\mathbf{b}\right]. \tag{14}$$

We first make note of a simple case, in which the channel symbols $\mathbf{b}$ are unconstrained (i.e., $\mathcal{B} = \{-1,+1\}^N$) and the signaling waveforms are orthogonal:

$$H_{m,n} = 0, \ \forall\ m \neq n. \tag{15}$$

In this case, the problem (14) simplifies to

$$\begin{aligned}\hat{\mathbf{b}} &= \text{argmax}_{\mathbf{b}\in\{-1,+1\}^N} \sum_{n=1}^{N} \left[2b_n y_n - H_{n,n} b_n^2\right] \\ &= \text{argmax}_{\mathbf{b}\in\{-1,+1\}^N} \sum_{n=1}^{N} b_n y_n,\end{aligned} \tag{16}$$

where we have made use of the fact that $b_n^2 = 1$. (This condition is not critical to the complexity issues that follow, which apply equally well to non-constant-modulus signaling alphabets.) This problem thus decouples to allow the symbol-by-symbol solution:

$$\hat{b}_n = \text{argmax}_{b_n\in\{-1,+1\}} b_n y_n = \text{sgn}\{y_n\}, \quad n = 1,2,\ldots,N, \tag{17}$$

which has $\mathcal{O}(N)$ complexity. So, the complexity per demodulated bit is constant with the sequence length.

Unfortunately, this low complexity quickly erodes if either the set $\mathcal{B}$ is constrained or the cross-correlation matrix $\mathbf{H}$ has nonzero off-diagonal elements. In the next subsection, we consider a case of the former difficulty, and in subsequent subsections we consider the latter difficulty.

3.1 THE VITERBI ALGORITHM

For now, we retain the assumption that the received signals satisfy the orthogonality condition (15), but impose the constraints of a rate-R code ($R < 1$) on the symbol sequence $\mathbf{b}$. In particular, we assume that $\mathbf{b}$ is a function of a shorter vector $\mathbf{d}$ of unconstrained data symbols from $\{-1, +1\}$:

$$\mathbf{d} \equiv \begin{pmatrix} d_1 \\ d_2 \\ \vdots \\ d_{RN} \end{pmatrix} \in \{-1, +1\}^{RN} \tag{18}$$

(For simplicity, we assume that RN is an integer.) In this case, the problem (14) reduces to

$$\max_{\mathbf{d} \in \{-1,+1\}^{RN}} \sum_{n=1}^{N} b_n(\mathbf{d})\; y_n, \tag{19}$$

where $b_n(\mathbf{d})$ denotes the functional dependence of b_n on the data symbols. In the general case, the problem (19) must be solved by exhaustion over the possible data sequences, which results in $\mathcal{O}\left(N\; 2^{RN}\right)$ complexity.

This complexity can be mitigated if the function $\mathbf{b}(\mathbf{d})$ is judiciously constrained. A key example of a complexity-reducing constraint is that in which the mapping from $\mathbf{d}$ to $\mathbf{b}$ is in the form of a convolutional code. In this case, we write R as the ratio of integers k/n and the encoder produces a n-long block of channel symbols for each k-long block of the data symbol sequence. A given channel symbol block is a function of the current data symbol block, plus a fixed number ν of previously generated data symbol blocks. That is, the data symbol sequence $\mathbf{d}$ is divided into a sequence $\mathbf{d}_1, \mathbf{d}_2, \ldots, \mathbf{d}_{N/n}$ of k-long blocks, and the channel symbol sequence $\mathbf{b}$ is divided into a sequence $\mathbf{b}_1, \mathbf{b}_2, \ldots, \mathbf{b}_{N/n}$ of n-long blocks, with the property that $\mathbf{b}_j$ depends only on $\mathbf{d}_j, \mathbf{d}_{j-1}, \ldots, \mathbf{d}_{j-\nu}$.

With this constraint, we can write the problem (14) as

$$\max_{\mathbf{d}\in\{-1,+1\}^{RN}} \sum_{j=1}^{N/n} \mathbf{b}_j' \left(\mathbf{d}_j, \mathbf{d}_{j-1}, \ldots, \mathbf{d}_{j-\nu}\right) \mathbf{y}_j \tag{20}$$

where we have written the $\mathbf{b}_j$'s as column vectors, and where $\mathbf{y}_j$ is a column vector containing those elements of $\mathbf{y}$ having the same indices as the elements of $\mathbf{b}_j$.

Now, ignoring end effects, the jth channel symbol block is generated as a function of the $(j-\nu)$th through jth data symbol blocks. These $\nu+1$ data symbol blocks form a state $\mathbf{x}_j$, which propagates forward to $\mathbf{x}_{j+1}$ by appending a new block on the leading edge, and deleting the oldest block on the trailing edge. That is, we have the recursive relationship

$$\mathbf{x}_{j+1} \equiv \begin{pmatrix} \mathbf{d}_{j+1} \\ \mathbf{d}_j \\ \mathbf{d}_{j-1} \\ \vdots \\ \mathbf{d}_{j-\nu+1} \end{pmatrix} = \begin{pmatrix} \mathbf{O} & \mathbf{O} & \mathbf{O} & \cdots & \mathbf{O} \\ \mathbf{I} & \mathbf{O} & \mathbf{O} & \cdots & \mathbf{O} \\ \mathbf{O} & \mathbf{I} & \mathbf{O} & \cdots & \mathbf{O} \\ \vdots & \vdots & \vdots & \ddots & \vdots \\ \mathbf{O} & \mathbf{O} & \mathbf{O} & \cdots & \mathbf{I} \end{pmatrix} \mathbf{x}_j + \begin{pmatrix} \mathbf{I} \\ \mathbf{O} \\ \mathbf{O} \\ \vdots \\ \mathbf{O} \end{pmatrix} \mathbf{d}_{j+1} \tag{21}$$

where $\mathbf{O}$ denotes the $k \times k$ matrix of all zeroes, $\mathbf{I}$ denotes the $k \times k$ identity matrix, and where we take $\mathbf{d}_j$ to be all zeroes for $j \leq 0$. Thus, this shift-register action forms a simple dynamical system whose state space has $2^{(\nu+1)k}$ elements. This gives a representation of (20) in a form amenable to dynamic programming. In particular, we wish to maximize the sum

$$\sum_{j=1}^{N/n} f_j(\mathbf{x}_j) \tag{22}$$

where

$$f_j(\mathbf{x}_j) = \mathbf{b}_j' \left(\mathbf{d}_j, \mathbf{d}_{j-1}, \ldots, \mathbf{d}_{j-\nu}\right) \mathbf{y}_j. \tag{23}$$

(The mappings from the data symbol state $\mathbf{x}_j$ to the corresponding channel symbol block $\mathbf{b}_j$ is typically in the form of a set of n linear (Boolean) functions on the binary symbols represented by $\mathbf{x}_j$.)

The recursive nature of (21) allows us to search for the ML data sequence with much lower complexity than the $\mathcal{O}\left(N\ 2^{RN}\right)$ required for exhaustion. In particular, for each j, we can evaluate all possible values of $f_j(\mathbf{x}_j)$ with $\mathcal{O}\left(n\ 2^{(\nu+1)k}\right)$ complexity. Moreover, because each value of the state $\mathbf{x}_j$ is connected to only 2^k values of the state $\mathbf{x}_{j-1}$, we can enumerate all values of the sum (22) with complexity $\mathcal{O}\left(N\ 2^{(\nu+2)k}\right)$ complexity. This dynamic programming technique [3] is the essence of the Viterbi algorithm [25]. In typical applications, the exponent $(\nu+2)k$ will be much smaller (less than 10) than the data length RN (more than 100), so this structure results in significant computational savings.

3.2 FORNEY'S MAXIMUM-LIKELIHOOD SEQUENCE DETECTOR

In the preceding section, we considered the version of the ML sequence detection problem (14) in which the signaling waveforms are orthogonal, but the channel symbols are constrained. In a classic 1972 paper [7], Forney observed that the alternative problem in which the channel symbols are unconstrained but the signaling waveforms are not orthogonal can be transformed into a problem of the former type. This observation suggests that complexity reduction similar to that gained by the Viterbi algorithm might be possible in the latter problem if a finite-memory property analogous to that induced by convolutional encoding can be invoked. In this section, we see that this suggestion is in fact valid, and arises in a very natural way in intersymbol-interference channels. The corresponding detector structure is Forney's maximum-likelihood sequence detector (MLSD).

To examine this issue, consider first the objective of the ML problem (14):

$$2\mathbf{b}'\mathbf{y} \;-\; \mathbf{b}'\mathbf{H}\mathbf{b}. \tag{24}$$

Assuming that the signals $\{s_n\}$ are linearly independent, the cross-correlation matrix $\mathbf{H}$ will be positive definite (and hence nonsingular). This implies that we can write a Cholesky factorization of $\mathbf{H}$:

$$\mathbf{H} \;=\; \mathbf{F}'\mathbf{F} \tag{25}$$

where $\mathbf{F}$ is a nonsingular lower-triangular matrix. On defining

$$\mathbf{c} \;=\; \mathbf{F}\mathbf{b} \;\text{ and }\; \mathbf{z} \;=\; \left(\mathbf{F}^{-1}\right)'\mathbf{y} \tag{26}$$

we can rewrite (24) as

$$2\mathbf{c}'\mathbf{z} \;-\; \mathbf{c}'\mathbf{c} \;=\; \sum_{n=1}^{N}\left[2c_n z_n - c_n^2\right]. \tag{27}$$

So, the ML sequence detection problem can be written as

$$\hat{\mathbf{b}} \;=\; \text{argmax}_{\mathbf{b}\in\{-1,+1\}^N} \sum_{n=1}^{N}\left[2c_n(\mathbf{b})z_n - c_n^2(\mathbf{b})\right] \tag{28}$$

where we have explicitly shown the dependence of **c** on **b**. Comparing (28) with analogous expression (19) in the preceding section, we can see that this nonorthogonal signaling channel is like a coded channel with orthogonal signals, where we interpret **b** as the underlying data symbols and **c** as the channel symbols. (The rate R of this code is unity.)

We now restrict attention to the case of a linearly dispersive *intersymbol interference* (ISI) channel, in which the signals $\{s_n\}$ are of the form

$$s_n(t) = p(t - nT), \tag{29}$$

where $p(t)$ is a basic received pulse shape and T is the symbol interval (i.e., $1/T$ is the signaling rate). In this case, the elements of the cross-correlation matrix $\mathbf{H}$ are given by

$$H_{m,n} = \int_{-\infty}^{\infty} s_m(t)\, s_n(t)\, dt = \int_{-\infty}^{\infty} p(t - mT)\, p(t - nT)\, dt. \tag{30}$$

Note that, if the duration of $p(t)$ is not greater than the symbol interval T, then the cross-correlation matrix satifies the orthogonality condition (15), and data detection in this problem reduces to the simple form (17). On the other hand, when the duration of $p(t)$ is greater than the symbol interval, then intersymbol interference results, and sequence detection is required to optimally detect the symbols. We now consider this latter case.

Typically, the pulse $p(t)$ will span only a few symbol intervals, say D, where the product DT is the maximal *delay spread* of the channel. In this case, we see from (30) that

$$H_{m,n} = 0, \quad |m - n| > D. \tag{31}$$

Thus, in this case $\mathbf{H}$ is a *banded matrix*; that is, all elements of $\mathbf{H}$ greater than a certain distance (in this case D) from the main diagonal are zero. It is easily seen that the Cholesky factor ($\mathbf{F}$ of (25)) of the banded matrix $\mathbf{H}$ is also banded, with the same bandwidth as $\mathbf{H}$. Thus, we also have

$$F_{m,n} = 0, \quad |m - n| > D. \tag{32}$$

It follows that the "encoding" **c** of (26) and (28) has finite memory:

$$c_n = \sum_{m=1}^{n} F_{n,m} b_m = \sum_{m=n-D}^{n} F_{n,m} b_m \equiv c_n(b_n, b_{n-1}, \ldots, b_{n-D}), \tag{33}$$

where we take $b_n = 0$ for $n \leq 0$. Using (33), we see that the ML data detection problem (28) can be written in a form analogous to (20) and

(22); namely,

$$\begin{aligned}\hat{\mathbf{b}} &= \text{argmax}_{\mathbf{b}\in\{-1,+1\}^N} \sum_{n=1}^{N} [2c_n(b_n, b_{n-1}, \ldots, b_{n-D})z_n - \\ &\qquad\qquad c_n^2(b_n, b_{n-1}, \ldots, b_{n-D})] \\ &= \text{argmax}_{\mathbf{b}\in\{-1,+1\}^N} \sum_{n=1}^{N} f_n(\mathbf{x}_n), \end{aligned} \tag{34}$$

where $\mathbf{x}_n$ now denotes the $(D+1)$-dimensional shift-register state

$$\mathbf{x}_n = \begin{pmatrix} b_n \\ b_{n-1} \\ \vdots \\ b_{n-D} \end{pmatrix}, \tag{35}$$

and where the functions f_n now become

$$f_n(\mathbf{x}_n) = 2c_n(b_n, b_{n-1}, \ldots, b_{n-D})z_n - c_n^2(b_n, b_{n-1}, \ldots, b_{n-D}). \tag{36}$$

Now, using the same argument as with (22) we see that the set of all possible symbol sequences can be searched for a maximum in (33) with a dynamic program using $\mathcal{O}\left(N\ 2^D\right)$ operations. This dynamic program, together with the transformation (26) is Forney's MLSD. As in the case of Viterbi decoding, the delay spread is typically such that D is much less than the framelength N, resulting in considerable computational savings.

Before leaving this structure, it is of interest to comment on the transformation (26). The sufficient statistic $\mathbf{y}$ of (9) is easily seen to satisfy the model

$$\mathbf{y} - \mathbf{Hb} + \mathbf{v}, \tag{37}$$

where the noise term $\mathbf{v}$ has the $\mathcal{N}(0, \sigma^2\mathbf{H})$ distribution. From this, it is straightforward to see that the transformed variables $\mathbf{z}$ satisfy the alternative model

$$\mathbf{z} \equiv (\mathbf{F}^{-1})'\,\mathbf{y} = \mathbf{c} + \mathbf{w} \tag{38}$$

where the transformed noise term $\mathbf{w} = (\mathbf{F}^{-1})'\,\mathbf{v}$ has the $\mathcal{N}(0, \sigma^2\mathbf{I})$ distribution. (Here, $\mathbf{I}$ denotes the $N \times N$ identity matrix.) So, the transformation $(\mathbf{F}^{-1})'\,\mathbf{y}$ results in an discrete-time observation model

$$z_n = c_n + w_n,\ n = 1, 2, \ldots, N, \tag{39}$$

in which the noise is white. Note that this sequence $\{z_n\}$ can be obtained from $y(\cdot)$ by passing it through a continuous-time linear filter followed by a symbol-rate sampler. This continuous-time linear filter is called a *whitened matched filter.*

We see from the above discussion that Forney's MLSD consists of a whitened matched filter, followed by a Viterbi decoder. In [20] Ungerboeck noted that the MLSD can be implemented directly without first passing through the whitened matched filter. To see this, we can use the facts that $b_n^2 = 1$ and $H_{n,m} = H_{m,n}$ to rewrite

$$\begin{aligned} 2\mathbf{b}'\mathbf{y} - \mathbf{b}'\mathbf{H}\mathbf{b} &= 2\sum_{n=1}^{N} b_n y_n - 2\sum_{n=1}^{N}\sum_{m=1}^{n-1} b_n b_m H_{n,m} - \sum_{n=1}^{N} H_{n,n}^2 \\ &= 2\sum_{n=1}^{N} b_n \left[y_n - \sum_{m=n-D}^{n-1} b_m H_{n,m} \right] - \sum_{n=1}^{N} H_{n,n}^2, \end{aligned} \quad (40)$$

where again we take $b_n = 0$ for $n \leq 0$. From (40) we see that (6) can be written in a form similar to (33) in terms of the shift-register state sequence of (35). Thus, ML symbol decisions can be found with $\mathcal{O}\left(N\ 2^D\right)$ complexity, as in the case of (33). Note that, in view of (30), we have in this ISI model that the cross-correlation $H_{m,n}$ depends only on the difference $m - n$, say

$$H_{m+k,m} = H_{1+k,1} \triangleq h_k, \quad (41)$$

from which the MLSD problem of (40) becomes

$$\max_{\mathbf{b}\in\{-1,+1\}^N} \sum_{n=1}^{N} b_n \left[y_n - \sum_{m=n-D}^{n-1} h_{n-m} b_m \right]. \quad (42)$$

We will return to this representation in the following subsection.

3.3 MAXIMUM-LIKELIHOOD MULTIUSER DETECTION

Maximization of the objective in (40) also arises in the problem of maximum-likelihood detection in multiple-access channels. To see this, consider the situation in which the received waveform $m_{\mathbf{b}}(t)$ is comprised of data signals of K active users superimposed in a single channel. Such a signal can be written as

$$m_{\mathbf{b}}(t) = \sum_{k=1}^{K}\sum_{i=1}^{B} b_k(i)\ p_k(t - iT - \tau_k), \quad (43)$$

where B is the number of symbols per user in the data frame of interest, T is the per-user symbol interval, and where τ_k, $\{b_k(i)\}$, and $p_k(t)$, denote, respectively, the received delay, symbol stream, and modulation waveform of the kth user. Here, we have a model of the form of (7) in which $N = KB$, and, with $i = \lceil \frac{n}{K} \rceil$ and $k = n - (i-1)K$, we take

$$b_n = b_k(i) \quad \text{and} \quad s_n(t) = p_k(t - iT - \tau_k). \tag{44}$$

The cross-correlation matrix is thus given by

$$H_{n,m} = \int_{-\infty}^{\infty} s_n(t)\, s_m(t)\, dt = \int_{-\infty}^{\infty} p_k(t - iT - \tau_k)\, p_\ell(t - jT - \tau_\ell)\, dt \tag{45}$$

with i and k as above, and with $j = \lceil \frac{m}{K} \rceil$ and $\ell = m - (j-1)K$. There are two sources of nonorthogonality in these cross-correlations. First, as in the preceding section, the individual waveforms p_k can be subject to dispersion; and secondly, pairs of waveforms $p_k(\cdot - iT - \tau_k)$ and $p_\ell(\cdot - jT - \tau_\ell)$ for $k \neq \ell$ can be nonorthogonal due either to nonorthogonal signaling (as, for example, in CDMA communications), or to channel effects (e.g., multipath, delay) on transmitted waveforms that were originally orthogonal. This latter situation gives rise to *multiple-access interference* (MAI) due to the correlation among different users' signals. Multiuser detection addresses the problem of data detection in the presence of such interference.

If we assume a nondispersive channel, so that $p_k(t) = 0$ for $t \notin [0, T]$, and (without loss of generality) that the users are numbered so that $0 \leq \tau_1 \leq \tau_2 \leq \cdots \leq \tau_K \leq T$, then the cross-correlations in (45) satisfy

$$H_{m,n} = 0, \quad |m - n| > K - 1. \tag{46}$$

Comparing with (31), we can conclude that the complexity of maximum-likelihood MUD can be reduced from the exhaustive $\mathcal{O}\left(KB\ 2^{KB}\right)$ to the more moderate $\mathcal{O}\left(KB\ 2^{K}\right)$. The corresponding dynamic programming algorithm based on the decomposition in (40) is the ML multiuser detector studied by Verdú in [21]. Note that a distinction between this ML MUD and Ungerboeck's MLSD for intersymbol-interference channels is that the cross-correlations for MUD are periodically time-varying:

$$H_{m+k,m} = H_{m+K+k,m+K}, \tag{47}$$

in contrast with the situation for the ISI channel (41).

Before leaving this topic, we make note of a few additional issues:

- The addition of length-D ISI to the MAI channel results in a cross-correlation matrix that is banded with bandwidth $(K-1)D$. Thus, the complexity of sequence detection for multiuser equalization becomes $\mathcal{O}\left(KB\ 2^{KD}\right)$.

- It is noteworthy that the complexity of maximum-likelihood MUD is not reduced by the additional restriction of synchrony: $\tau_1 = \tau_2 = \cdots = \tau_K$. In particular, the complexity in this case is still $\mathcal{O}\left(KB\ 2^K\right)$. The main difference between this case and the general asynchronous case is that sequence detection is not required in the synchronous case; that is, each symbol interval can be demodulated independently of the others. The given complexity in this case essentially corresponds to exhaustive search on a symbol-interval by symbol-interval basis.

- The complexity in the synchronous case can be mitigated if the signaling waveforms can be chosen judiciously. Obviously, if they can be chosen to be orthogonal then the complexity collapses to that of (17). But, more generally, particular choices of signaling waveforms give polynomial-in-K or even logarithmic-in-K complexity. (For some results of this type, see [12], [17], [18], and [19].)

4. BACKWARD IN TIME: MAXIMUM POSTERIOR SEQUENCE DETECTION

The sequence detectors of the preceding section were based on the principle of choosing the sequence of symbols that make the received observations most likely. Underlying this criterion is the philosophy that the symbols themselves are not random in the precise probabilistic sense, but rather are deterministic but unknown. Alternatively, in some applications it is desirable to adopt a Bayesian philosophy in which the symbols are endowed with a prior probabilistic structure. In such cases, the role of detection can be viewed as that of extracting, from the prior model and the observations, the posterior probabilistic structure of the symbols. This gives rise to so-called "soft-input/soft-output" detection which is useful in applications involving iterative or decision-feedback detector structures (such as turbo decoding [13] and detection [26]).

The maximum-likelihood detectors of the preceding section have their counterparts in this Bayesian framework. For example, the BCJR algorithm [2] solves the convolutional decoding problem in a Bayesian setting, while the minimum-error-probability multiuser detector [21, 24] solves the MUD problem in a Bayesian setting. Like ML sequence detection these problems can be solved using forward dynamic programming [8]; however, they can be solved more naturally and with lower complexity using a backward-forward dynamic programming structure [1, 5, 24]. In this section, we will discuss this framework briefly, referring the reader to the cited literature for further details and for specific algorithms.

We continue with the general signaling model introduced in Section 2, but now we adopt a Bayesian philosophy in which the symbol vector $\mathbf{b}$ is assumed in the uncoded case to be a random vector, drawn uniformly from $\{-1,+1\}^N$; or in which the data-symbol vector $\mathbf{d}$ is assumed in the rate-R coded case to be drawn uniformly from $\{-1,+1\}^{RN}$. For the purpose of exposition, we will restrict attention here to the uncoded case, but the convolutionally coded case has essentially the same structure.

The fact that the vector $\mathbf{y}$ is a sufficient statistic for $\mathbf{b}$ is unchanged by the assumption of a prior distribution on $\mathbf{b}$, so the differences between Bayesian detection and maximum-likelihood detection lie in the "software" aspects of the problem; i.e., in the mapping from $\mathbf{y}$ to $\mathbf{b}$.

In this case of uncoded symbols $\mathbf{b}$, we would like to compute the per-symbol *a posteriori* probabilities:

$$P(b_n = +1|\mathbf{y}) \equiv 1 - P(b_n = -1|\mathbf{y}), \; n = 1, 2, \ldots, N. \tag{48}$$

This computation provides the soft outputs needed for applications such as turbo decoding, and also allows for detecting the symbols in a way that minimizes the probability of symbol error:

$$P(\hat{b}_n \neq b_n), \tag{49}$$

the minimum (over $\hat{b}_n$) of which is achieved via the MAP detector[1]

$$\hat{b}_n = \text{argmax}_{a\in\{-1,+1\}} \, P(b_n = a|\mathbf{y}). \tag{50}$$

Recall from the preceding section that the maximum-likelihood detection problems reduce to the integer quadratic program (14); namely,

$$\max_{\mathbf{b}\in\mathcal{B}} \left[2\mathbf{b}'\mathbf{y} - \mathbf{b}'\mathbf{H}\mathbf{b}\right]. \tag{51}$$

The essence of the results of the preceding section is that significant complexity reductions are possible in either the case in which $\mathbf{H}$ is diagonal and $\mathcal{B}$ admits a finite-state model, or the case in which $\mathcal{B} = \{-1,+1\}^N$ and $\mathbf{H}$ is banded. The latter case can be viewed as being reducible to the former via the whitened matched-filter, or as being directly formalizable with a finite-state model. These same principles apply in the Bayesian case.

We first note that the quantity of interest, $P(b_n = a|\mathbf{y})$, can be written via Bayes formula (and using the fact that $\mathbf{b}$ is uniformly distributed) as

$$P(b_n = a|\mathbf{y}) = \frac{\ell(\mathbf{y}|b_n = a)}{\ell(\mathbf{y}|b_n = -1) + \ell(\mathbf{y}|b_n = +1)} \tag{52}$$

where $\ell(\mathbf{y}|b_n = a)$ denotes the likelihood of $\mathbf{y}$ given that $b_n = a$. This latter quantity can be written as

$$\ell(\mathbf{y}|b_n = a) \;=\; \sum_{\{\mathbf{b}\in\{-1,+1\}^N \;|\; b_n=a\}} \ell(\mathbf{y}|\mathbf{b}) \tag{53}$$

where $\ell(\mathbf{y}|\mathbf{b})$ denotes the likelihood of $\mathbf{y}$ conditioned on all the symbols $\mathbf{b}$, given via (37) as

$$\ell(\mathbf{y}|\mathbf{b}) \;=\; \exp\left\{\frac{1}{\sigma^2}\left[2\mathbf{y}'\mathbf{b} - \mathbf{b}'\mathbf{H}\mathbf{b}\right]\right\}. \tag{54}$$

We see from (53) that the raw complexity of this computation is essentially exponential in N. However, just as in the case of the maximum-likelihood problems of the preceding section, this complexity can be reduced to exponential-in-L for the case in which the matrix $\mathbf{H}$ is L-banded.

Also as in the maximum-likelihood case, there are two ways of viewing this complexity reduction. One view of the complexity reduction is analogous to Forney's MLSD. In particular, after applying the whitened matched-filter transformation (26) to the sufficient statistic $\mathbf{y}$ we obtain, as before, the observation model (39)

$$z_n \;=\; c_n(\mathbf{x}_n) \;+\; w_n, \; n = 1, 2, \ldots, N, \tag{55}$$

with state equation

$$\mathbf{x}_n \;=\; \begin{pmatrix} 0 & 0 & 0 & \cdots & 0 \\ 1 & 0 & 0 & \cdots & 0 \\ 0 & 1 & 0 & \cdots & 0 \\ \vdots & \vdots & \vdots & \ddots & \vdots \\ 0 & 0 & 0 & \cdots & 1 \end{pmatrix} \mathbf{x}_{n-1} \;+\; \begin{pmatrix} 1 \\ 0 \\ 0 \\ \vdots \\ 0 \end{pmatrix} b_n, \; n = 1, 2, \ldots, N, \tag{56}$$

where we take $b_n \;=\; 0$ for $n \;\leq\; 0$. With $\mathbf{b}$ drawn uniformly (and hence such that $b_1, b_2, \ldots, b_N$ are independent and identically distributed (i.i.d.)), this model will be recognized as a hidden-state Markov model (HMM) similar to the classical Kalman-Bucy model [10]. As in the Kalman-Bucy model, the conditional distribution of the state $\mathbf{x}_n$ of (56), given the measurements $z_1, z_2, \ldots, z_N$ of (55), can be computed via the *HMM smoother* which can be implemented using a pair of finite-dimensional recursive filters, one running forward in time and the other running backward in time (see, for example, [6]). The complexity of these filters is $\mathcal{O}(N\ 2^L)$ and the complexity of computing $P(b_n = a|\mathbf{y})$ from $P(\mathbf{x}_n|\mathbf{y})$ is $\mathcal{O}(2^L)$. So the overall complexity is $\mathcal{O}(N\ 2^L)$, as with the maximum-likelihood solution.

An alternative view of complexity reduction in (53) is a direct dynamic programming solution, that is analogous to the Ungerboeck MLSD. This algorithm [24] makes use of a generalization to computations such as (53) of the classic dynamic programming solution for the minimum of a sum of a function of a finite state. In particular, the main distinction between (53) and the corresponding ML problem (13) is that a *maximization* over a very large set ($\mathbf{b} \in \{-1,+1\}^N$) has been replaced by a *summation* over a similarly large set ($\{\mathbf{b} \in \{-1,+1\}^N \mid b_n = a\}$). It happens that both of these computations admit dynamic programming solutions in the case of banded $\mathbf{H}$, which corresponds in turn to the case in which a finite-state model can be used to compute the summands. As with the HMM smoother discussed above, this solution can be implemented with low complexity using a backward-forward algorithm. (Of course, the backward-forward structure of the HMM smoother is really just a manifestation of this same decomposition.) (See [24] for further details.)

5. CONCLUDING REMARKS

In this chapter, we have provided a brief overview of the problem of optimal sequence detection as it arises in three applications: decoding of convolutional codes, equalization of dispersive channels, and multiuser detection. With optimality defined in either the maximum-likelihood or maximum *a posteriori* probability sense, we have seen that the complexity of these problems is greatly reduced by dynamic programming for the cases of practical interest, in which the data to be detected enters the sufficient statistic via a relatively short shift-register state.

It should be noted that, despite their substantial complexity savings over brute-force computational methods, the algorithms discussed here are still too complex for many applications. Thus, in practice, sequence detection is often accomplished by any of a number of approximations to the algorithms presented here, including sequential and stack algorithms for the decoding of convolutional codes, linear and decision-feedback algorithms for channel equalization, and linear and iterative algorithms for multiuser detection. Nevertheless, the algorithms discussed here provide a fundamental basis for solving and understanding the optimality issues in these important applications.

Notes

1. Note that the minimum probability of *sequence* error $P(\hat{\mathbf{b}} \neq \mathbf{b})$ is achieved in this case of uniform prior by the ML sequence detector.

References

[1] K. Abend and B. D. Fritchman, "Statistical Detection for Communication Channels with Memory," *Proceedings of the IEEE*, Vol. 58, No. 5, pp. 779–785, May 1970.

[2] L. R. Bahl, J. Cocke, F. Jelinek, J. Raviv, "Optimal Decoding of Linear Codes for Minimizing Symbol Error Rate," *IEEE Transactions on Information Theory*, Vol. IT-20, No. 2, pp. 284–287, March 1974.

[3] R. Bellman, *Dynamic Programming.* (Princeton University Press: Princeton, NJ, 1957)

[4] R. H. Cameron and W. T. Martin, "Transformation of Wiener integrals under translation," *Annals of Mathematics*, Vol. 45, no. 3, pp. 386–396, 1944.

[5] D. Chang and J. Hancock, "On Receiver Structures for Channels Having Memory," *IEEE Transactions on Information Theory*, Vol. 12, No. 4, pp. 463–468, October 1966.

[6] R. J. Elliott, L. Aggoun, and J. B. Moore, *Hidden Markov Models: Estimation and Control.* (Springer-Verlag: New York, 1995).

[7] G. D. Forney, Jr., "Maximum-Likelihood Sequence Estimation of Digital Sequences in the Presence of Intersymbol Interference," *IEEE Transactions on Information Theory*, Vol. IT-18, No. 3, pp. 363–378, May 1972.

[8] J. Hayes, T. Cover and J. Riera, "Optimal Sequence Detection and Optimal Symbol-by-Symbol Detection: Similar Algorithms," *IEEE Transactions on Communications*, Vol. 30, No. 1, pp. 152–157, January 1982.

[9] T. Kailath and H. V. Poor, "Detection of Stochastic Processes," *IEEE Transactions on Information Theory*, Vol. 44, No. 6, pp. 2230–2259, October 1998.

[10] R. E. Kalman and R. S. Bucy, "New Results in Linear Filtering and Prediction Theory," *Transactions of the ASME. Ser. D, Journal of Basic Engineering*, Vol. 83, pp. 94–107, December 1961.

[11] H. Kobayashi, "Correlative Level Coding and Maximum-Likelihood Decoding," *IEEE Transactions on Information Theory*, Vol. 17, No. 1, pp. 586–594, January 1971.

[12] R. E. Learned, A. S. Willsky, and D. M. Boroson "Low Complexity Optimal Joint Detection for Oversaturated Multiple Access Communications," *IEEE Transactions on Signal Processing*, Vol. 45, No. 1, pp. 113–123, January 1997.

[13] R. J. McEliece, D. J. C. MacKay, and J.-F. Cheng, "Turbo Decoding as an Instance of Pearl's 'Belief Propagation' Algorithm," *IEEE Journal on Selected Areas in Communications*, Vol. 16, no. 2, pp. 140–152, February 1998.

[14] J. K. Omura, "On the Viterbi Decoding Algorithm," *IEEE Transactions on Information Theory*, Vol. IT-15, pp. 177–179, January 1969.

[15] H. V. Poor, *An Introduction to Signal Detection and Estimation - Second Edition.* (Springer-Verlag: New York, 1994)

[16] H. V. Poor and G. W. Wornell, Eds. *Wireless Communications: Signal Processing Perspectives.* (Prentice-Hall: Upper Saddle River, NJ, 1998)

[17] C. Sankaran and A. Ephremides and , "Solving a Class of Optimum Multiuser Detection Problems with Polynomial Complexity," *IEEE Transactions on Information Theory*, Vol. 44, No. 5, pp. 1958–1961, September 1998.

[18] C. Schlegel and A. Grant, "Polynomial Complexity Detection of Certain Multiple Access Systems," *IEEE Transactions on Information Theory*, Vol. 46, No. 6, pp. 2246–2248, September 2000.

[19] S. Ulukus and R. D. Yates, "Optimum Multiuser Detection is Tractable for Synchronous CDMA Systems Using m-Sequences," *IEEE Communications Letters*, Vol. 2, No. 4, pp. 89–91, April 1998.

[20] G. Ungerboeck, "Adaptive Maximum-Likelihood Receiver for Carrier Modulated Data Transmission Systems," *IEEE Transactions on Information Theory*, Vol. 22, No. 5, pp. 624–636, May 1974.

[21] S. Verdú, *Optimum Multiuser Signal Detection.* Ph.D. dissertation, Department of Electrical and Computer Engineering, University of Illinois at Urbana-Champaign, August 1984.

[22] S. Verdú, "Computational Complexity of Optimum Multiuser Detection," *Algorithmica*, Vol. 4, pp. 303–312, 1989.

[23] S. Verdú, *Multiuser Detection.* (Cambridge University Press: Cambridge, UK, 1998)

[24] S. Verdú and H. V. Poor, "Abstract Dynamic Programming Models Under Commutativity Conditions," *SIAM Journal on Control and Optimization*, Vol. 25, No. 4, pp. 990–1006, July 1987.

[25] A. J. Viterbi, "Error Bounds for Convolutional Codes and an Asymptotically Optimum Decoding Algorithm," *IEEE Transactions on Information Theory*, Vol. IT-13, pp. 260–269, April 1967.

[26] X. Wang and H. V. Poor, "Iterative (Turbo) Soft Interference Cancellation and Decoding for Coded CDMA," *IEEE Transactions on Communications*, Vol. 47, No. 7, pp. 1046–1061, July 1999.

Chapter 7

LEAST SQUARES AND KALMAN FILTERING ON FORNEY GRAPHS

H.-A. Loeliger
ISI, ETH Zürich, Switzerland

Abstract General versions of Kalman filtering and recursive least-squares algorithms are derived as instances of the sum(mary)-product algorithm on Forney-style factor graphs.

1. INTRODUCTION

Factor graphs [3, 6] are a unifying framework for a wide variety of system models, and the generic sum(mary)-product algorithm, which works by message passing in the factor graph, subsumes a wide variety of algorithms in coding, signal processing, and artificial intelligence [3]. It was pointed out in [3] that Kalman filtering can also be viewed as an instance of the sum-product algorithm, but this was explicitly demonstrated only for the scalar case; in the present paper, we discuss the vector case. We also demonstrate the equivalence of Kalman filtering and general recursive least-squares algorithms in this context. (For traditional state space models, this equivalence was shown in [4].)

In [2], Forney introduced a variation of factor graphs with a number of attractive properties. We will use these Forney graphs rather than the original factor graphs of [3].

The main result of the present paper are Tables 1 and 2, which state the message update rules for the building blocks of the classical linear state space models. Of course, these rules are just reformulations of the well established equations of Kalman filtering (e.g., [1]). Nevertheless, the whole paper is devoted to explain and to prove these tables.

The paper is structured as follows. Forney-style factor graphs are introduced in Section 2. In Section 3 some pertinent properties of Gaussian distributions are reviewed and the equivalence of the min-sum and the sum-product algorithm for Gaussian networks is pointed out. Kalman filtering as an instance of the sum(mary)-product algorithm is discussed in Section 4. The proofs of the message update rules of Tables 1 and 2, as well as some comments on these rules, are given in Section 5. Some conclusions are offered in Section 6. There is also an appendix with some background material in linear algebra, especially on the pseudo-inverse and on nonnegative definite matrices.

The following notation will be used. If z is a complex number, then $\overline{z}$ is its conjugate complex. If A is a matrix or a vector, then A^T is its transpose, $A^H \triangleq \overline{A^T}$, and $A^\#$ is the Moore-Penrose pseudo-inverse (see the appendix). The symbol "$\propto$" denotes equality of functions up to a scale factor. All vectors are column vectors.

2. FORNEY GRAPHS

A Forney-style factor graph (FFG) or "normal graph" [2] represents a factorization of a function of several variables. For example, assume that some function $f(u, w, x, y, z)$ can be factored as

$$f(u, w, x, y, z) = f_A(u, w, x) f_B(x, y, z) f_C(z). \tag{1}$$

This factorization is expressed by the graph of Figure 1. In general, an FFG consists of nodes, edges, and "half edges", where "half edges" are connected to only one node. The rules are as follows:

- There is a node for every factor.
- There is an edge (or half edge) for every variable.
- The node representing some factor g is connected with the edge (or half edge) representing some variable x if and only if x is an argument of g.

Implicit in these rules is the assumption that no variable appears in more than two factors. We will see below that this condition is far less restrictive than might appear at first sight.

The factors of the factorization expressed by the FFG are also called *local functions*; the overall function (i.e., the product of all local functions) is called the *global function*.

We will now rephrase some basic facts about factor graphs for FFGs; for more details, see [3] and [2]. We will first assume that all variables take values in finite sets; the modifications for continuous variables are given at the end of this section.

In probability theory, factorizations of joint probability measures are expressions of independence. E.g., let X, Y, and Z be discrete random

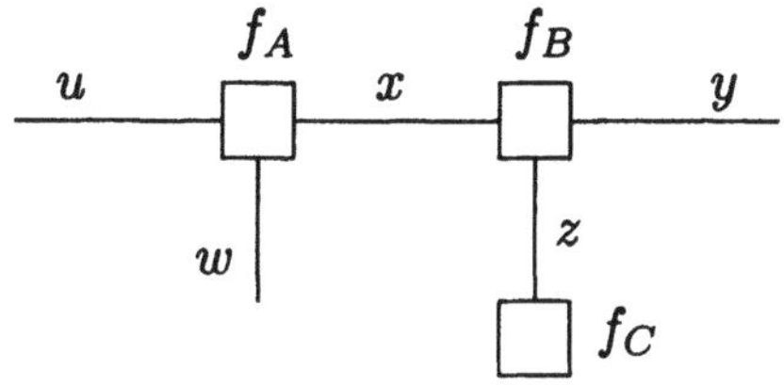

Figure 1 A Forney-style factor graph (FFG).

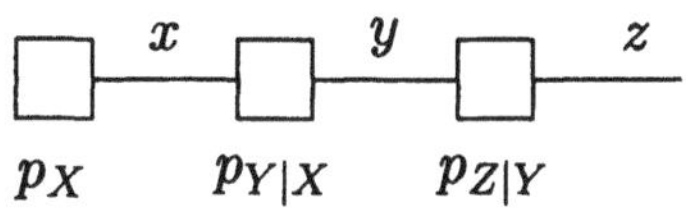

Figure 2 FFG of a Markov chain.

variables with joint probability mass function $p(x, y, z)$. Then X, Y, Z form a Markov chain if and only if $p(x, y, z)$ can be factored as

$$p(x, y, z) = p(x)p(y|x)p(z|y). \tag{2}$$

This factorization is shown in Figure 2. Upon removal of the edge y, the graph falls into two disconnected components, with x and z in different components, which expresses the conditional independence of X and Z given Y. It is easy to see that this generalizes to any FFG of a joint probability mass function: conditioned on the variables in any cut set of the graph, the variables in the two resulting components are independent.

A block diagram as in Figure 3 may also be viewed as an FFG. A function block $x = g(u, v)$ in the block diagram is then interpreted as representing the factor $\delta(x - g(u, v))$, where δ is the Kronecker delta function. When viewed as an FFG, the block diagram of Figure 3 thus represents the function

$$\delta(x - g(u, v))\, \delta(y - h(x, w)) = \begin{cases} 1, & \text{if } x = g(u, v) \text{ AND } y = h(x, w) \\ 0, & \text{else.} \end{cases} \tag{3}$$

In other words, the global function evaluates to 1 if and only if the values of all variables are consistent with the equations of the block diagram. Note that the arrows in the block diagram have no influence on its interpretation as an FFG.

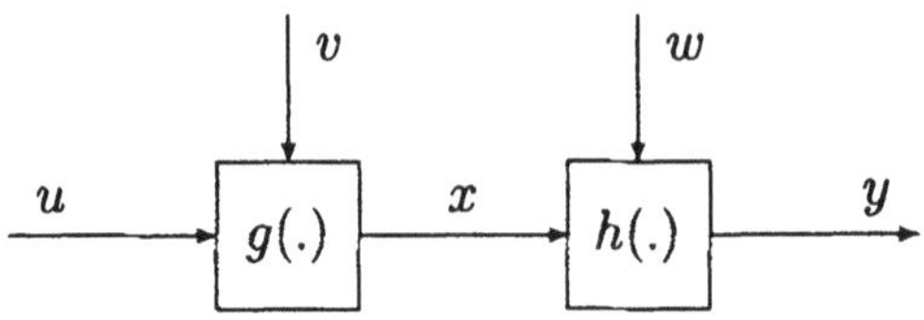

Figure 3 A block diagram.

Figure 4 Branching point (left) becomes a replication node (right).

As illustrated by these examples, an FFG can be used to express the structure of a "system" or "model". In this context, the domain of the global function f is called the *configuration space.* A *configuration* is an element of the configuration space, i.e., a particular assignment of values to all variables. A configuration ω is *valid* if $f(\omega) \neq 0$.

In a block diagram, we usually find also branching points as in Figure 4 (left). In an FFG, such branching points must be treated as factor nodes on their own, as is illustrated in Figure 4 (right). In doing so, there arise new variables (x_1 and x_2 in Figure 4) and a new factor

$$f_=(x, x_1, x_2) \triangleq \delta(x - x_1)\, \delta(x - x_2). \tag{4}$$

Note that, in every valid configuration, the new auxiliary variables have the same value as the original variable. By this device of variable "cloning", it is always possible to enforce the condition that a variable may appear in at most two factors (local functions).

The fact that ordinary block diagrams can be viewed as FFGs is one of the advantages of FFGs over factor graphs as defined in [3]. With this comes also the advantage of a clear distinction between external ("visible") variables and internal ("latent" or "state") variables: the former are represented by half edges, the latter by normal edges. Moreover, the operations of dividing a system into subsystems ("tearing") and of "zooming" into the interior of some subsystem — both central to Willems' system theory [7] — are naturally expressed in an FFG (cf. Figures 5 and 6).

Another attraction of FFGs is that the sum-product algorithm [3, 6] takes on a particulary simple form. Two messages are transmitted along each edge, one in each direction. (In practice, it often happens that only a subset of all these messages is actually needed.) Each message is, or represents, a *function* of the variable associated with that edge. Consider a node that represents some factor $f(x_1, \ldots, x_n)$. The message $\mu_{f\to x_k}$ out of this node along the edge x_k is the function

$$\mu_{f\to x_k}(x_k) = \sum_{x_1} \cdots \sum_{x_{k-1}} \sum_{x_{k+1}} \cdots \sum_{x_n} f(x_1, \ldots, x_n)\mu_{x_1\to f}(x_1) \cdot \ldots \cdot \mu_{x_{k-1}\to f}(x_{x-1})\mu_{x_{k+1}\to f}(x_{k+1}) \cdot \ldots \cdot \mu_{x_n\to f}(x_n), \quad (5)$$

where $\mu_{x_j\to f}$ is the message incoming on edge x_j. In words, the message $\mu_{f\to x_k}$ is the product of f and all messages towards f along all edges except x_k, summed over all variables except x_k. In practice, the update rule (5) is often modified to include a scale factor.

The messages in the graph are computed, or iteratively recomputed, according to some schedule. As is well known, if the graph is cycle free, the (final) message out of some half edge representing some variable x is the marginal function

$$\mu(x) \triangleq \sum_{\omega:\ x \text{ fixed}} f(\omega), \quad (6)$$

where the sum goes over all configurations ω with fixed x; for details see [3].

In this paper, we will be primarily interested in real or complex variables, or variables that are real or complex vectors. In this case, the Kronecker delta in equations such as (3) and (4) should be replaced by the Dirac delta and the summation in (5) and (6) should be replaced by integration.

3. GAUSSIAN DISTRIBUTIONS AND QUADRATIC COST FUNCTIONS

In Kalman filtering, the factors and messages are of the form

$$f(x) = e^{-q(x)} \quad (7)$$

with

$$\begin{aligned} q(x) &\triangleq (x - m_x)^H W_x (x - m_x) + c_x & (8) \\ &= x^H W_x x - 2\mathrm{Re}(x^H W_x m_x) + m_x^H W_x m_x + c_x, & (9) \end{aligned}$$

where x is a real or complex vector, where W_x is a nonnegative definite $n \times n$ matrix, and where the constant $c_x \in \mathbb{R}$ amounts to a scale factor in (7) which can often be ignored. The case where all quantities in (8) are real-valued and $q(.)$ is a function $\mathbb{R}^n \to \mathbb{R}$ will be referred to as "the real case"; the case where $q(.)$ is a function $\mathbb{C}^n \to \mathbb{R}$ will be referred to as "the complex case".

If W_x is positive definite, (7) may be viewed as a multidimensional Gaussian probability density with mean m_x. In the real case, the corresponding covariance matrix is $\frac{1}{2}W_x^{-1}$ and the appropriate scale factor is $e^{-c_x} = \sqrt{\det(W_x)/\pi^n}$; in the complex case, the covariance matrix is W_x^{-1} and the scale factor is $e^{-c_x} = \det(W_x)/\pi^n$. Note that the integral $\int_{-\infty}^{\infty} e^{-q(x)+c_x}\,dx$ depends on W_x but neither on m_x nor on c_x.

The sum of terms of the form (8) is again of that form (cf. the appendix), and thus the product of terms of the form (7) is also of the form (7). Moreover, integrating a function of the form (7) over some of the variables also preserves this form. Indeed, if $q(x, y)$ is the quadratic form

$$q(x,y) \triangleq \left((x-m_x)^H, (y-m_y)^H\right) \begin{pmatrix} W_{1,1} & W_{1,2} \\ W_{2,1} & W_{2,2} \end{pmatrix} \begin{pmatrix} x - m_x \\ y - m_y \end{pmatrix}, \tag{10}$$

then, considered as a function of x with parameter y, $q(x, y)$ is of the form (9) with $W_x = W_{1,1}$ and with both m_x and $c_x = \min_x q(x, y)$ depending on y. Assuming that $W_{1,1}$ is positive definite, we then have

$$\int_{-\infty}^{\infty} e^{-q(x,y)}dx = e^{-c_x(y)} \int_{-\infty}^{\infty} e^{-\left(q(x,y)-c_x(y)\right)}dx \tag{11}$$

$$\propto e^{-c_x(y)} \tag{12}$$

$$= e^{-\min_x q(x,y)} \tag{13}$$

because the integral on the right side of (11) does not depend on y.

It is often convenient to view Dirac delta functions as limits of Gaussian distributions: in the real case as

$$\delta(x) = \lim_{\beta\to\infty} \frac{\sqrt{\beta}}{(2\pi)^{n/2}} e^{-\frac{\beta}{2}||x||^2} \tag{14}$$

and in the complex case as

$$\delta(x) = \lim_{\beta\to\infty} \frac{\beta}{\pi^n} e^{-\beta||x||^2}. \tag{15}$$

If all factors in some FFG are of the form (7), then both (5) and (6) (with summation replaced by integration) are of the form (13) and yield results of the form (7). In this case, all computations, and indeed the FFG itself, may be interpreted in the logarithmic domain. The FFG then represents a "global" quadratic cost function $\varphi(\omega) = -\log f(\omega)$ of the form (8) as the sum of "local" quadratic cost functions of the same form, the messages are also quadratic cost functions of this form, and the sum-product rule (5) becomes the min-sum rule

$$\begin{aligned} q_{f\to x_k}(x_k) &= \min_{x_1}\ldots\min_{x_{k-1}}\min_{x_{k+1}}\ldots\min_{x_n}\Big(\varphi(x_1,\ldots,x_n) + \\ &\quad q_{x_1\to f}(x_1) + \ldots + q_{x_{k-1}\to f}(x_{x-1}) + \\ &\quad q_{x_{k+1}\to f}(x_{k+1}) + \ldots + q_{x_n\to f}(x_n)\Big) + c, \end{aligned} \tag{16}$$

with $\varphi(x_1,\ldots,x_n) \triangleq -\log f(x_1,\ldots,x_n)$, $q_{f\to x_k}(.) \triangleq -\log \mu_{f\to x_k}(.)$, $q_{x_j\to f}(.) \triangleq -\log \mu_{x_j\to f}(.)$, and where $c \in \mathbb{R}$ is a normalizing constant that corresponds to a scale factor in (5) and may be chosen freely. The sum-product algorithm thus becomes the min-sum algorithm [3]; if the graph has no cycles, the final message $q(x)$ out of some half edge x is

$$q(x) = \min_{\omega:\ x \text{ fixed}} \varphi(\omega) + \text{const.} \tag{17}$$

In other words, the marginalization (6) is equivalent to the least-squares problem (17) and the sum-product algorithm becomes the min-sum algorithm, which, in this case, is a general least-squares algorithm.

If the graph has cycles, the min-sum algorithm may fail to converge; however, if it converges, Weiss and Freeman [5] have shown that the mean vector of the message $q(x)$ is correct, i.e., it agrees with the mean vector of the quadratic form on the right side of (17).

4. LEAST SQUARES AND KALMAN-FILTERING ON AN FFG

Kalman filtering may be viewed as the sum-product algorithm applied to the state space model of Figure 5 with composite nodes as in Figure 6:

$$x[k] = Ax[k-1] + Bu[k] \tag{18}$$
$$y[k] = Cx[k] + w[k], \tag{19}$$

where for $k \in \mathbb{Z}$, $u[k]$, $w[k]$, $x[k]$, and $y[k]$ are real or complex vectors and where A, B, and C are real or complex matrices of appropriate dimensions.

In the traditional setup, it is assumed that $y[.]$ is observed and that both $u[.]$ and $w[.]$ are white Gaussian noise. In its most narrow sense, Kalman filtering is then only the forward sum-product recursion though the graph of Figure 5 (cf. Figure 7 left) and yields the a posteriori probability distribution of the state $x[k]$ given the observation sequence $y[.]$

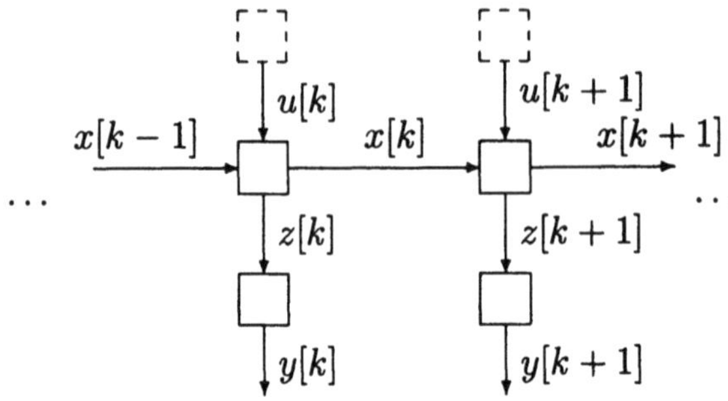

Figure 5 State space model.

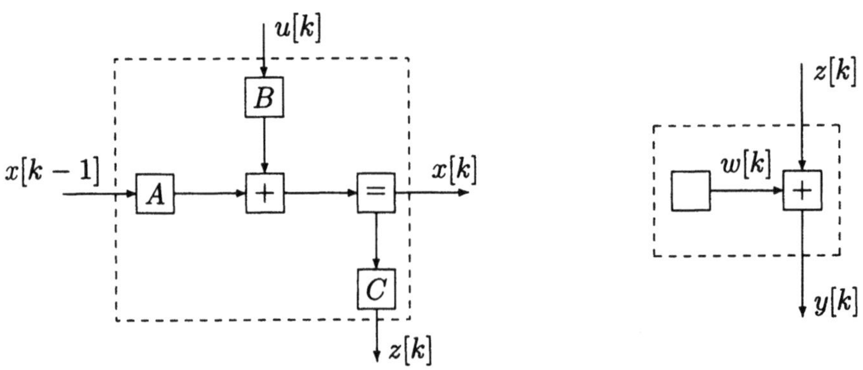

Figure 6 Details of linear state space model.

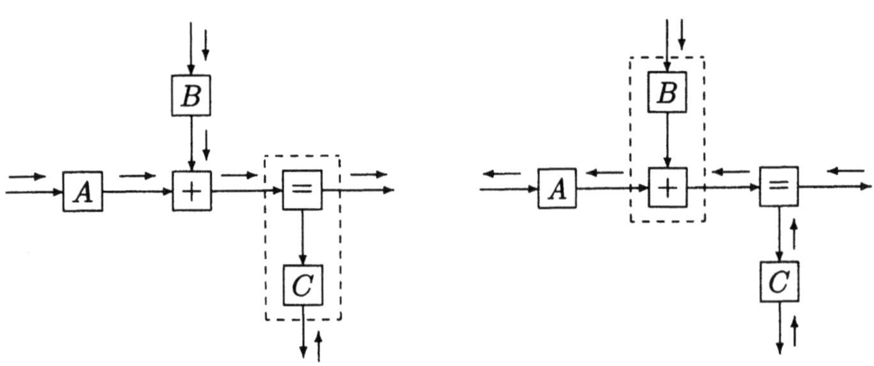

Figure 7 Use of composite-block update rules of Table 2.

1	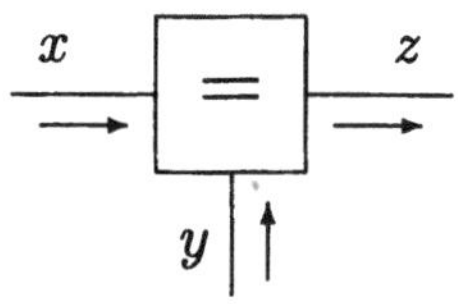 $\delta(x-y)\delta(x-z)$	$m_z = (W_x + W_y)^{\#}\,(W_x m_x + W_y m_y)$ $W_z = W_x + W_y$ $V_z = V_x(V_x + V_y)^{\#}\,V_y$
2	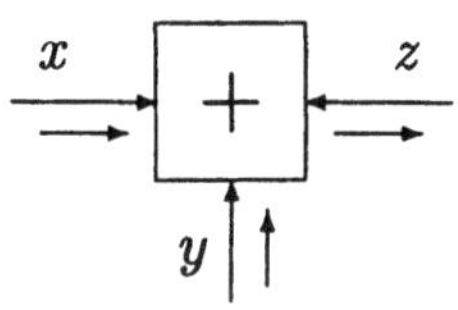 $\delta(x+y+z=0)$	$m_z = -m_x - m_y$ $V_z = V_x + V_y$ $W_z = W_x(W_x + W_y)^{\#}\,W_y$
3	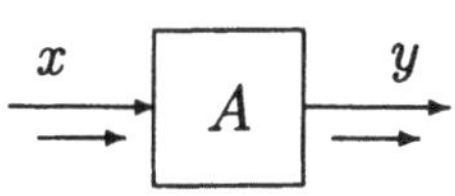 $\delta(y - Ax)$	$m_y = Am_x$ $V_y = AV_xA^H$ Problem with W_y if A has not full row rank!
4	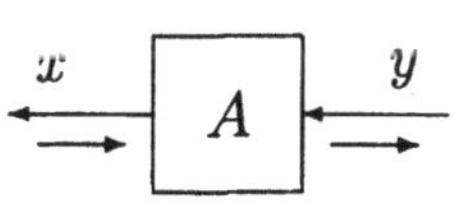 $\delta(x - Ay)$	$m_y = (A^H W_x A)^{\#} A^H W_x m_x$ $W_y = A^H W_x A$ If A has full row rank: $m_y = A^H (AA^H)^{-1} m_x$

Table 1 Update rules for messages consisting of mean vector m and covariance matrix V or $W = V^{-1}$.

5	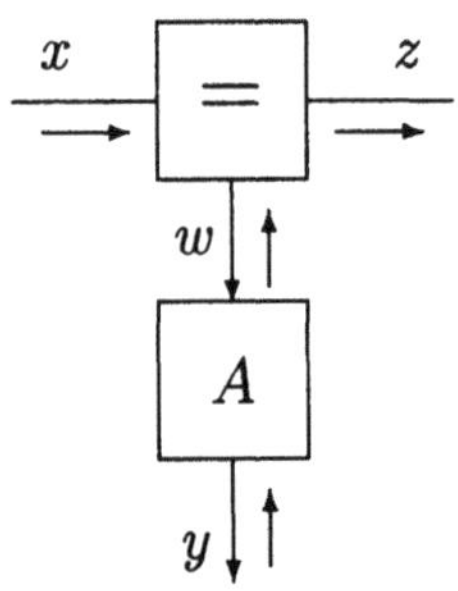	$m_z = m_x + V_x A^H G\,(m_y - Am_x)$ $V_z = V_x - V_x A^H G A V_x$ with $G \triangleq \left(V_y + AV_x A^H\right)^{-1}$
6	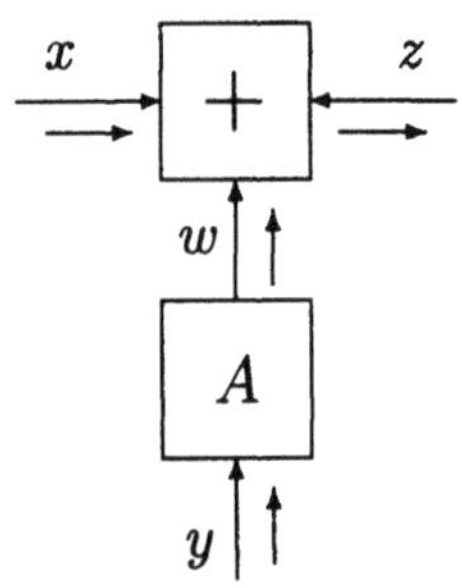	$m_z = -m_x - Am_y$ $W_z = W_x - W_x A H A^H W_x$ with $H \triangleq \left(W_y + A^H W_x A\right)^{-1}$

Table 2 Update rules for composite blocks.

up to time k. By computing also the backwards messages (cf. Figure 7 right), the posterior probability of all quantities given the whole observation sequence $y[.]$ may be obtained.

More generally, Kalman filtering amounts to the sum-product algorithm on any FFG (or part of an FFG) that consists of the linear building blocks listed in Table 1. As discussed in Section 3, any message along some edge x represents a quadratic cost function of the form (8) or, equivalently, a function of the form (7); for the actual computation, each such message consists of a mean vector m_x and a nonnegative definite "cost" matrix W_x or its inverse $V_x = W_x^{-1}$. As pointed out in Section 3. Kalman filtering may thus be viewed as, and is equivalent to, a general least-squares algorithm.

The rules for the computation of such messages are given in Table 1. As only one of the two messages along any edge, say x, is considered, the corresponding mean vectors and matrices are simply denoted m_x,

W_x, etc.. The derivation of these rules from the min-sum update rule (16) is given in the next section.

In general, we allow the matrices W or V to be nonnegative definite, which allows to express certainty in V and complete ignorance in W. However, whenever the inversion of such a matrix is required, it had better be positive definite; in practice, this may be achieved by the innocent device of adding a small multiple of an identity matrix.

The direct application of the update rules in Table 1 may lead to frequent matrix inversions. A key observation in Kalman filtering is that the inversion of large matrices can often be avoided. In the FFG, such simplifications may be achieved by using the update rules for the composite blocks given in Table 2. E.g., the vectors $u[k]$ and $z[k]$ in Figure 5 have usually much smaller dimensions than the state vector $x[.]$; in fact, they are often scalars. By working with composite blocks as in Figure 7, the forward recursion (left in Figure 7) using the covariance matrix $V = W^{-1}$ then requires no inversion of a large matrix and the backward recursion (right in Figure 7) using the cost matrix W requires only one such inversion for each discrete time index.

5. DERIVATION OF UPDATE RULES

We now derive the update rules of Tables 1 and 2 from the min-sum rule (16). As only one of the two messages along any edge, say x, is considered, the corresponding quadratic cost function will be denoted simply by $q_x(x) = (x - m_x)^H W_x(x - m_x)$. Frequent use will be made of facts stated in the appendix.

5.1 EQUALITY CONSTRAINT

Rule 1 of Table 1 is proved as follows. Assume that both W_x and W_y are nonnegative definite. According to the min-sum rule (16), we have for $\beta \to \infty$

$$\begin{aligned} q_z(z) &= \min_x \min_y \left(q_x(x) + q_y(y) + \beta \left(\|x - y\|^2 + \|x - z\|^2 \right) \right) && (20) \\ &= q_x(z) + q_y(z) && (21) \\ &= (z - m_x)^H W_x(z - m_x) + (z - m_y)^H W_y(z - m_y). && (22) \end{aligned}$$

This is of the form (95) with $x = z$, $A = W_x$, $B = W_y$, $a = m_x$, and $b = m_y$. Thus

$$q_z(z) = (z - m_z)^H W_z(z - m_z) + \textbf{const} \qquad (23)$$

with

$$W_z = W_x + W_y \qquad (24)$$

$$m_z = (W_x + W_y)^{\#} (W_x m_x + W_y m_y). \tag{25}$$

5.2 SUM CONSTRAINT

Rule 2 of Table 1 is proved as follows. The rule $V_z = V_x + V_y$ for the covariance matrices follows directly from elementary probability theory. An alternative derivation from the min-sum rule goes as follows. Assume that both W_x and W_y are nonnegative definite. According to the min-sum rule (16), we have for $\beta \to \infty$

$$\begin{aligned} q_z(z) &= \min_x \min_y \left(q_x(x) + q_y(y) + \beta \|x + y + z\|^2 \right) &(26)\\ &= \min_{x,y:\, x+y+z=0} \left(q_x(x) + q_y(y) \right) &(27)\\ &= \min_{x,y:\, x+y+z=0} \Big((x - m_x)^H W_x (x - m_x) + \\ &\qquad (y - m_y)^H W_y (y - m_y) \Big) &(28) \end{aligned}$$

Substituting $x+y = -z$ and $x-y = s$, i.e., $2x = s-z$ and $2y = -(s+z)$ yields

$$\begin{aligned} q_z(z) &= \min_{x,y:\, x+y+z=0} \frac{1}{4} \Big((2x - 2m_x)^H W_x (2x - 2m_x) + \\ &\qquad (2y - 2m_y)^H W_y (2y - 2m_y) \Big) &(29)\\ &= \min_s \frac{1}{4} \Big((s - z - 2m_x)^H W_x (s - z - 2m_x) + \\ &\qquad (s + z + 2m_y)^H W_y (s + z + 2m_y) \Big) &(30) \end{aligned}$$

This is of the form (95) with $x = s$, $A = W_x$, $B = W_y$, $a = z + 2m_x$, and $b = -(z + 2m_y)$. The minimum of (95) over x is c as in (107). With $a - b = 2z + 2m_x + 2m_y$, we thus have

$$q_z(z) = (z + m_x + m_y)^H W_x (W_x + W_y)^{\#} W_y (z + m_x + m_y) + \text{const.} \tag{31}$$

If both W_x and W_y are invertible, then $W_x + W_y$ is invertible and (87) implies $W_x (W_x + W_y)^{-1} W_y = (W_x^{-1} + W_y^{-1})^{-1}$.

5.3 MATRIX MULTIPLICATION: FORWARD MESSAGE

Rule 3 of Table 1 is proved as follows. The rule $V_y = A V_x A^H$ for the covariance matrices follows directly from elementary probability theory. An alternative derivation from the min-sum rule goes as follows. Assume that W_x is nonnegative definite. According to the min-sum rule (16), we have for $\beta \to \infty$

$$q_y(y) = \min_x \left(q_x(x) + \beta \|y - Ax\|^2 \right) \tag{32}$$

Clearly, if A does not have full row rank (i.e., if the rank of A is less than the number of rows), any vector y that is not in span(A) should be assigned infinite cost; the term $\beta\|y - Ax\|^2$ cannot be made to vanish in this case. Noting that $(I - AA^{\#})y$ is the projection of y onto span$(A)^{\perp}$, this case can be handled in practice by including in W_y a penalty term $\beta(I - AA^{\#})^H(I - AA^{\#})$ where β is some large positive real number.

Having settled that, we now consider the computation of $q_y(y)$ for those y that are in span(A). With this restriction in mind, we have

$$\begin{aligned} q_y(y) &= \min_{x:\, Ax=y} q_x(x) && (33) \\ &= \min_{x:\, Ax=y} (x - m_x)^H W_x (x - m_x) && (34) \end{aligned}$$

In general, this minimization is not easily handled. We therefore focus on two important special cases.

Case 1: If A is invertible, (34) becomes

$$\begin{aligned} q_x(A^{-1}y) &= (A^{-1}y - m_x)^H W_x (A^{-1}y - m_x) && (35) \\ &= (y - Am_x)^H (A^{-1})^H W_x A^{-1} (y - Am_x). && (36) \end{aligned}$$

Case 2: If W_x is invertible, it is positive definite, and so is its inverse $V_x = W_x^{-1}$. We then can write $V_x = BB^H$ for some matrix B that is itself invertible. Thus

$$\begin{aligned} q_x(x) &= (x - m_x)^H W_x (x - m_x) && (37) \\ &= (x - m_x)^H (B^{-1})^H B^{-1} (x - m_x) && (38) \\ &= (B^{-1}x - B^{-1}m_x)^H (B^{-1}x - B^{-1}m_x). && (39) \end{aligned}$$

With $w \triangleq B^{-1}x - B^{-1}m_x$, we obtain

$$\begin{aligned} \min_{x:\, Ax=y} q_x(x) &= \min_{w:\, A(Bw+m_x)=y} \|w\|^2 && (40) \\ &= \min_{w:\, ABw=y-Am_x} \|w\|^2 && (41) \\ &= \|(AB)^{\#}(y - Am_x)\|^2 && (42) \\ &= \left((AB)^{\#}(y - Am_x) \right)^H (AB)^{\#} (y - Am_x) && (43) \\ &= (y - Am_x)^H \left((AB)^{\#} \right)^H (AB)^{\#} (y - Am_x) && (44) \end{aligned}$$

$$= (y - Am_x)^H (ABB^H A^H)^{\#} (y - Am_x) \quad (45)$$
$$= (y - Am_x)^H (AW^{-1} A^H)^{\#} (y - Am_x), \quad (46)$$

where the step to (45) uses (81).

5.4 MATRIX MULTIPLICATION: BACKWARD MESSAGE

Rule 4 of Table 1 is proved as follows. Assume that W_x is nonnegative definite. According to the min-sum rule (16), we have for $\beta \to \infty$

$$\begin{aligned} q_y(y) &= \min_x \left(q_x(x) + \beta \|x - Ay\|^2 \right) && (47)\\ &= q_x(Ay) && (48)\\ &= (Ay - m_x)^H W_x (Ay - m_x) && (49)\\ &= (Ay)^H W_x Ay - 2\mathrm{Re}\,(y^H A^H W_x m_x) + m_x^H W m_x && (50)\\ &= y^H A^H W_x Ay - 2\mathrm{Re}\left(y^H (A^H W_x A)(A^H W_x A)^{\#} A^H W_x m_x\right) + \\ &\quad m_x^H W_x m_x && (51) \end{aligned}$$

where the equality $A^H W_x = (A^H W_x A)(A^H W_x A)^{\#} A^H W_x$ used in the last step follows from (76) and Property 9 of nonnegative definite matrices.

If the rank of A equals the number of rows, then $AA^{\#} = I$ and

$$\begin{aligned} A^H W_x &= A^H W_x A A^{\#} && (52)\\ &= \left(A^H W_x A\right) A^H \left(AA^H\right)^{-1}. && (53) \end{aligned}$$

5.5 COMPOSITE BLOCKS: RULE 5

Rule 5 of Table 2 is the combination of rules 1 and 4 of Table 1. We assume that both V_x and V_y (and thus W_x and W_y) are positive definite.

$$\begin{aligned} W_z &= W_x + W_w && (54)\\ &= W_x + A^H W_y A && (55) \end{aligned}$$

With the matrix inversion lemma (89), we obtain

$$\begin{aligned} V_z &= W_z^{-1} && (56)\\ &= V_x - V_x A^H \left(V_y + A V_x A^H\right)^{-1} A V_x. && (57) \end{aligned}$$

According to the rules 1 and 4, we further have

$$m_z = (W_x + W_w)^{-1} (W_x m_x + W_w m_w) \quad (58)$$

$$= V_z \left(W_x m_x + A^H W_y A \left(A^H W_y A \right)^{\#} A^H W_y m_y \right) \quad (59)$$

$$= V_z \left(V_x^{-1} m_x + A^H V_y^{-1} m_y \right). \quad (60)$$

Inserting (57) yields

$$m_z = \left(I - V_x A^H \left(V_y + A V_x A^H \right)^{-1} A \right) \left(m_x + V_x A^H V_y^{-1} m_y \right) \quad (61)$$

$$= m_x + V_x A^H V_y^{-1} m_y - V_x A^H \left(V_y + A V_x A^H \right)^{-1} A \left(m_x + V_x A^H V_y^{-1} m_y \right) \quad (62)$$

$$= m_x + V_x A^H \left(V_y + A V_x A^H \right)^{-1} \quad (63)$$

$$\cdot \left(\left(V_y + A V_x A^H \right) V_y^{-1} m_y - \left(A m_x + A V_x A^H V_y^{-1} m_y \right) \right) \quad (64)$$

$$= m_x + V_x A^H \left(V_y + A V_x A^H \right)^{-1} (m_y - A m_x). \quad (65)$$

5.6 COMPOSITE BLOCKS: RULE 6

Rule 6 of Table 2 is the combination of rules 2 and 3 of Table 1. We assume that both W_x and W_y are positive definite.

$$V_z = V_x + V_w \quad (66)$$

$$= V_x + A V_y A^H \quad (67)$$

With the matrix inversion lemma (89), we obtain

$$W_z = V_z^{-1} \quad (68)$$

$$= W_x - W_x A \left(W_y + A^H W_x A \right)^{-1} A^H W_x. \quad (69)$$

According to the rules 2 and 3, we further have

$$m_z = -m_x - m_w \quad (70)$$

$$= -m_x - A m_y. \quad (71)$$

6. CONCLUSIONS

Forney-style factor graphs ("normal graphs") are a natural extension of block diagrams that allows for both probabilistic and behavioral modelling. Kalman filtering and recursive least-squares algorithms may both be viewed as message passing in the graph that represents the underlying state space model. In this paper, we have discussed the corresponding message update rules for the building blocks of such models. In particular, we have shown in detail how these rules can be derived from the update rules of the general min-sum or sum-product algorithm.

Recursive least-squares and Kalman-filtering are important tools in problems like combined decoding, equalization, channel estimation, etc. The open-system spirit of Forney-style factor graphs makes it easy to use these tools for suitable subtasks of an overall iterative algorithm to solve such problems.

In 1967, Dave Forney introduced trellis diagrams, which have long since become deeply ingrained in the mind of every communications engineer. This writer finds Forney-style factor graphs equally helpful and wishes them an equally happy future.

Appendix

A Singular Value Decomposition and Pseudo-Inverse

A square complex matrix A is *unitary* if $AA^H = I$, i.e., $A^{-1} = A^H$. Any matrix A (not necessarily square) can be written as

$$A = USV^H, \tag{72}$$

where U and V are unitary matrices, and where the matrix S (not necessarily square) is real and diagonal with positive entries only:

$$S = \begin{pmatrix} S^{\dagger} & 0 \\ 0 & 0 \end{pmatrix} \tag{73}$$

with $S^{\dagger} \triangleq \mathrm{diag}(\sigma_1, \ldots, \sigma_\ell)$, $\sigma_k > 0$, $k = 1 \ldots \ell$. The decomposition (72) is called the *singular value decomposition* of A.

The *Moore-Penrose generalized inverse* or *pseudo inverse* of a complex matrix A (not necessarily square) with singular value decomposition (72) is the matrix

$$A^{\#} \triangleq VS^{\#}U^H \tag{74}$$

with

$$S^{\#} \triangleq \begin{pmatrix} \mathrm{diag}(\sigma_1^{-1}, \ldots, \sigma_\ell^{-1}) & 0 \\ 0 & 0 \end{pmatrix}. \tag{75}$$

Note that $S^{\#}$ has the same dimensions as S^H and $A^{\#}$ has the same dimensions as A^H. It is easy to see that

$$AA^{\#}A = A. \tag{76}$$

and

$$A^{\#}AA^{\#} = A^{\#}. \tag{77}$$

Proof: $AA^{\#}A = USV^HVS^{\#}U^HUSV^H = USS^{\#}SV^H = USV^H = A$.
$A^{\#}AA^{\#} = VS^{\#}U^HUSV^HVS^{\#}U^H = VS^{\#}SS^{\#}U^H = VS^{\#}U^H = A^{\#}$.
□

It is also easy to see that

$$A^{\#}x = 0 \Longleftrightarrow x^HA = 0. \tag{78}$$

Proof: $x^HA = 0 \Leftrightarrow x^HUS = 0 \Leftrightarrow S^{\#}U^Hx = 0 \Leftrightarrow A^{\#}x = 0.$ □

It follows from (76) and (78) that $AA^{\#}$ is the projection onto span(A), the space spanned by the columns of A.

If the rank of A equals the number of rows, then $AA^{\#} = I$ and

$$A^{\#} = A^H(AA^H)^{-1}; \tag{79}$$

if the rank of A equals the number of columns, then $A^{\#}A = I$ and

$$A^{\#} = (A^H A)^{-1}A^H. \tag{80}$$

Proof: We prove only (79); the proof of (80) is analogous. Under the stated assumptions, we have
$A^H(AA^H)^{-1} = VS^HU^H(USV^HVS^HU^H)^{-1} = VS^HU^H(USS^HU^H)^{-1}$
$= VS^HU^HU(SS^H)^{-1}U^H = VS^H(SS^H)^{-1}U^H = VS^{\#}U^H = A^{\#}$. □

Another useful formula is

$$(AA^H)^{\#} = (A^{\#})^H A^{\#}, \tag{81}$$

which follows from
$(AA^H)^{\#} = (USV(USV)^H)^{\#} = (USVV^HS^HU^H)^{\#} = (USS^HU^H)^{\#}$
$= U(SS^H)^{\#}U^H = U(S^{\#})^HS^{\#}U^H = U(S^{\#})^HV^HVS^{\#}U^H = (A^{\#})^HA^{\#}$.

B Nonnegative Definite Matrices

A square complex matrix A is *Hermitian* if $A^H = A$. If A is Hermitian, then $x^HAx = (x^HAx)^H = \overline{x^HAx}$ is real for any vector x. A Hermitian matrix A is *positive definite (nonnegative definite)* if $x^HAx > 0$ ($x^HAx \geq 0$) for all complex vectors $x \neq 0$. Some basic facts:

1. If A is positive definite (nonnegative definite), then so is aA for any real number $a > 0$.

2. If A and B are nonnegative definite matrices of the same dimensions, the $A + B$ is also nonnegative definite. If, in addition, A (or B) is positive definite, then $A + B$ is positive definite.

3. If A is positive definite, then A is invertible and A^{-1} is also positive definite. (See also Property 10.)

4. The unit matrix I (of any dimension) is positive definite.

5. If B is nonnegative definite, then so is ABA^H for any matrix A (of suitable dimensions, not necessarily square). In particular, AA^H is nonnegative definite.

6. If A is nonnegative definite, then it has a singular value decomposition $A = USU^H$ with unitary U and $S = \text{diag}(\sigma_1, \ldots, \sigma_n)$. $A = USU^H$ is positive definite if and only if all singular values σ_k, $k = 1, \ldots, n$, are nonzero (i.e., positive).

7. If A is nonnegative definite, then $x^H A x = 0$ implies $Ax = 0$.

8. If A and B are nonnegative definite matrices of the same dimensions, then we have $\text{span}(A + B) = \text{span}(A, B)$. (Here $\text{span}(W)$ denotes the linear space spanned by the columns of W; $\text{span}(A, B)$ is the space spanned by the columns of A and B.)

9. If B is nonnegative definite, then for any matrix A (of suitable dimensions, not necessarily square), $\text{span}(A^H B A) = \text{span}(A^H B)$.

10. If A is nonnegative definite, then so is its pseudo-inverse $A^\#$. Moreover, $AA^\# = A^\# A$.

Proof:

1. $x^H (aA) x = a(x^H A x)$.

2. $x^H (A + B) x = x^H A x + x^H B x$.

3. Let A be positive definite. Clearly, A is invertible if and only if $Ax = 0$ implies $x = 0$. But $Ax = 0$ implies $x^H A x = 0$, which in turn implies $x = 0$; thus A is invertible. It remains to show that A^{-1} is also positive definite. Assume $x^H A^{-1} x = 0$. With $y \triangleq A^{-1} x$, we then have $x^H A^{-1} x = y^H A y = 0$. Thus $y = 0$, which implies $x = 0$.

4. $x^H I x = x^H x = \|x\|^2$.

5. $x^H (ABA^H) x = (A^H x)^H B (A^H x) \geq 0$.

7. Let $A = USU^H$ be the singular value decomposition of A according to Property 6 and let $y \triangleq U^H x$. Then $x^H A x = (U^H x)^H S (U^H x) = y^H S y = \sum_{k=1}^n \sigma_k |y_k|^2$. Now assume $x^H A x = 0$. Since all σ_k are real and nonnegative, this implies $\sigma_k y_k = 0$, $k = 1, \ldots, n$, and thus $Sy = 0$. We therefore have $Ax = USU^H x = USy = 0$.

8. Clearly, we always have $\text{span}(A + B) \subseteq \text{span}(A, B)$. We prove equality by showing that the orthogonal complements of these spaces are equal. Some vector x is orthogonal to $\text{span}(A + B)$ if and only if $x^H (A + B) = 0$, and orthogonal to $\text{span}(A, B)$ if and only if $x^H A = x^H B = 0$. But with Property 7, we have

$x^H(A+B)=0 \iff x^H(A+B)x=0 \iff x^HAx+x^HBx=0 \iff x^HAx=x^HBx=0 \iff x^HA=x^HB=0.$

9. Clearly, we always have $\text{span}(A^HBA) \subseteq \text{span}(A^HB)$. As above, we prove equality by showing that the orthogonal complements of these spaces are equal. This is established by the following chain of equivalent conditions, where Property 7 is used twice: $x \in \text{span}(A^HB)^\perp \iff x^HA^HB = 0 \iff (Ax)^HB = 0 \iff (Ax)^HB(Ax)=0 \iff x^H(A^HBA)x=0 \iff x^HA^HBA=0 \iff x \in \text{span}(A^HBA)^\perp$.

10. Let $A = USU^H$ be the singular value decomposition of A according to Property 6. Then $A^\# = US^\#U^H$ and $x^HA^\#x = x^HUS^\#U^Hx = (U^Hx)^HS^\#(U^Hx) \geq 0$. Moreover, $AA^\# = USU^HUS^\#U^H = USS^\#U^H = US^\#SU^H = A^\#A$. □

If A and B are nonnegative definite matrices of the same dimensions, we also have

$$(A+B)(A+B)^\#A = A \tag{82}$$
$$= A(A+B)^\#(A+B) \tag{83}$$

Proof: The two equalities are obtained from each other by Hermitian transposition; it thus suffices to prove one of them. Now (76) implies $(A+B)(A+B)^\#x = x$ for any vector $x \in \text{span}(A+B)$. But according to property 8 above, $\text{span}(A+B) = \text{span}(A,B) \supseteq \text{span}(A)$. Thus $(A+B)(A+B)^\#x = x$ holds for any vector $x \in \text{span}(A)$, which implies (82). □

From (83) follows $A(A+B)^\#A + A(A+B)^\#B = A$ and thus

$$A(A+B)^\#B = A - A(A+B)^\#A; \tag{84}$$

from (82) follows $A(A+B)^\#A + B(A+B)^\#A = A$ and thus

$$B(A+B)^\#A = A - A(A+B)^\#A. \tag{85}$$

In particular, $A(A+B)^\#B = B(A+B)^\#A = \left(A(A+B)^\#B\right)^H$.

If both A and B are invertible, then $(A+B)$ is invertible and

$$A(A+B)^{-1}B = \left(B^{-1}(A+B)A^{-1}\right)^{-1} \tag{86}$$
$$= \left(A^{-1}+B^{-1}\right)^{-1}. \tag{87}$$

Finally, we have the

Matrix Inversion Lemma. Let B and D be positive definite matrices. If, for some matrices A and C, we have

$$A = B^{-1} + CD^{-1}C^H, \tag{88}$$

then

$$A^{-1} = B - BC(D + C^H BC)^{-1} C^H B. \tag{89}$$

Proof:

$$\begin{aligned}
&\left(B^{-1} + CD^{-1}C^H\right)\left(B - BC\left(D + C^H BC\right)^{-1} C^H B\right) && (90)\\
&= I + CD^{-1}C^H B - \left(B^{-1} + CD^{-1}C^H\right) BC\left(D + C^H BC\right)^{-1} C^H B && (91)\\
&= I + CD^{-1}C^H B - \left(C + CD^{-1}C^H BC\right)\left(D + C^H BC\right)^{-1} C^H B && (92)\\
&= I + CD^{-1}C^H B - CD^{-1}\left(D + C^H BC\right)\left(D + C^H BC\right)^{-1} C^H B && (93)\\
&= I. && (94)
\end{aligned}$$

□

C Sum of Quadratic Forms

Let both A and B be nonnegative definite matrices (which implies that they are Hermitian). Using (82) in the step to (98), we have

$$\begin{aligned}
&(x-a)^H A(x-a) + (x-b)^H B(x-b) && (95)\\
&= x^H Ax - 2\mathrm{Re}\,(x^H Aa) + a^H Aa + x^H Bx - 2\mathrm{Re}\,(x^H Bb) + b^H Bb && (96)\\
&= x^H (A+B)x - 2\mathrm{Re}\left(x^H(Aa+Bb)\right) + a^H Aa + b^H Bb && (97)\\
&= x^H (A+B)x - 2\mathrm{Re}\left(x^H(A+B)(A+B)^{\#}(Aa+Bb)\right) + \\
&\quad a^H Aa + b^H Bb && (98)\\
&= x^H Wx - 2\mathrm{Re}\,(x^H Wm) + m^H Wm + c && (99)
\end{aligned}$$

with

$$\begin{aligned}
W &= A + B && (100)\\
m &= (A+B)^{\#}(Aa + Bb) && (101)
\end{aligned}$$

and the scalar

$$\begin{aligned}
c &= a^H A a + b^H B b - m^H W m && (102)\\
&= a^H A a + b^H B b - \\
&\quad (Aa + Bb)^H (A+B)^{\#} (A+B)(A+B)^{\#}(Aa+Bb) && (103)\\
&= a^H A a + b^H B b - (a^H A + b^H B)(A+B)^{\#}(Aa+Bb) && (104)\\
&= a^H \left(A - A(A+B)^{\#} A\right) a - a^H A(A+B)^{\#} Bb + \\
&\quad b^H \left(B - B(A+B)^{\#} B\right) b - b^H B(A+B)^{\#} Aa && (105)\\
&= a^H A(A+B)^{\#} Ba - a^H A(A+B)^{\#} Bb + \\
&\quad b^H B(A+B)^{\#} Ab - b^H B(A+B)^{\#} Aa && (106)\\
&= (a-b)^H A(A+B)^{\#} B(a-b), && (107)
\end{aligned}$$

where the step to (106) follows from (84).

References

[1] B. D. O. Anderson and J. B. Moore, *Optimal Filtering.* Prentice-Hall, Englewood Cliffs, NJ, 1979.

[2] G. D. Forney, Jr., "Codes on Graphs: Normal Realizations," *IEEE Transactions on Information Theory*, vol. 47, pp. 520–548, Feb. 2001.

[3] F. R. Kschischang, B. J. Frey, and H.-A. Loeliger, "Factor graphs and the Sum-Product Algorithm," *IEEE Transactions on Information Theory*, vol. 47, pp. 498–519, Feb. 2001.

[4] A. H. Sayed and T. Kailath, "A State-Space Approach to Adaptive RLS Filtering," *IEEE Signal Processing Magazine*, vol. 11, pp. 18–60, 1994.

[5] Y. Weiss and W. T. Freeman, "On the Optimality of Solutions of the Max-Product Belief-Propagation Algorithm in Arbitrary Graphs," *IEEE Transactions on Information Theory*, vol. 47, pp. 736–744, Feb. 2001.

[6] N. Wiberg, *Codes and Decoding on General Graphs.* Linköping Studies in Science and Technology, Ph.D. Thesis No. 440, University of Linköping, Sweden, 1996.

[7] J. C. Willems, "Paradigms and Puzzles in the Theory of Dynamical Systems," *IEEE Transactions on Automatic Control*, vol. 36, pp. 259–294, March 1991.

Chapter 8

ITERATIVE ALGORITHMS FOR MAXIMUM LIKELIHOOD SEQUENCE DETECTION

J. A. O'Sullivan

Electronic Systems and Signals Research Laboratory

Department of Electrical Engineering

Washington University

St. Louis, MO 63130

Abstract The success of iterative decoding algorithms for turbo codes, for low density parity check codes, and for related codes has motivated studying connections to other potentially related iterative algorithms in information theory and in estimation theory. Important iterative algorithms in information theory include the Blahut-Arimoto algorithm for computing the distribution that achieves channel capacity and Blahut's algorithm for computing the rate-distortion function. In estimation theory, the expectation-maximization algorithm is widely used, especially in high-dimensional inference problems such as in image estimation. A study of these algorithms as alternating minimization algorithms lays the foundation for deriving new alternating minimization algorithms for a variety of problems. One family of such algorithms includes iterative algorithms for maximum-likelihood sequence detection. These algorithms and some of their properties are presented.

1. INTRODUCTION

The success of iterative decoding techniques has generated a great deal of interest in the underlying optimization theory. Conditions for the convergence of iterative decoding algorithms are of interest as are formalisms for deriving alternative iterative decoding algorithms. Some iterative decoding algorithms are closely connected to specific objective functions. Some algorithms are not directly connected to global objective functions.

At the heart of many algorithms is the message passing paradigm that has emerged from a generalization of several algorithms including the BCJR algorithm [3]. Bahl, Cocke, Jelinek, and Raviv derived the BCJR algorithm to exactly compute the posterior bit probabilities for a finite length noisy observation of a codeword from a convolutional code. This algorithm extends directly to codes defined on graphs that are trees, with the computations becoming messages passed along edges in the graphs; exact computations of posterior bit probabilities are achieved [16].

This algorithm has been extended to graphs with loops. Notably, the turbo decoding algorithm [4] can be viewed as a message passing algorithm, where the messages are analogous to the messages passed in the BCJR algorithm [16, 19]. Message passing algorithms have a long history in the computer science and artificial intelligence literature where graphical representations of data models have been exploited. Many connections between inference algorithms (in decoding, in estimation theory, and in learning theory) and graphical representations of data are described by Frey [16] and by various authors in the book edited by Jordan [19].

Message passing decoding has been particularly effective in decoding low density parity check codes [17, 22]. These codes are defined in terms of their parity check matrices which are designed to have a desired sparsity, typically a constant (or bounded) number of nonzero entries per column and row. The graph defining a code has two types of nodes: code bit nodes and parity check nodes. Messages are alternately passed from code bit nodes to parity check nodes, then from parity check nodes to code bit nodes. Recent work by Richardson and Urbanke [28] has found analytical bounds for the regions in which message passing decoding is optimal.

The Viterbi algorithm [13] can also be viewed as a message passing algorithm on a graph, the same graph used by Bahl, Cocke, Jelinek, and Raviv [3]. Instead of computing posterior bit probabilities, the Viterbi algorithm finds the most likely codeword. The computation of the most likely codeword can be achieved by a message passing algorithm on the graph, where the messages are viewed as being over the max-sum algebra rather than the sum-product algebra used for the BCJR algorithm [16]. These two algorithms are special cases of algorithms that compute the marginal of a joint function. While this provides opportunities for deriving new algorithms based on different types of marginalization (Aji and McEliece [1] describe many such generalizations), the algorithms have theoretical and computational difficulties when graphs have loops [29].

Iterative decoding algorithms are useful algorithms for inference in high dimensional spaces. Low density parity check codes have been studied with lengths greater than 1,000,000 code bits. Decoding is then over a discrete set with greater than $2^{1,000,000}$ elements. Practical iterative decoding algorithms have complexity linear in the number of code bits. Message passing algorithms over sparse graphs whose numbers of nodes and edges are proportional to the number of code bits have such complexity.

In image estimation problems, there is a similar need for algorithms for inference over high dimensional spaces. In a typical three-dimensional problem, the volume of interest may be discretized with a dimension greater than 100 in each direction, yielding greater than 1,000,000 voxels (volume elements). Image estimation is then over a continuous space larger than $\mathbf{R}^{1,000,000}$. Practical algorithms have complexity $O(n \log n)$ where n is the number of voxels. Iterative algorithms based on the expectation-maximization algorithm or variants of it such as the ordered-subset expectation-maximization (OSEM) algorithm [18] can have this complexity if the hidden data can be selected appropriately; essentially the graph describing the hidden data and the observed (or incomplete) data must be sparse with number of edges linear or $O(n \log n)$ in the number of voxels.

One of the goals of this paper is to draw on the experience in the image estimation literature to derive new algorithms for decoding. By providing common derivations for algorithms in these different applications, it is hoped more cross-fertilization of ideas may be obtained. Frey [16] describes many common ideas in iterative decoding, estimation theory, and Bayesian inference, emphasizing the use of graphical models, especially factor graphs. The chapters in the book edited by Jordan [19] describe theory that underlies inference using Bayesian networks and Markov random fields, and learning theory based on such models.

Alternating minimization algorithms provide a context in which many potentially good algorithms may be derived. We review the general framework for deriving alternating minimization algorithms, including the most important examples in information theory and estimation theory. A general tool is introduced that relaxes optimization problems defined on discrete sets to optimization problems on probability vectors. The optimizations over probability vectors become alternating minimization algorithms. One application is a new family of iterative decoding algorithms. While this new family has not found practical use to date, it contains many algorithms which are potentially useful.

The new family of iterative decoding algorithms are designed for maximum-likelihood sequence detection, the log-likelihood function being the global objective function. The iterative algorithms monotonically increase the log-likelihood function, a desirable property that not many iterative algorithms achieve. It is hoped that other desirable properties of algorithms used in estimation theory may be achieved in algorithms for iterative decoding, but this analysis is incomplete.

In the next section, we describe the decoding problems of interest, MAP decoding and maximum-likelihood sequence detection. In Section 3, we review variational methods used in information theory to derive alternating minimization algorithms, including algorithms for computing channel capacity, for computing rate-distortion functions, and for decomposing coupled problems into uncoupled problems. In Section 4, we return to the problem of maximum-likelihood sequence detection. The problem of inference over the vertices of a hypercube is relaxed to a problem of inference over probabilities that effectively index locations in the interior of the hypercube. An iterative algorithm for estimating these probabilities is obtained, and some of its properties are discussed. Conclusions and future work are described last.

2. TWO DECODING PROBLEMS

We describe a problem in decoding of binary data. The methods described in this paper extend directly to other finite alphabets. A general description of the overall problem is sketched and two decoding methods are defined: MAP decoding and maximum-likelihood seqence detection.

We envision problems in which there are data bits, unobservable states from a finite alphabet, and observations. Denote the data bits by $i_k, k = 1, 2, \ldots, K$ and the (vector) bit sequence with entries i_k by $\mathbf{i}$. Denote the states by $s_m, m = 1, 2, \ldots, M$, and the (vector) state sequence by $\mathbf{s}$. Denote the observations by $y_n, n = 1, 2, \ldots, N$, and the (vector) observation sequence by $\mathbf{y}$.

There is an underlying model for the data that is assumed to be known. The model is captured in terms of conditional probabilities for the discrete-valued variables and conditional probability density functions for the continuous-valued variables. For this paper, the data $\mathbf{y}$ are assumed to be continuous-valued; for discrete-valued observations, simply substitute the probabilities for probability densities. The triple $\mathbf{i} \rightarrow \mathbf{s} \rightarrow \mathbf{y}$ is assumed to form a Markov chain in that order. Thus, the prior (if known) on the bit sequence $P(\mathbf{i})$ and $P(\mathbf{s}|\mathbf{i})$ and $p(\mathbf{y}|\mathbf{s})$ completely specify the joint distribution.

In many decoding problems, the functions $P(\mathbf{s}|\mathbf{i})$ and $p(\mathbf{y}|\mathbf{s})$ are specified by a graphical model. Nodes in the graph correspond to entries of the vectors $\mathbf{s}$, $\mathbf{i}$, and $\mathbf{y}$. Edges correspond to conditional dependence between variables. In low density parity check codes, rather than specifying the conditionals $P(\mathbf{s}|\mathbf{i})$ (corresponding to the encoding), the parity check equations define the graph.

2.1 MAP DECODING

In MAP decoding, the posterior probabilities $P(i_k|\mathbf{y})$ on each bit are computed. For each k, it is equivalent to know any one of the three quantities:

$$P(i_k = 1|\mathbf{y}), \text{ or } L_k = \frac{P(i_k = 1|\mathbf{y})}{P(i_k = 0|\mathbf{y})}, \text{ or } l_k = \log\left(\frac{P(i_k = 1|\mathbf{y})}{P(i_k = 0|\mathbf{y})}\right). \quad (1)$$

There are differences in implementation considerations for computing these. The third, the log-likelihood ratios, are often preferred. The computation of the posterior probability relies on computation of the marginal

$$p(\mathbf{y}|i_k = 1)P(i_k = 1) = \sum_{\mathbf{s}} \sum_{\mathbf{i}:i_k=1} p(\mathbf{y}|\mathbf{s})P(\mathbf{s}|\mathbf{i})P(\mathbf{i}). \quad (2)$$

Note that

$$l_k = \log\left(\frac{p(\mathbf{y}|i_k = 1)}{p(\mathbf{y}|i_k = 0)}\right) + \log\left(\frac{P(i_k = 1)}{P(i_k = 0)}\right), \quad (3)$$

the second term on the right being the prior log-likelihood ratio. The first term on the right side of Equation (3) quantifies the amount of information provided by the observation sequence $\mathbf{y}$ and is usually referred to as the intrinsic information provided by $\mathbf{y}$. Extrinsic information (relative to $\mathbf{y}$) is provided by other sources including the prior.

Hard thresholding on the posterior bit probabilities yields

MAP Decoding:
Decide $\hat{i}_k = 1$ if $P(i_k = 1|\mathbf{y}) > 0.5$.

2.2 MAXIMUM-LIKELIHOOD SEQUENCE DETECTION

In maximum-likelihood sequence detection, the most likely bit sequence $\mathbf{i}$ is selected. For each bit sequence $\mathbf{i}$, compute

$$p(\mathbf{y}|\mathbf{i})P(\mathbf{i}) = \sum_{\mathbf{s}} p(\mathbf{y}|\mathbf{s})P(\mathbf{s}|\mathbf{i})P(\mathbf{i}). \quad (4)$$

Maximum-Likelihood Sequence Detection:
Select the sequence $\mathbf{i}$ to maximize $p(\mathbf{y}|\mathbf{i})P(\mathbf{i})$.

As in MAP decoding, computation of log-likelihood ratios is often more attractive. Here, the log-likelihood ratios are for the sequence. Assume a nominal bit sequence, such as the all zeros sequence, is specified and denoted $\mathbf{i}_0$. Then the log-likelihood ratios are

$$\log\left(\frac{p(\mathbf{y}|\mathbf{i})P(\mathbf{i})}{p(\mathbf{y}|\mathbf{i}_0)P(\mathbf{i}_0)}\right) = \log\left(\frac{p(\mathbf{y}|\mathbf{i})}{p(\mathbf{y}|\mathbf{i}_0)}\right) + \log\left(\frac{P(\mathbf{i})}{P(\mathbf{i}_0)}\right). \tag{5}$$

The first term on the right is the intrinsic information provided by the observations $\mathbf{y}$ for the bit sequence. The second term on the right is the prior log-likelihood ratio. In the absence of prior information, all bit sequences are considered equally likely and the prior log-likelihood ratios are zero.

3. VARIATIONAL METHODS IN INFORMATION THEORY

Suppose that a problem is specified as a minimization of an objective function $\phi(q)$ over $q \in \mathcal{Q}$, where $\mathcal{Q}$ is an appropriately defined set of interest. A variational representation of ϕ is a representation of ϕ as the result of an auxiliary minimization

$$\phi(q) = \min_{p\in\mathcal{P}} J(p,q), \tag{6}$$

where J is the auxiliary objective function and $\mathcal{P}$ is an appropriately defined auxiliary set. This yields the equivalent statement of the original problem as

$$\min_{q\in\mathcal{Q}} \phi(q) = \min_{q\in\mathcal{Q}} \min_{p\in\mathcal{P}} J(p,q). \tag{7}$$

The double minimization in Equation (7) immediately gives rise to an alternating minimization algorithm

1. Select $q^{(0)}$, set $r = 1$
2. Compute $p^{(r)} = \text{argmin}_{p\in\mathcal{P}} J(p, q^{(r-1)})$
3. Compute $q^{(r)} = \text{argmin}_{q\in\mathcal{Q}} J(p^{(r)}, q)$
4. If converged, stop, else $r = r + 1$, go to 2.

Often there are choices in the selection of J in Equation (6). In these cases, J should be selected to achieve two goals: the minimizations over p and q are easy; and the function J has desirable convexity properties. For any J, the algorithm monotonically decreases J since for all iterations r,

$$J(p^{(r-1)},q^{(r-1)}) \geq J(p^{(r)},q^{(r-1)}) \geq J(p^{(r)},q^{(r)}) \geq J(p^{(r+1)},q^{(r)}). \tag{8}$$

The variational representation immediately implies that $\phi(q^{(r-1)}) \geq \phi(q^{(r)})$. Choice of J influences the convergence rate. Convexity properties of J determine the region of convergence of a local minimum. I. Csiszár and G. Tusnády present general properties of alternating minimization algorithms [8]. O'Sullivan examined the connection between alternating minimization algorithms in information theory and in estimation theory [26].

3.1 VARIATIONAL METHODS IN INFORMATION THEORY: CHANNEL CAPACITY

In independent papers that shared the prize paper award in information theory in 1972, Blahut and Arimoto derived an algorithm for computing the distribution that achieves channel capacity [5, 2]. Suppose that a channel is given with probability transition matrix $\mathbf{Q} = [Q_{k|j}]$. The problem is to compute the input probability distribution $\mathbf{p} \in \mathcal{P}$ that maximizes the mutual information between the input and the output of the channel. Here, $\mathcal{P}$ is the set of probability vectors. The mutual information, written as a function of $\mathbf{p}$ and $\mathbf{Q}$ is

$$I(\mathbf{p}, \mathbf{Q}) = \sum_j \sum_k p_j Q_{k|j} \log \frac{Q_{k|j}}{\sum_{j'} p_{j'} Q_{k|j'}}. \tag{9}$$

Blahut and Arimoto introduced the variational representation of mutual information as

$$I(\mathbf{p}, \mathbf{Q}) = \max_{\mathbf{P}} \sum_j \sum_k p_j Q_{k|j} \log \frac{P_{j|k}}{p_j}, \tag{10}$$

where the maximization is over all conditional probability matrices $\mathbf{P}$ (that is, $\sum_j P_{j|k} = 1$ for all k).

The channel capacity computation is then written as the double maximization

$$\begin{aligned} C &= \max_{\mathbf{p} \in \mathcal{P}} I(\mathbf{p}, \mathbf{Q}) \\ &= \max_{\mathbf{p} \in \mathcal{P}} \max_{\mathbf{P}} \sum_j \sum_k p_j Q_{k|j} \log \frac{P_{j|k}}{p_j}. \end{aligned}$$

The resulting alternating maximization algorithm alternates between updates of $\mathbf{p}$ and $\mathbf{P}$.

3.2 VARIATIONAL METHODS IN INFORMATION THEORY: RATE-DISTORTION

Blahut introduced another variational representation of mutual information that is used to derive an algorithm in rate-distortion computations [5]. The distribution of a source is known to be **p**. Define the set of probability distributions

$$\mathcal{Q} \;=\{\mathbf{Q}=[Q_{k|j}]:\;\; Q_{k|j}\geq 0, \sum_{k} Q_{k|j}=1,\; \forall\, j\}.$$

The set $\mathcal{Q}_D$ of distributions satisfying a distortion constraint is

$$\mathcal{Q}_D \;=\; \{\mathbf{Q}\in\mathcal{Q}: \sum_{j} p_j \sum_{k} Q_{k|j} d(j,k) \leq D\},$$

where $d(j,k)$ is a distortion function. The rate-distortion function results from the minimization

$$R(D) \;=\; \min_{\mathbf{Q}\in\mathcal{Q}_D} I(\mathbf{p},\mathbf{Q}).$$

This becomes a double minimization through the variational representation of mutual information as

$$I(\mathbf{p},\mathbf{Q}) = \min_{\mathbf{q}\in\mathcal{P}} \sum_{j}\sum_{k} p_j Q_{k|j} \log \frac{Q_{k|j}}{q_k},$$

where the minimization is over all **q** that are probability vectors. Then

$$R(D) \;=\; \min_{\mathbf{Q}\in\mathcal{Q}_D} \min_{\mathbf{q}\in\mathcal{P}} \sum_{j}\sum_{k} p_j Q_{k|j} \log \frac{Q_{k|j}}{q_k}.$$

The standard implementation uses a Lagrange multiplier to enforce the distortion constraint. The iterations are performed for each choice of this Lagrange multiplier, parametrically sweeping out pairs of achievable rate-distortion pairs (R,D). For each Lagrange multiplier s, the double minimization is

$$R(D) \;=\; \min_{\mathbf{Q}\in\mathcal{Q}} \min_{\mathbf{q}\in\mathcal{P}} \sum_{j}\sum_{k} p_j Q_{k|j} \log \frac{Q_{k|j}}{q_k \exp(-s d(j,k))}.$$

3.3 LEGENDRE-FENCHEL TRANSFORMS

A variational representation of a function may be derived from the Legendre-Fenchel transform. A convex function f may be written as a maximization of a linear functional minus its dual function f^*. Essentially this transformation takes the argument of the convex function and places it in a linear functional, often significantly simplifying the original problem. In particular, the linear functional over the elements of a vector effectively decouples the elements of the vector in an alternating minimization algorithm. This method has been described for inference over problems with Markov priors or Bayesian networks models by Jordan, et al. [19]. It is fundamental to the understanding and application of large deviations theory [9, 6].

We describe one setting for the statement of the Legendre-Fenchel transform. This can be generalized in several ways including to Hausdorff topological vector spaces [9, Sec. 4.5].

A function $f : \mathbf{R}^n \to \mathbf{R}$ is convex over a convex set $\mathcal{D} \subset \mathbf{R}^n$ if for all $\mathbf{x}_1, \mathbf{x}_2 \in \mathcal{D}$,

$$f(\alpha\mathbf{x}_1 + (1-\alpha)\mathbf{x}_2) \leq \alpha f(\mathbf{x}_1) + (1-\alpha) f(\mathbf{x}_2), \tag{11}$$

for all $0 \leq \alpha \leq 1$; f is strictly convex if the inequality is strict for all $0 < \alpha < 1$. The conjugate set $\mathcal{D}^*$ is defined to be

$$\mathcal{D}^* = \{\lambda : \sup_{\mathbf{x}\in\mathcal{D}} [\langle \mathbf{x}, \lambda\rangle - f(\mathbf{x})] < \infty\}. \tag{12}$$

Definition 1 *The Legendre-Fenchel transform of f, defined on $\mathcal{D}^*$ is*

$$f^*(\lambda) = \sup_{\mathbf{x}\in\mathcal{D}} [\langle \mathbf{x}, \lambda\rangle - f(\mathbf{x})]. \tag{13}$$

The function f^* is also called the conjugate functional.

Lemma 1 *Given a convex function $f(\mathbf{x})$ defined on a convex set $\mathcal{D} \subset \mathbf{R}^n$, the conjugate set $\mathcal{D}^*$ and the conjugate functional f^* are convex. The set $[f^*, \mathcal{D}^*] = \{(r, \lambda) : \lambda \in \mathcal{D}^*, r \geq f^*(\lambda)\}$ is a closed convex set.*

For the proof, see Luenberger [21, p. 196]. Define the set

$$[f, \mathcal{D}] = \{(r, \mathbf{x}) : \mathbf{x} \in \mathcal{D}, r \geq f(\mathbf{x})\}. \tag{14}$$

Luenberger also shows that if $[f, \mathcal{D}]$ is closed, then f and $\mathcal{D}$ are conjugate to f^* and $\mathcal{D}^*$ and thus for $\mathbf{x} \in \mathcal{D}$,

$$f(\mathbf{x}) = \sup_{\lambda\in\mathcal{D}^*} [\langle \mathbf{x}, \lambda\rangle - f^*(\lambda)]. \tag{15}$$

In this case, the functions f and f^* (along with their convex domains $\mathcal{D}$ and $\mathcal{D}^*$) form a Legendre-Fenchel tranform pair. Table 1 lists common

$f(x)$	$\mathcal{D}$	$f^*(\lambda)$	$\mathcal{D}^*$
$-\ln x$	$\mathbf{R}_+$	$-1-\ln(-\lambda)$	$\mathbf{R}_-$
$-\ln(1+e^{-x})$	$\mathbf{R}$	$H(\lambda)$	$[0,1]$
e^x	$\mathbf{R}$	$\lambda\ln\lambda-\lambda$	$\mathbf{R}_+$
$\frac{1}{p}x^p,\ 1<p<\infty$	$\mathbf{R}_+$	$\frac{1}{q}\lambda^q,\ \frac{1}{p}+\frac{1}{q}=1$	$\mathbf{R}_+$
$\frac{1}{p}\lvert x\rvert^p,\ 1<p<\infty$	$\mathbf{R}$	$\frac{1}{q}\lvert\lambda\rvert^q,\ \frac{1}{p}+\frac{1}{q}=1$	$\mathbf{R}$

Table 1 Common Legendre-Fenchel Transform Pairs.

Legendre-Fenchel tranform pairs that occur in optimization problems. The binary entropy function is $H(\lambda) = -\lambda\log\lambda - (1-\lambda)\log(1-\lambda)$.

Other Legendre-Fenchel transform pairs may be obtained by simple manipulations of those in the table. For example, suppose $f(\mathbf{x}) = \sum_i f_i(x_i)$, where each f_i is a convex function on a convex domain $\mathcal{D}_i$. Suppose also that each f_i has conjugate functional f_i^* with conjugate domain $\mathcal{D}_i^*$. Then the conjugate functional for $f(\mathbf{x})$ is

$$f^*(\lambda) = \sum_i f_i^*(\lambda_i), \tag{16}$$

where $\lambda_i \in \mathcal{D}_i^*$.

More generally, a sum of convex functions is convex. The conjugate functional for such an arbitrary sum may be found in terms of the conjugate functionals of the functions in the sum. This expression may be complicated. A simple expression results for the sum of a linear function and a convex function. For example, the functional conjugate to $\langle\mathbf{x},\mathbf{a}\rangle + f(\mathbf{x})$ is $f^*(\lambda-\mathbf{a})$, with conjugate set $\mathbf{a}+\mathcal{D}^*$.

Legendre-Fenchel transforms may not be the best tool to use in deriving a variational representation of a function. One reason is readily apparent. Suppose the optimization criterion is expressed in terms of a convex function that should be minimized. Substitution of $\sup_{\lambda\in\mathcal{D}^*}[\langle\mathbf{x},\lambda\rangle - f^*(\lambda)]$ for the convex function yields a minimax optimization problem instead of a double minimization. This minimax or game-theoretic problem has optimization properties that can differ dramatically from properties of alternating minimization problems.

3.4 VARIATIONAL METHODS IN INFORMATION THEORY: CONVEX DECOMPOSITION

An alternative to the application of Legendre-Fenchel transforms comes from a simple application of Jensen's inequality to convex functions. Let $\mathcal{P} = \{\mathbf{p} : p_i \geq 0, \sum_i p_i = 1\}$. We refer to the following lemma as the convex decomposition lemma.

Lemma 2 *Suppose that f is a convex function defined on a convex domain $\mathcal{D}$. Given $\mathbf{x}_i \in \mathcal{D}$, $i = 1, 2, \ldots$, such that $\sum_i \mathbf{x}_i \in \mathcal{D}$,*

$$f(\sum_i \mathbf{x}_i) \leq \sum_i p_i f\left(\frac{1}{p_i}\mathbf{x}_i\right), \tag{17}$$

for all $\mathbf{p} \in \mathcal{P}$, with $p_i > 0$ and $\frac{1}{p_i}\mathbf{x}_i \in \mathcal{D}$ for all i. If f is strictly convex, equality holds if and only if $\frac{1}{p_i}\mathbf{x}_i = \mathbf{x}$ is independent of i.

Proof: As noted, the proof follows from Jensen's inequality. For any $\mathbf{p} \in \mathcal{P}$, Jensen's inequality gives

$$\begin{aligned} f(\sum_i \mathbf{x}_i) &= f\left(\sum_i \frac{p_i}{p_i}\mathbf{x}_i\right) \\ &\leq \sum_i p_i f\left(\frac{1}{p_i}\mathbf{x}_i\right). \end{aligned} \tag{18}$$

For f strictly convex, equality implies the arguments of f are constant. For a more thorough discussion of Jensen's inequality in an information theory context, see Cover and Thomas [7, p. 25]. □

We refer to the use of this lemma as a convex decomposition because the function of a sum of terms is decomposed as a convex combination of the function of each term. The separation can lead to a decoupling of the objective function. A standard use of the convex decomposition forms the basis for the expectation-maximization algorithm and other alternating minimization algorithms.

Lemma 3 *Given summable $\alpha_i \geq 0$, with one positive,*

$$-\log\left(\sum_i \alpha_i\right) = \min_{\mathbf{p}\in\mathcal{P}} \sum_i p_i \log \frac{p_i}{\alpha_i}. \tag{19}$$

Proof: For probabilities $\mathbf{p}$, Jensen's inequality gives

$$\log\left(\sum_i \frac{p_i}{p_i}\alpha_i\right) \geq -\sum_i p_i \log\frac{p_i}{\alpha_i} \tag{20}$$

with equality if and only if p_i/α_i = constant. That is, $p_i = \alpha_i/(\sum_{i'} \alpha_{i'})$. □

This lemma can be applied in several ways. For the MAP decoding problem, this lemma applied to the posterior probabilities $P(i_k = 1|\mathbf{y})$ yields

$$\log\left(\sum_{\mathbf{s}}\sum_{\mathbf{i}:i_k=1} p(\mathbf{y}|\mathbf{s})P(\mathbf{s}|\mathbf{i})P(\mathbf{i})\right) \geq$$
$$-\sum_{\mathbf{s}}\sum_{\mathbf{i}:i_k=1} q(\mathbf{s},\mathbf{i}|\mathbf{y},i_k)\log\left(\frac{q(\mathbf{s},\mathbf{i}|\mathbf{y},i_k)}{p(\mathbf{y}|\mathbf{s})P(\mathbf{s}|\mathbf{i})P(\mathbf{i})}\right)$$

for any probabilities $q(\mathbf{s},\mathbf{i}|\mathbf{y},i_k)$. For maximum-likelihood sequence detection,

$$\begin{aligned}\log\left(p(\mathbf{y}|\mathbf{i})\right) &= \log\left(\sum_{\mathbf{s}} p(\mathbf{y}|\mathbf{s})P(\mathbf{s}|\mathbf{i})\right)\\ &\geq -\sum_{\mathbf{s}} q(\mathbf{s}|\mathbf{y},\mathbf{i})\log\left(\frac{q(\mathbf{s}|\mathbf{y},\mathbf{i})}{p(\mathbf{y}|\mathbf{s})P(\mathbf{s}|\mathbf{i})}\right)\end{aligned}$$

for any probabilities $q(\mathbf{s}|\mathbf{y},\mathbf{i})$. Here this inequality for the maximum-likelihood sequence detection problem is of greater interest than for MAP decoding because it helps form the basis for the family of algorithms discussed below.

The convex decomposition lemma is used by Lange and Fessler in the derivation of their convex algorithm for transmission tomography [20], using ideas motivated by De Pierro [10, 11].

3.5 EXPECTATION-MAXIMIZATION ALGORITHM

The expectation-maximization algorithm can be viewed as resulting from an application of the convex decomposition lemma from the previous section. Suppose that the parameters to be estimated form the vector θ. The maximum-likelihood estimation problem is to compute the parameter vector θ that, given the observations $\mathbf{y}$, maximizes the log-likelihood function $\log p(\mathbf{y} : \theta)$. Assume that the triple $\theta \rightarrow \mathbf{s} \rightarrow \mathbf{y}$

forms a Markov chain in that order, where $\mathbf{s}$ is a vector of hidden data. Then the convex decomposition lemma yields

$$\log [p(\mathbf{y} : \theta)] = \log \left[\sum_{\mathbf{s}} p(\mathbf{y}|\mathbf{s}) p(\mathbf{s} : \theta) \right] = \\ - \min_{q(\mathbf{s}|\mathbf{y}:\theta)} \sum_{\mathbf{s}} q(\mathbf{s}|\mathbf{y} : \theta) \log \frac{q(\mathbf{s}|\mathbf{y} : \theta)}{p(\mathbf{y}|\mathbf{s}) p(\mathbf{s} : \theta)},$$

where $q(\mathbf{s}|\mathbf{y} : \theta)$ are probabilities. This is the desired variational representation of the log-likelihood function. The maximum-likelihood estimation problem becomes a double minimization

$$\hat{\theta}_{ML} = \underset{\theta}{\text{argmin}} \min_{q(\mathbf{s}|\mathbf{y}:\theta)} \sum_{\mathbf{s}} q(\mathbf{s}|\mathbf{y} : \theta) \log \frac{q(\mathbf{s}|\mathbf{y} : \theta)}{p(\mathbf{y}|\mathbf{s}) p(\mathbf{s} : \theta)}. \tag{21}$$

The expectation-maximization algorithm is simply the corresponding alternating minimization algorithm

$$\begin{aligned} \hat{q}^{(r+1)} &= q(\mathbf{s}|\mathbf{y} : \hat{\theta}^{(r)}) \\ &= \frac{p(\mathbf{y}|\mathbf{s}) p(\mathbf{s} : \hat{\theta}^{(r)})}{\sum_{\mathbf{s}'} p(\mathbf{y}|\mathbf{s}') p(\mathbf{s}' : \hat{\theta}^{(r)})} \\ \hat{\theta}^{(r+1)} &= \underset{\theta}{\text{argmax}} \sum_{\mathbf{s}} \hat{q}^{(r+1)}(\mathbf{s}|\mathbf{y} : \hat{\theta}^{(r)}) \log p(\mathbf{s} : \theta). \end{aligned}$$

The second step simply states that the new estimate for θ maximizes the expected value of the complete data log-likelihood given the last estimate and the incomplete data $\mathbf{y}$ (see the seminal paper by Dempster, Laird, and Rubin [12] and the paper by Miller and Snyder [23] for more complete derivations and discussion).

Note that the computational complexity of $E\{\log p(\mathbf{s} : \theta)|\hat{\theta}^{(r)}, \mathbf{y}\}$ can be large. Many of the algorithms used to speed up the computations are equivalent to either approximating this computation or simplifying it in other ways. Within the framework here, some of these simplifications correspond to different choices for $q(\mathbf{s}|\mathbf{y} : \theta)$ other than the global minimum.

4. ALTERNATING MINIMIZATIONS FOR MAXIMUM-LIKELIHOOD SEQUENCE DETECTION

For this section, let $\mathcal{P}_1$ be the set of probabilities over $\mathbf{i}$:

$$\mathcal{P}_1 = \{P(\mathbf{i}) : P(\mathbf{i}) \geq 0, \sum_{\mathbf{i}} P(\mathbf{i}) = 1\}. \tag{22}$$

Lemma 4 *For any function of* $\mathbf{i}$,

$$\max_{\mathbf{i}} g(\mathbf{i}) = \max_{P \in \mathcal{P}_1} \sum_{\mathbf{i}} g(\mathbf{i}) P(\mathbf{i}). \tag{23}$$

For example, for any two real numbers b and a,

$$\max(a, b) = \max_{\lambda \in [0,1]} [\lambda a + (1 - \lambda) b]. \tag{24}$$

This lemma states that maximization of a function of the discrete set of values for $\mathbf{i}$ (the corners of a hypercube) is equivalent to maximization of the convex combination of the values of the function. This formulation is typically not very helpful because the set of probabilities is unrestricted. In order to reduce the complexity, define the subset $\mathcal{P}_2 \subset \mathcal{P}_1$ of product distributions

$$\mathcal{P}_2 = \{P \in \mathcal{P}_1 : P(\mathbf{i}) = \prod_{k=1}^{K} P_k(i_k)\}, \tag{25}$$

where for each k, $P_k(1) + P_k(0) = 1$. Note that this set is completely parameterized by K values (one of $P_k(1)$, $P_k(1)/P_k(0)$, or $\log[P_k(1)/P_k(0)]$). We immediately have the following lemmas.

Lemma 5 *For any function of* $\mathbf{i}$,

$$\max_{\mathbf{i}} g(\mathbf{i}) = \max_{P \in \mathcal{P}_2} \sum_{\mathbf{i}} g(\mathbf{i}) \prod_{k} P_k(i_k). \tag{26}$$

When applied here, this lemma effectively transforms the maximum-likelihood sequence detection problem into a continuous-valued estimation problem over K variables defining $P \in \mathcal{P}_2$.

Lemma 6 *For any positive-valued function of* $\mathbf{i}$,

$$\begin{aligned} \max_{\mathbf{i}} \log g(\mathbf{i}) &= \max_{P \in \mathcal{P}_2} \log \left(\sum_{\mathbf{i}} g(\mathbf{i}) \prod_{k} P(i_k) \right) \\ &= -\min_{P \in \mathcal{P}_2} \min_{q \in \mathcal{P}_1} \sum_{\mathbf{i}} q(\mathbf{i}) \log \frac{q(\mathbf{i})}{g(\mathbf{i}) \prod_k P(i_k)}. \end{aligned}$$

This lemma applies the convex decomposition lemma that leads to the expectation-maximization algorithm to obtain a double minimization representation of the maximum-likelihood sequence detection problem.

4.1 ITERATIVE ALGORITHM FOR MAXIMUM-LIKELIHOOD SEQUENCE DETECTION

The double minimization form of the maximum-likelihood sequence detection problem is

$$-\max_{\mathbf{i}} \log p(\mathbf{y}|\mathbf{i}) = \min_{\mathbf{P}\in\mathcal{P}_2} \min_{q\in\mathcal{Q}} \sum_{\mathbf{i}} \sum_{\mathbf{s}} q(\mathbf{s},\mathbf{i}|\mathbf{y}) \log \frac{q(\mathbf{s},\mathbf{i}|\mathbf{y})}{p(\mathbf{y}|\mathbf{s})P(\mathbf{s}|\mathbf{i})\prod_k P_k(i_k)},$$

where $\mathcal{Q}$ is the unconstrained set of probabilities over $(\mathbf{s},\mathbf{i})$. In this formulation, all bit sequences are assumed to be equally likely; this can be removed by the incorporation of the priors above. The way it is written here is more intuitive, however, in that the P_k play the role of prior probabilities in the expression. One possible interpretation is that the problem becomes one of estimating the most likely priors, with the information bits modeled as being independent.

The resulting algorithm for maximum-likelihood sequence detection is as follows. Initialize $\hat{P}_k^{(0)}(i_k)$, and set $r = 0$. Perform the iterations

$$\begin{aligned} \hat{q}^{(r+1)}(\mathbf{s},\mathbf{i}|\mathbf{y}) &= \frac{1}{Z^{(r)}(\mathbf{y})} p(\mathbf{y}|\mathbf{s})P(\mathbf{s}|\mathbf{i}) \prod_k \hat{P}_k^{(r)}(i_k) \\ \hat{P}_k^{(r+1)}(i_k = 1) &= \hat{q}^{(r+1)}(i_k = 1|\mathbf{y}) \\ &= \sum_{\mathbf{s}} \sum_{\mathbf{i}:i_k=1} \hat{q}^{(r+1)}(\mathbf{s},\mathbf{i}|\mathbf{y}). \end{aligned}$$

Check for convergence and iterate if necessary.

The iterations for q essentially compute the posterior distribution on $(\mathbf{s},\mathbf{i})$ given that the prior is specified by the last estimate for P. The function $Z^{(r)}(\mathbf{y})$ never needs to be computed. The iterations for P_k essentially compute the marginal determined from q, the posterior probabilities. At the next iteration, the prior probabilities are set equal to these posterior probabilities and the iterations continue.

This sequence is guaranteed to monotonically increase the log-likelihood function as a result of the derivation based on the convex decomposition.

For convolutional codes, the BCJR algorithm exactly computes posteriors from the priors and can thus be used to implement the computations. The result is the following due to O'Sullivan, Snyder, and Duan [27], and Moher and Gulliver [24]. Let $BCJR(\mathbf{p})$ be the mapping of the BCJR algorithm from prior bit probabilities to posterior bit probabilities.

Lemma 7 *Suppose that the observations result from independent measurements of the code bits from a terminated convolutional code. The iterations* $\hat{\mathbf{p}}^{(r+1)} = BCJR(\hat{\mathbf{p}}^{(r)})$ *monotonically increase the log-likelihood.*

4.2 DISCUSSION OF IMPLEMENTATION ISSUES

The issues in implementation include the computational cost relative to MAP decoding, the convergence rate, and potential simplifications of the iterations.

The iterative algorithm as derived in the last subsection relies on the computations of the posterior bit probabilities. If the computations of these are difficult, as they are in graphs with loops, then the iterations defined above have exactly the same implementation issues as do algorithms for MAP decoding (such as standard message passing algorithms). Furthermore, as written, the computations require multiple iterations of those needed for MAP decoding. These are valid concerns, and an important part of a response to them is the appropriate or desired objective function. If performance is measured in bit error rate, then the objective function should attempt to minimize that. Choosing each information bit to maximize the marginal posterior bit probability arguably is an objective function matched to this criterion. If performance is measured in word error rate, then maximum-likelihood sequence detection is the desired problem. The computational cost for solving the correct problem may be higher.

In many problems it may not be fair to compare the computational cost as in the previous paragraph. The computational cost may in fact be identical, depending on implementation.

Our experience to date with convergence experimentally has been mixed. In many implementations, the algorithm converges rapidly to a bit sequence that does not achieve the maximum. Our interpretations are that the restriction of P to $\mathcal{P}_2$ yields a log-likelihood function that has local maxima and that the algorithm tends to converge too quickly in the early iterations. All vertices of the hypercube for $\mathbf{i}$ correspond to fixed points of the algorithm, many being local minima, local maxima, or saddlepoints (relative to changing $\mathbf{i}$ in one entry). One potential approach to ameliorate this situation would be to decrease the rate of convergence by taking a step in the direction of the next $\hat{P}_k^{(r)}$; our implementations of this have been encouraging, but the analysis is not yet complete.

Potential simplifications of the algorithm may be motivated by corresponding literature in imaging problems. For graphs with many loops, such as in Markov random field models and roughness penalties for images, updating only a subset of the variables at a time (as in a graph coloring algorithm) helps to decouple the maximization [25]. In the ordered subset algorithm of Hudson and Larkin [18], the observations $\mathbf{y}$ are separated into subsets. In subiterations of the algorithm, the iterations are performed assuming that the data from a particular subset are the only data available. This increases the speed of convergence dramatically (roughly reducing the number of iterations by a factor equal to the number of subsets while maintaining roughly the same complexity). Another approach may be to constrain the subset over which the q is updated, corresponding to constraining the algorithm to pass messages on a graph. Other algorithms are described in Jordan's book [19].

5. CONCLUSIONS

In this paper, we describe a new family of algorithms for use in maximum-likelihood sequence detection. The derivation uses a variational representation of the objective function, the log-likelihood function, as its basis. Variational methods in information theory and in estimation theory are reviewed in some detail. The variational representation for the log-likelihood function has two parts. The first part is a relaxation of the objective function from being defined on the discrete length-K information bit sequence to being defined on a set of probabilities parameterized by K continuous values. The second part uses the convex decomposition lemma to transform the logarithm of a sum into a sum of logarithms, with a minimization over a convex combination in the sum.

The methods used here are closely related to those used in imaging problems. More cross-fertilization of ideas between these fields may be fruitful for both.

Acknowledgments

This paper is dedicated to G. David Forney, Jr. on the occasion of his sixtieth birthday. Dave Forney's contributions to information theory and communication theory have been immeasurable, not just in content but in style as well. These high standards help set the tone for researchers in these fields. I am privileged to know Dave Forney both through his work and through social gatherings. I look forward to many more years of contributions and interactions. The topic of this paper was selected in part because two of Dave Forney's papers [14, 15] helped introduce me to iterative decoding.

Discussions with D. L. Snyder, L. Duan, B. Rimoldi, and R. Urbanke are gratefully acknowledged. This work was supported in part by NSF 9900159 and by NIH R01CA75371.

References

[1] S. M. Aji and R. J. McEliece, "The Generalized Distributive Law," *IEEE Transactions on Information Theory*, vol. 46, no. 2, pp. 325–343, 2000.

[2] S. Arimoto, "An Algorithm for Computing the Capacity of an Arbitrary Discrete Memoryless Channel," *IEEE Transactions on Information Theory*, vol. IT-18, pp. 14–20, 1972.

[3] L. Bahl, J. Cocke, F. Jelinek, and J. Raviv, "Optimal Decoding of Linear Codes for Minimizing Symbol Error Rate," *IEEE Transactions on Information Theory*, vol. IT-20, pp. 284–287, 1974.

[4] C. Berrou, A. Glavieux, and P. Thitimajshima, "Near Shannon Limit Error-Correcting Coding and Decoding: Turbo Codes (1)," *Proceedings of the IEEE International Conference on Communications 1993*, Geneva, pp. 1064–1070, June 1993.

[5] R. E. Blahut, "Computation of Channel Capacity and Rate Distortion Functions," *IEEE Transactions on Information Theory*, vol. IT-18, pp. 460–473, 1972.

[6] J. A. Bucklew, *Large Deviation Techniques in Decision, Simulation, and Estimation*, New York: Wiley and Sons, 1990.

[7] T. M. Cover and J. A. Thomas, *Elements of Information Theory*, New York: Wiley and Sons, 1991.

[8] I. Csiszár and G. Tusnády, "Information Geometry and Alternating Minimization Procedures," *Statistical Decisions*, Suppl. issue #1, pp. 205–207, 1984.

[9] A. Dembo and O. Zeitouni, *Large Deviations Techniques and Applications*, Boston, MA, Jones and Bartlett, Publishers, 1993.

[10] A. R. De Pierro, "On the Relation Between the ISRA and EM Algorithm for Positron Emission Tomography," *IEEE Transactions on Medical Imaging*, vol. 12, pp. 328–333, 1993.

[11] A. R. De Pierro, "A Modified Expectation Maximization Algorithm for Penalized Likelihood Estimation in Emission Tomography," *IEEE Transactions on Medical Imaging*, vol. 14, pp. 132–137, 1995.

[12] A. P. Dempster, N. M. Laird, and D. B. Rubin, "Maximum Likelihood from Incomplete Data via the EM Algorithm," *Journal of the Royal Statistical Society B*, vol. 39, pp. 1–37, 1977.

[13] G. D. Forney, Jr., "The Viterbi Algorithm," *Proceedings of the IEEE*, vol. 61, pp. 268–278, Mar. 1973.

[14] G. D. Forney, Jr., "The Forward-Backward Algorithm," *Proceedings of the 34th Allerton Conference on Communications, Control and Computing*, pp. 432–446, Monticello, IL, 1996.

[15] G. D. Forney, Jr., "On Iterative Decoding and the Two Way Algorithm," *Proceedings International Symposium on Turbo Codes and Related Topics*, Brest, France, pp. 12–25, Sep. 1997.

[16] B. J. Frey, *Graphical Models for Machine Learning and Digital Communication*, Cambridge, MA: MIT Press, 1998.

[17] R. G. Gallager, *Low-Density Parity-Check Codes.* Cambridge, MA: MIT Press, 1963.

[18] H. M. Hudson and R. S. Larkin, "Accelerated Image Reconstruction Using Ordered Subsets of Projection Data," *IEEE Transactions on Medical Imaging*, vol. 13, no. 4, pp. 601-609, 1994.

[19] M. I. Jordan, *Learning in Graphical Models*, MIT Press, Cambridge, MA, 1999.

[20] K. Lange and J. A. Fessler, "Globally Convergent Algorithms for Maximum a Posteriori Transmission Tomography," *IEEE Transactions on Image Processing*, vol. 4, no. 10, pp. 1430-1438, 1995.

[21] D. G. Luenberger, *Optimization by Vector Space Methods*, New York, NY, John Wiley & Sons, Inc., 1969.

[22] D. J. C. MacKay, "Good Error-Correcting Codes Based on Very Sparse Matrices," *IEEE Transactions on Information Theory*, vol. 45, pp. 399-431, Mar. 1999.

[23] M. I. Miller and D. L. Snyder, "The Role of Likelihood and Entropy in Incomplete-Data Problems: Applications to Estimating Point-Process Intensities and Toeplitz and Constrained Covariances," *Proceedings of the IEEE*, vol. 75, pp. 892-907, 1987.

[24] M. Moher and T. A. Gulliver, "Cross-Entropy and Iterative Decoding," *IEEE Transactions on Information Theory*, vol. 44, no. 7, pp. 3097–3104, 1998.

[25] J. A. O'Sullivan, "Roughness Penalties on Finite Domains," *IEEE Transactions on Information Theory*, vol. 4, no. 9, pp. 1258–1268, 1995.

[26] J. A. O'Sullivan, "Alternating Minimization Algorithms: From Blahut-Arimoto to Expectation-Maximization," in A. Vardy, ed., *Codes, Curves, and Signals: Common Threads in Communications*, Boston, Kluwer Academic, 1998.

[27] J. A. O'Sullivan, D. L. Snyder, and L. Duan, "Iterative Bahl Decoding is Maximum Likelihood Decoding," presented at *the IEEE 1998 Communication Theory Workshop*, Captiva Island, FL, 1998.

[28] T. Richardson and R. Urbanke, "The Capacity of Low-Density Parity Check Codes under Message-Passing Decoding," *IEEE Transactions on Information Theory*, vol. 47, pp. 599-618, Feb. 2001.

[29] Y. Weiss, "Correctness of local probability propagation in graphical models with loops," *Neural Computation*, vol. 12, pp. 1–41, 2000.

III

MODEMS

Chapter 9

V.92: THE FINAL CHAPTER IN THE NEVER-ENDING STORY OF DIAL-UP MODEMS

D.-Y. Kim
Airvana, Inc.

P. A. Humblet
Institute Eurecom, France

M. V. Eyuboglu
Airvana, Inc.

L. Brown
Conexant

G. D. Forney, Jr.
MIT, Cambridge, MA

S. Mehrabanzad
Airvana, Inc.

Abstract Ever since the first dial-up modems appeared in 1960s, many have repeatedly predicted their obsolescence. However, contrary to such predictions, dial-up modems actually thrived in the 80's and 90's as a result of slow roll-out of residential digital services, and the unprecented growth of Internet and remote access.

Since the first 300 b/s dial-up modem V.21, modem speeds increased steadily, most recently approaching download speeds of 56 kb/s in ITU

Recommendation V.90, finalized in 1998. V.90 takes advantage of the direct-digital PCM network connection of an Internet Service Provider's remote access server, to achieve such high downstream (from ISP to a user) speeds. However, for upstream transmission (from a user to ISP), V.90 employs the older V.34 technology which typically delivers 28.8 kb/s.

A new ITU modem standard called V.92 will increase upstream rates to above 40 kb/s, again taking advantage of PCM connections. In this paper, we present a new transmission scheme which is adopted in V.92. The proposed scheme is based on a generalization of Tomlinson-Harashima precoding.

As we predict that V.92 will be the last dial-up modem standard, we wonder whether we might be falling into the same trap that many others have fallen in the past. You can be the judge!

1. INTRODUCTION

Figure 1(a) gives a network-oriented view of a conventional voiceband

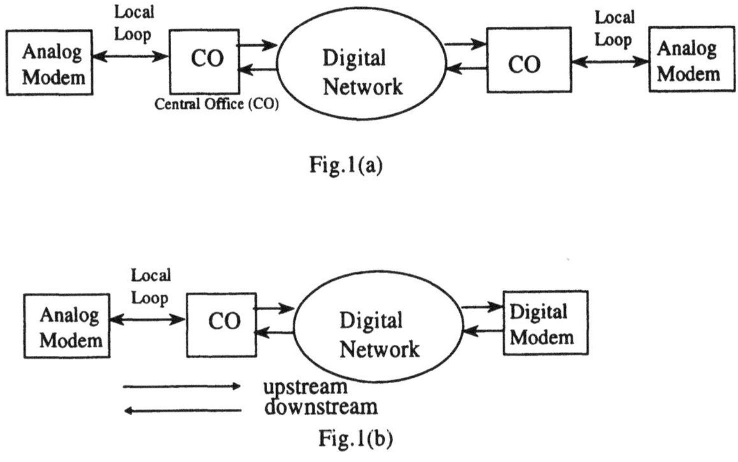

Figure 1 Modem Connections through a PSTN, (a) Analog end-to-end connection, (b) PCM modem connection

modem connection. Conventional modems, including V.34 [11], do not take account of A/D and D/A conversion processes in the telephone network [1]. Therefore, their performance is limited by the quantization noise that is introduced by pulse code modulation (PCM).

If the quantization noise is treated as additive white Gaussian noise, then the capacity of the telephone network for conventional voiceband modems can be calculated from the Shannon capacity formula [3]

$$C = W \log_2(1 + SNR), \tag{1}$$

where W is the channel bandwidth and SNR is the signal to noise ratio. Since W is about 3 to 3.5 kHz and the SNR due to quantization noise is about 33 to 35 dB, the capacity C is about 33 to 40 kb/s.

Pulse code modulation (PCM) modems [1] allow data transmission at over 50 kb/s downstream, e.g., from a service provider digitally connected to the Public Switched Telephone Network (PSTN) to a user who is connected through an ordinary telephone line. Figure 1(b) shows such a connection through the PSTN for a PC modem.

The ITU-T completed PCM modem Recommendation V.90 in 1998 [2]. V.90 takes into account the D/A converter in the downstream, and thereby eliminates quantization noise for downstream transmission. As described by Humblet and Troulis [1], V.90 uses a subset of the A/μ-law quantization levels as a pulse amplitude modulation (PAM) signal constellation, at a symbol rate of 8000 symbols/s. In contrast to conventional PAM systems, in PCM modems signal point selection and transmitter pulse shaping occur inside the network, in a PCM D/A converter. This restricts the PAM signal constellation to a subset of the A/μ-law quantizer levels. Therefore, the V.90 downstream rate is limited to 64 kbps (8 bits/symbol, 8000 symbols/s). In reality, the rate is limited to less than 56 kbps due to a FCC-mandated power constraint [1], digital impairments such as Robbed Bit Signaling (RBS), and channel noise.

For upstream transmission, V.90 uses the same modulation as in V.34, and therefore can achieve a maximum data transmission rate of only up to 33.6 kb/s. To achieve higher rates, the A/D quantizer in the central office (CO) must be taken into account. To avoid quantization distortion in the upstream PCM, the analog loop should be equalized at the transmitting modem and transmit timing should be precisely matched to the sampling timing of the A/D converter. Such timing can be acquired from the timing of the D/A on the downstream link.

E. Ayanoglu *et al.* [4, 5] made an earlier proposal for a new transmission scheme for PCM upstream. In this proposal, the sampling rate is less than 8 kHz, and multi-input-multi-output (MIMO) system theory is used for pre-equalization.

In this paper, we present a new transmission scheme to achieve data rates higher than 33.6 kb/s for PCM upstream. This scheme was adopted in June 2000 by the ITU-T in the next dial-up modem Recommendation V.92. Our scheme employs a new precoding scheme that is a generalization of Tomlinson-Harashima precoding (THP) [6]. This scheme is less complex and more flexible than that of reference [5], and achieves higher data rates. Our scheme also includes a design for optimal constellations for PCM upstream transmission.

We introduce generalized THP in Section 2. Optimal constellation design is explained in Section 3. Section 4 discusses the trellis coding scheme used in V.92. Section 5 discusses similarities and differences between our proposed transmission scheme and that of reference [5], and gives some simulation results.

2. NEW TRANSMISSION SCHEME

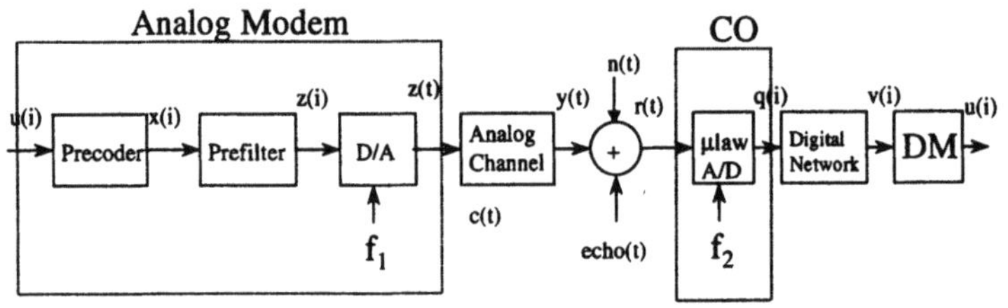

Figure 2 Block diagram for upstream

Figure 2 shows the block diagram of our PCM upstream transmission system. In our system, an analog PCM modem is connected to a central (CO) over a local analog loop. Channel noise and echo from downstream transmission are added to the received signal before it is quantized by the A/μ-law quantizer at a sampling frequency of $f_2 = 8$ kHz. The quantized octets [16], denoted by $q(i)$, are transmitted over the digital network, where they may be affected by various digital impairments such as Robbed Bit Signaling (RBS) [16].

Before data can be transmitted upstream, the clock (f_1) in the analog PCM modem must be synchronized to the clock (f_2) of the A/D. This can be achieved by learning the clock from the downstream PCM signal and synchronizing the clocks using techniques such as those described by Ling [7]. In this paper, we assume perfect clock synchronization. Once the clocks are synchronized, i.e., $f_1 = f_2$[1], the PCM upstream block diagram can be represented by an equivalent discrete-time block diagram, as shown in Figure 3. In Figure 3, i is the time index for 8 kHz samples.

As we have said, the signal at the A/D converter includes echo from the downstream transmission in a full-duplex system. Although this echo can be estimated in the receiver, it prevents us from ensuring that the received signal at the A/D will fall on a quantization level, even in absence of noise. Thus, in contrast to the downstream case, quantization noise cannot be avoided in the upstream direction.

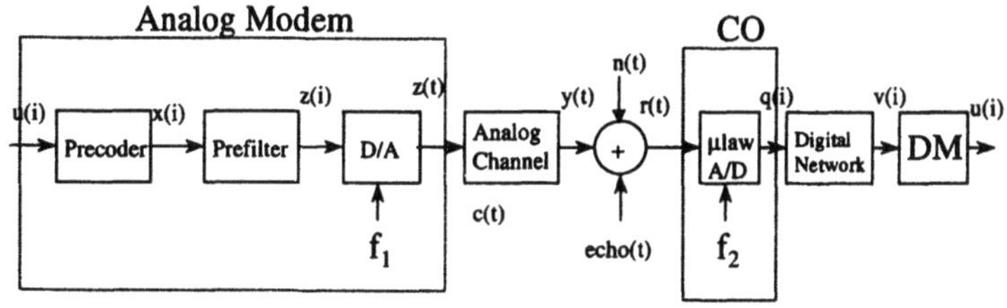

Figure 3 When $f_1 = f_2$, i is the time index for 8kHz samples.

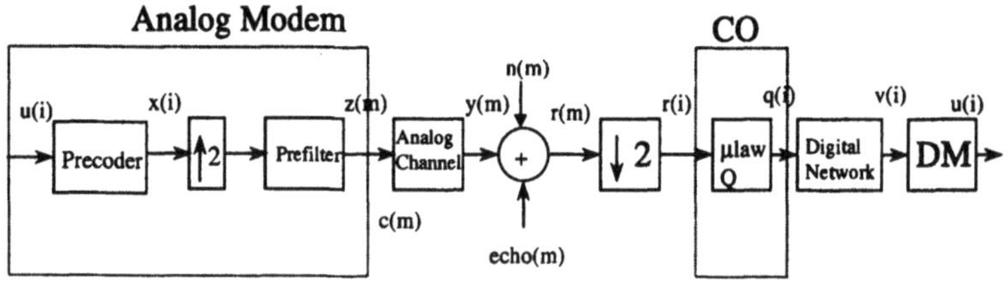

Figure 4 When $f_1 = 2f_2$, m and i are the time index for 16kHz and 8kHz samples, respectively.

The analog channel should be equalized at the transmitter because equalization at the receiver is not compatible with computation of proper likelihoods (see Section 3). Because the analog channel has a spectral null at DC [1], pre-equalization by a linear filter is not possible. Therefore we use a pre-equalization system based on a precoder and a prefilter, as shown in Figure 3.

The prefilter $g(i)$ is designed so that the overall response $p(i) = g(i) * c(i)$ is causal and monic. The precoder acts as a decision feedback circuit to remove the remaining causal interference. The precoder and prefilter are designed to transmit the signal $z(i)$ over the analog channel such that predetermined constellation points $y(i)$ (see Section 3) corresponding to digital data symbols $u(i)$ are produced at the input of the A/μ-law quantizer, if there is no noise and no echo. In the presence of noise and echo, the input of the A/μ-law quantizer will be $y(i) + n(i) + echo(i)$. Here $echo(i)$ denotes the echo, while the noise term $n(i)$ includes additive channel noise, intersymbol interference from imperfect equalization, and noise from imperfect echo estimation.

To explain the operation of the precoder and prefilter, Figure 3 is redrawn in Figure 5 on the assumption that there is no echo, no quantizer, and no digital network. We assume that the analog channel $c(i)$ has been accurately estimated during training. The prefilter $g(i)$ and the target causal, monic response $p(i)$ (where $p(0) = 1$) can be derived

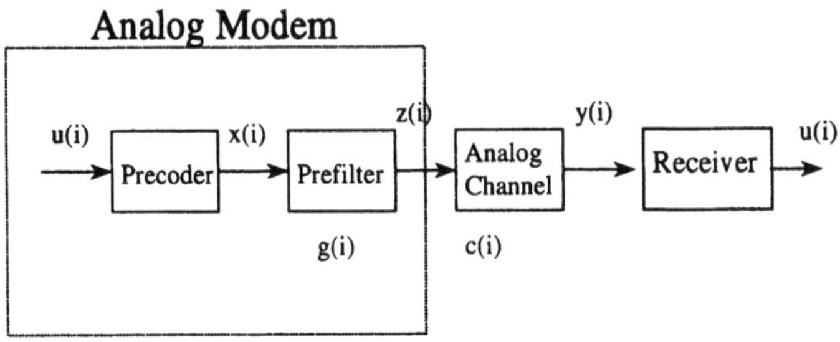

Figure 5 The precoding scheme used for upstream. To make the explanation simple, we assumed that there is no echo, no quantizer, and no digital network.

by minimizing the cost function

$$\zeta = \|g(i) * c(i) - p(i)\|^2 + \alpha\|g(i)\|^2. \tag{2}$$

The first term ensures small intersymbol interference (ISI), while the second term ensures small transmit power through the Lagrange multiplier α. The larger α, the lower will be the transmit power, but the larger will be the ISI.

Minimization of the cost function ζ is similar to determining the optimal feedforward and feedback filters for a decision feedback equalizer (DFE). A computationally efficient algorithm can be found in the paper of Al-Dhahir and Cioffi [8]. Note that in the limit of infinitely long filters, with a zero-forcing design, $g(i)$ will simply be an all-pass filter. We observe in practice that it is usually almost all-pass. The initially determined $p(i)$ and $g(i)$ can be used for as long as the analog channel $c(i)$ is time invariant. In practice, the analog channel is almost time invariant, and a fixed $p(i)$ and $g(i)$ can maintain initial performance for several hours.

Assuming that $p(i) = g(i) * c(i)$, the following relationship holds

$$y(i) = p(i) * x(i). \tag{3}$$

Since $p(0)$ is designed to be 1 and $p(i) = 0$ when $n < 0$ or $n > N_p$, Equation (3) can be rewritten as

$$x(i) = y(i) = \sum_{k=1}^{N_p} p(k)x(i-k). \tag{4}$$

Because $p(i)$ and the past values $x(n-i)$, $i = 1, 2, \ldots, N_p$, are known, the value of $x(i)$ needed to produce a given $y(i)$ at the output of the channel can be easily derived from Equation (4).

The system given by Equation (4) is unstable; however, we have some freedom in choosing $y(i)$ as a function of the data symbol $u(i)$ to send information over the channel. The given constellation points[2] for $y(i)$ are grouped into equivalence classes labeled by $u(i)$, and, given $u(i)$, one of the points in the corresponding equivalence class will be used as $y(i)$. The signal point $x(i)$ can then be computed according to Equation (4). As the signal sequence $x(i)$ is almost white and $g(i)$ is almost all-pass, choosing the point in the equivalence class to minimize $x(i)$ also tends to minimize the transmitted power $\|z(i)\|^2$ at the prefilter output[3].

An equivalence class is a set of one or more constellation points that represent the same data symbol $u(i)$. Let the N constellation points in A be denoted by $a(j)$, $-N/2 \le j < N/2$, where the indices[4] are in the same order as the levels. Thus negative points have negative indices, and positive points have nonnegative indices. Let the data $u(i)$ have M possible values, say $0 \le u \le M-1$. The equivalence class $E(u)$ corresponding to u is then defined as the set of all constellation points that have indices congruent to u modulo M; i.e., $E(u) = \{a(j) | j = u + zM, z \text{ an integer}\}$.

For example, if $N = 10$ and $M = 4$, then the four equivalence classes are $E(0) = \{a(-4), a(0), a(4)\}$, $E(1) = \{a(-3), a(1)\}$, $E(2) = \{a(-2), a(2)\}$, and $E(3) = \{a(-5), a(-1), a(3)\}$, as shown in Figure 6.

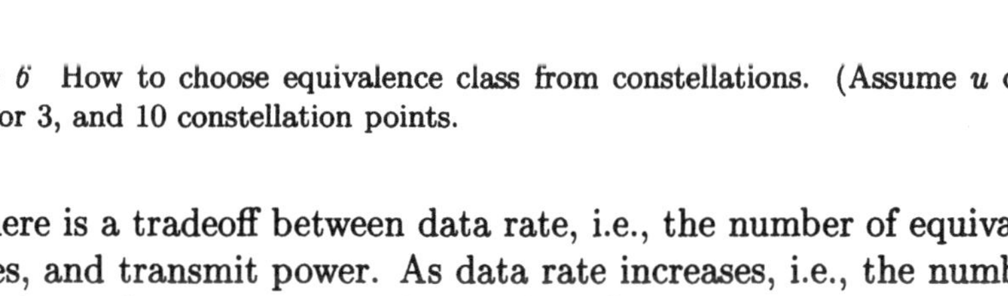

Figure 6 How to choose equivalence class from constellations. (Assume u can be 0,1,2, or 3, and 10 constellation points.

There is a tradeoff between data rate, i.e., the number of equivalence classes, and transmit power. As data rate increases, i.e., the number of equivalence classes increases, the number of points in an equivalence class decreases and the distance between them increases. Thus the smallest corresponding $x(i)$ and the transmit power get bigger.

The constellation for $y(i)$ will be called the *base constellation*. As shown in the next section, the base constellation can be designed to satisfy a certain target decoder error rate for given echo and noise characteristics on the line. From this base constellation, the equivalence class should be chosen to satisfy the transmit power constraint. This will give the supportable bit rate for a given transmit power constraint and line conditions.

We note at this point that although we have defined equivalence classes for one-dimensional constellations, equivalence classes could alternatively be defined for multidimensional constellations, or even for trellis codes, using techniques similar to those of Forney and Eyuboglu [14, 15]. This could give us a "shaping gain" which would help to reduce the transmit power, although at the expense of increased computational complexity.

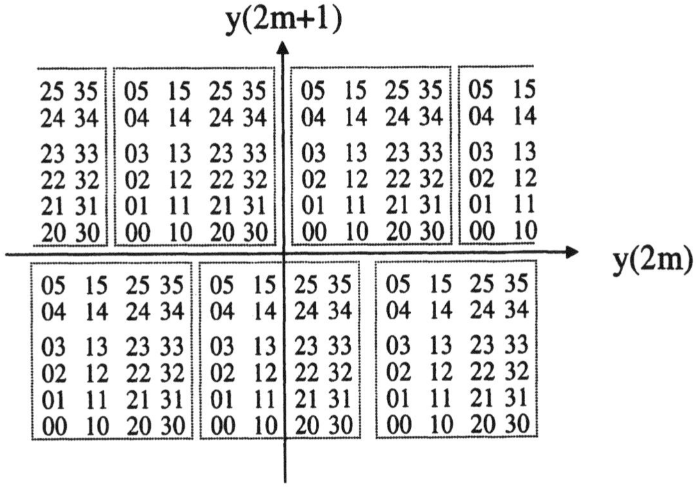

Figure 7 2 Dimensional Equivalence Class, $u(2m) = 0, 1, 2, 3$, and $u(2m + 1) = 0, 1, 2, 3, 4, 5$.

For example, Figure 7 gives an equivalence class defined for a two-dimensional constellation, where the number of points in the first and second dimension is issued to be 12. The first data symbol, $u(2m)$, is assumed to have four values, and the second data symbol, $u(2m + 1)$, six values. The equivalence class for input data symbols $u(2m) = i$ and $u(2m + 1) = j$ is denoted by the index pair "ij".

In the rest of this paper, we assume one-dimensional equivalence classes, as are used in V.92.

Tomlinson-Harashima precoding (THP) [6] can be seen to be a special case of the proposed one-dimensional equivalence class definition in which the base constellation is a uniform PAM constellation with an infinite number of points. Therefore, our precoding scheme will be called generalized THP (GTHP).

Figure 8 summarizes the operation of the precoder and the prefilter. Given digital data $u(i)$, the equivalence class of $u(i)$ is passed to the

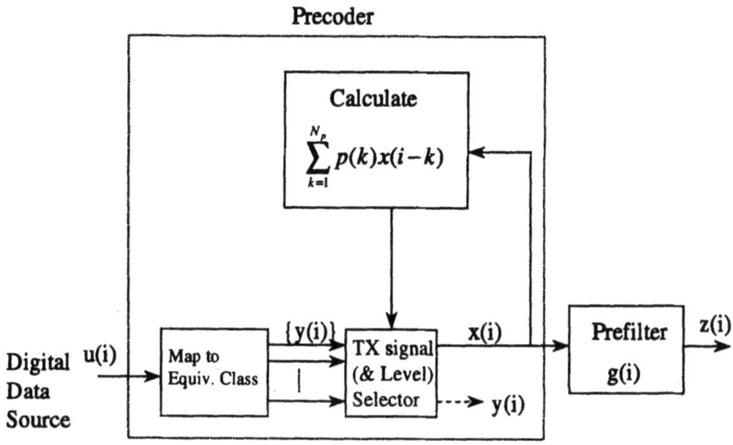

Figure 8 Block diagram how precoder andprefilter works.

transmit signal point selector. One of the points in this equivalence class is chosen as $y(i)$, and then $x(i)$ is calculated from Equation (4) and passed through a prefilter prior to transmission over the channel. The signal points $y(i)$ are chosen to minimize the energy of $x(i)$.

3. CONSTELLATION DESIGN

The explanation of the precoder and prefilter in the previous section ignored echo and quantization. In this case, $y(i)$ is affected only by additive noise, and therefore the receiver can decode the data sequence $y(i)$ straightforwardly, using a trellis decoder in the presence of channel coding or a symbol-by-symbol slicer in the absence of channel coding. Then $u(i)$ can be recovered from $y(i)$ based on the equivalence class definition.

In this section, we first show the receiver decodes the data sequence in the presence of noise, echo, and quantization. With this understanding, we will design an optimal base constellation for $y(i)$.

Our constellation design method is not restricted to the precoding scheme proposed here, but may be utilized with various other precoding schemes including that of [5].

Other papers, including [10], assume that the constellation should be a subset of the A/μ-law quantization levels. However, it will shortly be evident that this constraint should be relaxed to achieve an optimal constellation.

Figure 3 shows the block diagram of PCM upstream in the presence of echo, quantization, and digital network impairments. The analog channel output $y(i)$ is affected by noise and echo before μ-law quantization. The digital network may distort the μ-law quantizer output $q(i)$ into $v(i)$ through digital impairments such as RBS and digital loss [16]. We will describe the design of optimal constellations assuming that there are no digital impairments in the digital network, i.e., $v(i) = q(i)$. These results can be easily extended to the case when $v(i)$ is not equal to $q(i)$ by defining a new quantizer that combines the μ-law quantizer and the digital impairments introduced by the digital network. Digital impairments may be identified by the technique given by Kim [12].

The optimal constellation depends on the decoding scheme utilized by the digital modem. It will first be shown how to design the constellation assuming symbol-by-symbol decoding. Then the constellation design will be generalized to the case where there is trellis coding and the decoder employs a sequence-based decoding scheme such as the Viterbi decoding algorithm [16]. Our constellations will be designed to achieve a predetermined target error probability.

Assume that the N constellation points are $\{y_0, y_1, \ldots, y_{N-1}\}$. When the digital modem receives $v(i)$, given an estimate of $echo(i)$, the digital modem can decode which $y(i)$ has been transmitted by finding the most probable y_j, i.e., the y_j that maximizes

$$\Pr\,(v(n)|y_j, echo(n)). \tag{5}$$

If the noise $n(i)$ is Gaussian with variance σ_n^2, then this probability is given by

$$Q((\nu^\ell(i) - y_j - echo(i))/\sigma_n) - Q((\nu^u(i) - y_j - echo(i))/\sigma_n) \tag{6}$$

where ν^ℓ and $\nu^u(i)$ are, respectively, the lower and upper limits of the quantization interval associated with $v(i)$, and $Q(x)$ is the Gaussian probability-of-error function. Note that the noise variance is often dominated by $echo(i)$.

The decoding process may be better understood by observing the symbol-by-symbol decoding example depicted in Figure 10. In this figure, the 'x' marks on the axis represent μ-law quantizer levels, the '|' marks represent μ-law threshold levels. As specified in G.711 [16], there are 255 μ-law quantized levels with predefined thresholds. Of course, Figure 10 depicts only a small number of the possible levels.

In this example, it is assumed that a certain base constellation has points $y_4 = 695$ and $y_5 = 730$, and that $echo(i) = 15.4$. If the digital modem receives $v(i) = 751$[5], then it will determine that y_5 has been

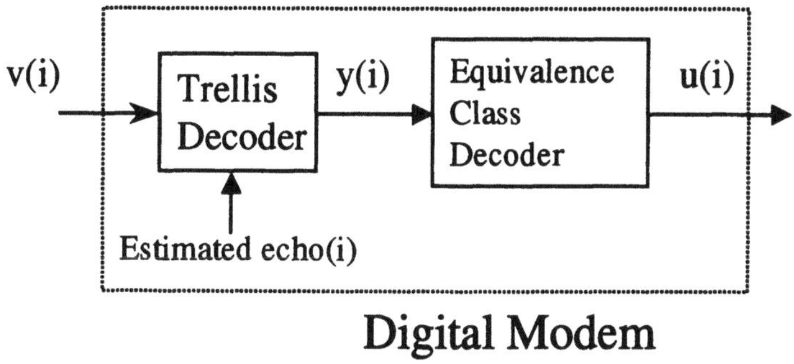

Figure 9 Receiver (Digial modem) in the presence of channel coding.

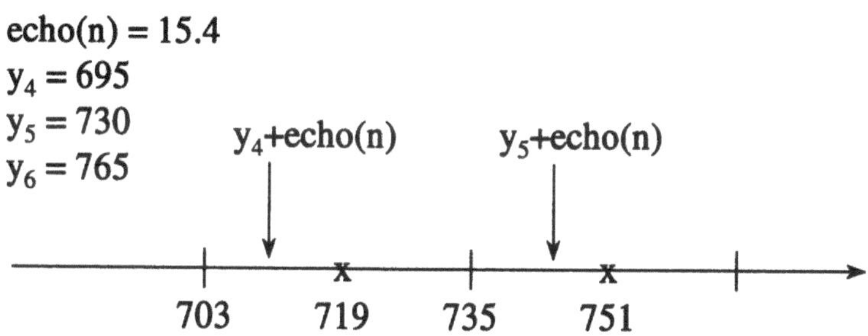

Figure 10 Symbol-by-Symbol Decoding Example

transmitted since, in the presence of noise, y_5 has the highest probability of having been the transmitted constellation point, given $echo(i)$, because it is the only point falling between the thresholds surrounding the μ-law level $v(i) = 751$.

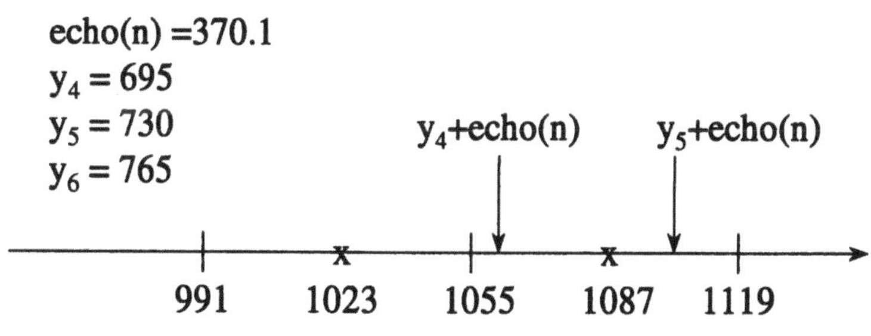

Figure 11 Symbol-by-Symbol Decoding Example

Another example with the same constellation points is shown in Figure 11. In this example, however *echo*(i) = 370.1. In this case, for either transmitted constellation point, y_4 or y_5, the digital modem will receive $v(i) = 1087$. As a result, the digital modem will have difficulty distinguishing between y_4 and y_5, and therefore these points will have a high error probability. The constellation should be designed so that this kind of error happens rarely enough to achieve some target symbol error probability (e.g., $P_e = 10^{-6}$). This could be achieved by increasing the distance between the constellation points; however, this will reduce the number of points that can be used in the constellation, which will reduce the data rate for fixed transmission power.

As shown by these examples, the constellation design depends heavily on the echo characteristics. As will be evident from the following algorithm, different constellations are optimal for different echo and noise characteristics.

We now give an algorithm to design the base constellation to achieve a certain target error probability for symbol-by-symbol decoding (i.e., no channel coding). We assume that the constellation has even symmetry, i.e., if $\{y_0, y_1, \ldots\}$ are the positive constellation points, then the negative points are $\{\ldots, -y_1, -y_0\}$. We also assume that the y_i are integers; this constraint may be relaxed, but we do not gain much.

The design algorithm is recursive. That is, if $y_0, y_1, \cdots, y_{k-1}$ have already been chosen, then y_k is chosen such that the following conditions are satisfied:

$$\begin{aligned} \Pr(\ldots, y_0, y_1, \cdots, y_{k-1} \text{ decoded } | y_k \text{ sent}) &< P_e/2, && (7) \\ \text{and } \Pr(y_k \text{ decoded } | y_{k-1} \text{ sent}) &< P_e/2 && (8) \end{aligned}$$

where P_e is the desired target symbol error probability for each constellation point. Note that by design the single-sided error probabilities (the left-hand error probability in Equation (7) and the right-hand error probability Equation (8)) are smaller than half of the target symbol error probability P_e. This ensures that the total error probability is than P_e. The recursion begins by finding the y_0 that satisfies $\Pr(-y_0 \text{ decoded } | y_0 \text{ sent }) < P_e/2$ and $\Pr(y_0 \text{ decoded } -y_0 \text{ sent}) < P_e/2$.

The left-hand error probability $\Pr(\ldots, y_0, y_1, \cdots, y_{k-1} \text{ decoded } | y_k \text{ sent })$ of Equation (7) may be calculated as follows:

$$\Pr(\ldots, y_0, y_1, \cdots, y_{k-1} \text{ decoded } | y_k \text{ sent }) =$$

$$\int_{-\infty}^{\infty} \Pr(\ldots, y_0, y_1, \cdots, y_{k-1} \text{ decoded } | y_k \text{ sent }, e) P_E(e) de, \qquad (9)$$

where e denotes the echo.

In Figure 11, we show an assumed Gaussian distribution for noise centered about the point $y_k + echo$. The area under the distribution beyond quantizer threshold is the left-hand error probability of Equation (7) or Equation (10). From this we can derive the following

$$\begin{aligned}\Pr(\ldots, y_0, y_1, \ldots, y_{k-1} \text{ decoded } |y_k \text{ sent }, e) \\ = Q((y_k + e - \text{ Threshold})/\sigma_n).\end{aligned} \tag{10}$$

Here "Threshold" is the μ-law quantizer threshold, where the μ-law level higher (resp. lower) than this threshold will be decoded as y_k (resp. y_{k-1}).

Equation (10) provides the term inside the integral in Equation (9). The integration over e in Equation (9) can be approximated by a sum over small intervals of e for calculation. For the probability of echo $P_E(e)$, we have assumed a Gaussian distribution. (The constellation would be different if we assumed a different distribution for echo.)

The right-hand error probability $\Pr(y_k, y_{k+1}, \ldots \text{ decoded } |y_{k-1} \text{ sent})$ of Equation (8) may be calculated similarly.

This recursive process continues until y_k reaches the largest quantization level.

In summary, the constellation design algorithm is as follows:

1. Find the minimum y_0 that satisfies the two conditions:

$$\Pr(-y_0 \text{ decoded } |y_0 \text{ sent }) < P_e/2$$
$$\Pr(y_0 \text{ decoded } | -y_0 \text{ sent }) < P_e/2.$$

2. Set $k = 1$.

3. Find the minimum y_k that satisfies

$$\Pr(\ldots, y_0, y_1, \ldots, y_{k-1} \text{ decoded } |y_k \text{ sent }) < P_e/2$$

and

$$\Pr(y_k, y_{k+1}, \ldots, \text{ decoded } |y_{k-1} \text{ sent }) < P_e/2.$$

4. If $y_k <$ largest A/μ-law value, set $k = k + 1$ and go to Step 3; else STOP.

For example, here is a constellation designed under the assumption that the noise and echo distributions are Gaussian with $\sigma_n = 7$ and $\sigma_e = 150$, and with no digital impairments (only positive values are shown);

{37, 113, 192, 275, 361, 450, 544, 646, 755. 870, 988, 1108, 1229, 1351, 1479, 1634, 1804, 1982, 2164, 2348, 2532, 2716, 2900, 3084, 3268, 3452, 3722, 4022, 4331, 4640, 4949, 5258, 5567, 5876, 6185, 6494, 6803, 7112, 7422 }.

We can use a similar algorithm for constellation design with channel coding. The only difference is in how to calculate the left-hand and right-hand error probabilities of Equation (7) and Equation (8). The exact error probabilities depend on the particular code being used and are hard to calculate precisely. In practice we use an error probability bound, such as that of [10], instead of the exact error probability.

Finally, because RBS impairments usually occur with period six and different digital impairments require different constellations, V.92 allows up to six constellations to be defined.

4. CHANNEL CODING

Channel coding may be combined with our precoding scheme to achieve higher data rates or greater noise margins. Since decoding is performed on the channel output sequence $y(i)$, the coded sequences $y(i)$ should have good distance properties.

While a novel coding scheme tailored to the PCM upstream application might be possible, we have found that the four-dimensional (4D) trellis codes used for Recommendation V.34 work well, although perhaps not optimally, when the constellation is designed as in the previous section. In V.92, the same trellis codes as in V.34 [11] were adopted to reduce the time and effort of both standardization and product development.

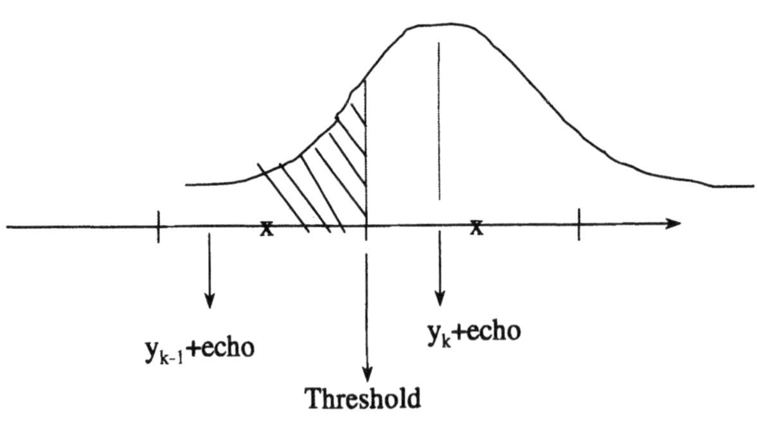

Figure 12 How to get the probability of error.

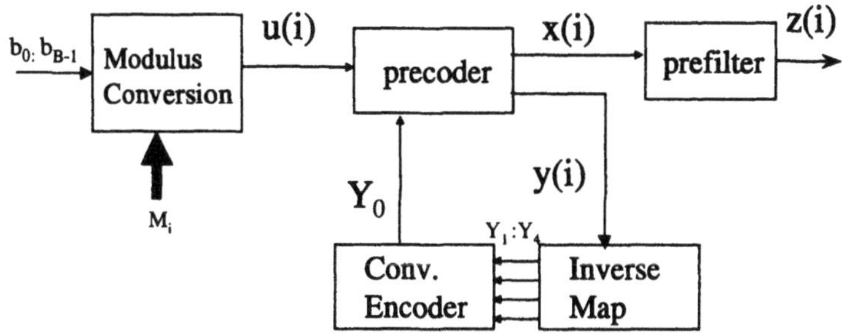

Figure 13 V.92 analog modern transmitter block diagram

Figure 13 is a block diagram of all major elements of the V.92 analog modem transmitter.

The data symbols $u(i)$ are generated from the user input bit sequence using twelve-symbol modulus conversion [2] with bases M_i, $0 \leq i \leq 11$. The block size of twelve was chosen to accommodate both period-six base constellations and four-dimensional trellis codes.

Our precoding scheme is based on the indices of the constellation points rather than their levels. It is therefore natural to combine our precoding scheme with trellis codes as in V.34, except using indices rather than levels.

The convolutional encoder operates exactly as in V.34 and produces the parity bit Y_0 [11] once every four-symbol period. The inverse map in Figure 13 operates similarly to the inverse map used in V.34, except that it is based on constellation indices rather than levels.

As explained in Section 2, each data symbol $u(i)$ is represented an equivalence class $E(u) = \{a(j) | j = u + zM_i, z \text{ and integer }\}$, where j is the constellation index.

To combine precoding with 4D trellis coding, the equivalence class definition is extended as follows. Let $k = 0, 1, 2, 3$ be a cyclic time index that indicates the symbol position in the four-symbol trellis frame. Then for $k = 0, 1, 2$, $E(u)$ is defined as

$$E(u) = \{a(j_k) | j_k = u + z_k M_i, z_k \text{ an integer}\}, \tag{11}$$

while for $k = 3$ (and z_k an integer), it is defined as

$$E(u) = \{a(j_k) | j_k = 2u + 2z_k M_i + (j_0 + j_1 + j_2 + Y_0) \bmod 2\}. \tag{12}$$

This definition ensures that $y(i)$, the output of the channel in the absence of echo and noise, will be a sequence in the trellis code. A four-tuple of indices $\{j_0, j_1, j_2, j_3\}$ is defined to be even or odd according to whether their sum is even or odd. The equivalence class definition of Equation (12) makes the four-tuple $\{j_0, j_1, j_2, j_3\}$ even or odd according to whether Y_0 is 0 or 1. Therefore, the proposed equivalence class definition together with feedback through the inverse map and the convolutional encoder guarantees that $y(i)$ is a valid trellis code sequence.

5. DISCUSSION AND SIMULATIONS

In this section, we give simulation results and compare the performance of our PCM upstream transmission system to that of [5].

We first show that both our system and the system given in [5] can be understood using equivalence classes. As shown in the previous section, our proposed transmission system defines an equivalence class at the output of the analog channel, i.e., for $y(i)$. Our proposed system defines an equivalence such that the constellation has multiple points for each symbol. We then use generalized THP to force a certain point at the output of the channel. Moreover, although V.92 defines the equivalence classes in a one-dimensional constellation, our scheme can be extended to multiple dimensions.

The system proposed in [5] only uses $N-1$ out of N symbols. If $N = 6$, for example, it can be thought as defining an equivalence class for the data $\{a_1, a_2, a_3, a_4, a_5\}$ in a six-dimensional constellation as $\{a_1, a_2, a_3, a_4, a_5, \text{ any}\}$, where "any" means the decoder does not care about the sixth value. In other words, to send the information $\{a_1, a_2, a_3, a_4, a_5\}$, the output of the analog channel can be $\{a_1, a_2, a_3, a_4, a_5, \text{ any}\}$. This can be thought as expanding the constellation by employing on extra symbol for every five symbols. To obtain a channel output of $\{a_1, a_2, a_3, a_4, a_5, \text{ any}\}$ when the data is $\{a_1, a_2, a_3, a_4, a_5\}$, a MIMO pre-equalization scheme is used in [5].

The theoretical performance of a precoding system thus depends only on the definition of equivalence classes. Note that the base constellation does not depend on what equivalence class definition is employed; therefore any precoding scheme can use an optimal constellation as defined in Section 3.

Without specifying what kind of pre-equalization scheme is used, the upstream precoding problem may be restated as follows: Minimize the transmit power, i.e., energy of z, assuming y is in the equivalence class of u and $y(i) = c(i) * z(i)$:

$$\text{Min}_{\vec{y}} \text{ Energy } (\vec{z}), \text{ s.t. } \vec{y} = \vec{c} * \vec{z} \text{ and } \vec{y} \in \text{ Equivalence class } (\vec{u}). \quad (13)$$

As can be seen from this equation, the only factor affecting this minimization is the equivalence class definition. Either type of pre-equalization —i.e., DFE-type pre-equalization, as in our system, or MIMO system, as in [5] —can then be used in the transmitter to force the resulting point $y(i)$ at the output of the channel.

Our equivalence class definition is more flexible, in that it allows an optimal amount of constellation expansion, whereas the equivalence class definition of [5] can provide only a discrete amount of constellation expansion, such as one out of six symbols.

Simulations have been performed to compare the performance of our proposed system to that of [5]. We assume that there is no channel coding, since both systems can achieve about the same amount of coding gain. The constellation given as an example in Section 3 is used. The analog channel is taken to be $(1-D)/\sqrt{2}$.

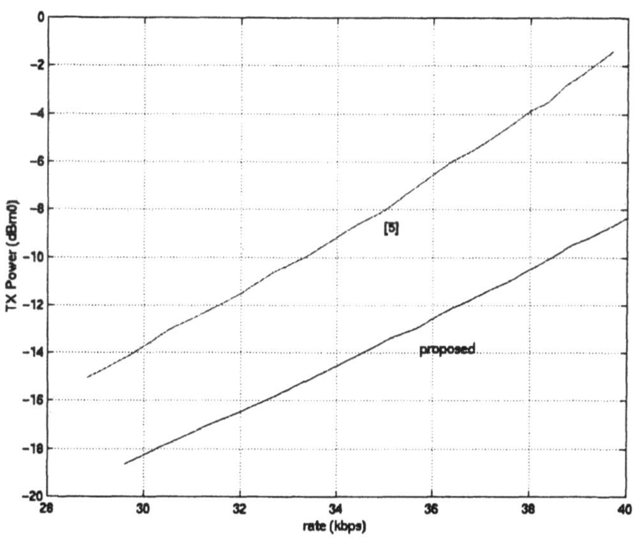

Figure 14 Transmit power versus data rate.

Figure 14 shows the transmit power versus bit rate for the proposed system and for [5].[6] In the figure, [5] is assumed to use one extra symbol every six symbols. As shown in Figure 13, our proposed system has about 4-6 dB power gain, or 4-5 kb/s rate over [5]. Furthermore, [5] requires five different equalization filters, each one of which must be longer than the feedforward filter of our proposed system. Our proposed system can therefore achieve the same reduction of ISI using far fewer taps for feedforward and feedback filters than the MIMO filters of [5].

6. CONCLUSIONS

We have proposed a new transmission scheme for PCM upstream that can achieve significantly higher rates than 33.6 kb/s. Our scheme employs generalized Tomlinson-Harashima precoding with nonuniform PAM constellations, and may be combined with trellis coding. An optimal constellation design algorithm has been developed. This scheme has been adopted for the forthcoming Recommendation V.92.

We have compared our scheme to the previously proposed scheme of [5]. We have showed how both can be expressed in a common equivalence class framework, and have concluded that our scheme is less complex and more flexible. Through simulations, we have shown that our scheme achieves higher bit rates than that of [5].

ACKNOWLEDGMENTS

The authors would like to thank E. Elias, S. Lundby, P. Maurer, J. Pilozzi, and J. Yang for their contributions to the development of this PCM upstream transmission system.

Notes

1. Note that f_1 and f_2 do not have to be the same, as long as they are synchronized. For example, the block diagram when $f_1 = 2f_2$ is given in Figure 4. In this paper, we assume $f_1 = f_2$ to simplify explanations.

2. The constellation design for $y(i)$ will be covered in Section 3. In this section, we assume the constellation points for $y(i)$ are given.

3. The selection could also look ahead to minimize $||x(i)||$ or even $||z(i)||$.

4. Index j is not a time index.

5. To be precise, $v(i)$ is a μ-law code which ranges from -127 to 128 [16]. However, in this paper, we use $v(i)$ to represent both the μ-law code and its decoder amplitude (which ranges from -8031 to 8031 [16]) since it will be evident from the context which it represents.

6. We assume that intersymbol interference (ISI) is small enough for both systems to have similar error probability. In fact, our proposed system can perfectly precode this channel, i.e., there is no ISI, whereas [5] has some residual ISI for this analog channel.

References

[1] P. A. Humblet and M. G. Troulis, "The Information Driveway," *IEEE Communications Magazine*, pp. 64–68, Dec. 1996.

[2] ITU-T Recommendation V.90, "A Digital Modem and Analog Modem Pair for Use on the Public Switched Telephone Network at Data Signaling Rates Up to 56000 Bit/s Downstream and Up to 33600 Bit/s Upstream," 1998.

[3] C. E. Shannon, "A Mathematical Theory of Communication," *Bell Systems Technical Journal*, vol. 27, pp. 379–423/623–656,

July/Oct. 1948.

[4] E. Ayanoglu, "Data Transmission When the Sampling Frequency Exceeds the Nyquist Rate," *IEEE Communications Letters*, vol. 1, pp. 157–159, Nov. 1997.

[5] E. Ayanoglu, N. R. Dagdeviren, G. D. Golden, and J. E. Mazo, "An Equalizer Design Technique for the PCM Modem: A New Modem for the Digital Public Switched Network," *IEEE Transactions on Communications*, vol. 46, pp. 763–774, June 1998.

[6] E. A. Lee and D. G. Messerschmitt, *Digital Communication*, 2nd Edition. Norwell, MA: Kluwer Academic, 1994.

[7] F. Ling, "First and Second Digital Rate Converter Synchronization Device and Method," U.S. patent 5,199,046, Mar. 1993.

[8] N. Al-Dhahir and J. M. Cioffi, "Efficient Computation of the Delay-Optimized Finite-Length MMSE-DFE," *IEEE Transactions on Signal Processing*, vol. 44, pp. 1288–1292, May 1996.

[9] D.-Y. Kim and M. V. Eyuboglu, "Convolutional Spectral Shaping," *IEEE Communications Letters*, vol. 3, pp. 9–11, Jan. 1999.

[10] H. Herzberg and B. R. Saltzberg, "Coding for a Channel with Quantization in the Presence of an Estimable Interference," *IEEE Transactions on Communications*, vol. 45, pp. 45–51, Jan. 1997.

[11] ITU-T Recommendation V.34, "A Modem Operating at Data Signaling Rates of Up to 33600 Bit/s for Use on the General Switched Telephone Network and on Leased Point-to-Point 2-Wire Telephone-Type Circuits," 1996.

[12] D.-Y. Kim, "Device and Method for Detecting PCM Upstream Digital Impairments in a Communication Network," U.S. patent filed Dec. 1997 (pending).

[13] K. Shanmugan and A. M. Breipohl, *Random Signals: Detection, Estimation, and Data Analysis.* Wiley, 1988.

[14] G. D. Forney, Jr., "Trellis Shaping," *IEEE Transactions on Information Theory*, vol. 38, pp. 281–300, Mar. 1992.

[15] M. V. Eyuboglu and G. D. Forney, Jr., "Trellis Precoding: Combined Coding, Precoding, and Shaping for Intersymbol Interference Channels," *IEEE Transactions on Information Theory*, vol. 38, pp. 301–314, Mar. 1992.

[16] J. Bellamy, *Digital Telephony*, 2nd Edition. Wiley, 1991.

[17] G. D. Forney and G. Ungerboeck, "Modulation and Coding for Linear Gaussian Channels," *IEEE Transactions on Information Theory*, vol. 44, pp. 2384–2415, Oct. 1998.

[18] R. F. H. Fischer, W. H. Gerstacker, and J. B. Huber, "Dynamics-Limited Precoding, Shaping, Blind Equalization for Fast Digital Transmission over Twisted-Pair Lines," *IEEE Journal on Selected Areas in Communications*, vol. 13, pp. 1622–1633, Dec. 1995.

[19] L.-F. Wei, "Trellis-Coded Modulation with Multidimensional Constellations," *IEEE Transactions on Information Theory*, vol. 33, pp. 483–501, July 1987.

[20] R. Laroia, "Coding for Intersymbol Interference Channels—Combined Coding and Precoding," *IEEE Transactions on Information Theory*, vol. 42, pp. 1053–1061, July 1996.

Chapter 10

MODEMS TO MARS

G. J. Pottie
Electrical Engineering Department
University of California, Los Angeles

Abstract The practice of modem engineering has generated many interesting theoretical problems, which now apply to a vast range of systems. The title of this paper has dual meaning: the engineering path leading from voiceband modems to communications in distributed sensor networks for Martian exploration, and preliminary plans for the establishment of an interplanetary Internet.

1. INTRODUCTION

Modems are one of the basic building blocks of the Internet, and voiceband modem technology has shown the way to a wide variety of popular access devices including cable TV modems and digital subscriber lines. Recent advances in integrated circuit fabrication technology now permit the construction of compact and low-cost devices that include signal processing, sensing, and wireless communication. This will enable the extension of the Internet to the physical world. Simultaneously, new extensions of the Internet are being planned for our solar system, with an entirely new vision of how communication infrastructure will support science missions and human exploration. This paper presents an overview of the technological path from voiceband modems to ubiquitous communication networks.

2. SOME REALLY COOL THINGS ABOUT VOICEBAND MODEMS

A bias must be acknowledged in that my graduate digital communications course is centered on the study of V.34 modems, with excursions into other technologies. The main point I make about such modems is that the channel resource was extremely limited and thus the signal rates low, which meant that many signal processing cycles can go into each bit transmitted and received. The result was that very advanced techniques were used that only later were applied to higher speed systems. It is a consumer application in which Shannon's limits are approached, over a medium never intended to support data communications, in a product that has become a low-priced commodity. Along the way, new concepts in line probing, adaptive equalization, constellation shaping, coded modulation, combined coded and shaped modulation and equalization, echo cancellation, and synchronization were developed. A backwards compatible family of devices was produced such that each succeeding generation could actually in a bug-free manner interoperate with preceding generations. A standards process took place that actually spurred rather than smothered technological change. These are all significant accomplishments that merit academic study on their own.

3. V.34 TO GIGABIT ETHERNET

The very fast or very cheap communication systems usually first have an analog solution, which is quite fortunate for digital designers. Since it is difficult to get anything analog to work that involves loops of degree larger than 2, the channels are inadvertently engineered in such a fashion that adaptive least squares techniques are nearly certain to converge. The obvious success in adaptive techniques in raising voiceband modem speeds from 1200 bits/s to 28.8 kb/s over "analog" lines has paved the way for many of these techniques to be applied to digital subscriber lines, cable modems, and high speed ethernet. However, the demands of higher speed were met not by simply having more general purpose DSPs, but by design of ASICs capable of carrying out the most common operations with a fraction of the chip area and thus cost and energy. Here the most impressive technological change from the design point of view is the improvement in computer aided design tools for ASICs, which enable experienced designers to very rapidly turn out new chips using standard cell libraries. New cells need to be created only for the most critical functions. These designers need to know both communications theory (what algorithms are suitable) and be versant in the tools. A recent example of the success of this approach is the very rapid growth of Broadcom, with a variety of high speed chips with extensive adaptive features being produced on short design cycles.

4. SENSOR NETWORKS

Wireless sensor networks are an example of systems in which energy resources are very highly constrained [1, 2, 3, 4, 5, 6]. In such systems, each node includes sensors, signal processing, and a radio, with the objective of detecting and identifying events with high probability. Prior architectures would typically send raw data to a central site for processing. However, given that millions of operations can be executed for the same energy as it takes to transport 1 kbit 100 m (for typical parameters) [6], it makes much more sense to process and make decisions in the nodes insofar as possible. Shannon's limit and Maxwell's law together dictate that there will always be a minimum energy cost to transmit the bits, while continuing advances in integrated circuit technology indicate we are far from the minimum energy costs for signal processing. Further, by coming to decisions rather than transporting the raw data, we ease bandwidth limitations in the network, promoting scalability. The resulting bit rates on any given link will be low, and many messages will be very short. Here advanced communications techniques are needed not for increasing the bandwidth efficiency, but for conserving transmit energy, and avoiding costly retransmissions. Early prototype systems [1, 2], later systems [5, 6], and a general overview of the principles of wireless sensor network design [6] are described in the literature.

The high cost of communications dictates distributed rather than centralized algorithms for a wide range of sensor functions, from networking to data fusion. The large number of nodes further implies that the network should self-organize for these tasks, rather than relying on human intervention to set-up or adjust to failures of nodes. A set of algorithms to accomplish this is presented, for example, by Sohrabi et al. [4]. The severe energy constraint is in part mitigated by the loose latency requirements for many of the applications. Indeed, progress on this problem is made only when considering the complete system from physical layer through application. Challenges include questions of network hierarchy (flat or multitiered), interaction of signal processing, storage, and communication, heterogeneity in the hardware and environment, and interaction with remote networks such as the Internet with their associated resources. The first commercial products that deal with these questions are beginning to appear, but we have not even constructed the theoretical scaffolding for dealing with issues of fundamental limits. Like voiceband modems, this seems to be an area where very interesting theory will be spurred by practical need.

Figure 1 illustrates how sensing, networking, signal processing, and database management are all intertwined with wireless sensor networks.

Figure 1: Information Processing Hierarchy

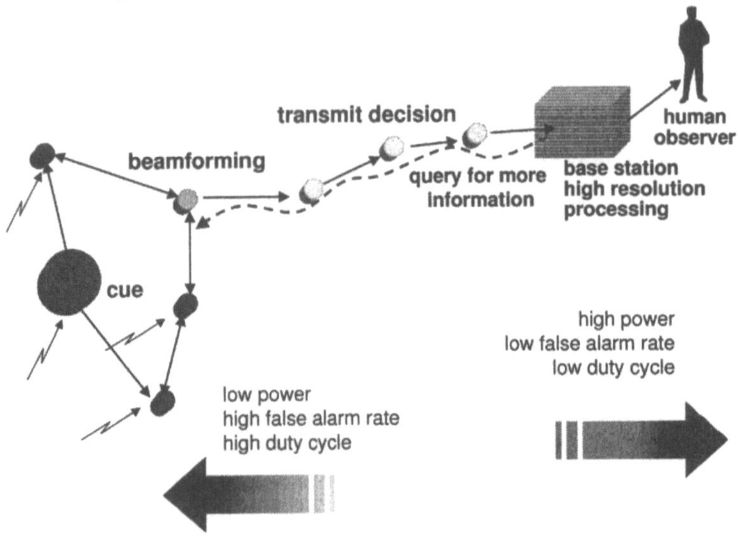

Consider a large network whose purpose is to detect and identify particular physical events. Individual nodes are powered off some combination of batteries and solar cells so that energy is at a premium. As noted above, this compels processing as much of the data at source as possible so as to limit communications. Further, there will be a signal processing hierarchy within the nodes themselves. Energy-detection circuits use little power, and can be constantly vigilant. However, to meet detection probability requirements they will produce an unacceptable rate of false alarms. Thus sampled data can be queued, and in the event that the energy detection threshold is exceeded some higher level of signal processing can be invoked, e.g., frequency analysis. By queuing data and proceeding through a sequence of operations of increased sophistication (lower false alarm rates, higher energy consumption), energy is expended only to the extent required while assuring performance targets are met. Yet a single node cannot on its own provide accurate position information for targets, and may have insufficient SNR for producing a reliable decision about an event. Thus, collections of nodes may form ad hoc networks to perform data fusion (exchange of likelihood function values) or coherent beamforming (raw data) [3]. This step is taken only when lower-energy procedures fail to produce the desired results. Finally, information can be multihopped back towards an end user, possibly being queued and aggregated along the way, for example to produce activity summary reports. This possibility of aggregation expands network scalability.

The end user may actually desire more information, and has the option to query the network to provide further details about particular events. Thus queued data may be recovered and transmitted. For example, for data on which the network has come to a murky decision a central site may perform highly sophisticated adaptive processing, the end result of which will be new detection table parameters being sent to all the nodes. For the highest detection/identification reliability, the end user must retain the ability to drill down to the lowest level in the network. At the same time, to avoid information overload, bandwidth contention, and massive energy usage the nodes must only rarely be in the situation where the transport of raw data to the end user is required.

Notice that in this example there is no clear division between signal processing, database issues, and networking. It is rather a unified process, in which design decisions at one level have a large impact on other levels.

5. ROBOT ECOLOGIES

Sensor nodes scattered upon the ground can end up in disadvantageous locations with respect to antenna elevation or sensor field of view. The thought naturally occurs that it would be nice if either the whole node or an appendage could move to the more favorable location. The advantage of extending an arm is that we do not have the energy or complexity cost of moving the whole unit. On the other hand, having some mobile elements may allow holes in the network to be filled, or afford the possibility of physical exchange of information and energy. In this scenario, the fixed robotic elements are the analog of plants—accumulating energy—and the mobile elements are the analog of animals—using high energy supplies. Such a robot ecology [7] provides the possibility of sustained autonomous operation. The plants provide the infrastructure that makes possible the long-term deployment of mobile objects, and in addition are useful in themselves in collecting information. A very wide variety of plants can be conceived: balloon-launchers (roots tap down to water, and generate hydrogen by electrolysis), burrs, tumbleweeds, lily pads, seed launchers, creeper vines (say propelled by mobots), etc. Plants could provide navigational assistance, in effect trading information against energy consumption. The number and complexity of the interactions that are possible makes clear the need for a more formal theory of cooperation. At the moment, cooperation is pursued on an ad hoc basis, and typically with either a very rigid hierarchy (master/slave) or actions among homogeneous peers. Extensions to large numbers of heterogeneous devices is very challenging. Nevertheless, a broad range of applications are possible with even relatively simple robotic ecologies.

Figure 2: Mars Geological Survey

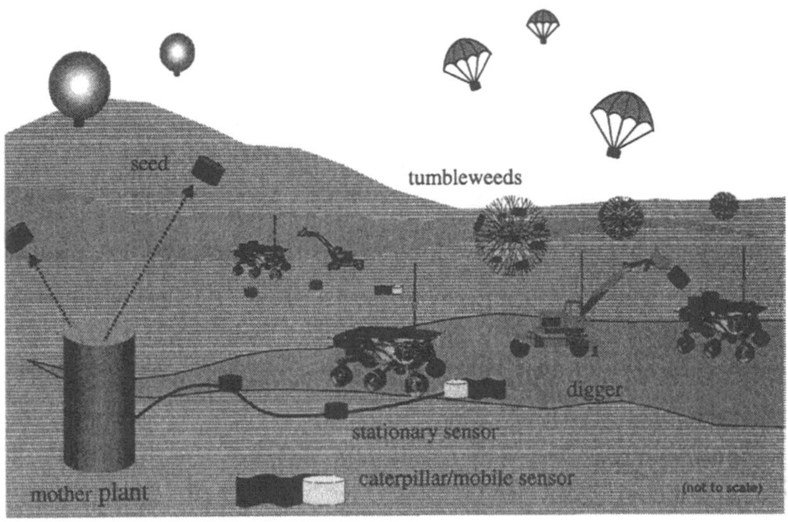

Figure 2 illustrates a fanciful ecology for performing a geological survey of Mars, with mother plants equipped with creeper vines and sensor seed launchers, mobile diggers and trucks, wind-propelled sensor tumbleweeds and balloons, and air-drop of fresh supplies. It is of interest that the rover in the celebrated Mars Pathfinder mission moved not all that much further than spring-launched sensor seeds could be sent. Energy constraints are even more important in planetary exploration applications than for Earth-based systems due to the huge expense of transporting heavy power-generation and storage units. Thus, unconventional systems which opportunistically use local resources (to live off the land) may be an interesting avenue for future research.

6. THE MARTIAN INTERNET

Plans are afoot for a sustained series of robotic missions to Mars. In contrast to the practice of the past, it is intended that missions will contribute to a common infrastructure with the goal of supporting robot and eventually human colonies on the red planet. Part of this task will be construction of an interplanetary network (IPN) [8]. Eventually, a network of communications/position location satellites will be created for Mars, in support of a local internet similar to the terrestrial internet. Management of traffic between Earth and Mars however will be challenging due to the large latency. Standard Internet protocols will not work; the signaling will be more akin to sending email than carrying on an interactive session. It is anticipated that the traffic flow will be highly asymmetric (from Mars to Earth, since the information consumers are terrestrial).

Presently, communications time is allocated according to committee decisions. This will clearly not be possible when there are many scientific missions in progress, as well as requests from the public at large. Market auction algorithms show some promise in reducing complicated allocation criteria to a single parameter that can determine access priority. Such mechanisms may also be used to provide incentives for missions to create the necessary infrastructure (e.g., awarding Mars phone cards which can then be sold for real money).

7. GRAND INFORMATION THEORY CHALLENGES

Distributed sensor and robotic networks bring large new challenges to information theory. In sensor networks, the basic problem is to identify classes of events to a required level of fidelity (missed/false detection) using the minimum resources (energy, bandwidth) within latency constraints. Neglecting latency, this can be cast as a network rate-distortion problem with the resources in the role of rate and fidelity in the role of distortion. Resource usage includes that needed for determining which set of sensor nodes will be involved in the decision, including inhibition of others from participating. There are also capacity problems, in that groups of nodes may cooperatively signal to overcome gaps in the network. How to form the transmitting and receiving groups, what cost is associated with maintenance of a particular level of synchronism, and how to optimally code are all unknown. Further, we may have to deal with heterogeneous devices.

The challenges in robotics are even greater, in that we must deal with both of the above problems in the context of mobility, with interactions between fixed and moving assets. Vast latencies are possible in attempting to control the operations of robot colonies on distant planets, so that adjustments can only be made to parameters controlling behaviors, rather than individual actions. A very interesting interaction between theory and practice is before us.

References

[1] K. Bult, A. Burstein, D. Chang, M. Dong, M. Fielding, E. Kruglick, J. Ho, F. Lin, T.-H. Lin, W. J. Kaiser, H. Marcy, R. Mukai, P. Nelson, F. Newberg, K. S. J. Pister, G. Pottie, H. Sanchez, O. M. Stafsudd, K. B. Tan, C. M. Ward, S. Xue, and J. Yao, "Low Power Systems for Wireless Microsensors," *Proceedings of International Symposium on Low Power Electronics and Design*, pp. 17–21, 1996.

[2] G. Asada, M. Dong, T. S. Lin, F. Newberg, G. Pottie, H. O. Marcy, and W. J. Kaiser, "Wireless Integrated Network Sensors: Low Power Systems on a Chip," *Proceedings of the 24th IEEE European Solid-State Circuits Conference*, 1998.

[3] K. Yao, R. E. Hudson, C. W. Reed, D. Chen, and F. Lorenzelli, "Blind Beamforming on a Randomly Distributed Sensor Array System," *IEEE Journal on Selected Areas in Communications*, vol. 16, no. 8, pp. 1555–1567, Oct. 1998.

[4] K. Sohrabi, J. Gao, V. Ailawadhi, and G. Pottie, "A Self-Organizing Sensor Network," *Proceedings of the 37th Allerton Conference on Communication, Control, and Computing*, Monticello, IL, Sept. 1999.

[5] J. R. Agre, L. P. Clare, G. J. Pottie, and N. P. Romanov, "Development Platform for Self-Organizing Wireless Sensor Networks," *Aerosense '99*, Orlando, FL, April 1999.

[6] G. J. Pottie and W. J. Kaiser, "Wireless Integrated Network Sensors," *Communications of the ACM*, vol. 43, no. 5, pp. 51–58, May 2000.

[7] G. J. Pottie and R. Brooks, "Towards a Robotic Ecology," *DARPA ISAT Study*, 1999.

[8] V. Cerf, et al., "Interplanetary Internet Seminar," http://www.ipnsig.org/reports/CalTechSeminar1.html, Oct. 1999.

Chapter 11

EXISTENCE OF GOOD CODES WITH LOW PEAK-TO-AVERAGE POWER RATIOS FOR MULTICARRIER COMMUNICATIONS

K. G. Paterson
Hewlett-Packard Laboratories
Filton Road, Stoke Gifford
Bristol BS34 8QZ, U.K.

V. Tarokh
Massachusetts Institute of Technology
77 Massachusetts Avenue
Cambridge, MA, USA

Abstract An appendum to a theorem of Shannon is proved. It is proved that that there exists asymptotically good codes of length n, whose peak-to-average power ratio for multicarrier communications is at most $8 \log n$.

1. INTRODUCTION

Multicarrier communications (OFDM) is an old communication technique that has recently seen rising popularity in wireless and wireline applications. International standards making use of OFDM for wireless LANs are currently being established by IEEE 802.11 and ETSI BRAN committees. For wireless applications, an OFDM-based system can be of interest because it provides a greater immunity to impulse noise and fast fades and eliminates the need for equalizers, while efficient hardware implementations can be realized using FFT techniques.

A problem associated with OFDM is the high peak-to-average power ratio (PAPR) of uncoded OFDM signals. If the peak transmit power is limited, either by regulatory or application constraints, this has the effect of reducing the average power allowed under OFDM relative to that under constant power modulation techniques. This in turn reduces the range of OFDM transmissions. Moreover, to prevent spectral growth of the OFDM signal in the form of intermodulation amongst subcarriers and out-of-band radiation, the transmit amplifier must be operated in its linear region (i.e., with a large input back-off), where the conversion from DC to RF power is inefficient. This may have a deleterious effect on battery lifetime in mobile applications. In many low-cost applications the drawbacks of high PAPR outweigh all the potential benefits of OFDM systems.

A number of approaches have been proposed to deal with this power control problem. One idea is to choose $\mathcal{C}$ for its error-correcting properties and the offset to reduce the PAPR of the resulting coded OFDM transmissions. This approach enjoys the twin benefits of power control and error correction and is simple to implement in practice, but requires extensive calculation to find good codes and offsets. Recently, Tarokh and Jafarkhani [15] introduced a geometrical approach to the offset selection problem. This approach leads to a computationally efficient algorithm making use of a maximum-likelihood decoder for the underlying code to find good offsets. As an example, the algorithm was used to obtain a reduction of 4.5 dB in the PAPR of the rate 1/2 BPSK code proposed in the ETSI BRAN Hiperlan-II standardization committee.

Another strand of work on codes with reduced PAPR can be found in a sequence of papers [2, 3, 10, 9, 11]. Davis and Jedwab [3] developed a powerful theory which yields a class of codes enjoying PAPR as low as 2, large minimum distance and possessing efficient soft-decision decoding algorithms [4, 12]. These codes are formed from unions of cosets of the classical Reed-Muller codes and new generalizations of them. Others gave special cases of these codes [10, 9] and further developed the underlying theory [11]. While these block codes reduce PAPR, they also result in reduced transmission rates, severely so for large numbers of carriers.

In this paper, we make a contribution to the use of coding for combined error correction and peak to average power reduction.

We prove that there exist asymptotically good sequences of codes achieving analogs of Varshamov-Gilbert bound, with PAPR growth of order only $8 \log n$.

The organization of the paper is as follows.

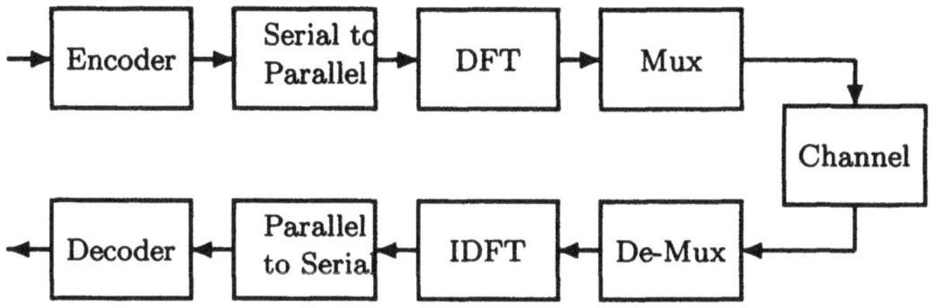

Figure 1 OFDM Block Diagram.

In Section 2, we introduce the communication model, formulate the PAPR problem that we study and establish our notation. We then prove our result in Section 3.

2. THE COMMUNICATION MODEL

Figure 1 shows a block diagram of an OFDM system. At each time $t = 0, \mathcal{T}, 2\mathcal{T}, \ldots$ blocks B_t of k bits arrive at the encoder. These k bits are encoded as a vector $\mathbf{c}$ of n constellation symbols from a constellation $\mathcal{Q} \subset \mathbb{C}$. The admissible sequences are called codewords, and the ensemble of all possible codewords is a code $\mathcal{C}$ of rate $R = k/n$. We denote the minimum Euclidean distance of the code $\mathcal{C}$ by d.

The vector $\mathbf{c}$ of n constellation symbols are provided to the input of a discrete Fourier transform by a serial to parallel block, producing a sequence of symbols $C_0, C_1, \ldots, C_{n-1}$, where

$$C_l = \sum_{i=0}^{n-1} c_i \exp(\frac{-j2\pi\, il}{n}), \tag{1}$$

for $l = 0, 1, 2, \ldots, n-1$. Here $j = \sqrt{-1}$.

This sequence is the input to the RF chain which produces the transmitted signal. This signal at time t is modelled by the real part of the complex envelope:

$$S_{\mathbf{c}}(t) = \left(\sum_{i=0}^{n-1} c_i \exp(-j2\pi(f_0 + if_s)t)\right) \tag{2}$$

for $0 \le t \le \frac{1}{f_s}$, where f_0 is the carrier frequency and f_s is the bandwidth of each tone. The relation between the quantities f_s and $\mathcal{T}$ depends on whether a guard time is assigned, or a cyclic prefix is used and these details have no bearing upon the bounds derived in this paper. However, we note that $f_s = \frac{1}{T}$ is commonly assumed in an ideal situation.

The receiver receives the signal $\mathrm{Re}[S_{\mathbf{c}}(t)]$ perturbed by noise and performs the inverse operations: the RF chain at the receiver down-converts, processes the received data and obtains estimates of the parameters $C_\ell, \ell = 0, 1, \ldots, n-1$. The receiver next applies an inverse discrete Fourier transform on these estimates and generates estimates of $c_0, c_1, \ldots, c_{n-1}$. The receiver then extracts the block B_T of input bits by applying a suitable error correction algorithm.

For any codeword $\mathbf{c}$, the instantaneous power of the corresponding transmitted signal $\mathrm{Re}[(S_{\mathbf{c}}(t))]$ is equal to $(\mathrm{Re}[S_{\mathbf{c}}(t)])^2$. This power is less than or equal to the function $|S_{\mathbf{c}}(t)|^2$, called the *envelope power* of the OFDM signal.

The average value of the envelope power is exactly equal to $||\mathbf{c}||^2$ while, for $f_0 >> f_s$, the average power of the actual OFDM signal is approximately equal to $\frac{1}{2}||\mathbf{c}||^2$. We define PAPR($\mathbf{c}$), the peak-to-average power ratio of the OFDM signal, to be the ratio of the peak power of $\mathrm{Re}(S_{\mathbf{c}}(t))$ to $||\mathbf{c}||^2$, the average envelope power. We write $\zeta = f_0/f_s$, and note that in a typical OFDM application, we have $\zeta >> 1$ (for example, in the ETSI BRAN standard, $f_0 = 5 \times 10^9$ and $f_s \simeq 300 \times 10^3$). Then we have

$$\mathrm{PAPR}(\mathbf{c}) = \max_{0 \leq t \leq 1} \frac{\left|\mathrm{Re}\left(\sum_{i=0}^{n-1} c_i \exp(-2\pi j(\zeta + i)t)\right)\right|^2}{||\mathbf{c}||^2}. \tag{3}$$

Note that in the literature, PAPR is often referred to as peak-to-mean power ratio (PMPR). Notice also that

$$\mathrm{PAPR}(\mathbf{c}) \leq \max_{0 \leq t \leq 1} \frac{|S_{\mathbf{c}}(t)|^2}{||\mathbf{c}||^2}.$$

The function on the right of the above inequality is called the peak-to-mean envelope power ratio (PMEPR) of the codeword or OFDM signal. We denote it by PMEPR($\mathbf{c}$). It is often more convenient to work with PMEPR than PAPR. It is straightforward to show that if

$$c(z) = c_0 + c_1 z + \cdots + c_{n-1} z^{n-1}$$

denotes the degree $n-1$ polynomial whose coefficients are derived from $\mathbf{c}$, then

$$\mathrm{PMEPR}(\mathbf{c}) = \frac{1}{||\mathbf{c}||^2} \max_{|z|=1} |c(z)|^2.$$

Thus, the PMEPR of a codeword is related to the maximum squared absolute value of the corresponding polynomial on the unit circle. This observation will be useful in the sequel.

For a code $\mathcal{C}$, we define

$$\mathrm{PAPR}(\mathcal{C}) = \max_{\mathbf{c}\in\mathcal{C}}(\mathrm{PAPR}(\mathbf{c}))$$

and refer to it as the peak-to-average power ratio (PAPR) of $\mathcal{C}$.

We assume throughout the remainder of this paper that all the codewords in our codes have average envelope power $\|\mathbf{c}\|^2$ equal to n (we say that the codes have constant energy). This assumption certainly holds for any signal constellation in which each symbol has absolute value 1, for example any PSK constellation.

For any subset S of the n-dimensional complex sphere of radius $\sqrt{n}$ and any $r \geq 0$, we define $H(S,r)$ to be the surface consisting of all those points of the sphere which are within distance r of S. We let $A(S,r)$ denote the area of $H(S,r)$. Clearly, for any point $\mathbf{x}$ on the sphere, $H(\mathbf{x},r)$ is a spherical cap. It is well-known that (see for instance [14]):

$$A(r) := A(\mathbf{x},r) = \frac{2\pi^{n-\frac{1}{2}}n^{n-\frac{1}{2}}}{(n-\frac{3}{2})!}\int_0^{2\arcsin(r/2\sqrt{n})}\sin^{2n-2}\theta d\theta. \quad (4)$$

3. ON THE EXISTENCE OF CODES WITH A GIVEN MINIMUM DISTANCE AND PAPR

In this section, we assume that d and d_* are given and prove the existence of codes of length n, minimum distance at least d and peak-to-average power ratio at most d_*. Our tool is a Varshamov-Gilbert type of argument. We require a technical lemma given below.

Lemma 1 *Let $c(z)$ be a polynomial of degree $n-1$ over $\mathbb{C}$. Let $W = \lceil 2\pi(n-1)\rceil$. Then*

$$\max_{|z|=1}|c(z)| \leq 2\cdot \max_{0\leq k<W}|c(\exp(-2\pi jk/W))|.$$

Proof: Our proof of Lemma 1 is obtained by correcting an argument of Beck [1]. We require an additional classical result from complex analysis known as Bernstein's inequality.

Lemma 2 *Let $c(z)$ be a polynomial of degree $n-1$ over $\mathbb{C}$. Then*

$$\max_{|z|=1}|c'(z)| \leq (n-1)\cdot \max_{|z|=1}|c(z)|.$$

Writing $z = \exp(2\pi j\beta)$, we can restate the above lemma as:

$$\max_{0\leq\beta<1}|c'(\exp(2\pi j\beta))| \leq 2\pi(n-1)\cdot \max_{0\leq\beta<1}|c(\exp(2\pi j\beta))|.$$

Proof: (of Lemma 1) Recall that $W = \lceil 2\pi(n-1) \rceil$. Suppose $M = \max_{|z|=1} |c(z)|$ is attained at $z = \exp(2\pi j\beta)$ where $0 \leq \beta < 1$. Then for some i w ith $0 \leq i < W$, we have

$$\left|\beta - \frac{-i}{W}\right| \leq \frac{1}{2W}$$

and then

$$\begin{aligned}
&|M - |c(\exp(-2\pi ji/W))|| = \\
&||c(\exp(2\pi j\beta))| - |c(\exp(-2\pi ji/W))|| \leq \\
&|c(\exp(2\pi j\beta)) - c(\exp(-2\pi ji/W))| = \\
&\left|\int_{-i/W}^{\beta} c'(\exp(2\pi j\theta))d\theta\right| \leq \\
&\int_{-i/W}^{\beta} |c'(\exp(2\pi j\theta))|d\theta \leq \\
&\left|\beta - \frac{-i}{W}\right| \max_{0\leq\theta<1} |c'(\exp(2\pi j\theta))| \leq \\
&\frac{1}{2W} \cdot 2\pi(n-1) \max_{0\leq\theta<1} |c(\exp(2\pi j\theta))| = \frac{\pi(n-1)M}{W}.
\end{aligned}$$

Hence

$$\begin{aligned}
M &\leq \frac{1}{1 - \frac{\pi(n-1)}{W}} \cdot |c(\exp(-2\pi ji/W))| \leq \\
&2 \max_{0\leq k<W} |c(\exp(-2\pi jk/W))|
\end{aligned}$$

and the proof is complete.

Theorem 3 *Let d and d_* be given and let 2^{nR} be the largest integer for which*

$$4\lceil 2\pi(n-1)\rceil A(d_*) + 2^{nR} A(d) \leq \frac{2n^{n-\frac{1}{2}}\pi^n}{(n-1)!}$$

holds. Then there exists a constant energy code C of rate R with minimum Euclidean distance at least d and peak-to-average power ratio at most $8n(1-\delta)^2$ where $\delta = d_*^2/2n$.

Proof: Let $W = \lceil 2\pi(n-1) \rceil$. We define the vectors

$$\mathbf{v}_k = \left(1, \exp(\frac{2\pi jk}{W}), \exp(\frac{2\pi j2k}{W}), \ldots, \exp(\frac{2\pi j(W-1)k}{W})\right)$$

for $0 \leq k < W$ and

$$\mathbf{v}_{k+W} = -\mathbf{v}_k, \quad 0 \leq k < W.$$

Consider the spherical caps $I(\mathbf{v}_k, d_*)$, $I(j\mathbf{v}_k, d_*)$, $k = 0, 1, \ldots, 2W-1$ on the n-dimensional complex sphere of radius $\sqrt{n}$. Suppose that the sum of the areas of these caps is less than the area of this sphere. Then we can choose a point $\mathbf{c}_0$ on the sphere which is not in the union of these caps.

Inductively, let us assume that the codewords $\mathbf{c}_0, \mathbf{c}_1, \ldots, \mathbf{c}_{\ell-1}$ ($\ell \geq 1$) are chosen. Then we consider the spherical caps $H(\mathbf{v}_k, d_*)$, $H(j\mathbf{v}_k, d_*)$, $k = 0, 1, \ldots, 2W-1$ and $H(\mathbf{c}_i, d)$, $i = 0, 1, \ldots, \ell - 1$. If the sum of areas of these caps is less than the area of the n-dimensional complex sphere of radius $\sqrt{n}$, then we can choose a codeword $\mathbf{c}_\ell$ which is not in the union of the caps. It is easy to see that this process can be repeated as long as

$$4\lceil 2\pi(n-1)\rceil A(d_*) + (\ell+1)A(d) \leq \frac{2n^{n-\frac{1}{2}}\pi^n}{(n-1)!}.$$

At the end of the induction process, we have chosen points $\mathbf{c}_0$, $\mathbf{c}_1$, ..., $\mathbf{c}_{2^{nR}-1}$. We define the code C to consist of these points. Clearly this code has minimum distance at least d by virtue of the manner in which the codewords were chosen. Furthermore, the distance between any codeword $\mathbf{c}$ and any of $\mathbf{v}_k$, $j\mathbf{v}_k$ $(0 \leq k < 2W-1)$ is at least d_*. We thus have

$$\begin{aligned} \|\mathbf{c} - \mathbf{v}_k\| &\geq d_*, \quad 0 \leq k < W, \\ \|\mathbf{c} + \mathbf{v}_k\| &\geq d_*, \quad 0 \leq k < W, \\ \|\mathbf{c} - j\mathbf{v}_k\| &\geq d_*, \quad 0 \leq k < W, \\ \|\mathbf{c} + j\mathbf{v}_k\| &\geq d_*, \quad 0 \leq k < W. \end{aligned}$$

which together imply that:

$$2\left|\mathrm{Re}\left(\sum_{i=0}^{n-1} c_i \exp(\frac{-2\pi jki}{W})\right)\right| \leq 2n - d_*^2, \quad 0 \leq k < W,$$

$$2\left|\mathrm{Im}\left(\sum_{i=0}^{n-1} c_i \exp(\frac{-2\pi jki}{W})\right)\right| \leq 2n - d_*^2, \quad 0 \leq k < W.$$

Writing $c(z) = \sum_{i=0}^{n-1} c_i z^i$, these can be written as

$$2\left|\mathrm{Re}\left(c(\exp(\frac{-2\pi jk}{W}))\right)\right| \leq 2n - d_*^2, \quad 0 \leq k < W,$$

$$2\left|\mathrm{Im}\left(c(\exp(\frac{-2\pi jk}{W}))\right)\right| \leq 2n - d_*^2, \quad 0 \leq k < W.$$

We deduce that

$$|c(\exp(\frac{-2\pi jk}{W}))| \leq \sqrt{2}(n - \frac{d_*^2}{2}), \quad 0 \leq k < W.$$

Applying Lemma 1, we conclude that for all z with $|z| = 1$,

$$|c(z)| \leq 2\sqrt{2}(n - \frac{d_*^2}{2}).$$

Thus

$$\text{PAPR}(\mathbf{c}) \leq \text{PMEPR}(\mathbf{c}) = \frac{1}{n} \max_{|z|=1} |c(z)|^2 \leq 8n(1-\delta)^2$$

where $\delta = d_*^2/2n$. This completes the proof.

An Inequality of Shannon: For our analysis, we will need to estimate the area of a spherical cap. For a completely different purpose, it transpires that Shannon has considered the function $A(r)$ and asymptotic approximations to it. Indeed, he establishes [14, p. 292] that, for $\theta_1 < \frac{\pi}{2}$

$$\frac{\sin^{2n-1}\theta_1}{\cos\theta_1}(1 - \frac{1}{2n}\tan^2\theta_1) \leq$$
$$(2n-1)\int_0^{\theta_1} \sin^{2n-2}\theta d\theta \leq \frac{\sin^{2n-1}\theta_1}{\cos\theta_1}.$$

This means in particular that

$$0 \leq \int_0^{\theta_1} \sin^{2n-2}\theta d\theta \leq \frac{\sin^{2n-1}\theta_1}{(2n-1)\cos\theta_1} \tag{5}$$

for any $n \geq 2$ and $0 \leq \theta_1 < \frac{\pi}{2}$.

We can now prove an existence result showing that asymptotically good codes with low peak-to-average power ratio exist. Our result is summarized in the following result.

Theorem 4 *Let $R \geq 0$, and $\Delta \geq 0$ be such that*

$$2^R\sqrt{2\Delta(1 - \frac{\Delta}{2})} < 1.$$

Then for all sufficiently large n, there exists a constant energy code C of length n, rate R and minimum distance $d = \sqrt{2\Delta n}$ having peak-to-average power ratio of at most $8\log n$.

Proof: Write $\delta = 1 - \sqrt{\frac{\log n}{n}}$. By Theorem 3, provided that

$$4\lceil 2\pi(n-1)\rceil A(\sqrt{2\delta n}) + 2^{nR} A(\sqrt{2\Delta n}) \leq \frac{2n^{n-\frac{1}{2}}\pi^n}{(n-1)!},$$

then there exists a code $\mathcal{C}$ of rate R with minimum Euclidean distance at least d and peak-to-average power ratio at most $8n(1-\delta)^2 = 8\log n$. We thus have to prove that the above inequality holds for all sufficiently large n.

We replace the expression for $A(r)$ given in Equation (4) in the above inequality. After simple manipulations, we observe that it suffices to prove that the inequality

$$4\lceil 2\pi(n-1)\rceil \int_0^{2\arcsin(\sqrt{2\delta}/2)} \sin^{2n-2}\theta d\theta +$$
$$2^{nR} \int_0^{2\arcsin(\sqrt{2\Delta}/2)} \sin^{2n-2}\theta d\theta \leq \frac{2\sqrt{\pi}(n-\frac{3}{2})!}{(n-1)!}$$

holds for all sufficiently large n. We now use the above inequality of Shannon. After further computations, and using the Wallis product identity to observe that for large n,

$$\frac{(n-\frac{3}{2})!}{(n-1)!} \approx \frac{2}{\sqrt{n}},$$

it can be shown that it suffices to prove that the inequality

$$4\pi\sqrt{n}\frac{2n-2}{(2n-1)}\frac{\sin^{2n-1}(\theta_1)}{\cos(\theta_1)} + \frac{\sqrt{n}}{(2n-1)}2^{nR}\frac{\sin^{2n-1}(\theta_2)}{\cos(\theta_2)} \leq 2\sqrt{\pi} \qquad (6)$$

holds asymptotically, where

$$\theta_1 = 2\arcsin(\sqrt{2\delta}/2),$$
$$\theta_2 = 2\arcsin(\sqrt{2\Delta}/2).$$

Because $2^R \sin(\theta_2) = 2^R\sqrt{2\Delta(1-\frac{\Delta}{2})} < 1$, we see that

$$\lim_{n\to\infty} \frac{\sqrt{n}}{(2n-1)}2^{nR}\frac{\sin^{2n-1}(\theta_2)}{\cos(\theta_2)} = 0.$$

It is also easy to show that $\sin(\theta_1) = \sqrt{2\delta(1-\frac{\delta}{2})} < 1$. It follows that $\sin(\theta_1) = \sqrt{1-\frac{\log n}{n}}$. Thus $\sin^2(\theta_1) = 1 - \frac{\log n}{n}$ and $\cos(\theta_1) = \sqrt{\frac{\log n}{n}}$.

Then we have

$$4\pi\sqrt{n}\frac{2n-2}{(2n-1)}\frac{\sin^{2n-1}(\theta_1)}{\cos(\theta_1)} = \frac{4\pi n}{\sqrt{\log n}}\frac{2n-2}{(2n-1)}(1-\frac{\log n}{n})^{(n-0.5)}.$$

But $n(1-\frac{\log n}{n})^n \to 1$ as $n \to \infty$ and so

$$\lim_{n\to\infty} 4\pi\sqrt{n}\frac{2n-2}{(2n-1)}\frac{\sin^{2n-1}(\theta_1)}{\cos(\theta_1)} = 0.$$

Thus inequality Equation (6) holds for all sufficiently large n and the result is proved.

References

[1] J. Beck, "Flat Polynomials on the Unit Circle—Note on a Problem of Littlewood," *Bulletin of the London Mathematical Society*, vol. 23, pp. 269–277, 1991.

[2] J. A. Davis and J. Jedwab, "Peak-To-Mean Power Control and Error Correction for OFDM Transmission Using Golay Sequences and Reed-Muller Codes," *Electronic Letters*, vol. 33, pp. 267–268, 1997.

[3] J. A. Davis and J. Jedwab, "Peak-To-Mean Power Control in OFDM, Golay Complementary Sequences and Reed-Muller Codes," *IEEE Transactions on Information Theory*, vol. 45, pp. 2397–2417, Nov. 1999.

[4] A. J. Grant and R. D. J. van Nee, "Efficient Maximum-Likelihood Decoding of Q-ary Modulated Reed-Muller Codes," *IEEE Communication Letters*, vol. 2, pp. 134–136, 1998.

[5] J. Jedwab, "Comment: M-Sequences for OFDM Peak-to-Average Power Ratio Reduction and Error Correction," *Electronic Letters*, vol. 33, pp. 1293–1294, 1997.

[6] A. E. Jones, T. A. Wilkinson, and S. K. Barton, "Block Coding Scheme for Reduction of Peak to Mean Envelope Power Ratio of Multicarrier Transmission Schemes, *Electronic Letters*, vol. 30, pp. 2098–2099, 1994.

[7] A. E. Jones and T. A. Wilkinson, "Combined Coding Error Control and Increased Robustness to System Nonlinearities in OFDM," *Proceedings of the IEEE 46th Vehicular Technology Conference*, pp. 904–908, Atlanta, GA, 1996.

[8] S. H. Müller, R. W. Bäuml, R. F. H. Fischer, and J. B. Huber, "OFDM with Reduced Peak-to-Average Power Ratio by Multiple

Signal Representation," *Annales des Télécommunications*, vol. 52, pp. 58–67, 1997.

[9] R. D. J. van Nee, "OFDM Codes for Peak-to-Average Power Reduction and Error Correction," *Proceedings of the IEEE Globecom 1996*, pp. 740–744, London, 1996.

[10] H. Ochiai and H. Imai, "Block Coding Scheme Based on Complementary Sequences for Multicarrier Signals," *IEICE Transaction Fundamentals*, pp. 2136–2143, 1997.

[11] K. G. Paterson, "Generalized Reed-Muller Codes and Power Control in OFDM Modulation," *IEEE Transactions on Information Theory*, vol. 46, pp. 104–120, 2000.

[12] K. G. Paterson and A. E. Jones, "Efficient Decoding Algorithms for Generalized Reed-Muller Codes," *IEEE Transactions on Communication*, to appear.

[13] P. Piret, "Bounds for Codes over the Unit Circle," *IEEE Transactions on Information Theory*, vol. 32, pp. 760–767, 1986.

[14] C. E. Shannon, "Probability of Error for Optimal Codes in a Gaussian Channel," *Claude Shannon Collected Papers*, IEEE Press, pp. 279–324, 1993.

[15] V. Tarokh and H. Jafarkhani, "On Reducing the Peak to Average Power Ratio in Multicarrier Communications," *IEEE Transactions on Communication*, vol. 48, pp. 37–44, 2000.

[16] T. A. Wilkinson and A. E. Jones, "Minimization of the Peak to Mean Envelope Power Ratio of Multicarrier Transmission Schemes by Block Coding," *Proceedings of the IEEE 45th Vehicular Technology Conference*, pp. 825–829, Chicago, IL, July 1995.

IV

PHYSICS AND INFORMATION THEORY

Chapter 12

ON SYNCHRONIZATION AND INTERSYMBOL INTERFERENCE FOR JAMMED TIMING CHANNELS

J. Giles
T.J. Watson Research Center
Hawthorne, New York

B. Hajek
Coordinated Science Laboratory
University of Illinois at Urbana-Champaign

Abstract This paper focuses on two aspects of timing channels with intelligent delay jammers that touch on the work of G. David Forney, Jr. First, the delay inserted by a jammer can complicate the task of synchronization at the decoder. A solution to this problem is proposed for the class of maximum delay jammers. Second, delay jammers inherently offer channels with memory. It is shown that incorporation of the channel memory into the decoding process can be used to increase the data rate through a buffer constrained jammer.

1. INTRODUCTION

An often overlooked technique for communicating via a packet communication system is to encode data in the timing of the packets in addition to encoding data in the bits of the packets. While potentially quite noisy, this packet timing channel provides an additional data path, which is available since the packet network is not operating at capacity. In the sense that encoding information into the packet timing is not the

*This work was completed while Dr. Giles was at the University of Illinois. The research was supported by a Department of Defense NDSEG Graduate Fellowship and by the National Science Foundation under contract NSF ANR 99-80594.

usual way of communicating with a packet communication system, such use of the system can be considered covert.

For example, as shown in Figure 1, Bob can transmit packets containing innocuous messages to Alice while transmitting secret messages using packet timing. While the channel may be naturally noisy due to delays in the underlying packet channel, it still might be possible for Bob and Alice to communicate at a high data rate using timing. Even if the communication is monitored, the secret message may be overlooked.

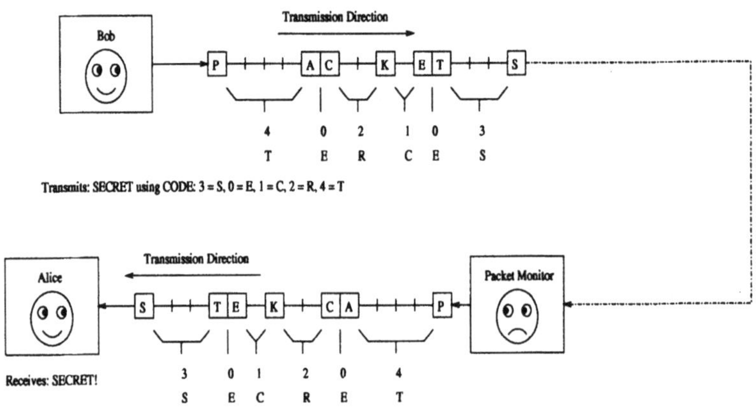

Figure 1 Motivating example

We define a covert channel to be any channel used for communication that is either not intended to be used for communication or that was intended to be used in a fundamentally different way. A *covert timing channel* is a covert channel that uses timing to convey information. See [11] for a collection of other definitions.

Covert timing channels can exist due to processor sharing on a single machine, shared message buffers, or standard network connections. In a typical covert channel scenario, a trusted user with access to sensitive data passes the data to a user who does not have access to the sensitive data by using a shared resource as a covert channel. The trusted user may be unaware that data is being passed if he is running a program containing a Trojan horse.

Taking the viewpoint of a security authority who expects to be able to monitor all communications on a given packet communications system, the existence of this covert packet timing channel may be undesirable. In particular, the complexity of monitoring all possible ways in which information can be conveyed through packet timing may make monitoring impractical. The security authority could be in the network at large (e.g. a government agency), or it could be associated with a security zone containing Bob.

Rather than trying to monitor all packet timing channels, one solution is to employ jamming devices which limit the usefulness of the packet timing channels, as shown in Figure 2. If the rate at which information can be conveyed using a packet timing channel in the presence of a jamming device is guaranteed to be acceptably low, the security authority will not need to monitor the packet timing channel. We would like to find an optimal jamming strategy and determine the maximum rate at which information can be conveyed using a packet timing channel in the presence of such a jammer.

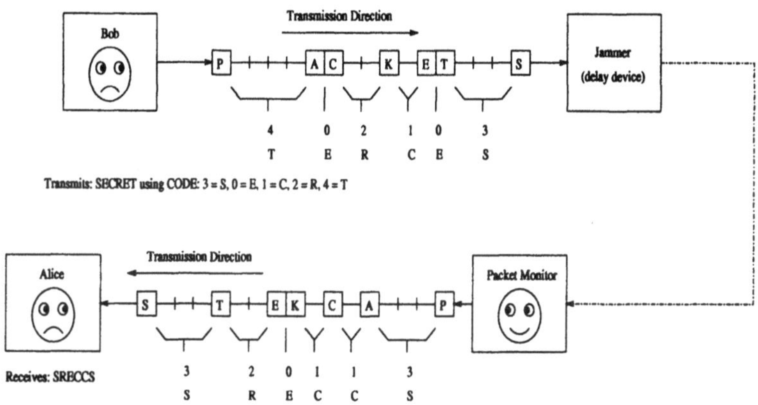

Figure 2 Motivating example with jammer

On the other hand, taking the viewpoint of a network user concerned with secrecy, the use of a packet timing channel to circumvent packet monitoring systems may be desirable. Of course, encryption is a straightforward way to attempt privacy, but encryption can draw attention and can be compromised with sufficient effort. The use of a packet timing channel, possibly with encryption, is not obvious to a casual observer and may be a way to communicate privately without drawing attention. The packet timing channel may have noise added by a jamming device and noise due to delay in the underlying packet communications system. Given constraints on the total noise in the packet timing channel, we would like to determine the minimum rate at which information can be conveyed using a packet timing channel in the presence of the noise and to find optimal coding strategies.

This paper highlights two portions of the authors' paper [8] which seem most related to the work and interests of G. David Forney, Jr. The status of covert communication via timing channels is far from that of the status of communication via highspeed modems over phone lines. Still, we can envision for jammed timing channels the use of many sophisticated tools such as generalized minimum distance decoding, geometrically uniform codes, Viterbi decoding, group codes, iterative methods based on factor graphs, channel sounding, trellis coded modulation, numerous methods for equalization, etc. In this paper, we consider two among many possible issues that arise in connection with covert timing channels. The first issue, raised by Dr. Forney as a question from the audience during the 1999 IEEE IT Workshop in Metsovo, Greece, is how synchronization can be achieved over jammed timing channels. The second issue, inspired by a landmark paper of Forney [7], is how to track the memory in the channel to reduce the effects of the jammer, which essentially causes intersymbol interference. The reader is referred to [8] for a summary of earlier work, study of a min-max information rate game, consideration of jammers with average delay constraints, and the study of related models in which packets are replaced by fluid inputs.

2. PROBLEM FORMULATION

We consider two waveform classes for covert packet timing channels: the continuous time packet waveforms and the discrete time packet waveforms. Input signals for each class are represented by cumulative packet arrival waveforms, as shown in Figure 3.

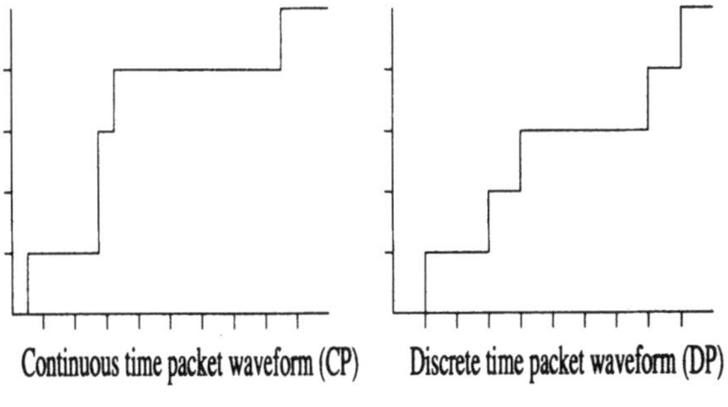

Figure 3 Typical waveforms

For continuous time packet waveforms, there are no lower bounds on the spacing between initiations of packet transmissions. In particular, such a waveform can have any positive integer-sized jump at any instant. The only restriction for continuous time packet waveforms is that packets cannot be split.

For discrete time packet waveforms, time is slotted and both the input and the jammer can transmit zero or one packet in each time slot.

Discrete time packet waveforms provide a tractable way to introduce a lower bound on the interpacket spacing. A packet that arrives at a jammer in a given time slot is available for departure in the same time slot (i.e., packets arrive at the beginning of time slots). As for continuous time packet waveforms, packets cannot be split.

Note that the waveforms do not include packet sequence numbers. This reflects our focus on packet timing rather than on bits within the packets. In particular, we do not permit the input to code covert information by using out-of-order sequence numbers, so the jammer has no incentive to re-order packets.

Ideally, we would like to say something about actual codes and jamming strategies. That is, for a particular class of jamming strategies, we would like to find codes of high rate, with arbitrarily small probability of decoding error when passed through a jammer. In addition, we would like to understand the worst case jamming strategies in a particular class. This view of the problem is called the *coding framework*.

An (M,T) code using a class $\mathcal{S}(T)$ of input waveforms and a class $\mathcal{Y}(T)$ of output waveforms is a pair of mappings (f,ϕ), where $f:\{1,\ldots,M\}\to\mathcal{S}(T)$, and $\phi:\mathcal{Y}(T)\to\{0,\ldots,M\}$ where 0 is included to denote an erasure. The rate of such a code is $\frac{1}{T}\log M$.

Let $\mathcal{W}$ represent a class of jammers and let $\lambda>0$. Say that a rate R is achievable by randomized codes for arrival rate λ codewords and a jammer class $\mathcal{W}$ if for any $\delta>0$ there exists a sequence of random (M,T) codes with $\frac{1}{T}\log M\geq R-\delta$ for T large enough, such that the maximal probability of decoding error over all jammers in the class and over all messages tends to zero as $T\to\infty$ and such that $P\left(\left|\frac{S_m(T)}{T}-\lambda\right|\leq\delta\right)=1$ for $1\leq m\leq M$. For a code with arrival rate λ, the capacity C is the supremum of such rates R.

We assume that a jammer can choose any causal delay strategy, including strategies that change packet ordering, subject to constraints on the delay. The jamming strategy can be deterministic, or it can be random with memory. However, waveforms, the jammer cannot delete packets or insert duplicates or additional packets since this might impact the underlying packet communication system. The jammer knows that the number of packets in each codeword is approximately the same, but in most cases the jammer does not know λ or the codewords. We use randomized code capacity, which implies that the coder and decoder have access to a source of common randomness, so that they can randomly select codes without the jammer's knowledge.

The constraints considered for jammers are a maximum delay constraint or a maximum buffer size constraint. A jammer is called a *maximum-delay-constrained* (MDC) jammer with delay parameter D if it delays no packet by D or more time units. A jammer is called a *maximum-buffer-constrained* (MBC) jammer with parameter B if it holds no more than B packets at any time in the continuous time case, or holds no more than B packets from one slot to the next slot in the discrete time case.

The coder and the decoder are aware of the particular delay constraints placed on the jammer, but are not aware of the actual strategy used by the jammer. In the case of a maximum delay constraint, for example, the coder and the decoder will know that there is a maximum delay constraint for the jammer and also know the value of the maximum delay. The coder does not receive feedback from the decoder. In practice, if the delay constraints are unknown to the coder and the decoder, they can make conservative assumptions. For example, in the case of a maximum delay constraint, the coder and the decoder can assume a particular maximum delay, and as long as the actual maximum delay is less than or equal to the assumed maximum delay, the communication will be reliable.

We comment briefly on the definition we have given for the capacity C. Our choice was guided by the theory of arbitrarily varying channels (AVC), initiated by Blackwell, Brieman, and Thomasian [3]. See [4] and [9] for recent surveys of the theory of AVC's, and [12] for a study of the related error exponents. Three aspects of the jammed timing channel do not fit the original formulation of AVC's: (i) The jammed timing channel has constraints, namely an arrival rate constraint on the coder and, in the case of average delay constraints, a delay constraint on the jammer, (ii) The jammed timing channel has memory, and (iii) The jammer can causally observe the particular input waveform used. Work of Csiszár and Narayan [5, 6] addresses point (i): the use of constraints for AVC's. The paper [5] in particular indicates that in defining the channel capacity C, it is more natural to impose constraints on each codeword and each received word, rather than on averages over codewords or over random codes. This suggests that our definition of channel capacity C is most appropriate. Work of Lapidoth and Teletar [10] addresses point (ii): the use of certain AVC's with memory. Unfortunately, there is no known extension of the AVC theory that addresses point (iii).

3. COMMUNICATION WITHOUT SYNCHRONIZATION

This section focuses on maximum delay constrained jammers using continuous time packet waveforms. A particular jamming strategy, called the *periodic dump strategy* with period D, works as follows. A sequence of dump times is fixed, with fixed spacing D between consecutive dump times. The jammer accumulates packets received between dump times and at each dump time releases all packets it has. Since this jammer is deterministic, the mutual information rate between the input and output is equal to the entropy rate of the output process. The information contained in the output process is all in the batch sizes. The mean batch size is λD. Therefore, the maximum output entropy rate would occur if the output batch sizes were mutually independent, having the geometric distribution with mean λD starting from zero. This distribution is denoted by $Geo_o(\lambda D)$. Setting $\mu = \lambda D$, the distribution is given by $p(x) = \frac{1}{1+\mu D}(1 - \frac{1}{1+\mu D})^x, x = 0, 1, 2, \ldots$ and it has entropy $H(Geo_o(\mu)) = (1 + \mu)\log(1 + \mu) - \mu \log(\mu)$. By standard arguments, this shows [8] that $C \leq H(Geo_o(\lambda D))/D$.

Next, we indicate an explicit coding strategy to show that $C = H(Geo_o(\lambda D))/D$. Moreover, we will show that the decoder does not need to have separate timing information. For many communication systems, a possible approach for achieving synchronization is to have the transmitter send some training signal. However, for many delay jammers, the jammer could delay the entire training segment by a fixed amount. The amount by which the jammer delayed the signal will not be learned at the receiver. The approach we suggest is analogous to noncoherent signaling, in that the codes work even though the receiver cannot always synchronize to the transmitter with high precision.

Consider a block timing code with codewords of duration $(N + 2)D$ time units, where each codeword is such that packets are transmitted only at times which are an integer multiple of D time units. Label the batch size transmitted at time jD for codeword i as X_j^i for $0 \leq j \leq N+1$. For each i, set $X_0^i = 1$ and $X_{N+1}^i = 0$ so that exactly one packet is always transmitted at time 0, and zero packets are transmitted at time $(N+1)D$. Take the set of codewords to be the set of all possible sequences of N positive integers, $(X_1^i, \ldots, X_N^i)$ such that $\sum_{j=1}^N X_j^i = \lceil \lambda D(N + 2) \rceil$ and $\sum_{j=1}^N jX_j^i = s$ for each i, where s is a fixed constant specified below, and $\sum_{j=1}^N jX_j$ is called the *delay sum* for the sequence $(X_1, \ldots, X_N)$.

We first show that for such codewords, error-free decisions can be made at the decoder, even without timing information. Next, we specify the delay sum s and show that there are enough codewords in this code to achieve the desired information rate, $H(Geo_o(\lambda D))/D$.

The decoder works as follows. Let time zero for the coder be the time that the first batch (of size 1) is transmitted and let time zero for the decoder be the time that the first packet arrives. To determine the

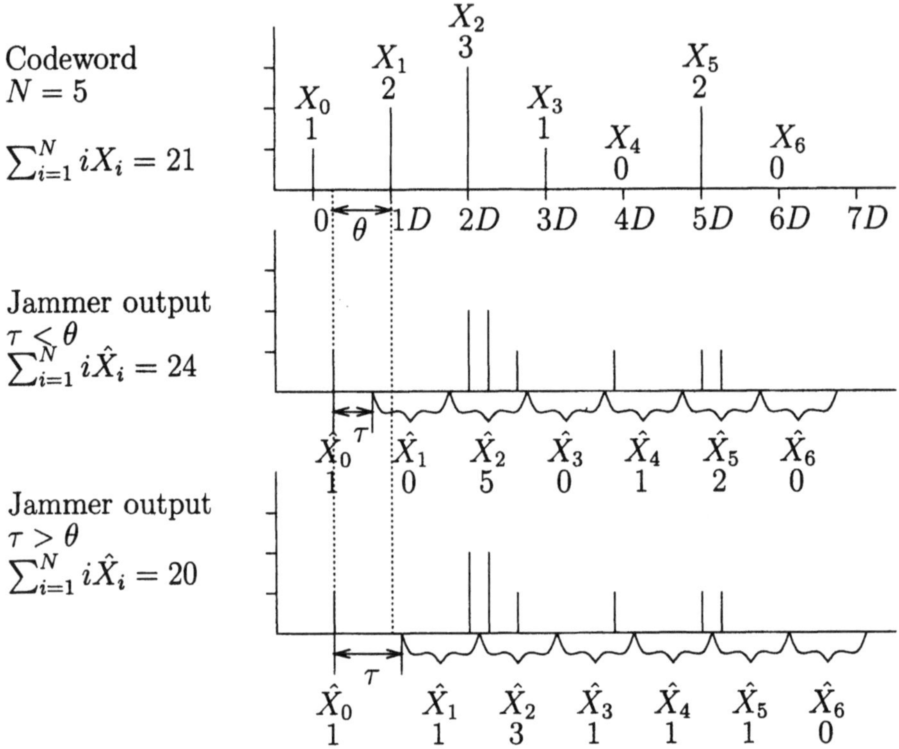

Figure 4 Coding and decoding: decoder does not know time that coder begins transmitting

transmitted batch sizes, $X_1, \ldots, X_N$, the decoder would like to know the offset, $\theta \in [0, D)$, from *its* time zero to *the coder's* time D as in Figure 4. By guessing an offset τ, the decoder can get an estimate of X_1, denoted $\hat{X}_1$, by counting packets in the interval $[\tau, \tau+D)$, $\hat{X}_2$ by counting packets in the interval $[\tau + D, \tau + 2D)$, and so on, as shown in Figure 4. The codeword in Figure 4 with $N = 5$ satisfies $\sum_{i=1}^{N} iX_i = s = 21$. Note that for $\tau = \theta$, the estimate of the codeword equals the codeword, and of course the estimate of the codeword has the same delay sum s as the codeword. As a function of τ, the delay sum is nonincreasing and its jumps mark changes in the estimated codeword. Thus, the decoder can correctly find the codeword by adjusting τ until $\sum_{j=1}^{N} j\hat{X}_j = s$. Therefore, a codeword can be transmitted without error every $(N+2)D$ time units.

Next we specify the delay sum s and show that there are enough codewords to achieve the saddle point information rate. Let $\mathcal{U}$ be the collection of sequences of positive integers, $(X_1^i \ldots, X_N^i)$, such that $\sum_{j=1}^N X_j^i = \lceil \lambda D(N+2) \rceil$ for each i. Then,

$$|\mathcal{U}| = \binom{\lceil \lambda D(N+2) \rceil + N - 1}{N} \tag{1}$$

$$\geq \frac{2^{(\lceil \lambda D(N+2) \rceil + N - 1) h\left(\frac{N}{\lceil \lambda D(N+2) \rceil + N - 1}\right) / \log 2}}{\lceil \lambda D(N+2) \rceil + N} \tag{2}$$

where $h(p) = -p \log p - (1-p)\log(1-p)$ and (2) is obtained from the inequality $\binom{n}{k} \geq \frac{1}{n+1} 2^{nh(k/n)/\log 2}$. For each $k \in \{0, 1, \ldots, N(\lceil \lambda D(N+2) \rceil)\}$, let

$$S_k = \left\{ (X^1, \ldots, X^N) : (X^1, \ldots, X^N) \in \mathcal{U} \text{ and } \sum_{j=1}^N j X^j = k \right\}.$$

We choose $s = \arg\max_k |S_k|$, so that s is the most common delay sum. Since exactly $\lceil \lambda D(N+2) \rceil$ packets are in each codeword, the largest possible delay sum is given by $N\lceil \lambda D(N+2) \rceil$, corresponding to all packets being transmitted in the last batch. Given any $\epsilon > 0$, for N large enough, the number of codewords having delay sum s is lower bounded by $\frac{|\mathcal{U}|}{N\lceil \lambda D(N+2) \rceil} \geq 2^{\frac{N}{\log 2}\left((\lambda D+1) h\left(\frac{N}{\lceil \lambda D(N+2) \rceil + N - 1}\right) - \epsilon\right)}$.

One such codeword can be transmitted without error every $(N+2)D$ time units, so the rate of the code is at least

$$\frac{N}{(N+2)D}\left((\lambda D + 1) h\left(\frac{N}{\lceil \lambda D(N+2) \rceil + N - 1}\right) - \epsilon\right) \xrightarrow{N \to \infty}$$
$$\frac{(\lambda D+1) h\left(\frac{1}{\lambda D+1}\right) - \epsilon}{D} = \frac{H\left(\mathrm{Geo}_o(\lambda D)\right) - \epsilon}{D}.$$

Since ϵ is arbitrary, for the MDC jamming channel with delay parameter D and continuous time packet waveforms, we have that $C = \frac{H(\mathrm{Geo}_o(\lambda D))}{D}$. The capacity C is shown in Figure 5.

4. TRACKING CHANNEL MEMORY

In this section, we consider a maximum-buffer-constrained (MBC) jammer with parameter B that can carry over at most B packets in its buffer. We assume that time is slotted and that at most one packet can enter the buffer and at most one packet can leave the buffer in a time slot. There can temporarily be up to $B+1$ packets in the buffer during

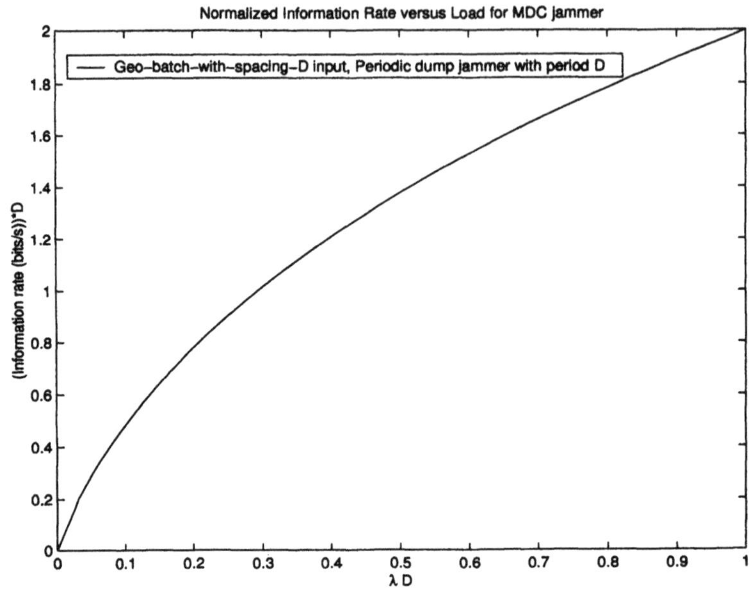

Figure 5 Channel capacity for the continuous time MDC channel

a time slot as long as no more than B packets are carried over to the next time slot.

To begin we obtain an upper bound on the channel capacity C for an MBC jammer by considering a specific jamming strategy, called the periodic quantized dump strategy. Both the period and the number of packets per quanta are taken to be $\lfloor \frac{B}{2} \rfloor + 1$. The jammer chooses dump slots $(\theta_1, \theta_2, \ldots)$ with $\theta_{i+1} = \lfloor \frac{B}{2} \rfloor + 1 + \theta_i$ for all $i \geq 1$. If the jammer has less than or equal to $\lfloor \frac{B}{2} \rfloor$ packets in its buffer at θ_i including the arrival in that slot, then the jammer holds all packets until θ_{i+1} and at most B packets will be in the jammer just before θ_{i+1}. If instead the jammer has more than $\lfloor \frac{B}{2} \rfloor$ packets in its buffer at θ_i, then the jammer will release one packet in each slot during the period from θ_i to just before θ_{i+1} and the total buffer size will not increase before θ_{i+1}. Hence this jamming strategy will never hold over more than B packets and it is an MBC jamming strategy. The jammer output is either 0 or $\lfloor \frac{B}{2} \rfloor + 1$ packets in a $\left(\lfloor \frac{B}{2} \rfloor + 1\right)$-interval, so the output can be considered a discrete binary process taking values 0 or $\lfloor \frac{B}{2} \rfloor + 1$ at each θ_i. The entropy rate for the output will be maximized if the outputs at the dump slots are independent. The mean number that should depart at each θ_i is $\lambda \left(\lfloor \frac{B}{2} \rfloor + 1\right)$ since the arrival rate is λ. Hence the output entropy rate

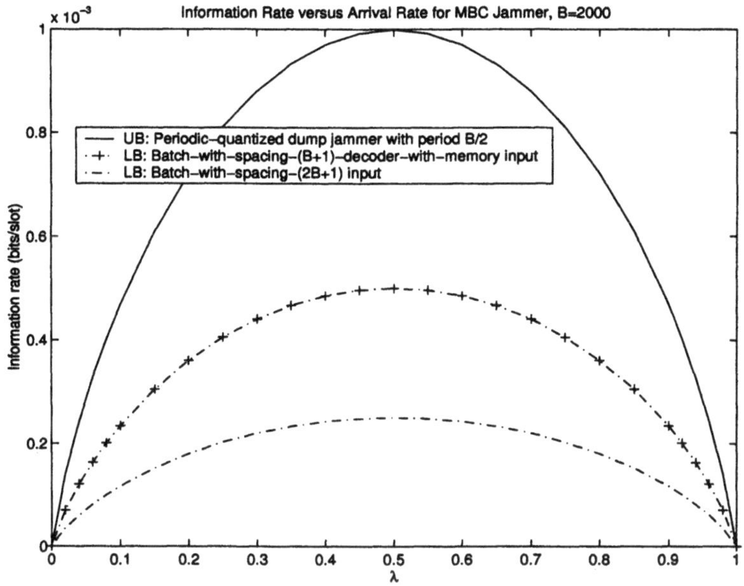

Figure 6 Bounds on information rate for MBC jammers

and therefore the capacity is upper bounded as $C \leq \frac{h(\lambda)}{\lfloor \frac{B}{2} \rfloor + 1} \leq \frac{2h(\lambda)}{B+1}$ which is illustrated in Figure 6 for $B = 2000$ packets and is labeled *periodic quantized dump jammer.*

A simple lower bound on the capacity C is obtained with the following batch-with-spacing-$2B+1$ input process. In each $(2B+1)$-interval, the input process will with probability $1 - \lambda$ transmit a binary zero by not sending any packets, and otherwise transmit a binary one by sending $2B+1$ packets, one slot at a time in contiguous slots starting with the dump slot. Since any MBC jammer can hold over at most B packets from slot to slot, the input is completely determined by the output, assuming the initial input transmission time is known at the output. In particular, if the number of packets at the output in each $(2B+1)$-interval is $B+1$ or more, then the input was a one, and if the number of packets is B or fewer, then the input was a zero. The entropy rate for this coding scheme is $\frac{h(\lambda)}{2B+1}$, which is in turn a lower bound on C for MBC jammers. This lower bound is within a factor of four of the upper bound described above and is illustrated in Figure 6 for $B = 2000$ packets and is labeled Batch-with-spacing-$(2B+1)$ input.

Taking inspiration from Forney [7], we can obtain a better lower bound on the capacity. The idea is to note that the jammer is a source of memory which can cause intersymbol interference, and therefore the decoder can improve its performance by tracking the channel state. A batch-with-spacing-$B+1$ coder similar to that for the previous input process is used, in which a binary one is transmitted by sending one packet in each of $B+1$ slots, and a binary zero is transmitted by remaining silent for $B+1$ slots. We will show that the input is still completely determined by the output. The decoder makes hard decisions at slots that are integer multiples of $B+1$. We define outstanding packets as any packets sent by the input process, but not yet seen at the output. At the beginning of slot 0, there are no outstanding packets. If the input sends a binary one, then all of the outstanding packets from the previous hard decision plus at least one additional packet will be observed at the output at a hard decision time since the buffer can only hold B packets and the input sends $B+1$ packets. If the input sends a zero, then no more than the number of outstanding packets will be observed at the output at a hard decision time. Thus, an error-free decision can be made about each symbol at the output. Based on that decision, the new number of outstanding packets can be determined. Since the decoder receives one error-free binary symbol every $B+1$ slots and the arrival rate is λ, we have that both $\underline{V}_{DP}$ and C are lower bounded by $\frac{h(\lambda)}{B+1}$. This lower bound differs from the upper bound by a factor of two and is illustrated in Figure 6 for $B = 2000$ and is labeled *Batch-with-spacing-(B+1)-decoder-with-memory input.*

To close this section, we explore another philosophy for jammers. The idea is for the jammer to try to make the output process as random as possible. For example, the jammer might like to output a stream of independent, mean λ Bernoulli variables, regardless of the input. Of course, this is not possible since the buffer can become full or empty. A reasonable fix would be for the jammer to adjust the output probability as a function of the buffer size. For example, consider a jammer that in any slot i outputs a packet with probability $\frac{k}{B+1}$, where k is the number of packets in the buffer at the beginning of the slot, including the new arrival, if any. However, this is apparently not a good jammer strategy, at least for B large. Indeed, this jammer is equivalent to the generalized billiard ball channel of T. Berger [1].

In Berger's model, the coder adds a red or white ball to a billiard table at the beginning of each slot, and the channel output is a ball selected from the table, with all balls having equal probability. The equivalence can be seen by mapping slots with packets to red balls, and idle slots to white balls. It can be shown [2] that for large B the capacity of the channel is at least as large as a constant times $B^{-\frac{2}{3}}$. For large B, this far exceeds the maximum capacity ($\Theta(1/B)$) described above. Roughly speaking, the generalized billiard ball channel can allow the coder to reliably convey information by causing small fluctuations in buffer size over short time intervals.

5. NUMERICAL EXAMPLE

In this section, we give a numerical example for discrete time packet waveform jammers to illustrate the effectiveness of the jamming and coding strategies. Consider a 100-Mbps link over which Bob is trying to communicate with Alice using packet timing (see Figure 7). For simplicity, we assume that the only traffic on the link is the traffic generated by Bob, and that discrete time packet waveforms are used so that time is slotted with at most one packet transmitted in each slot. In addition, we assume that any delay introduced beyond that introduced by a jammer is known to both Bob and Alice.

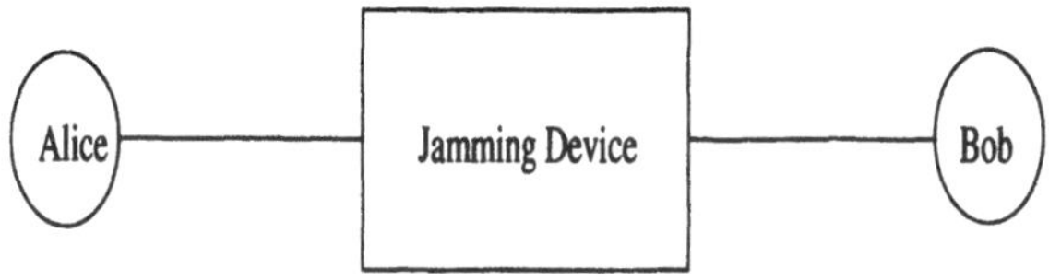

Figure 7 Numerical example for discrete time packet waveforms

We assume that Bob transmits packets with length 1,250 bytes (10,000 bits) and that the packets are not fragmented on the network. We choose a slot size of 10^{-4} seconds so that exactly one packet can be transmitted in each slot.

First, note that the capacity of this timing channel *without a jammer* is given by $h(\lambda)/10^{-4}$ where λ is the long-term average packet transmission rate corresponding to a scheme that transmits a binary one with probability λ by sending a packet in a slot and a binary zero with probability $1 - \lambda$ by remaining silent in the slot. For $\lambda = 1/2$ (one packet transmitted every two slots on average), such a coding strategy can successfully communicate data at a rate of 10,000 bps. For comparison, we will assume that on average Bob transmits packets at a rate of one packet every three slots ($\lambda = 1/3$) which corresponds to the base packet channel bit rate being about 33.3-Mbps.

We display in Table 1 upper and lower bounds on C for MDC and MBC jammers for discrete time packet waveforms. The bounds for MBC are from this paper, while the bounds for MDC can be found in [8], and are obtained for discrete time in a manner similar to that used in this paper for the continuous time model. We consider rate 1/3 input processes and MDC jammers with maximum delays 2 ms, 20 ms, 200 ms, and 2 s and MBC jammers with maximum buffer sizes 20, 200, 2,000, and 20,000

Note that the delays correspond to 20, 200, 2,000, and 20,000 slots since the slots are 10^{-4} seconds. The results show, for example, that for an MDC jammer with maximum delay 20 ms, jamming strategies exist for which no coding strategy can exceed 358 bps through the jammer. Additionally, for rate 1/3 coders and MDC jammers with maximum delay 20 ms, coding strategies exist for which at least 188 bps can be transmitted, regardless of the particular choice of MDC jamming strategy. Thus, the capacity is between 188 and 358 bps for MDC jammers with maximum delay 20 ms and rate 1/3 coders.

Table 1 Discrete time numerical example (100-Mbps link, 33.3-Mbps host stream bit rate, 10,000 bits/packet)

Jammers \ Delay (Buffer Size)	D=2 ms (B=20 packets)		D=20 ms (B=200 packets)		D=200 ms (B=2,000 packets)		D=2,000 ms (B=20,000 packets)	
	UB	LB	UB	LB	UB	LB	UB	LB
No Jammer	9183 bps							
Maximum-delay-constrained	1,985 bps	1,073 bps	358 bps	188 bps	52 bps	28 bps	6.9 bps	3.7 bps
Maximum-buffer-constrained	875 bps	437 bps	91 bps	45 bps	9.2 bps	4.5 bps	0.92 bps	0.45 bps

As can be seen from the table, for relatively small delay or small buffers, the maximum rate at which information can be transmitted using the timing channel in the presence of good jammers is reduced significantly.

References

[1] T. Berger, "Generalized Billiard Ball Channels," *Proceedings of the 1999 Information Theory Workshop*, Metsovo, Greece, June 1999.

[2] T. Berger and Z. Zhang (private communication), June 1999.

[3] D. Blackwell, L. Breiman, and A. J. Thomasian, "Proof of Shannon's Transmission Theorem for Finite-State Indecomposable Channels," *Annals of Mathematical Statistics*, vol. 31, pp. 558–867, 1958.

[4] I. Csiszár and J. Körner, *Information Theory*, Akadémiai Kiadó, Budapest, 1981.

[5] I. Csiszár and P. Narayan, "Arbitrarily Varying Channels with Constrained Inputs and States," *IEEE Transactions on Information Theory*, vol. 34, no. 1, pp. 27–34, January 1988.

[6] I. Csiszár and P. Narayan, "Capacity and Decoding Rules for Classes of Arbitrarily Varying Channels," *IEEE Transactions on Information Theory*, vol. 35, no. 4, pp. 752–769, July 1989.

[7] G. D. Forney, Jr., "Maximum-Likelihood Sequence Estimation of Digital Sequences in the Presence of Intersymbol Interference," *IEEE Transactions on Information Theory*, vol. IT-18, pp. 363–378, May 1972.

[8] J. Giles and B. Hajek, "An Information-Theoretic and Game-Theoretic Study of Timing Channels," Preprint, September 2000.

[9] A. Lapidoth and P. Narayan, "Reliable Communication Under Channel Uncertainty," *IEEE Transactions on Information Theory*, vol. 44, no. 6, pp. 2148–2177, October 1998.

[10] A. Lapidoth and I. E. Teletar, "The Compound Channel Capacity of a Class of Finite-State Channels," *IEEE Transactions on Information Theory*, vol. 44, pp. 973–983, May 1998.

[11] National Computer Security Center, "A Guide to Understanding Covert Channel Analysis of Trusted Systems," National Computer Security Center, Ft. George G. Meade, MD, Tech. Rep. NCSG-TG-030, November 1993.

[12] I. Stiglitz, "Coding for a Class of Unknown Channels," *IEEE Transactions on Information Theory*, vol. 12, no. 2, pp. 189–195, April 1966.

Chapter 13

DECODING ONLY THE STRONGEST CDMA USERS

S. Shamai (Shitz)
Department of Electrical Engineering, Technion
Israel Inst. of Technology, Haifa, 32000, ISRAEL

S. Verdú
Department of Electrical Engineering
Princeton University
Princeton NJ08544, USA

Dedicated to George David Forney, Jr.
a scholar, and a ripe and good one;
exceeding wise, fair-spoken, and persuading.

Abstract We consider the standard randomly spread CDMA channel where users are subject to independent flat fading. We analyze the outage probability of a joint receiver that decodes only the users that are received with the strongest power. All users are assumed to transmit at the same rate which is adjusted so that all users that are decoded by the receiver have vanishing block error probability. For an arbitrary number of users, we show that the capacity region achieved by this decoder is amenable to significant simplification. When the number of users grow without bound, we give asymptotic readily-computable outage formulas based on our previous application of the theory of asymptotic singular values of large matrices to CDMA systems.

1. INTRODUCTION

Ranging from the single-user receiver that neglects the presence of multiaccess interference to the maximum likelihood joint receiver, a wide variety of techniques have been proposed for demultiplexing digital streams that overlap in both time and frequency [1]. Linear and nonlinear receivers that take into account the presence of multiaccess interference can achieve substantial gains in spectral efficiency, as evidenced by the information theoretic analysis of the capacity of randomly spread CDMA [2, 3]. The analysis in [2, 3] obtains the maximum throughput (total rate) in a multiaccess channel where all active users are reliably decoded regardless of their received powers assuming that the transmitters can adjust their rates (and possibly also their powers) as a function of the channel fade levels they experience. An alternative strategy, motivated by practical applications, is to rule out even a modicum of feedback and have all users transmit at the same rate. Then, due to fading, the receiver can no longer guarantee with overwhelming probability that it will decode all active users. The appropriate performance measure in that scenario is the probability that a randomly selected user will not be successfully decoded, commonly referred to as the outage probability. References [4, 3] show how to compute the outage probability of linear CDMA receivers as a function of the rate, E_b/N_0, and fading distribution.

The receiver analyzed in this paper ranks all K active users by their received powers and then demodulates the J strongest users. J is chosen as the largest integer for which the demodulation is successful, a number that depends not only on the fade levels but on the noise realization. Thus, in general, the receiver may need to attempt to demodulate several population sizes until it ascertains the maximal decodable subset of users. However, since different users experience independent fades, as the number of users grows, the percentage of successfully decodable users converges to a deterministic constant. Thus, in many practical scenarios the receiver can be easily simplified to attempt the demodulation of a fixed percentage of users dictated by either complexity constraints or by a specified tolerable probability of joint decoding failure. Furthermore, if the rate is communicated to the users by the receiver, it can be adjusted so that all the decoded users are reliably decoded. If the fading evolves ergodically, then such feedback to the transmitters is not required. Underpinning the rationale for this receiver is the unsurprising property of randomly spread CDMA that for any $L = 1, \ldots K$, the probability of decoding failure for an arbitrary subset of L users is greater or equal than the probability of decoding failure for the strongest L users. While the $K - J$ weakest users are not decoded, their presence is not ignored by the decoder; Indeed, it takes into account their spreading sequences and received powers (but not the structure of their error control codes, which are assumed to have empirical Gaussian statistics in accordance with well-known multiaccess channel capacity principles). With that information, the decoder then makes an optimum decision about the J strongest users. Clearly, the capabilities of such a receiver serve as an upper bound to those of less complex receivers such as the group detectors of [6] where the effect of the unwanted users is mitigated by means of a linear front end.

Following the above referenced papers on the capacity of randomly spread CDMA, our analysis of the strongest-subset decoder focuses on the asymptotic limit of $K \to \infty$ and $N \to \infty$ with $K/N = \beta$, where N is the spreading gain and β is a design parameter.

The analysis of outage probability assumes that all users are subject to the same flat-fading distribution and that the fade levels are independent from user to user, stay constant for the duration of the transmitted codeword, and are perfectly known at the receiver.

Power ranking has played a longstanding role in multiuser detection. For example, in successive cancellation, it is common (although not leading to minimum uncoded bit error rate) to demodulate users in decreasing order of received powers [1]. Recently proposed receivers that involve sorting of powers include [8] and [5], the latter of which analyzes decoding only the strongest users in an unspread narrowband channel.

2. MODEL

Let K and N denote the number of users and the spreading gain, respectively. Following the standard notation [1, 3] for the synchronous nondispersive CDMA channel, the receiver obtains an N-dimensional observable

$$\mathbf{y} = \mathbf{S}\mathbf{A}\mathbf{x} + \mathbf{m} \tag{1}$$

where the kth column of the matrix $\mathbf{S}$ represents the unit-norm kth random signature sequence of the kth user, the diagonal matrix of received complex-valued fading coefficients is denoted by

$$\mathbf{A} = \mathsf{diag}\{\mathsf{A}_1, \ldots \mathsf{A}_K\}, \tag{2}$$

$$\mathbf{x} = [x_1, \ldots, x_K]^T \tag{3}$$

is the K-vector of information-carrying channel symbols, and $\mathbf{m}$ is a complex Gaussian vector all whose real and imaginary components have variance 1/2 and are independent.

All transmitted codewords of blocklength n are subject to the average power constraint:

$$\frac{1}{n}\sum_{i=1}^{n} |x_k[i]|^2 \leq \mathsf{SNR}. \tag{4}$$

The number of bits transmitted in each symbol $x_k[i]$ is common to all users. We denote the number of bits per second per Hertz transmitted by each user by R. If the codebook has M messages, then

$$R = \frac{\log_2 M}{nN}. \tag{5}$$

3. ORDERED DISTRIBUTION OF FADING POWERS

Let i_k denote the index of the kth strongest user at the receiver, i.e.,

$$|\mathsf{A}_{i_1}|^2 \geq |\mathsf{A}_{i_2}|^2 \geq \ldots \geq |\mathsf{A}_{i_K}|^2. \tag{6}$$

Although the definition of i_k is nonunique, it is immaterial how ties are broken. The (power) fade level experienced by the kth strongest user is denoted by

$$\nu_k = |\mathsf{A}_{i_k}|^2. \tag{7}$$

The set of the J strongest users is denoted by

$$\mathcal{D}_J = \{i_1, \ldots, i_J\}. \tag{8}$$

Furthermore, we use overbar/underbar to denote the subset of strong/weak users as in

$$\bar{\mathbf{X}}_J = \{X_{i_1}, \ldots, X_{i_J}\}, \tag{9}$$
$$\underline{\mathbf{X}}_J = \{X_{i_{J+1}}, \ldots, X_{i_K}\}, \tag{10}$$
$$\bar{\mathbf{A}}_J = \mathsf{diag}\{\mathsf{A}_{i_1}, \ldots, \mathsf{A}_{i_J}\}, \tag{11}$$
$$\underline{\mathbf{A}}_J = \mathsf{diag}\{\mathsf{A}_{i_{J+1}}, \ldots, \mathsf{A}_{i_K}\}. \tag{12}$$

Analogously, $\bar{\mathbf{S}}_J$ denotes the $N \times J$ matrix of signatures of the J strongest users and $\underline{\mathbf{S}}_J$ denotes the $N \times (K-J)$ matrix of signatures of the remaining users.

Let $F(x)$ designate the cumulative probability distribution function (cdf) of the fading power $|\mathsf{A}_k|^2$. Then, the distribution of ν_k is given by (e.g. [1, p.388])

$$F_{\nu_k}(x) = \int_0^x \frac{K!}{(K-k)!(k-1)!}\left(1 - F(z)\right)^{k-1} F(z)^{K-k}\, dF(z) \tag{13}$$

which for $K \to \infty$ yields [5]

$$\lim_{K\to\infty} F_{\nu_{k(K)}}(x) - \left\{ \begin{array}{ll} 0 \;, & 1 - F(x) > \frac{k(K)-1}{K-1} \\ 1 \;, & 1 - F(x) \leq \frac{k(K)-1}{K-1} \end{array} \right\} = 0, \tag{14}$$

where we have explicitly denoted the fact that we are interested in a strongest-user index $k(K)$ which depends on the total number of users. The asymptotic expectation of the kth strongest power is

$$\begin{aligned} E(\nu_{k(K)}) &= \int_0^\infty \left(1 - F_{\nu_{k(K)}}(x)\right) dx & (15)\\ &= \int_0^\infty 1\left\{1 - F(x) > \frac{k(K)-1}{K-1}\right\} dx + o(1) & (16)\\ &= F^{-1}\left(1 - \frac{k(K)-1}{K-1}\right) + o(1)\,, & (17)\end{aligned}$$

where $1\{\cdot\}$ designates the indicator function. Regarding (17) note that for a continuous and differentiable function $g(x)$

$$\begin{aligned} \frac{1}{K}\sum_{k=w_LK}^{k=w_HK} g(\nu_k) &\stackrel{K\to\infty}{\longrightarrow} \int_{w_L}^{w_H} g\left(F^{-1}(1-x)\right) dx \\ &= \int_{F^{-1}(1-w_H)}^{F^{-1}(1-w_L)} g(y)\, dF(y)\,. & (18)\end{aligned}$$

4. NONASYMPTOTIC CAPACITY REGION

In principle, the fact that some users are decoded whereas the code structure of some users is not exploited could lead to analytical difficulties. However, it is possible to sidestep those difficulties crisply by using Kolmogorov's identity. The mutual information between the channel output and the information transmitted by the users to be demodulated can be decomposed as

$$\begin{aligned} I(\bar{\mathbf{X}}_J;\mathbf{Y}) &= I(\bar{\mathbf{X}}_J,\underline{\mathbf{X}}_J;\mathbf{Y}) - I(\underline{\mathbf{X}}_J;\mathbf{Y}|\bar{\mathbf{X}}_J) & (19)\\ &= I(\mathbf{X};\mathbf{Y}) - I(\underline{\mathbf{X}}_J;\mathbf{Y}|\bar{\mathbf{X}}_J) & (20)\end{aligned}$$

where the input and output random vectors in (1) are denoted by the upper case symbols $\mathbf{X}$ and $\mathbf{Y}$. Both mutual informations appearing in (20) find closed form expressions. It was shown in [9] that

$$I(\mathbf{X};\mathbf{Y}) = \log\det\left(\mathbf{I} + \mathsf{SNR}\mathbf{A}^*\mathbf{S}^*\mathbf{S}\mathbf{A}\right). \quad (21)$$

Analogously, since the information transmitted by different users is independent

$$I(\underline{\mathbf{X}}_J;\mathbf{Y}|\bar{\mathbf{X}}_J) = \log\det\left(\mathbf{I} + \ \mathsf{SNR}\ \underline{\mathbf{A}}_{\mathrm{J}}^*\underline{\mathbf{S}}_{\mathrm{J}}^*\underline{\mathbf{S}}_{\mathrm{J}}\underline{\mathbf{A}}_{\mathrm{J}}\right). \quad (22)$$

It should be emphasized that the fact that the signature sequences are known also for the weak $K - J$ users that are not decoded is fully exploited in (20)-(22). Unlike [5] where no spreading is considered, it would be suboptimal to have a front end of J matched filters which neglects the $K - J$ weak users.

Conditioned on the channel fades and signature waveforms, the capacity region determining the rates achievable by the J strongest users is given by

$$\mathbf{C} = \bigcap_{\mathcal{D} \subset \mathcal{D}_J} \left\{ \sum_{i \in \mathcal{D}} R_i \leq I(\bar{\mathbf{X}}_J; \mathbf{Y} \mid \bar{\mathbf{X}}_{J-\mathcal{D}}) \right\} \tag{23}$$

As we see in the following result, due to the particular structure of our problem, J of the rate constraint equations among the $2^J - 1$ equations in (23) are sufficient in order to describe the capacity region.

Theorem 1 *The capacity region achievable by the J strongest users is equal to (bits/sec)*

$$\mathbf{C} = \bigcap_{j=1}^{J} \left\{ \sum_{\ell=j}^{J} R_{i_\ell} \leq I(\underline{\mathbf{X}}_{j-1}; \mathbf{Y} \mid \bar{\mathbf{X}}_{j-1}) - I(\underline{\mathbf{X}}_J; \mathbf{Y} \mid \bar{\mathbf{X}}_J) \right\} \tag{24}$$

Proof.

First we need to show that J equations are sufficient:

$$\mathbf{C} = \bigcap_{j=1}^{J} \left\{ \sum_{\ell=j}^{J} R_{i_\ell} \leq I(X_{i_j}, \ldots, X_{i_J}; \mathbf{Y} \mid \{X_{i_1}, \ldots X_{i_{j-1}}\}). \right\} \tag{25}$$

To that end, proceeding as in (20), note that all the mutual informations arising in (23) take the form

$$\begin{aligned} I(\bar{\mathbf{X}}_J; \mathbf{Y} \mid \mathbf{X}_{\mathcal{D}_J - \mathcal{D}}) &= I(\bar{\mathbf{X}}_{\mathcal{D}}, \underline{\mathbf{X}}_J; \mathbf{Y} \mid \bar{\mathbf{X}}_{J-\mathcal{D}}) \\ &- I(\underline{\mathbf{X}}_J; \mathbf{Y} \mid \bar{\mathbf{X}}_J) \qquad (26) \\ &= \log\det(\mathbf{I} + \mathsf{SNR}\, \underline{\mathbf{A}}_{\mathcal{B}}^* \underline{\mathbf{S}}_{\mathcal{B}}^* \mathbf{S}_{\mathcal{B}} \mathbf{A}_{\mathcal{B}}) \\ &- I(\underline{\mathbf{X}}_J; \mathbf{Y} \mid \bar{\mathbf{X}}_J) \qquad (27) \end{aligned}$$

where $\underline{\mathbf{A}}_{\mathcal{D}}$ and $\underline{\mathbf{S}}_{\mathcal{D}}$ denote the fade coefficients and signature matrices of the users in the subset

$$\mathcal{B} = \mathcal{D} \cup \mathcal{D}_J^c = (\mathcal{D}_J - \mathcal{D})^c. \tag{28}$$

The right side of (28) is always lower bounded by the choice of $\mathcal{D}$ corresponding to the l-weakest users among the J strongest users, i.e. $J - l + 1, \ldots, J - 1, J$. Thus, the corresponding equation dominates.

Note that the interfering part $I(\underline{\mathbf{X}}_J; \mathbf{Y} \mid \bar{\mathbf{X}}_J)$ is fixed for fixed J and thus is not affected by the specific subset of users.

Finally, generalizing the decomposition in (20) we can write every mutual information appearing in (25) as

$$\begin{aligned} I(X_{i_k}, \ldots, X_{i_J}; \mathbf{Y} \mid \{X_{i_1}, \ldots X_{i_{k-1}}\}) &= I(X_{i_k}, \ldots, X_{i_J}, \underline{\mathbf{X}}_J; \mathbf{Y} \mid \{X_{i_1}, \ldots X_{i_{k-1}}\}) \\ &- I(\underline{\mathbf{X}}_J; \mathbf{Y} \mid \{X_{i_1}, \ldots X_{i_{k-1}}\}, \{X_{i_k}, \ldots, X_{i_J}\}) \\ &= I(\underline{\mathbf{X}}_{k-1}; \mathbf{Y} \mid \bar{\mathbf{X}}_{k-1}) - I(\underline{\mathbf{X}}_J; \mathbf{Y} \mid \bar{\mathbf{X}}_J) \quad (29) \end{aligned}$$

for $k = 1, \ldots J$.

Naturally, the explicit expression in (22) not only can be used to compute the second mutual information in the rate constraints of (24) but also the first, since they both can be viewed as coming from a channel where there are no users other than the (j or J) weakest ones.

5. ASYMPTOTIC CAPACITY

From (22) and (24), we need to focus on the behavior of

$$\log\det(\mathbf{I} + \mathsf{SNR}\ \underline{\mathbf{A}}_L^* \underline{\mathbf{S}}_L^* \underline{\mathbf{S}}_L \underline{\mathbf{A}}_L),$$

for values of L ranging from 1 to J. Using results on the asymptotic behavior of the singular values of random matrices, the main result in [3] can be used to show that as $K \to \infty$, $N \to \infty$, $K/N = \beta$, $L/K = \delta$ (where $0 \le \delta \le 1$ and $\beta > 0$ are arbitrary parameters) we obtain the asymptotic behavior:

$$\begin{aligned} \frac{1}{N} \log\det(\mathbf{I} + \mathsf{SNR}\ \underline{\mathbf{A}}_L^* \underline{\mathbf{S}}_L^* \underline{\mathbf{S}}_L \underline{\mathbf{A}}_L) &= -\log \eta_K^L + (\eta_K^L - 1)\log e + o(1) \\ &\quad \frac{1}{N} \sum\nolimits_{m=L+1}^{K} \log(1 + \mathsf{SNR}\, \nu_m\, \eta_K^L), \quad (30) \end{aligned}$$

where η_K^L solves the equation

$$\frac{1}{\eta_K^L} = 1 + \frac{1}{N} \sum\nolimits_{m=L+1}^{K} \frac{\nu_m}{\frac{1}{\mathsf{SNR}} + \eta_K^L \nu_m}. \tag{31}$$

Using (18), we can see that

$$\lim_{K \to \infty} \eta_K^{\delta K} = \eta_\delta \tag{32}$$

where η_δ is the solution to

$$1 - \beta(1 - \delta) = \eta - \beta\Theta(\delta;\ \mathsf{SNR}\, \eta), \tag{33}$$

and

$$\Theta(w; a) = \int_{F^{-1}(0)}^{F^{-1}(1-w)} \frac{1}{1 + a\phi} dF(\phi) \tag{34}$$

Analogously, in order to deal with the summation in (30), we define

$$\Gamma(w;a) = \int_{F^{-1}(0)}^{F^{-1}(1-w)} \log(1+a\phi)\, dF(\phi) \tag{35}$$

and

$$\mathsf{C}(\delta) = \beta\,\Gamma(\delta;\mathsf{SNR}\eta_\delta) - \log\eta_\delta + (\eta_\delta - 1)\log e. \tag{36}$$

It follows from the above definitions that (30) converges to

$$\lim_{K\to\infty} \frac{1}{N} \log\det\left(\mathbf{I} + \mathsf{SNR}\ \underline{\mathbf{A}}_L^* \underline{\mathbf{S}}_L^* \underline{\mathbf{S}}_L \underline{\mathbf{A}}_L\right) = \mathsf{C}(\delta). \tag{37}$$

It is convenient to realize that the functions defined in (34)-(35) are related by

$$\Theta(w;a) = \frac{-a\partial\Gamma(w;a)}{\partial a} + \frac{1}{a}\left(F^{-1}(1-w) - F^{-1}(0)\right). \tag{38}$$

Now we proceed to plug the asymptotic expressions in the capacity region (24). Because of the equal-rate assumption, and recalling (5), the capacity region becomes

$$\mathbf{C} = \bigcap_{j=1}^{J} \left\{ (J-j+1)R \le \frac{1}{N} I(\underline{\mathbf{X}}_j;\mathbf{Y} \mid \bar{\mathbf{X}}_j) - \frac{1}{N} I(\underline{\mathbf{X}}_J;\mathbf{Y} \mid \bar{\mathbf{X}}_J) \right\} \tag{39}$$

$$= \bigcap_{j=1}^{J} \left\{ (J-j+1)R \le \mathsf{C}\left(\frac{j-1}{K}\right) - \mathsf{C}\left(\frac{J}{K}\right) \right\}. \tag{40}$$

We are now ready to state the main result. Letting $1-q$ and $1-x$ take the asymptotic role of J/K and j/J respectively, leads to the following result on the total achievable rate RJ in the notation of (40):

Theorem 2 *Let the tolerable outage probability be q, then the total achievable rate (bits per second per Hz) is*

$$R_T(q) = \inf_{0<x\le 1} \frac{\mathsf{C}\left((1-q)(1-x)\right) - \mathsf{C}(1-q)}{x} \tag{41}$$

where the function C *is defined in (36).*

Note that in the limit with ordered received power the fraction $\frac{J}{K}$ of reliably decoded user hardens and is not a random variable anymore (see [5]). It is only the identity of the J operating users, which remains absolutely random.

The results can also be cast in form of an outage of an arbitrary user signaling at either zero rate (if it is not strong enough) or at a rate r times the expected throughput [3], that is:

$$R = \frac{r}{K} \mathcal{C}^{\text{opt}} (\beta, \ \mathsf{SNR}) \,, \tag{42}$$

where $\mathcal{C}^{\text{opt}} (\beta, \ \mathsf{SNR})$ is as in [3]. Thus $R_T(q) = RJ$ (41) should satisfy

$$R_T(q) = (1-q) r \, \mathcal{C}^{\text{opt}} (\beta, \ \mathsf{SNR}) \,. \tag{43}$$

The outage as a function of the parameter r can be simply obtained by solving for q in (43).

6. RAYLEIGH FADING

For Rayleigh fading,

$$F(\nu) = 1 - e^{-\nu} \,, \ F^{-1}(x) = -\ln(1-x) \,. \tag{44}$$

Using the exponential integral function

$$- - \mathcal{E}_i(-x) = \int_x^\infty \frac{e^{-t}}{t} \, dt \,, \qquad x \geq 0 \tag{45}$$

we can write the key functions defined in (34) and (35) explicitly:

$$\begin{aligned}
\Gamma(w;a) &= \int_0^{-\ln w} \log(1+a\phi)\, e^{-\phi} \, d\phi \\
&= -w \log(1 - a \ln w) + w e^{\left(\frac{1}{a} - \ln w\right)} \mathcal{E}_i\left(-\left(\frac{1}{a} - \ln w\right)\right) \log e \\
&\quad -e^{\left(\frac{1}{a}\right)} \mathcal{E}_i\left(-\frac{1}{a}\right) \log e = -w \log\left(1 - a \ln w\right) \\
&\quad +e^{\left(\frac{1}{a}\right)} \left[\mathcal{E}_i\left(-\left(\frac{1}{a} - \ln w\right)\right) - \mathcal{E}_i\left(-\frac{1}{a}\right)\right] \log e \,.
\end{aligned} \tag{46}$$

$$\begin{aligned}
\Theta(w;a) &= \int_0^{-\ln w} \frac{1}{1+a\phi} e^{-\phi} \, d\theta \\
&= \frac{e^{\left(\frac{1}{a}\right)}}{a} \left\{\mathcal{E}_i\left(-\left(\frac{1}{a} - \ln w\right)\right) - \mathcal{E}_i\left(-\frac{1}{a}\right)\right\} \,.
\end{aligned} \tag{47}$$

7. DISCUSSION

In this analysis, we approached the outage issue of a joint detector by first ranking the received powers, following the approach in [5]. At the limit of large systems, the randomness manifests itself only through the actual identity of the successfully operating users. The fraction of successfully decoded user actually converges to a deterministic constant. Hence, the detector does not have to seek what the surrounding of the set of the reliably decoded users is, as the proportion of successfully decoded users becomes deterministic. Yet, the variance can be an important factor as $J \to \infty$, so the receiver needs to carry on with the elimination process at the vicinity of $J = y \cdot K$. This view can also be adopted for all the linear receivers considered in [3]. See also [4]. Clearly, the results of [3] will follow but now with the interpretation of $y = J/K$ designating the availability probability. (See [5] for a matched-filter, single-user receiver in case of no spreading.)

We have assumed throughout that the fading channel state information is ideally provided to the receiver. Clearly, in the case of no dynamics this can be relaxed as this information could in principle be estimated with arbitrary accuracy.

Following the conclusions in [5, 10] we expect a dramatic effect in terms of rate-outage performance of this joint detector, even when compared to say DFE-MMSE (cancellation based) detectors capable of achieving the same overall throughput in a variety of models [3], [5], [2].

It is questionable whether this detector is optimal in a more general setting where codes are designed to account for the fact that some of the users are not supposed to be decoded (a given outage design goal), which resembles a broadcast setting [7]. Regardless of whether random spreading is used, that more general problem can be cast within the class of problems addressing combined multiaccess broadcast and interference channels—a class of channels whose solution remains elusive.

8. ACKNOWLEDGEMENTS

This work was supported by the US-Israel Binational Science Foundation. We are grateful to Mr. Benny Zaidel for a careful reading of the manuscript.
Dedication after King Henry VIII, Act iv. Sc. 2. by William Shakespeare.

References

[1] S. Verdú, *Multiuser Detection,* Cambridge University Press, 1998.

[2] S. Verdú and S. Shamai (Shitz), "Spectral Efficiency of CDMA with Random Spreading," *IEEE Transactions on Information Theory,* Vol. 45, No. 2, pp. 622–640, March 1999.

[3] S. Shamai (Shitz) and S. Verdú, "The Impact of Frequency-Flat Fading on the Spectral Efficiency of CDMA," *IEEE Transactions on Information Theory,* Vol.47, pp. 1302-1327, May 2001.

[4] E. Biglieri, G. Caire, G. Taricco and E. Viterbo, "How Fading Affects CDMA: An Asymptotic Analysis with Linear Receivers," *IEEE Journal on Selected Areas in Communications,* pp. 191-201, February 2001.

[5] S. Shamai (Shitz) and I. Bettesh, "Outages, Expected Rates and Delays in Multiple-User Fading Channels," *2000 Conference on Information Sciences and Systems (CISS'2000),* pp. WA4.7–WA4.15, 15–17, March 2000, Princeton University, USA.

[6] M. Varanasi, Group Detection for Synchronous Gaussian CDMA Channels. *IEEE Transactions on Information Theory,* 41:1083–1096, July 1995.

[7] S. Shamai, "A Broadcast Approach for the Multiple-Access Slow Fading Channel," *Proceedings of the 2000 IEEE International Symposium on Information Theory,* p. 128, Sorrento, Italy, 25–30, June 2000.

[8] B. R. Vojcic, A. AlRustamani, and A. D. Damnjanovic, "Greedy Iterative Multiuser Detection for Turbo Coded Multiuser Communications," *International Conference on Telecommunications, ICT2000,* King's College, London, May 2000.

[9] S. Verdú, Capacity Region of Gaussian CDMA channels: The Symbol-Synchronous Case. *Proceedings 24th Allerton Conference on Communications, Control and Computing,* pages 1025–1034, Oct. 1986.

[10] S. Shamai (Shitz), S. Verdú and B.M. Zaidel, "Information Theoretic Aspects of "Strongest-User" Only Detectors for Randomly Spread CDMA", Technical Report (CCIT 354), Technion (ITT), *To appear.*

Chapter 14

POWER LIMITED CHANNELS: CODING, MULTIACCESS, AND SPREAD SPECTRUM

R. G. Gallager
Laboratory for Information and Decision Systems
Massachusetts Institute of Technology

Abstract

This is a slightly edited version of a paper written, but not published, in the mid-eighties. Its age is symbolic of my much older friendship and respect for Dave Forney, and it is dedicated to that friendship. It's age should not be taken as any claim for precedence of the results, but merely to excuse the lack of up-to-date references.

We consider a class of communication channels modelling situations in which there is an energy constraint but almost an unlimited number of degrees of freedom per unit time. Examples of such channels are broadband additive Gaussian noise channels, fading dispersive channels, and quantum optical channels. We start by restricting such channels to binary inputs and then find the reliability function. In the limit of infinite bandwidth, the reliability function for these channels can be found exactly for all rates if there is a finite capacity in terms of bits per unit energy. Particular attention is paid to how this limiting reliability is approached as bandwidth increases. We then show that the restriction to binary inputs is essentially optimal in the broadband limit. Finally we apply these results to multiaccess use of such channels. Multiaccess coding for these broadband channels provides us with an abstraction of spread spectrum communication.

1. INTRODUCTION

There are many communication systems in which the primary constraint on the transmitted sequences arises from power limitations rather than bandwidth limitations. These systems have been widely studied for additive Gaussian noise channels and to a lesser extent for fading dispersive channels and optical fiber channels. There is no cohesive treatment of the class of such channels, and this paper is an initial effort at such a treatment. For multiaccess channels, in which a large number of transmitters share the same physical medium, it is usually desirable to limit the power of each transmitter so as to limit interference. In this case again, there is usually more bandwidth than any one transmitter could use effectively, and the primary limit is from the imposed power constraint. For example, in local area networks of the Ethernet type, transmitters use high energy while transmitting, but transmit only a small fraction of time, thus leading to small overall power. In spread spectrum communication, transmitters spread their power over a broad bandwidth, thus resulting in low power per unit bandwidth and not too much interference with other transmitters.

The above situations can be modelled by a discrete time memoryless channel in which the transmitted and received discrete time symbols can be regarded as inputs and outputs over incremental intervals of time and frequency. The memoryless assumption is not very good for fading channels, but becomes reasonable if the time-frequency intervals of interest include enough degrees of freedom. Initially, we consider a single transmitter and receiver and we restrict the discrete time input to be binary, the symbol one representing a nonzero energy input, and the symbol zero representing a zero energy input. Consider block codes of blocklength N with M codewords; given a constraint level δ, each codeword is constrained to contain at most $D = \lfloor N\delta \rfloor$ nonzero symbols (where $\lfloor x \rfloor$ denotes the integer part of x). The rate R of such a code, in nats per letter, is $R = (\ln M)/N$. Let $P_e(N, R, \delta)$ be the minimum probability of decoding error achievable with codes of blocklength N, constraint δ, and at least $\exp(NR)$ codewords. Define the corresponding reliability function as

$$E(R, \delta) = \limsup_{N \to \infty} \frac{-\ln P_e(N, R, \delta)}{N} \tag{1}$$

We shall also be interested in the rate per energy symbol, $\widetilde{R} = R/\delta$ and in the normalized reliability function $\widetilde{E}(\widetilde{R}, \delta) = E(\widetilde{R}\delta, \delta)/\delta = \limsup_{N\to\infty} \frac{-\ln P_e(N, \widetilde{R}\delta, \delta)}{N\delta}$. We will show that the limit $\widetilde{E}(\widetilde{R}) = \lim_{\delta \to 0} \widetilde{E}(\widetilde{R}, \delta)$ exists and show that the limit is given by

$$\widetilde{E}(\widetilde{R}) = \max_{0 \le \rho \le 1} [\widetilde{E}_0(\rho) - \rho \widetilde{R}] \quad \text{where} \tag{2}$$

$$\widetilde{E}_0(\rho) = -(1+\rho) \ln \int_y p_0(y)^{\rho/(1+\rho)} p_1(y)^{1/(1+\rho)} dy \tag{3}$$

Here, $p_0(y)$ and $p_1(y)$ are the transition probability densities defining the discrete time channel. This result is similar to the well known results for reliability functions for discrete memoryless channels (DMC) except for the following two differences. First, the reliability function here can be calculated exactly for all rates, whereas for the DMC it can only be evaluated exactly for rates greater than the critical rate [1]. Second, the capacity here (in nats per energy symbol) can be infinite and in fact is infinite whenever there is a range of outputs that can be reached from input symbol one but not zero.

We next consider a more general discrete time, point-to-point channel with an input alphabet $\{0, 1, \ldots, K\}$. The zero input again corresponds to zero energy, and there is some arbitrary energy associated with each of the other input letters. We again impose a constraint on the aggregate energy per codeword. A reliability function can be defined in the limit as the allowable energy per degree of freedom approaches zero. For all rates between the critical rate and capacity, the optimal input distribution for this limiting reliability function turns out to be binary, using the zero energy symbol as one of the binary symbols.

Except for some pathological cases, the optimal input distribution is similarly binary for all rates below the critical rate, and thus the reliability function can be calculated exactly for all rates except in these pathological cases. Note, however, that the nonzero energy symbol that is used in the optimal binary distribution can change with rate. This provides some guidance into the amount of energy that should be used in those time-frequency intervals that are used with nonzero energy.

Finally, we consider multiaccess channels with the same type of energy constraint on each transmitter's input. We derive a converse to the multiaccess coding theorem for channels with such an energy constraint. This is not an obvious extension of the converse without an energy constraint, and apparently this is new. We then restrict attention to the binary input case and model the channel as an 'or' channel followed by an arbitrary binary input memoryless point-to-point channel. The achievable rate region can then be evaluated as a function of the energy constraint. We also derive a random coding bound for this class of multiaccess channel for three different cases, one in which the codewords from the different transmitters are independently decoded, one in which the codewords are decoded in sequence, using the earlier decoded words to help in decoding, and one in which the decoding of all codewords is simultaneous.

2. BINARY INPUT CHANNELS

Consider a discrete time memoryless communication channel with binary inputs, $x = \{0, 1\}$. For simplicity of notation, assume the outputs are real numbers with transition probability densities $p_0(y)$ and $p_1(y)$. We consider a block code of constraint length N with M codewords, $\vec{x}_1, \vec{x}_2, \ldots, \vec{x}_M$. We consider the input 0 to be a zero energy input and the input one to be a positive energy input; the energy constraint that we impose is then

$$\sum_{n=1}^{N} x_{mn} \leq N\delta \tag{4}$$

for each codeword $\vec{x}_m = (x_{m1}, x_{m2}, \ldots, x_{mN})$. For a given code, we define the probability of error as the average, over all codewords, of the probability of incorrect decoding using maximum-likelihood decoding. Let $P_e(N, R, \delta)$ be the minimum probability of error over all block codes of blocklength N with at least $\exp[NR]$ codewords, each satisfying the energy constraint in Equation (4). The reliability function $E(R, \delta)$ is then defined in Equation (1).

The coding theorem asserts that $E(R, \delta)$ is positive for R less than the capacity (in nats per unit energy) of the channel, subject to the constraint in Equation (4). What we consider in this section is what happens in the limit as $\delta \to 0$. For many physical channels, the discrete time elements here can be viewed as inputs and outputs over incremental intervals of both time and frequency. Increasing N and decreasing δ while holding $D = \lfloor N\delta \rfloor$ constant can be viewed as increasing the frequency band of the transmission without changing the time duration or power of the codewords and without changing the discrete time channel model.

Note that if N is increased while holding D and the number of code words M fixed, then $P_e(N, R, \delta)$ either decreases or remains constant, i.e.,

$$P_e(N, \frac{\ln M}{N}, \frac{D}{N}) \geq P_e(N', \frac{\ln M}{N'}, \frac{D}{N'}) \quad \text{for } N' \geq N \tag{5}$$

The reason for this is that a code for the smaller blocklength could be converted into a code of larger blocklength simply by using zeros for the extra letters of each codeword. Thus the limit of $P_e(N, R, \delta)$ as $N \to \infty$ (with M and $N\delta$ fixed) must exist, and must be reached (and in fact must be reached for $N \leq MD$, since that is the maximum number of positions in which one or more codewords can contain ones). It is intuitively plausible that the optimal code in this limit is a code in which no two codewords have ones in any of the same positions, thus requiring a blocklength $N = MD$. We refer to this code in what follows as the orthogonal code; Massey [8] has demonstrated the optimality of this code. Our major concern is not with the error probability of the orthogonal code, which is very wasteful of blocklength, but rather with how this limiting error probability is approached with increasing N for a given δ. Given our interest in the limit $N \to \infty$, it is desirable to define a normalized rate, $\widetilde{R} = (\ln M)/D = R/\delta$ and a normalized reliability function,

$$\widetilde{E}(\widetilde{R}, \delta) = \limsup_{N\to\infty} \frac{-\ln P_e(N, \widetilde{R}\delta, \delta)}{N\delta} = \frac{E(\widetilde{R}\delta, \delta)}{\delta} \tag{6}$$

In the next subsection, we analyze the random coding bound [1] for these channels, using a fixed value of δ. After that, we find the reliability function at zero rate and use this, along with known results about the relationship of lower and upper bounds to error probability, to find upper and lower bounds to the reliability function $E(R, \delta)$. We shall find that the upper and lower bounds come together in the limit $\delta \to 0$, and thus $\lim_{\delta\to 0} \widetilde{E}(\widetilde{R}, \delta)$ is specified for all $\widetilde{R}$. This is rather surprising, since the only other channels for which the reliability function is completely known are a few channels that have a positive zero error capacity, and very noisy channels [1]. Another surprising aspect of these channels is that the capacity (in bits per unit D) can be infinite in the limit $\delta \to 0$, and we find the conditions under which this can happen.

3. THE RANDOM CODING BOUND

In applying the random coding bound, we can regard Equation (4) as an energy constraint and then use a shell constraint in the random coding bound as developed in Section 7.3 of [1]. An alternative would be to use a fixed composition random coding argument, using $N\delta$ ones in each codeword. We take the first alternative here since it is analytically simpler. The result (from Theorem 7.3.2 of [1]) is that the reliability function $E(R, \delta)$ satisfies

$$E(R, \delta) \geq E_r(R, \delta) \tag{7}$$

where the random coding exponent $E_r(R, \delta)$ is given by

$$E_r(R, \delta) = \max_{0\leq\rho\leq 1,\ 0\leq\phi} E_0(\rho, \phi, \delta) - \rho R \tag{8}$$

$$E_0(\rho, \phi, \delta) = -\ln \int_y \left[(1-\delta)p_0(y)^{\frac{1}{1+\rho}} e^{-\phi\delta} + \delta p_1(y)^{\frac{1}{1+\rho}} e^{\phi(1-\delta)}\right]^{1+\rho} dy \tag{9}$$

In this expression, we have used the obvious input distribution, using input 1 with probability $P_1 = \delta$. For large δ, the optimal distribution could have $P_1 < \delta$, but, for taking the limit $\delta \to 0$, there is no essential loss of generality in assuming $P_1 = \delta$. We want to factor out the term $p_0(y)$ from the sum in Equation (9), but observe that there may be a range of y over which $p_0(y) = 0$. To take care of this case, define q as

$$q = \int_{y:\, p_0(y)=0} p_1(y)dy \tag{10}$$

We shall soon see that whether $q = 0$ or not plays a critical role in whether or not the limiting capacity of the channel, in bits per unit energy, is finite or not. We can now rewrite Equation (9) in the form

$$E_0(\rho,\phi,\delta) = -\ln\left\{q\delta^{1+\rho}e^{\phi(1-\delta)(1+\rho)} + \int_y p_0(y)\left[e^{-\phi\delta}[1-\delta+\delta f(y)]\right]^{1+\rho} dy\right\} \tag{11}$$

$$\text{where} \quad f(y) = e^{\phi}\left[\frac{p_1(y)}{p_0(y)}\right]^{\frac{1}{1+\rho}} \tag{12}$$

We next find an upper bound to E_0 that is independent of δ. We will later see that this yields the asymptotic infinite bandwidth result and will find a lower bound to E_0 so as to discover how the asymptotic result is approached. For the upper bound, we must lower bound the expression inside the log. We first lower bound q by zero, and then, inside the bracketed term in the integral, we use the bound $(1+x)^{1+\rho} \geq 1+x(1+\rho)$; this is valid for $x \geq -1$ and $\rho \geq 0$. This yields

$$E_0(\rho,\phi,\delta) \leq \phi\delta(1+\rho) - \ln\int_y p_0(y)\left\{1+\delta(1+\rho)\left[f(y)-1\right]\right\}dy \tag{13}$$

The right side of Equation (13) can now be maximized over the free parameter $\phi \geq 0$, which appears both in the first term and implicitly in $f(y)$. The maximizing ϕ occurs where the integral above is one; this maximizing ϕ depends on ρ but not on δ and is given by

$$\phi(\rho) = -\ln\int_y [p_0(y)]^{\frac{\rho}{1+\rho}}[p_1(y)]^{\frac{1}{1+\rho}}dy \tag{14}$$

The integral above exists and is convex in ρ for $\rho > 0$ except in the special case $q = 1$ in which $p_0(y)p_1(y) = 0$ for all y. This special case is equivalent to a noiseless binary symmetric channel (BSC) and will be treated separately as an example later. We take $\phi(0)$ to be $\lim_{\rho\to(0+)}\phi(\rho)$, i.e., $\phi(0) = -\ln(q)$. We assume in what follows (except for the noiseless BSC) that $q < 1$ so that $\phi(\rho)$ exists for $\rho > 0$. We then define $\widetilde{E}_0(\rho)$ as

$$\widetilde{E}_0(\rho) = (1+\rho)\phi(\rho) = -(1+\rho)\ln\int_y [p_0(y)]^{\frac{\rho}{1+\rho}}[p_1(y)]^{\frac{1}{1+\rho}}dy \tag{15}$$

so that

$$\max_{\phi} E_0(\rho, \phi, \delta) \leq \delta \widetilde{E}_0(\rho) \tag{16}$$

Substituting this into Equation (8), we have

$$E_r(R, \delta) \leq \max_{0 \leq \rho \leq 1} \left[\delta \widetilde{E}_0(\rho) - \rho R\right] \tag{17}$$

On the surface, this is not an entirely satisfactory result, since Equation (17) gives an upper bound to $E_r(R, \delta)$, which, from Equation (7), is a lower bound to the reliability function $E(R, \delta)$. We shall show later that the right side of Equation (17) is also an upper bound to $E(R, \delta)$. First, however, we evaluate a lower bound to $E_0(\rho, \phi, \delta)$ and show that the bounds in Equations (16) and (17) are tight in the limit of small δ. The following lemma, which is proven in the appendix, gives us the necessary result.

Lemma 1 *For* $0 < \rho \leq 1,\ \phi \geq 0,\ 0 \leq q < 1,$

$$\max_{\phi} E_0(\rho, \phi, \delta) \geq \delta \left[\widetilde{E}_0(\rho) - \delta^{\rho} \exp \widetilde{E}_0(\rho)\right] \tag{18}$$

The next lemma relates our upper and lower bounds. Recall that we are assuming that $\phi(\rho)$ exists (i.e., that the channel is not a noiseless BSC).

Lemma 2 *For* $0 < \rho \leq 1,\ 0 \leq q < 1$

$$\lim_{\delta \to 0} \delta^{-1} \max_{\phi} E_0(\rho, \phi, \delta) = \widetilde{E}_0(\rho) \tag{19}$$

Proof: From Equation (16), $\delta^{-1} \max_{\phi} E_0(\rho, \phi, \delta) \leq \widetilde{E}_0(\rho)$. From Equation (18), $\delta^{-1} \max_{\phi} E_0(\rho, \phi, \delta) \geq \widetilde{E}_0(\rho) - \delta^{\rho} \exp \widetilde{E}_0(\rho)]$. As $\delta \to 0$, δ^{ρ} also approaches zero. Thus the upper and lower bounds approach equality, completing the proof.

Define

$$\widetilde{E}_r(\widetilde{R}) = \max_{0 \leq \rho \leq 1} \left[\widetilde{E}_0(\rho) - \rho \widetilde{R}\right] \tag{20}$$

We show in the next theorem that this is the limiting random coding exponent as δ approaches zero. First, however we discuss the maximization over ρ above. The function $\widetilde{E}_0(\rho)$ is nondecreasing and concave in ρ (see the lemma in appendix 5B of [1]). Thus the maximization in Equation (20) is given parametrically by

$$\begin{aligned}
\widetilde{E}_r(\widetilde{R}) &= \widetilde{E}_0(1) - \widetilde{R} && \text{for } \widetilde{R} < \widetilde{E}_0'(1) \\
\widetilde{E}_r(\widetilde{R}) &= \widetilde{E}_0(\rho) - \rho\widetilde{R} && \text{for } \widetilde{R} = \widetilde{E}_0'(\rho)\,;\ 0 < \rho \le 1 \\
\widetilde{E}_r(\widetilde{R}) &= \widetilde{E}_0(0) && \text{for } \widetilde{R} \ge \widetilde{E}_0'(0)
\end{aligned} \tag{21}$$

In the last part of the above equation, $\widetilde{E}_0(0)$ and $\widetilde{E}_0'(0)$ are taken as limits as $\rho \to 0$. What is quite surprising here, however, is that $\widetilde{E}_0(0)$ need not be zero. In fact, evaluating the limit from Equations (14) and (15), we find that

$$\widetilde{E}_0(0) = -\ln(1-q) \tag{22}$$

This means that if $q > 0$, then $\widetilde{E}_r(\widetilde{R})$ is positive for all $\widetilde{R} \ge 0$; we shall soon see that this implies that the capacity, in bits per unit energy, is infinite. This might seem less surprising if we observe that, for the orthogonal code, if message m is sent, correct decoding is assured if, in any position where the mth codeword has a one, the output falls in the set of probability q for which input 0 is impossible. Thus the probability of error for the orthogonal code is lower bounded by $(1-q)^D$. In general, $\widetilde{E}_r(\widetilde{R})$ is decreasing with $\widetilde{R}$ for $\widetilde{R} < \widetilde{E}_0'(0)$ and is constant for all larger $\widetilde{R}$.

Theorem 1 *For $\widetilde{R} \ge 0$ and $0 \le q < 1$,*

$$\lim_{\delta \to 0} \delta^{-1} E_r(\widetilde{R}\delta, \delta) = \widetilde{E}_r(\widetilde{R}) \tag{23}$$

Proof: From Equation (17) and the definition of $\widetilde{E}_r$ in Equation (20), we have

$$\delta^{-1} E_r(\widetilde{R}\delta, \delta) \le \widetilde{E}_r(\widetilde{R})$$

Thus, we need only establish that, for any $\epsilon > 0$, we have

$$\delta^{-1} E_r(\widetilde{R}\delta, \delta) \ge \widetilde{E}_r(\widetilde{R}) - \epsilon \tag{24}$$

for sufficiently small δ. First consider $\widetilde{R}$ satisfying $\widetilde{R} = \widetilde{E}_0'(\rho)$ for some ρ, $0 < \rho \le 1$. Using this value of ρ to lower bound the right side of Equation (8) and combining with Equation (18),

$$\begin{aligned}
E_r(\widetilde{R}\delta, \delta) \ge \delta \left[\widetilde{E}_0(\rho) - \delta^{\rho} \exp[\widetilde{E}_0(\rho)] - \rho\widetilde{R}\right] &= \\
\delta \left[\widetilde{E}_r(\widetilde{R}) - \delta^{\rho} \exp \widetilde{E}_0(\rho)\right]&
\end{aligned} \tag{25}$$

Dividing both sides by δ, we see that Equation (24) is satisfied for small enough δ. The same argument applies for $\widetilde{R} < \widetilde{E}_0'(1)$, using one for ρ. Finally consider the last case where $\widetilde{R} \geq \widetilde{E}_0'(0)$. If $\widetilde{E}_0(0) = 0$, then $\widetilde{E}_r(\widetilde{R}) = 0$ and Equation (24) is obviously satisfied for all δ. If $\widetilde{E}_0(0) > 0$, then choose ρ to satisfy

$$\widetilde{E}_0(\rho) - \rho\widetilde{R} = \widetilde{E}_0(0) - \epsilon/2 \tag{26}$$

Using the first half of Equation (24) with this value of ρ, and combining with Equation (26),

$$E_r(\widetilde{R}\delta, \delta) \geq \delta\left[\widetilde{E}_0(0) - \epsilon/2 - \delta^\rho \exp\widetilde{E}_0(\rho)\right] \tag{27}$$

Again Equation (24) is satisfied for small enough δ and the proof is complete.

4. LOWER BOUNDS TO ERROR PROBABILITY

We next show that $\widetilde{E}_r(\widetilde{R})$ is also an upper bound on the actual reliability function $\delta^{-1}E(\widetilde{R}\delta, \delta)$. Since the random coding exponent is a lower bound to the reliability function, this and Theorem 1 will show that $\widetilde{E}_r(\widetilde{R})$ is also the limiting reliability function as $\delta \to 0$. We start by evaluating error probability for a code with two codewords. Assume that the code is orthogonal, since for two codewords this clearly minimizes error probability. This is a classical binary hypothesis testing problem, made even simpler by the symmetry between the two codewords. The error probability (see, for example, Section 5.4 of [1]) is given by

$$P_e = o\left(\frac{1}{\sqrt{N}}\right)\exp\left\{-D\left[-2\ln\int_y \sqrt{p_0(y)p_1(y)}dy\right]\right\} \tag{28}$$

Note that the quantity in brackets is $\widetilde{E}_0(1) = \widetilde{E}_r(0)$. Also, this error probability is a lower bound to the error probability of any code with two or more codewords, so this exponent is an upper bound to the reliability function at $R = 0$, $\widetilde{E}(0)$. Thus $\widetilde{E}_r(0) = \widetilde{E}(0)$. It is well known [6] that for any DMC (and thus any value of $\delta > 0$), the reliability function is upper bounded by the sphere packing exponent and by the straight line exponent. The sphere packing exponent is given by

$$E_s(R, \delta) = \max_{\rho\geq 0}\left[\max_{\phi\geq 0} E_0(\rho, \phi, \delta) - \rho R\right] \tag{29}$$

It follows from Equation (16) that

$$E_s(\widetilde{R}\delta, \delta) \leq \delta \max_{\rho\geq 0} \widetilde{E}_0(\rho) - \rho\widetilde{R} \tag{30}$$

Letting $\widetilde{E}_s(\widetilde{R})$ be $\max_{\rho\geq 0}[\widetilde{E}_0(\rho) - \rho\widetilde{R}]$, we have

$$\widetilde{E}_s(\widetilde{R}) = \widetilde{E}_0(\rho) - \rho\widetilde{E}_0'(\rho) \qquad \text{for } \widetilde{R} = \widetilde{E}_0'(\rho),\ \rho > 0 \tag{31}$$

$$\widetilde{E}_s(\widetilde{R}) = \widetilde{E}_0(0) \qquad \text{for } \widetilde{R} \geq \lim_{\rho\to 0} \widetilde{E}_0'(\rho) \tag{32}$$

The straight line exponent is tangent to the sphere packing exponent (as a function of R) and equals the zero rate exponent, $\widetilde{E}(0,\delta)$ at $R = 0$. The straight line exponent is then upper bounded as the tangent to the upper bound to E_s in Equation (30); i.e., as $\delta[\widetilde{E}_0(1) - \widetilde{R}]$. Thus combining the upper bound to the sphere packing exponent with this bound to the straight line exponent, we have our earlier upper bound to the random coding exponent; we have thus proven the following theorem:

Theorem 2 *For all* $\widetilde{R} \geq 0$,

$$\delta^{-1}E(\widetilde{R}\delta, \delta) \leq \widetilde{E}_r(\widetilde{R}) \tag{33}$$

$$\lim_{\delta\to 0} \delta^{-1}E(\widetilde{R}\delta, \delta) = \widetilde{E}_r(\widetilde{R}) \tag{34}$$

We have seen in the above theorem that the reliability function $\widetilde{E}_r(\widetilde{R})$ in the limit $\delta \to 0$ is similar to the usual reliability functions for discrete memoryless channels with the following two major differences. First the reliability function here is known for it all $\widetilde{R} > 0$. Second, if $0 < q < 1$, $\widetilde{E}_r(\widetilde{R})$ is positive for *all* $\widetilde{R} > 0$; that is, the capacity in bits per unit energy of the channel is infinite. A special case of this phenomenon was first observed for a model of quantum optical channels [7,8,9]. It does not suggest that there is anything wrong with the concept of capacity for such channels and in no way gives added importance to the computational cutoff rate (see [9] for a more complete discussion); it does suggest that models for which $q > 0$ are not completely realistic, especially for large $\widetilde{R}$.

Another peculiar phenomenon that does not occur for the DMC is that it is possible for $\lim_{\rho\to 0} \widetilde{E}_0'(\rho)$ to be infinite. If $q = 0$ and this limit is infinite, then the capacity (in bits per unit energy) is infinite, but the exponent approaches zero as $\widetilde{R} \to \infty$.

5. ARBITRARY NUMBER OF INPUTS

Next consider an arbitrary input alphabet $\{0, 1, \ldots, K\}$ and let h_k be the energy associated with input letter k. Assume that $h_0 = 0$ and

$h_k > 0$ for $1 \leq k \leq K$. Assume further that each codeword $\vec{x}_m = (x_{m1}, \ldots, x_{mN})$ must satisfy the constraint

$$\sum_{n=1}^{N} h_{X_{mn}} \leq N\delta \tag{35}$$

For simplicity we assume that $\delta < h_k$ for $1 \leq k \leq K$ so that some inputs in each codeword must be zero. The random coding bound of Equation (7) to Equation (9) is as before, but now Equation (9) generalizes to

$$E_0(\rho, \phi, \delta, \vec{Q}) = -\ln \int_y \left\{ \sum_{k=0}^{K} Q_k e^{\phi[h_k - \delta]} p_k(y)^{\frac{1}{1+\rho}} \right\}^{1+\rho} dy \tag{36}$$

where $p_k(y)$ is the transition probability density given input k and where $\vec{Q} = (Q_0, Q_1, \ldots, Q_k)$ satisfies, first, $\sum_k Q_k h_k \leq \delta$, second, $\sum_k Q_k = 1$, and third, $Q_k \geq 0$ for $0 \leq k \leq K$. We now lower bound the integral in Equation (36), and thus upper bound E_0, by integrating only over y such that $p_0(y) > 0$. We can then factor $p_0(y)$ out of the term in braces and substitute $1 - \sum_{k \geq 1} Q_k$ for Q_0 to get

$$E_0(\rho, \phi, \delta, \vec{Q}) \leq \phi\delta(1+\rho) - \ln \int_y p_0(y) \left\{ \sum_{k=1}^{K} Q_k [f_k(y) - 1] \right\}^{1+\rho} dy \tag{37}$$

where

$$f_k(y) = e^{\phi h_k} \left[\frac{p_k(y)}{p_0(y)} \right]^{\frac{1}{1+\rho}} \tag{38}$$

Equation (37) can be further upper bounded by the inequality $(1 + x)^{1+\rho} \geq 1 + x(1+\rho)$ for $x \geq -1, \rho \geq 0$, yielding

$$E_0(\rho, \phi, \delta, \vec{Q}) \leq \phi\delta(1+\rho) - \ln \int_y p_0(y) \left\{ 1 + (1+\rho) \sum_{k=1}^{K} Q_k [f_k(y) - 1] \right\} dy \tag{39}$$

The integral above is a linear function of $\vec{Q} = (Q_1, \ldots, Q_K)$, subject to the linear constraints $\sum_k Q_k h_k \leq \delta, Q_k \geq 0$, and thus the right side of Equation (39) is maximized over $\vec{Q}$ for any given ρ, ϕ, δ by $Q_k = 0$ for all but one k, $1 \leq k \leq K$ (i.e., by a binary input). It follows that the right side of Equation (39) can be jointly maximized over ϕ and $\vec{Q}$ by maximizing over ϕ for each binary choice and then selecting the largest of these K possibilities. Using the result of Equation (13) to Equation (15) for a binary input, then, we have

$$\max_{\phi,\vec{Q}} E_0(\rho,\phi,\delta,\vec{Q}) \le \delta \widetilde{E}_0(\rho) \tag{40}$$

where

$$\widetilde{E}_0(\rho) = \max_{k\neq 0} \frac{-(1+\rho)}{h_k} \ln \int_y p_0(y)^{\frac{\rho}{1+\rho}} p_k(y)^{\frac{1}{1+\rho}} \tag{41}$$

Note that Equation (41) differs from the earlier binary result in Equation (15) by including the maximization over which nonzero input to use and by observing that input k can be used only with a relative frequency δ/h_k. We can also obtain a lower bound on $\max_{\phi,\vec{Q}} E_0(\rho,\phi,\delta,\vec{Q})$ by restricting $\vec{Q}$ to have only a single nonzero component, chosen to maximize Equation (41). In this case, the lower bound in Equation (18) applies, so that

$$\max_{\phi,\vec{Q}} E_0(\rho,\phi,\delta,\vec{Q}) \ge \delta \left[\widetilde{E}_0(\rho) - \left(\frac{\delta}{h_k}\right)^{\rho} \exp \widetilde{E}_0(\rho)\right] \tag{42}$$

It is seen that Lemma 2 is again valid for this more general case, and Theorem 1, which we restate for this case as Theorem 3, follows as before.

Theorem 3 *For all $\widetilde{R} \ge 0$,*

$$\lim_{\delta\to 0} \delta^{-1} E_r(\widetilde{R}\delta,\delta) = \widetilde{E}_r(\widetilde{R}) \tag{43}$$

where

$$E_r(\widetilde{R}\delta) = \max_{0\le\rho\le 1,\, 0\le\phi,\, \vec{Q}} E_0(\rho,\phi,\delta,\vec{Q}) - \rho R \tag{44}$$

$$\widetilde{E}_r(\widetilde{R}) = \max_{0\le\rho\le 1} E_0(\rho) - \rho\widetilde{R} \tag{45}$$

The sphere packing exponent $E_s(R,\delta)$ also follows as before with

$$\lim_{\delta\to 0} \delta^{-1} E_s(\widetilde{R}\delta,\delta) \le \widetilde{E}_s(\widetilde{R}) \tag{46}$$

$$\widetilde{E}_s(\widetilde{R}) = \max_{\rho\ge 0} E_0(\rho) - \rho\widetilde{R} \tag{47}$$

Recall now that in the binary case we showed that $\lim_{\delta\to 0} \delta^{-1}E(\widetilde{R}\delta, \delta)$ was equal to $\widetilde{E}_r(\widetilde{R})$. This was done by evaluating the zero rate exponent, i.e.,

$$\widetilde{E}(0) = \lim_{\delta\to 0}\left[\lim_{\widetilde{R}\to 0} \frac{E(\widetilde{R}\delta, \delta)}{\delta}\right]$$

This was shown to be $\widetilde{E}_0(1)$ and the desired result followed by combining the straight line exponent with the sphere packing exponent.

In this more general case, we shall find that the zero rate exponent is not *always* equal to $\widetilde{E}_0(1)$. This equality holds for all cases that appear to be of real interest (and thus $\lim \delta^{-1}E(\widetilde{R}\delta, \delta) = \widetilde{E}_r(\widetilde{R})$ in these cases) but in cases of less interest, the zero rate exponent exceeds $\widetilde{E}_0(1)$.

As an example, consider the DMC in figure 1. From Equation (41), $\widetilde{E}_0(\rho) = \rho \ln 2$. Thus $\widetilde{E}_r(\widetilde{R}) = \ln 2 - \widetilde{R}$ for $0 \le \widetilde{R} \le \ln 2$ and $\widetilde{E}_s(\widetilde{R}) = \infty$ for $0 \le \widetilde{R} \le \ln 2$. Both are zero for $\widetilde{R} > \ln 2$.

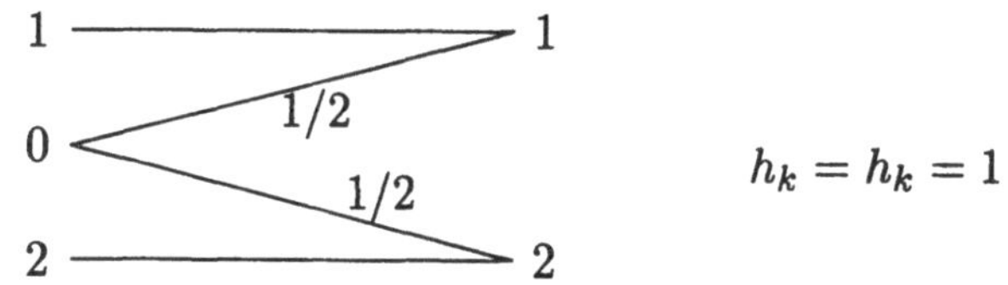

Figure 1 An example of a channel with infinite error probability exponent at zero rate

For any blocklength $N \ge D$, this channel can be used as a noiseless binary channel, using inputs 1 or 2 for the first D components of each codeword and using input 0 for the final $N - D$ components. Thus the reliability function $\delta^{-1}E(\widetilde{R}\delta, \delta)$ is infinite for $0 \le \widetilde{R} \le \ln 2$ and zero for $\widetilde{R} > \ln 2$.

This example can be generalized by adding very small transition probabilities, ϵ, from input 1 to output 2 and from input 2 to output 1. In this case, there is no zero error capacity and the random coding and sphere packing exponents are equal over a range of rates. For ϵ small enough, however, the zero rate exponent is achieved by using inputs 1 and 2 exclusively on the first D components of the blocklength and is unequal to $\widetilde{E}_0(1)$. In what follows, we assume that the channel has no zero error capacity equal to zero. We shall prove the following theorem.

Theorem 4 *If the zero error capacity is zero, then*

$$\widetilde{E}_0(1) = \lim_{\delta\to 0} \frac{E(0,\delta)}{\delta} = \sup_{\vec{Q}} \frac{\sum_{i,k} -Q_i Q_k \mu_{ik}(1/2)}{\sum_i Q_i h_i} \qquad (48)$$

where

$$\mu_{ik}(s) = \ln \int_y p_i(y)^{1-s} p_k(s)^s \, dy \tag{49}$$

and the sup *is over probability vectors* $\vec{Q} = (Q_0, \ldots, Q_K)$ *with* $Q_0 < 1$.

Before proving the theorem, we discuss the supremum in Equation (48) and find the conditions under which it is equal to $\widetilde{E}_0(1)$. For any $\vec{Q}$, define $q_i, 1 \leq i \leq k$, by $q_i = Q_i/(1-Q_0)$. Thus $\vec{q}$ is a probability vector over the nonzero inputs. The expression on the right side of Equation (48), which we denote $F(\vec{Q})$, is then

$$F(\vec{Q}) = \frac{-2Q_0 \sum_i q_i \mu_{i0}(1/2) - (1 - Q_0) \sum_{i,k} q_i q_k \mu_{ik}(1/2)}{\sum_i q_i h_i} \tag{50}$$

Note that for any given $\vec{q}$, $F(\vec{Q})$ is linear in Q_0, and is either maximized at $Q_0 = 0$, or as a supremum in the limit where $Q_0 \to 1$. Thus the supremum of $F(\vec{Q})$ can be found by first maximizing the right side of Equation (50) over $\vec{q}$ with $Q_0 = 0$ and then with $Q_0 = 1$. With $Q_0 = 1$, this maximum occurs with some $q_i = 1$, and is given by

$$\max_k \frac{-2\mu_{k0}(1/2)}{h_k} = \widetilde{E}_0(1) \tag{51}$$

We thus have proved the following corollary to Theorem 4:

Corollary 1 *If*

$$\widetilde{E}_0(1) \geq \frac{-\sum_{i,k} q_i q_k \mu_{ik}(1/2)}{\sum_i q_i h_i} \tag{52}$$

Then

$$\widetilde{E}(0) = \widetilde{E}_0(1) \tag{53}$$

This corollary establishes the condition (i.e., Equation (52)) under which the reliability function is equal to the random coding exponent for all rates. If Equation (52) is violated, on the other hand, the zero rate exponent is equal to the right side of Equation (52). What the violation of Equation (52) means is that at rates approaching zero, the best codes of a given energy use only nonzero energy inputs for all codewords over a blocklength just long enough to use the available energy. If Equation (52) is satisfied, however, good codes have orthogonal codewords and a binary input is sufficient.

Proof of Theorem 5: We first use the expurgated random coding bound to lower bound the reliability $E(0, \delta)$. From [1], the expurgated exponent at rate zero with constraint δ is given by

$$E_{ex}(0, \delta) = \max_{\vec{Q}: \sum_i Q_i h_i \leq \delta} - \sum_{i,k} Q_i Q_k \mu_{ik}(1/2) \tag{54}$$

An alternative, given blocklength N and total energy constraint $N\delta$, is to use a shorter effective constraint length fN $(0 < f \leq 1)$, using energy δ/f per letter for the first fN letters of the block and using all zeros on the last $(1-f)N$ letters of each codeword. Thus, for any $f, 0 < f \leq 1$ and any $\vec{Q}$ satisfying $\sum_i Q_i h_i \leq \delta/f$, we can achieve an exponent (on a blocklength of N basis) that is at least

$$-f \sum_{i,k} Q_i Q_k \mu_{ik}(1/2) \tag{55}$$

For any given $\vec{Q}$ such that $Q_0 < 1$, and for any $\delta < \sum_i Q_i h_i$, we can choose $f, 0 < f \leq 1$ to satisfy $\sum_i Q_i h_i = \delta/f$. Substituting this for f in Equation (55) yields

$$\lim_{\delta \to 0} \frac{E(0, \delta)}{\delta} \geq \frac{\sum_{i,k} -Q_i Q_k \mu_{ik}(1/2)}{\sum_i Q_i h_i} \tag{56}$$

Since this is valid for any $\vec{Q}, Q_0 < 1$, the right side of Equation (48) is a lower bound to the left side. In order to establish the opposite inequality, and complete the proof, we use the results in [6 part II]. The results there establish the zero rate exponent for the unconstrained case, and, rather than repeating the entire (lengthy and difficult) argument for the constrained case, we simply point out the required modifications, using a fixed value δ for the constraint throughout. Theorem 1 of [6 part II] must be modified in a straightforward way (i.e., using Eqs. (1.01) and (1.02) rather than (1.07) and (1.08) of [6 part II]) because of the continuum of the alphabet here. Theorem 2 and Lemmas 4.1 and 4.2 of [6 part II] follow without change, but Lemma 4.3 must be modified. Lemma 4.3 develops an upper bound on the exponent (there called D_{min}) of an ordered code of $2M$ codewords (the definition of ordered is not relevant to our modification) of blocklength N. It is shown in (1.36) of [6 part II] that

$$d_{\min} \leq \frac{1}{N} \sum_{n=1}^{N} \sum_{i} {}^{,} k Q_i^t(n) Q_k^b(n) \left[-\mu_{ik}(1/2) - \frac{1}{2} \mu'_{ik}(1/2) \right] \tag{57}$$

where $\mu_{ik}(s)$ is defined in Equation (49), $Q_i^t(n)$ is the relative frequency of input letter i in position n of the top M codewords, and $Q_i^b(n)$ is the relative frequency of i in position n of the bottom M codewords. Next $Q_i(n)$ and $r_i(n)$ are defined by

$$Q_i(n) = \frac{1}{2}\left[Q_i^t(n) + Q_i^b(n)\right] \tag{58}$$

$$r_i(n) = \frac{1}{2}\left[Q_i^t(n) - Q_i^b(n)\right] \tag{59}$$

Thus

$$Q_i^t(n)Q_i^b(n) = Q_i(n)Q_k(n) + r_i(n)Q_k(n) - Q_i^t(n)r_k(n) \tag{60}$$

When Equation (60) is substituted into Equation (57), one term is $-(1/N)\sum_n \sum_{i,k} Q_i(n)Q_k(n)\mu_{ik}(1/2)$. In contrast to Equation (1.40) of [6 part II], we have, for each n,

$$\sum_{i,k} -Q_i(n)Q_k(n)\mu_{i,k}(1/2) \leq \left[\sum_i Q_i(n)h_i\right] \sup_{\vec{Q}} \frac{\sum_{i,k} -Q_iQ_k\mu_{ik}(1/2)}{\sum_i Q_ih_i} \tag{61}$$

Summing over n and using the energy constraint, we then have

$$\frac{1}{N}\sum_{n=1}^{N}\sum_{i,k} -Q_i(n)Q_k(n)\mu_{i,k}(1/2) \leq \delta \sup_{\vec{Q}} \frac{\sum_{i,k} -Q_iQ_k\mu_{ik}(1/2)}{\sum_i Q_ih_i} \tag{62}$$

For a given δ, the other terms resulting from substituting Equation (60) into Equation (57) go to zero with N as shown in Lemmas 4.3, 4.4, and 4.5 of [6 part II]. Thus for any $\delta > 0$,

$$\frac{E(0,\delta)}{\delta} \leq \sup_{\vec{Q}} \frac{\sum_{i,k} -Q_iQ_k\mu_{ik}(1/2)}{\sum_i Q_ih_i} \tag{63}$$

This completes the proof of Theorem 4

6. MULTIACCESS COMMUNICATION

Now suppose that, instead of one transmitter sending a codeword to a receiver, there are J transmitters, each of which simultaneously sends a codeword to a single receiver. Assume that each transmitter has the same channel input alphabet $\{0, 1, ..., K\}$ and uses a block code of the same constraint length N. Let M_j be the number of codewords and $R_j = (\ln M_j)/N$ be the rate of the jth transmitter, $1 \leq j \leq J$.

Let $\{\mathbf{x}_m(j), 1 \leq m \leq M_j\}$ be the codewords of the jth code, where $\mathbf{x}_m(j) = (x_{m1}(j), ..., x_{mN}(j))$. Finally, let $h_k(j)$ be the energy associated with transmitter j using letter k and assume that each codeword from each transmitter must satisfy the constraint

$$\sum_{n=1}^{N} h_{x_{mn}(j)}(j) \leq N\delta_j \quad ; \quad 1 \leq m \leq M_j, \; 1 \leq j \leq J \tag{64}$$

We suppose that the transmitters are all synchronized in the sense that the receiver gets a channel output y_n corresponding to the nth channel input from each transmitter. We also assume that the channel is memoryless in the sense that if $\mathbf{x}(j) = x_1(j), \ldots, x_N(j)$, the channel output density is given by

$$p_N(\mathbf{y}|\mathbf{x}(1), \ldots \mathbf{x}(J)) = \prod_{n=1}^{N} p(y_n|x_n(1), \ldots, x_n(j)) \tag{65}$$

where $p(y|x(1), \ldots, x(J))$ is the transition probability density defining the channel.

Given a code for each transmitter, and given a channel transition probability density, we assume that each source j independently produces a message $m(j)$, uniformly distributed from one to M_j; the encoder at each transmitter j generates codeword $\mathbf{x}_{m(j)}(j)$, and the receiver maps the received $\mathbf{y}$ into J decoded outputs, $m^*(1), \ldots, m^*(J)$. An error occurs if $m(j) \neq m^*(j)$ for any $j, 1 \leq j \leq J$. There are two different definitions of error probability of interest here. The first, block error probability, P_e, is simply the probability that an error occurs over the ensemble of source inputs and channel outputs for a given set of codes. For the second, bit error probability, we assume that M_j is a power of 2 for each j and that the input to encoder j is a sequence of $\log_2 M_j$ IID equiprobable binary digits. The bit error probability for source j, $P_e^*(j)$ is then defined as the expected number of binary digits into encoder j that are decoded in error. The overall bit error probability P_e^* is then defined as $\max_j P_e^*(j)$.

We shall use P_e (as in the previous sections) to discuss the direct part of the coding theorem and exponential error bounds, and use bit error probability, P_e^*, for the converse to the coding theorem. For the converse, we view N as the number of uses of the channel over its lifetime rather than as blocklength for any given code. Thus, showing that P_e^* is bounded away from zero for certain transmission rates applies to block codes, convolutional codes, block codes in which the blocks at different transmitters are not synchronized, block codes in which different transmitters have different blocklengths, etc. Note that showing that P_e is bounded away from zero, and even showing that P_e approaches one with increasing N, does not imply that P_e^* is bounded away from zero and does not show that reliable communication in some reasonable sense is not possible.

We now define the achievable region for a multiaccess channel with energy constraints and establish the converse to the coding theorem. The coding theorem and converse for multiaccess channels without an energy constraint was established by Ahlswede [10] and Liao [11] and has been extended in many subsequent papers, for example [12-14]. However, none of the converses use bit error probability and none include general energy constraints. The inclusion of an energy constraint into the converse is nontrivial and forces us to look at some subtleties that appear to have been overlooked in the simpler existing proofs of the converse. We will not bother to prove the forward part of the coding theorem in general since it is a trivial extension of the result without an energy constraint.

Let $Q_k(j)$, $0 \le k \le K$ be a probability assignment on the channel input alphabet for transmitter j, $1 \le j \le J$, and consider the ensemble $X(1)$, $X(2), \ldots, X(J), Y$ in which $X(1), \ldots, X(J)$ are independent, $X(j)$ has the alphabet $\{0, 1, \ldots, K\}$ with probability assignment $Q_k(j)$ and Y is described by $p(y|x(1), \ldots, x(J))$.

Let S be an arbitrary nonempty subset of the set $\{1, 2, \ldots, J\}$ of transmitters, let $\mathbf{X}(S)=X(i_1), X(i_2), \ldots, X(i_j)$ where $S=\{i_1, i_2, \ldots, i_j\}$, and let $\mathbf{X}(S^c) = \{X(i_{j+1}) \cdots X(i_J)\}$ where $\{i_{j+1}, i_{j+2}, \ldots, i_J\}$, is the complement S^c of S. Finally let $I_{\vec{Q}}(\mathbf{X}(S); Y|\mathbf{X}(S^c))$ be the average mutual information between $\mathbf{X}(S)$ and Y conditional on $\mathbf{X}(S^c)$ for the ensemble defined by $\vec{Q} = \{Q_k(j), 0 \le k \le K, 1 \le j \le J\}$. A vector $\mathbf{R} = (R(1), ..., R(J))$ of input rates to the J transmitters is called *achievable* for the joint probability assignment $\vec{Q}$ if

$$0 \le \sum_{i \in S} R(i) \le I_{\vec{Q}}(\mathbf{X}(S); Y|\mathbf{X}(S^c)) \quad \text{for all } S \tag{66}$$

Let $\vec{\delta} = \delta_1, \ldots, \delta_J$ be a vector of allowable energies for the J transmitters. The pair $(\mathbf{R}, \vec{\delta})$ is defined to be *directly achievable* if there is some $\vec{Q}$ such that $\mathbf{R}$ is achievable for $\vec{Q}$ and such that $\vec{Q}$ satisfies the energy constraint, i.e.,

$$\sum_k h_k(j) Q_k(j) \le \delta_j \quad \text{for all } j;\ 1 \le j \le J \tag{67}$$

Finally the pair $(\mathbf{R}, \vec{\delta})$ is defined to be *achievable* if it is in the convex hull of a set of directly achievable pairs $(\mathbf{R}^{(m)}, \vec{\delta}^{(m)})$. The set of $\mathbf{R}$ such that $(\mathbf{R}, \vec{\delta})$ is achievable for a given $\vec{\delta}$ is denoted $\mathbf{R}(\vec{\delta})$. Note that $\mathbf{R}(\vec{\delta})$ is not the same as the convex hull of the set of **R** such that **R** is achievable for some $\vec{Q}$ satisfying Equation (67) for that $\vec{\delta}$; we shall shortly see an example of this difference.

Before proceeding, we must understand Equation (66) a little better; for a given $\vec{Q}$, Equation (66) is a set of $2^J - 1$ linear inequalities on the vector **R**. The set of **R** that satisfy Equation (66) is a convex polytope. Let $i_1, i_2, \ldots, i_J$ be a permutation of the integers one to J, and let $S_j = \{i_1, i_2, \ldots, i_j\}$ for $1 \leq j \leq J$. Then we claim that the solution to

$$\sum_{i \in S_j} R(i) = I_{\vec{Q}}(\mathbf{X}(S_j); Y|\mathbf{X}(S_j^c)) \quad \text{for all } j;\ 1 \leq j \leq J \tag{68}$$

is an extreme point of the set of **R** that satisfy Equation (66) for a given $\vec{\delta}$ and $\vec{Q}$. It is easy to see that there is a unique **R** satisfying Equation (68) for a given permutation, and it is not hard, by manipulating information inequalities, to see that this satisfies all the other inequalities in Equation (66). This, however, is sufficient to guarantee that **R** is an extreme point. It is also not hard to see that the $J! - 1$ extreme points (not necessarily distinct) formed in this way constitute all the extreme points other than those with one or more zero rates.

Lemma 3 *Let $\vec{Q}^{(1)}, \ldots, \vec{Q}^{(M)}$ be M joint distributions on the input alphabets, where for $1 \leq m \leq M$, $\vec{Q}^{(m)} = \{Q_k^{(m)}(j);\ 1 \leq j \leq J,\ 0 \leq k \leq K\}$ and let $\mathbf{X}^{(m)} = X^{(m)}(1), \ldots, X^{(m)}(J)$ be the corresponding ensembles. Let $\lambda_m \geq 0$, $\sum_m \lambda_m = 1$. If*

$$0 \leq \sum_{i \in S} R(i) \leq \sum_{m=1}^{M} \lambda_m I(\mathbf{X}^{(m)}(S); Y^{(m)}|\mathbf{X}^{(m)}(S^c)) \quad \text{for all } S \tag{69}$$

$$\text{and} \quad \sum_{m=1}^{M} \lambda_m \sum_k h_k(j) Q_k^{(m)}(j) \leq \delta_j \quad \text{for all } j,\ 1 \leq j \leq J \tag{70}$$

then $(\mathbf{R}, \vec{\delta})$ is achievable where $\mathbf{R} = (R(1), \ldots, R(J))$ and $\vec{\delta} = \{\delta_1, \ldots, \delta_J\}$.

Proof: We prove the lemma for $M = 2$; the extension to arbitrary M follows immediately by induction on M. The set of rate vectors that satisfy Equation (69) for a given $\vec{Q}^{(1)}, \vec{Q}^{(2)}, \lambda_1$, and $\lambda_2 = 1 - \lambda_1$, forms a convex polytope. As before, each extreme point of this polytope with all $R(i) > 0$ satisfies the J equations, $1 \leq j \leq J$,

$$\sum_{i \in S_j} R(i) = \lambda_1 I_{\vec{Q}^{(1)}}(\mathbf{X}(S_j); Y|\mathbf{X}(S_j^c)) + \lambda_2 I_{\vec{Q}^{(2)}}(\mathbf{X}(S_j); Y|\mathbf{X}(S_j^c)) \tag{71}$$

where $S_j = \{i_1, \ldots, i_j\}$ for some permutation $i_1, \ldots, i_J$ of the integers one to J. For a given permutation (and thus a given extreme point $\mathbf{R}$), there is an extreme point $\mathbf{R}^{(1)}$ of the set of $\mathbf{R}$ that satisfies Equation (68) for $\vec{Q}^{(1)}$ and an extreme point $\mathbf{R}^{(2)}$ of the set of $\mathbf{R}$ that satisfies Equation (68) for $\vec{Q}^{(2)}$. Clearly $\mathbf{R} = \lambda_1 \mathbf{R}^{(1)} + \lambda_2 \mathbf{R}^{(2)}$. This same result holds for all the extreme points of Equation (71) involving some $R(i) = 0$. Thus every extreme point of Equation (69) (for $N = 2$) is λ_1, times an extreme point of the solutions of Equation (66) for $\vec{Q}^{(1)}$, plus λ_2, times an extreme point of the solutions for Equation (66) for $\vec{Q}^{(2)}$. Since every point satisfying Equation (69) is a convex combination of the extreme points of Equation (69), it is also equal to λ_1 times a point achievable for $\vec{Q}^{(1)}$ plus λ_2 times a point achievable for $\vec{Q}^{(2)}$. Thus $(\mathbf{R}, \vec{\delta})$ is a convex combination of a point $(\mathbf{R}^{(1)}, \vec{\delta}^{(1)})$ directly achievable with $\vec{Q}^{(1)}$ and a point $(\mathbf{R}^{(2)}, \vec{\delta}^{(2)})$ directly achievable with $\vec{Q}^{(2)}$.

We are now ready to state the coding theorem and converse; these show that 'reliable communication is possible if all components of $\mathbf{R}$ are less than those for some $\mathbf{R}^*$ for which $(\mathbf{R}^*, \vec{\delta})$ is in the achievable rate region and that reliable communication is impossible if $(\mathbf{R}, \vec{\delta})$ is outside of the achievable region.

Theorem 5 *Assume that* $(\mathbf{R}, \vec{\delta})$ *is achievable for a given memoryless multiaccess channel with* J *transmitters. Then for any* $\epsilon > 0$, $\mu > 0$, *there is an* $N(\epsilon, \mu)$ *such that for all blocklengths* $N \geq N(\epsilon, \mu)$, *there exists a code for each transmitter* j, *satisfying the energy constraint* δ_j *with* $M_j \geq \exp[N(R_j - \mu)]$ *codewords, and there exists a decoding rule such that* $P_e \leq \epsilon$.

We shall not prove this theorem, since, as stated before, it is a minor extension adding energy constraints to the theorems of [10] and [11]. We shall prove the following converse, however, since it is a nontrivial extension. The theorem states that, for a given energy constraint $\vec{\delta} = \{\delta_1, \ldots, \delta_J\}$, if the bit error probability P_e^* is small, then the rate vector $\mathbf{R} = (R(1), \ldots, R(J))$ must be close to an achievable rate vector.

Theorem 6 *Consider a given memoryless multiaccess channel with* J *transmitters and with energy constraint* $\vec{\delta} = \{\delta_1, \ldots, \delta_J\}$. *Assume that, over an arbitrary blocklength* N, *the input to each transmitter* j *is* $NR(j)/\ln 2$ *IID equiprobable binary digits and that the bit error probability satisfies* $P_e^* \leq \epsilon$. *Let* $H_2(\epsilon) = -\epsilon \log_2 \epsilon - (1-\epsilon)\log_2(1-\epsilon)$ *and let* $R^*(j) = R(j)(1 - H_2(\epsilon))$. *Then* $(\mathbf{R}^*, \vec{\delta})$ *is achievable where* $\mathbf{R}^* = \{R^*(1), \ldots, R^*(J)\}$.

Proof: Let $S = \{j_1, j_2, \ldots, j_i\}$ be an arbitrary nonempty subset of the transmitters. The number of binary digits encoded and transmitted by transmitters in S over time N is given by

$$L = \frac{N}{\ln 2} \sum_{j \in S} R(j) \tag{72}$$

Let U^L be the ensemble of these binary digits (assumed IID and equally likely) and let V^L be the ensemble of the corresponding decoded binary digits. The average error probability over these L binary digits is at most $P_e^* \leq \epsilon$, so from Fano's inequality extended to a sequence of letters (Theorem 4.3.2 in [1]), we have

$$H_2(\epsilon) \geq \frac{H(U^L|V^L)}{L \ln 2} \tag{73}$$

where $H_2(\epsilon)$ is in binary units and $H(U^L|V^L)$ is in natural units. Since $H(U^L) = L \ln 2$, this can be rewritten, with the help of Equation (72) as

$$[1 - H_2(\epsilon)] \sum_{j \in S} R(j) \leq \frac{I(U^L; V^L)}{N} \tag{74}$$

From the data processing theorem, we have

$$I(U^L; V^L) \leq I(\mathbf{X}^N(S); \mathbf{Y}^N) \tag{75}$$

where $\mathbf{X}^N(S) = \mathbf{X}^N(j_1), \mathbf{X}^N(j_2), \ldots, \mathbf{X}^N(j_i)$ is the ensemble of transmitted letters from the transmitters in S for the given codes and $\mathbf{Y}^N$ is the ensemble of received letters. Using the independence of the codewords transmitted by the different transmitters,

$$I(\mathbf{X}^N(S); \mathbf{Y}^N) \leq I(\mathbf{X}^N(S); \mathbf{Y}^N | \mathbf{X}^N(S^c)) \tag{76}$$

Since the channel is memoryless,

$$I(\mathbf{X}^N(S); Y^N | \mathbf{X}^N(S^c)) \leq \sum_{n=1}^{N} I(\mathbf{X}_n(S); Y_n | \mathbf{X}_n(S^c)) \tag{77}$$

Combining Equation (74) to Equation (77) and recalling that $R^*(j) = [1 - H_2(\epsilon)]R(j)$,

$$\sum_{j \in S} R(j)^* \;\leq\; \frac{1}{N} \sum_{n=1}^{N} I(\mathbf{X}_n(S); Y_n | \mathbf{X}_n(S^c)) \tag{78}$$

Since Equation (78) is valid for all nonempty sets S, the previous lemma asserts that $(\mathbf{R}^*, \vec{\delta})$ is achievable, completing the proof.

7. SIMPLE MULTIACCESS EXAMPLE

Consider the special case in which each transmitter has a binary alphabet, using unit energy on input 1 and zero energy on input 0. Each transmitter j has the same energy constraint δ, which means that, over blocklength N, each transmitter sends at most $N\delta$ one's. Finally assume that the channel output depends statistically only on whether all transmitters send zero or whether one or more transmitters transmit one. Let $p_0(y)$ be the transition probability density given that all transmitters send zero and let $p_1(y)$ be the density given that one or more transmitters send one.

This multiaccess channel can be viewed as the cascade of two channels, the first of which is a multiaccess 'or' channel for which the output is binary and equal to the logical 'or' of the set of inputs; the second is a point-to-point channel with binary inputs and transition probability densities $p_1(y)$ and $p_0(y)$. This is not a very realistic model in practice, since the output of most multiaccess channels depends on how many transmitters are simultaneously transmitting a one. The model does capture the essence of multiaccess interference, however, and is suitable for achieving several simple insights.

Assume that the J transmitters are numbered so that the rate vector $\mathbf{R} = \{R(1), \ldots, R(J)\}$ satisfies $R(1) \geq R(2) \geq \cdots \geq R(J) \geq 0$. Let $\vec{Q}$ be the input probability assignment in which each transmitter independently sends a one with probability δ and a zero with probability $1-\delta$. The set of rate vectors $\mathbf{R}$, subject to the above ordering, that are feasible for $\vec{Q}$, is given by the following inequalities, $l \leq j \leq J$:

$$0 \leq \sum_{i \in S_j} R(i) \;\leq\; I_{\vec{Q}}(\mathbf{X}(S_j); Y|\mathbf{X}(S_j^c)) \tag{79}$$

where $S_j = \{1, \ldots, j\}$. From the symmetry between the transmitters, and the monotonicity of $\{R(i)\}$, it can be seen that all the inequalities in Equation (66) are satisfied if these inequalities are satisfied. Let $I(\epsilon)$ be the mutual information on the point-to-point part of the multiaccess channel given that input 1 is used with probability ϵ, i.e.,

$$I(\epsilon) = \int_y \left[\epsilon p_1(y) \ln \frac{p_1(y)}{\epsilon p_1(y) + (1-\epsilon)p_0(y)} + (1-\epsilon)p_0(y) \ln \frac{p_0(y)}{\epsilon p_1(y) + (1-\epsilon)p_0(y)}\right] dy \tag{80}$$

Since ϵ is an input probability assignment to a DMC, $I(\epsilon)$ is concave in ϵ. We can evaluate the right side of Equation (79) in terms of $I(\epsilon)$ to yield

$$0 \leq \sum_{i \in S_j} R(i) \;\leq\; (1-\delta)^{J-j} I(1-(1-\delta)^j) \tag{81}$$

The right hand side of this equation is concave in j, as can be seen by noting, from Equation (79), that the increment from $j-1$ to j is $I(X_j; Y|\mathbf{X}(S_j^c))$. From the symmetry between inputs, this increment depends only on the conditioning, which is decreasing with j. The conditioning is independent of X_j, so the increment is decreasing in j, and thus concavity is established. Thus, in the important special case where all the transmitter rates are equal to R, the most stringent inequality above is $j = J$, where we have

$$JR \leq I(1-(1-\delta)^J) \tag{82}$$

Let $\epsilon_{\max}$ be the value of ϵ that maximizes $I(\epsilon)$; i.e., $I(\epsilon_{\max})$ is the capacity of the point-to-point channel. If $1-(1-\delta)^J > \epsilon_{\max}$, then an input distribution should be used that reduces the probability of using input 1 to the δ' that satisfies $1-(1-\delta')^J = \epsilon_{\max}$. Our primary interest here, however, is where the power limit requires δ to be smaller than this limit, and this is assumed in what follows.

A more interesting situation is that in which some rates are much larger than others. In this case, the interference from users of very low rate can be reduced by using those inputs with probability smaller than δ. In other words, the directly achievable region for this multiaccess channel requires looking at a more general class of input probability vectors in which transmitter j uses input 1 with an arbitrary probability $\delta_j \leq \delta$. For such a $\vec{Q}$, the average mutual informations in Equation (79) can be evaluated to yield

$$0 \leq \sum_{i \in S_j} R(i) \leq \left[\prod_{i=j+1}^{J} (1-\delta_j)\right] I\left(1-\prod_{i=1}^{j}(1-\delta_i)\right) \tag{83}$$

There appears to be no simple characterization of the set of directly achievable rates, although it is not hard to verify whether or not a given rate vector is directly achievable. Note that (under the low power assumption) the mutual information term in Equation (83) is increasing in each δ_i, so there is no loss of generality in setting $\delta_1 = \delta$. This places a lower bound, $B_1 = R(1)/I(\delta)$ on the product $\prod_{j>1}(1-\delta_j)$. Next, for $j = 2$ and a given choice of δ_2, note that

$$\prod_{j>2}(1-\delta_j) \geq \frac{R(1)+R(2)}{I[1-(1-\delta)(1-\delta_2)]}\,; \qquad \prod_{j>2}(1-\delta_j) \geq \frac{B_1}{1-\delta_2}$$

The bound is on the left decreasing in δ_2 and the bound on the right is increasing. We define B_2 as the larger of these bounds, minimized over $\delta_2 \leq \delta$. For each successively larger value of i, we use the corresponding pair of equations to establish a lower bound B_i to $\prod_{j\geq i}(1-\delta_J)$. We continue until either all constraints are met, or some constraint can not be met.

The directly achievable rate region above is typically smaller than the entire achievable region. Consider a convex combination of J pairs of rate/energy vectors in which the jth, say $(\mathbf{R}^{(j)},\vec{\delta}^{(j)})$ has $R^{(j)}(i)=0$, $\delta_i^{(j)}=0$ for all $i\neq j$. This corresponds to frequency or time division multiplexing between the transmitters, as will be discussed later. The achievability inequalities of Equation (66) for each (j) reduce to the single inequality

$$0 \leq R^{(j)}(j) \leq I_{\vec{Q}^{(j)}}(X_j;Y) = I(\delta_j^{(j)}) \tag{84}$$

Letting $\lambda_1,\ldots,\lambda_J$ be nonnegative numbers summing to one, consider the convex combination $\mathbf{R}=\sum_j \lambda_j \mathbf{R}^{(j)}$, $\vec{\delta}=\sum_j \lambda_j \vec{\delta}^{(j)}$. This means that $R(j)=\lambda_j R^{(j)}(j)$ and $\delta_j=\lambda_j\delta_j^{(j)}$. From Equation (84), then, $(\mathbf{R},\vec{\delta})$ is achievable if

$$0 \leq R(j) \leq \lambda_j I(\delta_j/\lambda_j) \tag{85}$$

We shall refer to the rate region defined by Equation (85) as the frequency division multiplexing (FDM) region. In the special case where all rates are equal to some fixed R and $\lambda_j=1/J$ for $1\leq j\leq J$, this reduces to

$$0 \leq JR \leq I(J\delta) \tag{86}$$

Note that $NJ\delta$ is an upper bound on the total number of ones that can enter the point-to-point channel. It follows (still assuming $J\delta \leq \epsilon_{\max}$) that $I(J\delta)$ is an upper bound on the mutual information per letter on the point-to-point channel. Thus, for equal rates, FDM achieves the outer limit of the achievable region. Note that the equal rate point on the boundary of the directly achievable region, as given by Equation (82), is strictly inside this FDM equal rate point, and thus strictly inside the achievable region.

Figure 2 illustrates the FDM region for $J=2$ and compares it with the directly achievable region using $Q_1(1)=Q_1(2)=\delta$. Note that the extreme points of the directly achievable region lie outside the FDM region for this example, so that neither dominates over the entire region. Finally, when R_1 is greater than its value at the extreme point on the right, then R_2 is maximized by choosing $Q_1(2)=\delta_2<\delta$.

Finally, we briefly investigate the use of coding to achieve rates in the interior of the feasible region. For points in the interior of the FDM region, each transmitter can be allocated a disjoint set of positions in a block of length N, allocating $\lambda_j N$ positions (which can be viewed as

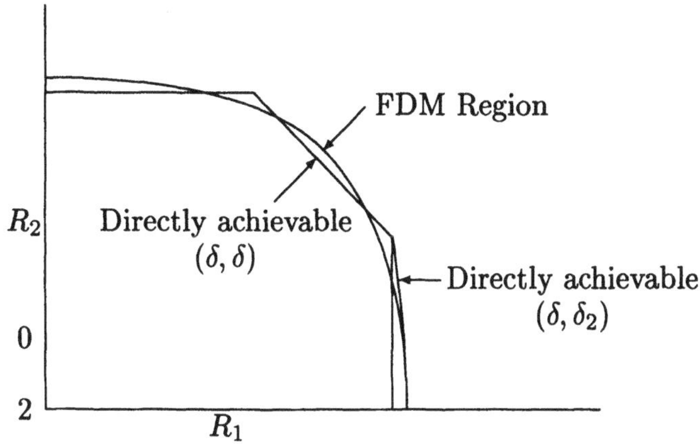

Figure 2 An example of the achievable region achieved by FDM, that achieved by directly achievable points using input 1 with probability δ for each transmitter, and that using $\delta_2 < \delta$ for the second transmitter.

a frequency band) to the jth transmitter. The jth transmitter then uses a code within its subblock of $\lambda_j N$ positions, using at most δN ones in each codeword, or δ/λ_j ones per allocated position. There is no interference between transmitters, and the decoder can decode each transmitter's codewords independently. Thus the system reduces to J independent point-to-point channels with energy constraints; conventional coding techniques can be used and the error probability results of Section 2 are directly applicable.

The FDM approach is often undesirable for data transmission. The major reason for this is not so much that it does not achieve the full achievable rate region but rather the burstiness of the sources. A transmitter needs enough bandwidth to send its messages with tolerable delay, but this leads to wasted bandwidth when there is nothing to send. A possible solution is dynamic bandwidth allocation, but this introduces scheduling delay and distributed scheduling algorthms. Another disadvantage of FDM is that it is susceptible to jamming, either intentional or unintentional.

Alternatively, if the desired set of rates is inside the directly achievable region for some $\{Q_1(j) = \delta_j\}$, then independently selected codes can be chosen, using a shell constraint with $\delta_j N$ ones in each codeword for each transmitter j. This can be viewed as an abstraction of spread spectrum coding (see, for example [16,17]). In practice, a spread spectrum transmitter pseudo-randomly chooses a subset of the available N components, and then uses algebraic or convolutional coding to select code words within that subset; all codewords use input 0 (i.e., don't transmit) for all components outside that subset.

One advantage of this approach is that if there is a very large set of transmitters, but only J of them are transmitting in a block (i.e., the others are sending a constant stream of zeros), then the error probability is the same as if only J transmitters existed. Thus no scheduling is required and there is no loss of efficiency due to bursty sources (other than that J becomes a random variable and the error probability becomes large as J becomes too large).

Our objective in this section has not been to provide a substantive guide to coding for multiaccess power limited channels, but merely to give some introduction to how power constraints can be incorporated into standard information theoretic arguments.

8. APPENDIX

Before proving Lemma 1, we first rewrite $E_0(\rho, \phi, \delta)$ from Equation (11).

$$\begin{aligned} E_0(\rho, \phi, \delta) &= -\ln\left\{ q(\delta e^{\phi})^{1+\rho} + \int_y p_0(y)\left[1-\delta+\delta f(y)\right]^{1+\rho} dy \right\} \\ &\quad +\phi\delta(1+\rho) \end{aligned} \tag{87}$$

The major task in establishing the Lemma 1 lower bound to $E_0(\rho, \phi, \delta)$ is to upper bound the integral in Equation (86). The following lemma upper bounds the critical component in that integral.

Lemma 4 *For any ρ, $0 \le \rho \le 1$, any $f \ge 0$, and any δ, $0 \le \delta \le 1$,*

$$(1-\delta+\delta f)^{1+\rho} \le 1 + (1+\rho-\rho\delta)(f-1)\delta + (\delta f)^{1+\rho} \tag{88}$$

Proof: The left side of Equation (88) is convex in ρ and thus can be bounded, for $0 \le \rho \le 1$, by its chord between those points, i.e.,

$$(1-\delta+\delta f)^{1+\rho} \le 1 + (1+\rho)(f-1)\delta + \rho\delta^2(f-1)^2 \tag{89}$$

Upper bounding $(f-1)^2$ by $f^2 - (f-1)$,

$$(1-\delta+\delta f)^{1+\rho} \le 1 + (1+\rho-\rho\delta)(f-1)\delta + \rho(\delta f)^2 \tag{90}$$

For $\delta f \le 1$ and $\rho \le 1$, we have $(\delta f)^2 \le (\delta f)^{1+\rho}$, completing the proof for $\delta f \le 1$.

For $\delta f > 1$, we have

$$(1-\delta+\delta f)^{1+\rho} = \left[1 + \frac{1-\delta}{\delta f}\right]^{1+\rho} (\delta f)^{1+\rho} \tag{91}$$

Using Equation (89) on the first factor of Equation (91), replacing δ by $1/f$ and δf by $1/(\delta f)$), we get

$$(1-\delta+\delta f)^{1+\rho} \le (\delta f)^{1+\rho} + (1+\rho)(1-\delta)(\delta f)^{\rho} + \rho(1-\delta)^2(\delta f)^{\rho-1} \quad (92)$$

Since $\delta f > 1$, we have $(\delta f)^{\rho} \le \delta f$ and $(1+\rho)(1-\delta) \le 1+\rho-\rho\delta$, so

$$(1-\delta+\delta f)^{1+\rho} \le (\delta f)^{1+\rho} + (1+\rho-\rho\delta)\delta f + \rho(1-\delta)^2 \quad (93)$$

Substituting $\rho(1-\delta)^2 \le (1-\rho\delta)(1-\delta) = 1-(1+\rho-\rho\delta)\delta$ into Equation (95), we get Equation (88), completing the proof of lemma 4.

Proof of Lemma 1

Substituting (88) into Equation (86),

$$E_0(\rho,\phi,\delta) \ \ge\ -\ln\left\{q\left[\delta e^{\phi}\right]^{1+\rho} + \int_y p_0(y)\left[1+(1+\rho-\rho\delta)(f(y)-1)\delta + [\delta f(y)]^{1+\rho}\right]dy\right\} + \phi\delta(1+\rho) \quad (94)$$

Let $F(\rho,\phi,\delta)$ denote the integral above. We evaluate it for $\phi=\phi(\rho)$, which is a lower bound to the maximum over ϕ. From Equation (13), $\int_y p_0(y)[f(y)-1]dy = 0$ for $\phi(\rho)$, so that

$$F(\rho,\phi(\rho),\delta) = 1 + \int_y p_0(y)[\delta f(y)]^{1+\rho}dy \quad (95)$$

We recall that the integral in Equation (11) is over y such that $p_0(y) > 0$. We then have

$$\begin{aligned} F(\rho,\phi(\rho),\delta) &= 1 + \int_{y:\,p_0(y)>0} p_0(y)\left[\delta e^{\phi}\right]^{1+\rho}\frac{p_1(y)}{p_0(y)}dy \\ &= 1 + (1-q)\left[\delta e^{\phi}\right]^{1+\rho} \end{aligned}$$

We have used the definition of q in Equation (10). Substituting Equation (95) into Equation (93),

$$\begin{aligned} E_0(\rho,\phi(\rho),\delta) &\ge \phi(\rho)\delta(1+\rho) - \ln\left\{1+\left[\delta e^{\phi(\rho)}\right]^{1+\rho}\right\} \\ &\ge \delta\phi(\rho)(1+\rho) - \delta^{1+\rho}e^{(1+\rho)\phi(\rho)} \end{aligned}$$

$$= \delta\left[\widetilde{E}_0(\rho) - \delta^{\rho}\exp\widetilde{E}_0(\rho)\right] \quad (96)$$

since, from Equation (15), $\widetilde{E}_0(\rho) = (1+\rho)\phi(\rho)$. This yields Equation (18), completing the proof.

References

[1] R. G. Gallager, *Information Theory and Reliable Communication*, New York: Wiley, 1968

[2] A. J. Viterbi and J. K. Omura, *Principles of Digital Communication and Coding*, New York: McGraw Hill, 1979.

[3] J. M. Wozercraft and I. M. Jacobs, *Principles of Communication Engineering*, New York: Wiley, 1965.

[4] R. S. Kennedy, *Fading Dispersive Channels*, New York: Wiley, 1969.

[5] J. S. Richters, "Communication over Fading Dispersive Channels," MIT *RLE Tech Report 464*, 1967.

[6] C. E. Shannon, R. G. Gallager, and E. R. Berlekamp, "Lower Bounds to Error Probability for Coding on Discrete Memoryless Channels," *Information and Control 10*, pp. 65-103 (part I), pp. 522-552 (part II).

[7] R. J. McEliece, "Practical Codes for Photon Communication," *IEEE Transactions on Information Theory*, Vol. IT-27, 4, July 1981.

[8] J. L. Massey, "Capacity, Cutoff Rate, and Coding for a Direct Detection Optical Channel," *IEEE Transactions on Communications*, Vol. COMM-29, 11, pp 1615-1621, Nov. 1981.

[9] P. A. Humblet, "Error Exponents for a Direct Detection Optical Channel,"M.I.T. LIDS Report LIDS-P-1337, Oct. 1983; also *Proceedings. 23rd Allerton Conference on Communication, Control and Computing*, Monticello, Ill., Oct. 1985.

[10] R. Ahlswede, "Multi-way Communication Channels," *Proceedings of the 1971 IEEE International Symposium on Information Theory* Tsahkadsor, Armenian S.S.R., 1971.

[11] H. Liao, "A Coding Theorem for Multiple Access Communications," in *Proceedings of the 1972 IEEE International Symposium on Information Theory*, Asilomar, CA, 1972; also "Multiple Access Channels," Ph.D. thesis, Univ. of Hawii, 1972.

[12] A. El Gamal and T. M. Cover, "Multiple user Information Theory," *Proceedings of the IEEE*, Vol. 68, pp. 1466-1483, Dec. 1980

[13] E. C. van der Meulen, "A Survey of Multi-way Channels in Information Theory 1961-1976," *IEEE Transactions on Information Theory*, Vol. IT-23, Jan 1977.

[14] I. Csiszar and J. Korner, *Information Theory: Coding Theorems for Discrete Memoryless Systems*, New York: Academic Press 1981.

[15] T. Kasami, S. Lin, V. K. Wei, and S. Yamamura, "Graph Theoretic Approaches to the Code Construction for the Two-user Multiple-access Binary Adder Channel," *IEEE Transactions on Information Theory*, Vol. IT-20, pp. 114-130, Jan. 1983.

[16] M. B. Pursley, "Frequency-Hop Transmission for Satellite Packet Switching and Terrestrial Packet Radio Networks," *IEEE Transactions on Information Theory*, Vol. IT32, pp 652-668, Sept. 1986.

[17] Special Issue on Spread Spectrum: *IEEE Transactions on Communications*, May 1982.

[18] A. J. Viterbi, "Spread Spectrum Multiple Access with Binary Modulation can Approach Shannon Capacity for the Aggregate Gaussian Channel" Manuscript.

[19] S. Verdu, "Maximum Likelihood Sequence Detection for Intersymbol Interference Channels: A New Upper Bound on Error Probability," *IEEE Transactions on Information Theory*, Vol. IT-33, pp. 62-69, Jan. 1987.

[20] E. Arikan, *Sequential Decoding for Multiple Access Channels*, Ph. D. Thesis, Dept. of EECS, M.I.T., Nov. 1985.

[21] D. Slepian and J. Wolf, "A Coding Theorem for Multiple Access Channels with Correlated Sources," *Bell System Technical Journal*, Vol. 52, pp. 1037-1076, Sept. 1973.

[22] R. G. Gallager, "A Perspective on Multiaccess Channels," *IEEE Transactions on Information Theory*, Vol. IT-31, pp. 124-143, Mar. 1985.

[23] J. Y. N. Hui and P. A. Humblet, "The Capacity region of the Totally Asynchronous Multiple-Access Channel," *IEEE Transactions on Information Theory*, Vol. IT-31, pp. 207-217, Mar. 1985.

[24] T. M. Cover, R. J. McEliece, and E. C. Posner, "Asynchronous Multi-access channel Capacity," *IEEE Transactions on Information Theory*, Vol. IT-26,pp. 291-298, Nov. 1980.

Chapter 15

RECURSIVE CONSTRUCTION OF MULTI-CHANNEL TRANSMISSION LINES WITH A MAXIMUM ENTROPY PROPERTY

A. H. Sayed
Department of Electrical Engineering
University of California, Los Angeles, CA 90024 USA

T. Constantinescu
Programs in Mathematical Sciences
University of Texas, Richardson, TX 75083 USA

T. Kailath
Department of Electrical Engineering
Stanford University, Stanford, CA 94305 USA

Abstract J. P. Burg gave a now-famous solution of the problem of finding the maximum-entropy extension of a section $\{c_0, c_1, \ldots, c_n\}$ of a scalar (stationary) covariance sequence, with a nice interpretation in terms of the reflection coefficients of an associated discrete transmission line: namely, that the maximum-entropy extension is obtained by adding trivial sections with zero reflection coefficients. This work inspired various extensions and generalizations, mostly without any associated physical interpretation. In this paper we shall use the generalized transmission lines associated with the basic triangular matrix factorization algorithms of the Displacement Structure Theory to obtain several interesting maximum entropy extension results. One is the perhaps surprising fact that Burg's zero-reflection-coefficient extension solution may not always hold when the coefficients $\{c_i\}$ are matrix-valued.

DEDICATED TO G. DAVID FORNEY WITH ADMIRATION AND APPRECIATION.

*This work was partially supported by the National Science Foundation under award CCR-9732376 and by the Army Research Office under grant DAAH-04-96-1-0176.

1. INTRODUCTION

The breadth of Dave Forney's abilities, interest and contributions is truly remarkable. He has studied, developed, learned, and applied a surprising diversity of mathematical tools to a variety of problems in information, communication and system theory. At another end of the spectrum, he has been closely involved with software and hardware development, and with standards issues in the conception, development and successful deployment of pioneering communications modems.

It is not possible, for several reasons, to contribute a paper directly overlapping some of Dave's research areas. The present contribution is not too far afield, however, and in fact relates to one of the most appealing, at least to us, features of several of Dave's mathematical contributions. Namely, that the development of many of his apparently quite abstract results has been grounded in a concrete physical implementation: for example, his very important contributions to the algebraic structure of rational matrices are intimately connected to the structure and properties of convolutional codes.

In our work, such a concrete foundation for several purely mathematical results has been provided by generalized transmission line realizations of a set of fundamental matrix factorization algorithms that we have called Generalized Schur algorithms; in essence, these provide fast algorithms for a large variety of mathematical and physical problems by incorporating various kinds of Displacement Structure into the fundamental and ubiquitous Gaussian elimination algorithm (see [1], [2]).

In this paper we show how this generalized transmission line structure can be used to generalize Burg's famous result [3] on maximum entropy extensions of a scalar covariance sequence, $\{1, c_1, \ldots, c_N\}$. In the transmission line interpretation, the classical Schur algorithm (Burg used the closely related Levinson algorithm) is used to construct N sections of a lossless transmission line, each section being characterized by a scalar reflection coefficient (of course, less than unity in magnitude). All extensions of the given covariance sequence can be obtained by adding additional sections to the line; the maximum entropy extension is obtained by adding (trivial) sections with reflection coefficients that are zero. Another characterization is that all extensions are obtained by terminating the constructed N-section line with a passive load; the maximum entropy section corresponds to using a zero load.

Burg's result has been further studied and generalized in several ways, see *e.g.*, [4], [5]. This paper presents some further extensions with some surprising features, especially that in the apparently straightforward case where the c_i are matrix-valued, the zero load termination may not give the maximum entropy based extension (Section 7). An outline of the rest of the paper is as follows: in the next section we present classical material about the trigonometric moment problem and the Schur test for its solution. Then we describe the array form of Schur's algorithm and its scattering interpretation. The maximum entropy extension is introduced in Section 5. The last two sections contain our results concerning the connection with structured matrices and the construction of the multichannel maximum entropy extension.

2. CLASSICAL BACKGROUND MATERIAL

Consider a sequence of covariance, (or autocorrelation) coefficients $\{c_k\}$ of a zero-mean stationary random process $\{y_i\}$, say

$$c_k \triangleq \mathrm{E}\, y_i y_{i-k}^* \qquad \text{for} \quad k = \ldots -2, -1, 0, 1, 2 \ldots$$

where the letter E denotes expectation. The z–spectrum of the process $\{y_i\}$ is defined as the z-transform of its covariance sequence,

$$S_y(z) \triangleq \sum_{k=-\infty}^{\infty} c_k z^k \,. \tag{1}$$

The z–spectrum $S_y(z)$ has two fundamental properties:

i) Para-Hermitian symmetry, which means that $S_y(z)$ satisfies the relation $S_y(z) = \left[S_y\left(\frac{1}{z^*}\right)\right]^*$. This can be easily established from the definition (1) and the fact that $c_{-k} = c_k^*$ for all k.

ii) Nonnegativity, which means that $S_y(e^{j\omega}) \geq 0$. This is a nontrivial property that characterizes any z–spectrum of a truly stochastic process.

More importantly perhaps, is that the converse statement also holds. That is, any function of z obeying the above two properties must be the z–transform of a covariance (also known as a moment) sequence.

Two classical problems that were extensively studied at the beginning of the last century are the so-called trigonometric moment problem and the extension problem (see, *e.g.*, [6] for an account of these earlier contributions).

Problem 1 (Trigonometric Moment Problem) *Given a sequence of complex numbers $\{\ldots c_{-2}, c_{-1}, c_0, c_1, c_2 \ldots\}$, satisfying $c_k = c_{-k}^*$, find necessary and sufficient conditions for the sequence to be a moment (or covariance) sequence.*

In other words, the above problem seeks conditions that guarantee that a given sequence $\{c_k\}$ can be regarded as the covariance sequence of a true stochastic process, *viz.*, that its entries can be recovered via a relation of the form

$$c_k = \frac{1}{2\pi}\int_{-\pi}^{\pi} e^{-j\omega k} d\mu(\omega) \tag{2}$$

for some positive measure μ on $[-\pi, \pi)$ (say $d\mu(\omega) = S_y(e^{j\omega})d\omega$, for some power spectral density function $S_y(e^{j\omega})$.

Problem 2 (Extension Problem) *Given a finite sequence of complex numbers $\{c_{-n}, \ldots, c_{-1}, c_0, c_1, \ldots, c_n\}$, satisfying $c_k = c^*_{-k}$, describe all possible extensions $\{c_k, |k| > n\}$ that result in a moment sequence.*

The two problems are related and, not surprisingly, have applications in many areas of statistical signal processing (*e.g.*, in the form of maximum entropy spectral estimation), estimation theory (*e.g.*, Wiener-Hopf theory and spectral factorizations), circuit theory (*e.g.*, lossless and passive circuit design), and control theory (*e.g.*, positive-real systems and the so-called Kalman-Yakubovich-Popov characterization of such systems).

The solutions of the above two problems were given in a series of works by Toeplitz, Carathéodory, and Schur. They were later extended, and at times significantly generalized, in several other directions.

In particular, Toeplitz (1907) showed that the sequence $\{c_k$, such that $c_k = c^*_{-k}\}_{-\infty}^{\infty}$ is a moment sequence if, and only if, the matrices (now known as *Toeplitz* matrices)

$$T_k \triangleq \begin{bmatrix} c_0 & c_{-1} & c_{-2} & \cdots & c_{-k} \\ c_1 & c_0 & c_{-1} & \cdots & c_{-k+1} \\ c_2 & c_1 & c_0 & \ddots & \vdots \\ \vdots & \vdots & \ddots & \ddots & c_{-1} \\ c_k & c_{k-1} & \cdots & c_1 & c_0 \end{bmatrix} \quad \text{for } k = \ldots, -1, 0, 1, \ldots \tag{3}$$

are positive semidefinite for all k, *i.e.*, $T_k \geq 0$. Checking this condition would, at first sight, involve checking the nonnegativity of large determinantal formulas. Carathéodory (1911) connected this nonnegativity condition with function theory. He defined the power series

$$c(z) \triangleq c_0 + 2\sum_{k=1}^{\infty} c_k z^k , \tag{4}$$

and showed that $\{c_k\}_{-\infty}^{\infty}$ is a moment sequence if, and only if,

$$c(z) \text{ is analytic and } \mathrm{Re}\,[c(z)] \geq 0 \text{ in } |z| < 1. \tag{5}$$

In other words, $c(z)$ needs to be a positive-real function (now known as a *Carathéodory* function). Following this work, Schur (1917) considered the bilinear transformation

$$s(z) \triangleq \frac{c(z) - c_0}{c(z) + c_0}, \tag{6}$$

and noted that $c(z)$ is positive-real if, and only if, $s(z)$ is analytic and bounded by unity ($|s(z)| \leq 1$) in $|z| < 1$ (such functions are now called *Schur functions*). Moreover, and strikingly for the times, Schur presented a recursive test (rather than large determinantal expressions) for checking if a scalar function $s(z)$ is of Schur type or not.

Theorem 1 (Schur's Test) *Consider the following recursive algorithm that starts with a given initial function $s(z)$,*

$$s_{i+1}(z) = \frac{1}{z}\frac{s_i(z) - \gamma_i}{1 - \gamma_i^* s_i(z)}, \quad s_0(z) = s(z), \quad \textit{and} \quad \gamma_i = s_i(0). \tag{7}$$

Then the following statements hold:

- $s(z)$ *is analytic and bounded by unity in* $|z| < 1 \iff |\gamma_i| \leq 1$ *for all* i.
- $|\gamma_i| < 1$ *for* $0 \leq i < n$ *and* $|\gamma_n| = 1$ *for some* n *if, and only if,* $s(z)$ *is a finite Blaschke product of degree* n *(i.e., a rational all-pass function). This means that* $s(z)$ *is analytic and bounded by unity in* $|z| < 1$ *and* $|s(e^{j\omega})| = 1$.
- *Starting with a Schur function* $s(z)$*, each function* $s_i(z)$ *is also of Schur type.*

The above theorem also shows that Schur functions can be of two types. One type is characterized by reflection coefficients that are always strictly bounded by one, $\{|\gamma_i| < 1\}$ for all i. The other type corresponds to all-pass functions and is characterized by a finite set of initial reflection coefficients that are strictly bounded by one, say $\{|\gamma_i| < 1,\ 0 \leq i < n\}$, followed by $|\gamma_n| = 1$. It can be shown that any Schur-type function $s(z)$ that satisfies

$$\int_{-\pi}^{\pi} \log[1 - |s(e^{j\omega})|^2]d\omega \; > \; -\infty \tag{8}$$

belongs to the first type. That is, all its reflection coefficients will be strictly less than one in magnitude. Now, in view of relation (10), the condition (8) is implied by the following so-called Paley-Wiener condition on the power spectral density function of the random process $\{y_i\}$,

$$\int_{-\pi}^{\pi} \log[S_y(e^{j\omega})]d\omega > -\infty. \tag{9}$$

A stationary stochastic process that satisfies this condition is said to be *purely nondeterministic* to indicate that even prediction from an infinite past leaves a nonzero residual error. This is because in prediction theory, the integral in (9) is related to the prediction error energy from an infinite-order prediction problem. More specifically, and briefly, let $\hat{y}_{i|i-1:i-m}$ denote the linear least-mean-squares estimate of y_i that is based on the past m data, and define

$$\sigma_m^2 \triangleq E\left|y_i - \hat{y}_{i|i-1:i-m}\right|^2 .$$

That is, σ_m^2 is the variance of the corresponding prediction error (its independence of i is due to the stationarity assumption on the random process $\{y_i\}$). The asymptotic behaviour of σ_m^2 as $m \to \infty$ has been studied in some detail in [7, 8]. In Chapter 2 of the first reference it is shown that

$$\lim_{m\to\infty} \sigma_m^2 = \exp\left[\frac{1}{2\pi}\int_{-\pi}^{\pi} \log[S_y(e^{j\omega})]d\omega\right],$$

which explains our earlier claim that for a purely nondeterministic process $\{y_i\}$, the prediction residual variance is finite even for an infinite-order prediction. In fact, it is further shown in [7] that this happens if, and only if, the sequence of reflection coefficients is square summable,

$$\sum_{i=1}^{\infty} |\gamma_i|^2 < \infty.$$

This is known as Szegö's formula and is a celebrated result in the prediction theory of stationary processes.

We therefore have a sequence of results that address the moment problem by working with the interrelated quantities $\{c(z), s(z), S_y(z)\}$, which can be seen to be related as follows:

$$1 - s(z)\left[s\left(\frac{1}{z^*}\right)\right]^* = \frac{4S_y(z)}{[c(z)+1][c^*(z^{-*})+1]}. \tag{10}$$

But what about the extension problem? It is well-known that the solution is highly nonunique and that, among all solutions, a so-called central solution with a maximum entropy property stands out. This central solution has received considerable attention in the literature, especially since Burg's original work on spectral analysis [3], and it can be associated with the construction of a special scattering cascade with a zero load.

The purpose of this article is to show how to construct more general scattering cascades in a multichannel scenario under a certain maximum entropy requirement. The cascades are more general not only in the sense that they involve multi-input/multi-output ports but also in that the pure delays that are associated with classical scattering media are now replaced by first-order Blaschke functions. In addition, and in order to meet the maximum entropy property, special care is needed in constructing the scattering cascade and in choosing the load.

3. ARRAY FORM OF SCHUR'S ALGORITHM

It turns out that a convenient framework to pursue these extensions is by the study of matrices with so-called displacement structure (see, e.g., [1, 9] and the references therein). A generalized form of Schur's recursion plays a significant role in this development and is therefore reviewed next.

To begin with, note that Schur's recursion (7) is nonlinear in $s_i(z)$. A linearized form can be obtained by expressing $s_i(z)$ as the ratio of two power series, say

$$s_i(z) \triangleq \frac{y_i(z)}{x_i(z)},$$

so from (7) we must have

$$\frac{y_{i+1}(z)}{x_{i+1}(z)} = \frac{y_i(z) - \gamma_i x_i(z)}{z\left[x_i(z) - \gamma_i^* y_i(z)\right]}.$$

We can therefore employ the following updates for the numerator and denominator of $s_{i+1}(z)$:

$$y_{i+1}(z) = \alpha_i\left[y_i(z) - \gamma_i x_i(z)\right] \qquad (11)$$

$$x_{i+1}(z) = \alpha_i z\left[x_i(z) - \gamma_i^* y_i(z)\right] \qquad (12)$$

for some arbitrary nonzero complex number α_i. If we choose (assuming $|\gamma_i| < 1$, which will be our focus in this paper for reasons to become clear as we progress),

$$\alpha_i \triangleq \frac{1}{\sqrt{1 - |\gamma_i|^2}},$$

then we can write (11)–(12) in compact matrix form as follows (after extracting a z term)

$$z\begin{bmatrix} x_{i+1}(z) & y_{i+1}(z) \end{bmatrix} = \begin{bmatrix} x_i(z) & y_i(z) \end{bmatrix} \underbrace{\Theta_i \begin{bmatrix} z & 0 \\ 0 & 1 \end{bmatrix}}_{\Theta_i(z)} \tag{13}$$

where Θ_i is the 2×2 elementary hyperbolic rotation,

$$\Theta_i \triangleq \frac{1}{\sqrt{1-|\gamma_i|^2}} \begin{bmatrix} 1 & -\gamma_i \\ -\gamma_i^* & 1 \end{bmatrix}, \quad \gamma_i = \lim_{z \to 0} \frac{y_i(z)}{x_i(z)},$$

and $\Theta_i(z)$ is a so-called 2×2 J–lossless elementary section,

$$\Theta_i(z) \triangleq \Theta_i \begin{bmatrix} z & 0 \\ 0 & 1 \end{bmatrix}. \tag{14}$$

It is J–lossless because it is analytic in $|z| < 1$ and is J–unitary on the unit circle,

$$\Theta_i(e^{j\omega}) J \left[\Theta_i(e^{j\omega})\right]^* = J, \quad J = \begin{bmatrix} 1 & 0 \\ 0 & -1 \end{bmatrix}.$$

[This follows from the fact that the hyperbolic rotation satisfies $\Theta_i J \Theta_i^* = J$.]

We further invoke the power series expansions of $x_i(z)$ and $y_i(z)$, say

$$\begin{aligned} x_i(z) &= x_{ii} + x_{i+1,i}z + x_{i+2,i}z^2 + \ldots \\ y_i(z) &= y_{ii} + y_{i+1,i}z + y_{i+2,i}z^2 + \ldots \end{aligned}$$

and collect the coefficients $\{x_{ik}, y_{ik}\}$ into an array:

$$G_i \triangleq \begin{bmatrix} x_{ii} & y_{ii} \\ x_{i+1,i} & y_{i+1,i} \\ x_{i+2,i} & y_{i+2,i} \\ \vdots & \vdots \end{bmatrix}.$$

Then recursion (13) can be interpreted as follows:

$$G_i = \begin{bmatrix} \times & \times \\ \times & \times \\ \times & \times \\ \vdots & \vdots \end{bmatrix} \xrightarrow{\Theta_i(\gamma_i)} \begin{bmatrix} \times' & 0 \\ \times' & \times' \\ \times' & \times' \\ \vdots & \vdots \end{bmatrix} \xrightarrow{shift} \begin{bmatrix} 0 & 0 \\ \times'' & \times' \\ \times'' & \times' \\ \vdots & \vdots \end{bmatrix} = \begin{bmatrix} 0 \quad 0 \\ G_{i+1} \end{bmatrix}.$$

In words,

1. We start with an array G_i and define γ_i as the ratio of the entries of its top row.

2. We apply Θ_i to G_i and thus annihilate the rightmost entry of its top row.

3. We then shift down the first column of $G_i\Theta_i$, while leaving the second column unaltered.

4. The resulting array has a top zero row. Its lower part is G_{i+1}, which contains the coefficients of the power series expansions of $\{x_{i+1}(z), y_{i+1}(z)\}$ that define $s_{i+1}(z)$.

In matrix language, (13) and the above array picture can be expressed as

$$\begin{bmatrix} 0 \quad 0 \\ G_{i+1} \end{bmatrix} = \mathcal{Z}G_i\Theta_i \begin{bmatrix} 1 & 0 \\ 0 & 0 \end{bmatrix} + G_i\Theta_i \begin{bmatrix} 0 & 0 \\ 0 & 1 \end{bmatrix}, \tag{15}$$

where $\mathcal{Z}$ denotes the (semi-infinite) lower triangular shift matrix with ones on the first subdiagonal and zeros elsewhere.

We therefore described an equivalent description of Schur's algorithm in terms of a procedure that involves only matrix operations on arrays of numbers. Such matrix-based descriptions are very useful and also very powerful; for one thing, they allow us to pursue generalizations that are difficult to develop (or perhaps envision) if we persist with the function language.

We can now state Schur's solution to the trigonometric moment problem in this array form and using the bilinear representation (6).

Algorithm 1 (Schur's Solution to the Moment Problem) *Apply Schur's test to $s(z) = y_0(z)/x_0(z)$ where,*

$$y_0(z) = \frac{1}{2}[c(z) - c_0] \quad \textit{and} \quad x_0(z) = \frac{1}{2}[c(z) + c_0],$$

or, equivalently, to the array

$$G_0 = \begin{bmatrix} c_0 & 0 \\ c_1 & c_1 \\ c_2 & c_2 \\ c_3 & c_3 \\ \vdots & \vdots \end{bmatrix},$$

and verify whether the resulting reflection coefficients $\{\gamma_i\}$ are bounded by one, $|\gamma_i| \leq 1$. More specifically,

1. *If* $|\gamma_j| > 1$ *for some* j, *then the sequence* $\{c_k\}$ *is not a moment sequence.*

2. *If* $|\gamma_j| < 1$ *for* $0 \leq j < n$ *and* $|\gamma_n| = 1$ *at some iteration* n, *then* $\{c_k\}$ *is a moment sequence.*

3. *If* $|\gamma_j| < 1$ *for all* j *then* $\{c_k\}$ *is also a moment sequence.*

4. THE SCATTERING FRAMEWORK

An important aspect of Schur's algorithm is that it admits a physical interpretation in terms of a cascade of elementary sections that combine together to form a layered-medium structure. Such structures play a fundamental role in modern signal processing.

Returning to the array recursion (15), note that it can be graphically depicted as a cascade of elementary sections as shown in Figure 1, where we defined

$$\gamma_i^c \triangleq \sqrt{1 - |\gamma_i|^2} \,.$$

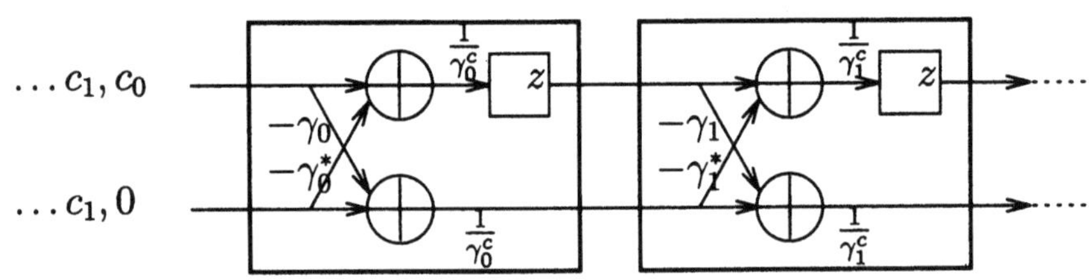

Figure 1 The feedforward structure associated with Schur's recursion.

Each section consists of a 2×2 hyperbolic rotation Θ_i followed by a unit-time delay (or storage) element denoted by z. The cascade represents a feedforward (and pipelineable) implementation of the array algorithm, where the entries of the two columns of G_0 are available at the input lines of the first section.

We can regard each section as a 2–input, 2–output system with transfer function $\Theta_i(z)$ as in (14). Moreover, our convention in this paper is that inputs to a transfer matrix function are applied to its left (they are row input vectors) and, correspondingly, outputs are also row vectors. Hence, we may write (see Figure 2)

$$\begin{aligned} \begin{bmatrix} o_1(z) & o_2(z) \end{bmatrix} &= \begin{bmatrix} i_1(z) & i_2(z) \end{bmatrix} \Theta_i(z) \\ &= \begin{bmatrix} i_1(z) & i_2(z) \end{bmatrix} \begin{bmatrix} 1/\gamma_i^c & -\gamma_i/\gamma_i^c \\ -\gamma_i^*/\gamma_i^c & 1/\gamma_i^c \end{bmatrix} \begin{bmatrix} z & 0 \\ 0 & 1 \end{bmatrix}, \end{aligned}$$

to denote the input-output map of each elementary section in the z-transform domain, with $\{i_1(z), i_2(z)\}$ generically denoting the input signals and $\{o_1(z), o_2(z)\}$ denoting the output signals.

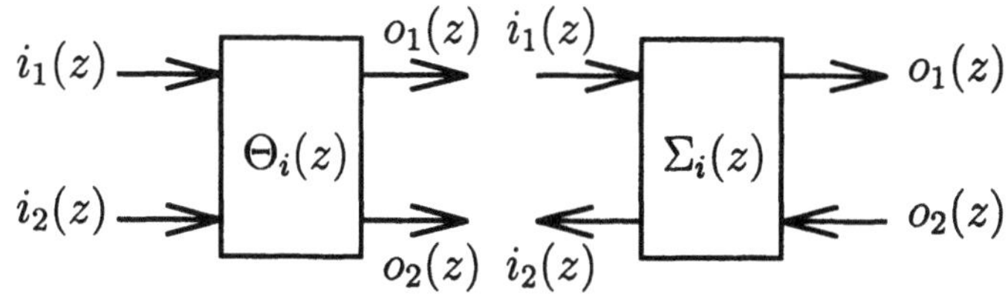

Figure 2 The scattering formulation for individual sections.

By reversing the direction of flow in a section we obtain the mapping

$$\left[\ o_1(z) \quad i_2(z)\ \right] = \left[\ i_1(z) \quad o_2(z)\ \right] \underbrace{\begin{bmatrix} \gamma_i^c & \gamma_i \\ -\gamma_i^* & \gamma_i^c \end{bmatrix} \begin{bmatrix} z & 0 \\ 0 & 1 \end{bmatrix}}_{\Sigma_i(z)}$$

where $\{i_1, o_2\}$ are now treated as input ports and $\{o_1, i_2\}$ are now treated as output ports (see Figure 2). Each section $\Sigma_i(z)$ is now composed of a unitary gain matrix Σ_i ($\Sigma_i \Sigma_i^* = I$) followed by a unit-time delay element.

By combining several elementary sections $\Theta_i(z)$ we obtain a feedforward cascade whose overall 2×2 transfer matrix we shall denote by $\Theta(z)$ and is given by (*cf.* Figure 1),

$$\Theta(z) = \Theta_0(z)\Theta_1(z)\Theta_2(z)\ldots$$

If we partition its entries accordingly with $J = (1 \oplus -1)$, say

$$\Theta(z) \triangleq \begin{bmatrix} \Theta_{11}(z) & \Theta_{12}(z) \\ \Theta_{21}(z) & \Theta_{22}(z) \end{bmatrix}, \tag{16}$$

where $\Theta_{ij}(z)$ are 1×1 transfer functions, then the entries of the scattering matrix $\Sigma(z)$ that corresponds to the combination of the corresponding scattering sections $\Sigma_i(z)$, i.e.,

$$\Sigma(z) = \Sigma_0(z) \star \Sigma_1(z) \star \Sigma_2(z) \star \ldots,$$

where $\star$ denotes the Redheffer product, are given by

$$\Sigma(z) = \begin{bmatrix} \Theta_{11}(z) - \Theta_{12}(z)\Theta_{22}^{-1}(z)\Theta_{21}(z) & -\Theta_{12}(z)\Theta_{22}^{-1}(z) \\ \Theta_{22}^{-1}(z)\Theta_{21}(z) & \Theta_{22}^{-1}(z) \end{bmatrix}. \tag{17}$$

The scattering cascade is depicted in Figure 3, where we are denoting the entries of the sequences at its left-most ports generically by $\{x_{k0}, y_{k0}\}$.

[In particular, if we use the special input sequences of Figure 1, then $\{x_{k0}\} = \{c_0, c_1, \ldots\}$ and $\{y_{k0}\} = \{0, c_1, \ldots\}$. The transfer function from the left-most input to output ports in the scattering cascades is seen to be $-\Theta_{12}(z)\Theta_{22}^{-1}(z)$.

We are now in a position to describe a solution to the extension problem

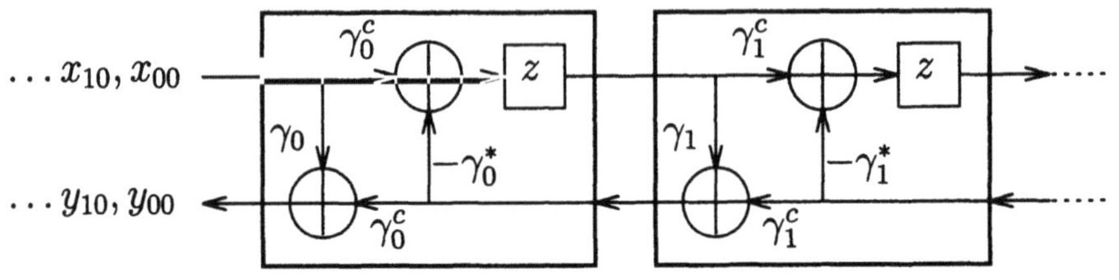

Figure 3 The feedback structure associated with Schur's algorithm.

that we posed earlier. Indeed, it is well-known that the above cascade can be used to solve the following slightly more specialized version of the extension problem, in which we require the Carathéodory function to be strictly positive-real.

Problem 3 (Specialized Extension Problem)
Given a finite sequence of complex numbers $\{c_{-n}, \ldots, c_{-1}, c_0, c_1, \ldots, c_n\}$, *satisfying* $c_k = c_{-k}^*$, *describe all possible extensions* $\{c_k, |k| > n\}$ *that result in a moment (or covariance) sequence such that the corresponding* $c(z)$ *is strictly positive-real.*

The strict positivity requirement on $c(z)$ guarantees that the corresponding Schur function $s(z)$ that is obtained via the bilinear relation (6) is strictly bounded by unity in $|z| \leq 1$. We say that it is a strict Schur function and, in view of (8), its reflection coefficients will all be strictly bounded by one. This further means that the extended moment sequence will be the covariance sequence of a purely nondeterministic stationary stochastic process.

Now given the first $n+1$ moment values $\{c_k, 0 \leq k \leq n\}$ (and their complex conjugate values), we can determine the coefficients $\{s_k, 0 \leq k \leq n\}$ from the bilinear relation (6). Likewise, given $\{s_k, 0 \leq k \leq n\}$ (and c_0), we can determine the corresponding $\{c_k, 0 \leq k \leq n\}$. In order to guarantee a one-to-one correspondence between the $\{c_k\}$ and the $\{s_k\}$, we shall assume without loss of generality that the moment sequence is normalized so that $c_0 = 1$. We therefore pose the following equivalent problem.

Problem 4 (Interpolation Problem) *Given $\{s_k, 0 \leq k \leq n\}$, satisfying $s_0 = 0$, what are the possible extensions $\{s_k, k > n\}$ that result in a strict Schur function $s(z)$?*

The reason we refer to the above equivalent formulation as an interpolation problem is that the given coefficients $\{s_k\}$ can be interpreted as interpolation conditions on a function $s(z)$, *viz.*, we seek strict Schur functions $s(z)$ that satisfy

$$\frac{1}{k!}\frac{d^k}{dz^k}s(z)\bigg|_{z=0} = s_k \,. \tag{18}$$

This is a special analytic interpolation problem that can be solved in many different ways. For our purposes in this article, it is convenient to consider a solution that is based on the scattering cascade of Figure 3 [10].

Given $\{s_k, s_0 = 0, 0 \leq k \leq n\}$, we form the matrix

$$G_0 = \begin{bmatrix} 1 & s_0 \\ 0 & s_1 \\ 0 & s_2 \\ \vdots & \vdots \\ 0 & s_n \end{bmatrix}, \tag{19}$$

and apply Schur's recursion (15) to these rows. Notice that in Section 3. we used the notation G_i (and G_0) to refer to a semi-infinite matrix. Here, we are using the same notation for a finite matrix (with $(n+1)$ rows). Referring to the array formulation (15) of Schur's algorithm we see that, for any k, the first k rows of the resulting array

$$\begin{bmatrix} 0 \;\; 0 \\ G_{i+1} \end{bmatrix}$$

are completely determined by the first k rows of the array G_i. Hence, when we are only interested in a finite number of Schur steps, we can use a finite-dimensional version of (15), which we write as

$$\begin{bmatrix} 0 \;\; 0 \\ G_{i+1} \end{bmatrix} = ZG_i\Theta_i \begin{bmatrix} 1 & 0 \\ 0 & 0 \end{bmatrix} + G_i\Theta_i \begin{bmatrix} 0 & 0 \\ 0 & 1 \end{bmatrix}, \tag{20}$$

where Z now denotes a finite-dimensional lower triangular shift matrix (of appropriate dimensions) with ones on the first subdiagonal and zeros elsewhere.

By applying (20) to the above G_0, we identity $(n+1)$ reflection coefficients $\{\gamma_i, i = 0, \ldots, n\}$. If the coefficients $\{s_k, 0 \leq k \leq n\}$ we started with arose from numbers $\{c_k, c_0 = 1, c_k = c^*_{-k}, 0 \leq k \leq n\}$ that are part of a valid strict moment sequence, then the result of Schur's test in Theorem 1 (and the corresponding array solution in Algorithm 1) indicates that all these $(n+1)$ reflection coefficients must be strictly bounded by one. If at any step k, $0 \leq k \leq n$, we find $|\gamma_k| \geq 1$, then the above interpolation problem is not solvable.

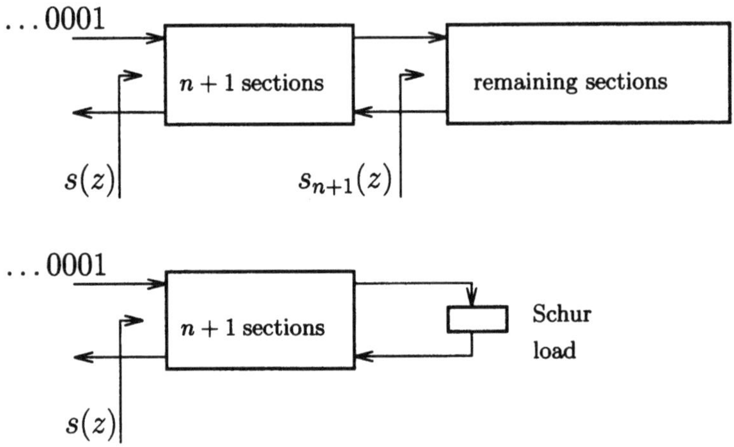

Figure 4 A scattering interpretation of the first $(n+1)$ steps.

Assuming that we succeed in getting $\{|\gamma_i| < 1, 0 \leq i \leq n\}$, we then obtain the $(n+1)$ sections of the feedforward cascade of Figure 1, with individual transfer functions $\{\Theta_i(z), 0 \leq i \leq n\}$. The corresponding overall transfer matrix is

$$\Theta(z) = \Theta_0(z)\Theta_1(z)\ldots\Theta_n(z) .$$

It is analytic in $|z| \leq 1$ and J–unitary on the unit circle. This property in turn implies that if we partition the entries of $\Theta(z)$ as in (16) then $-\Theta_{12}(z)\Theta_{22}^{-1}(z)$ is a strict Schur function.

The corresponding scattering cascade $\Sigma(z)$ is shown schematically in Figure 4. According to the partitioning (17), the above strict Schur function is the $(1,2)$ entry of $\Sigma(z)$ and is equal to the transfer function at the input of the scattering cascade of these $(n+1)$ sections. This transfer function is denoted by $s(z)$ in Figure 4 because, besides being a strict Schur function, it also satisfies the required interpolation conditions. There are many ways to see this. Here we follow [2, Sec. 3.4]. Let $G_0(z)$ denote the z–transform of the initial array G_0,

$$G_0(z) \triangleq \begin{bmatrix} 1 & z & z^2 & \ldots & z^n \end{bmatrix} G_0 = \begin{bmatrix} 1 & \sum_{k=1}^{n} s_k z^k \end{bmatrix} . \quad (21)$$

Then recall from the linearized function form (13) that the array algorithm (20) creates successive arrays G_i whose z-transforms satisfy

$$zG_{i+1}(z) = G_i(z)\Theta_i(z) .$$

This means that each step of Schur's algorithm produces a new row function $zG_{i+1}(z)$ with one more zero at the origin than the previous function $G_i(z)$. Therefore, after $(n+1)$ such steps and starting with $G_0(z)$ we must have

$$G_0(z)\Theta_0(z)\Theta_1(z)\ldots\Theta_n(z) = G_0(z)\Theta(z) = O(z^{n+1}) .$$

The notation $O(z^{n+1})$ means that the powers of z^i, $0 \leq i \leq n$, in the Taylor series expansion of $G_0(z)\Theta(z)$ around $z = 0$ are all zero. Using the partitionings (16) and (21) and defining $s(z) = -\Theta_{12}(z)\Theta_{22}^{-1}(z)$, we obtain that

$$s(z) = \left[\sum_{k=1}^{n} s_k z^k\right] + O(z^{n+1}) .$$

This establishes that $s(z)$ satisfies (18).

In summary, we showed that the $(1,2)$ entry of the scattering cascade that is obtained from G_0 gives us one solution $s(z)$ in the form of the transfer function at the input of the cascade. But what about other solutions? It turns out that all of these can be obtained by attaching to the right side of $\Sigma(z)$ any strict Schur-type function (see, e.g., [5]). This is indicated in Figure 4, where $s_{n+1}(z)$ denotes the transfer function at the input terminals of the load and the block denoted "remaining sections" would correspond to a scattering implementation of this Schur load, if desired. The sequence $\{\ldots, 0001\}$ in the figure is used to indicate a unit pulse sequence and, therefore, that the impulse response of such a cascade will be $\{\boxed{s_0}, s_1, \ldots, s_n,\}$ followed by valid $\{s_k, k > n\}$ such that the corresponding $s(z)$ is a Schur function that satisfies the interpolation conditions.

In the above, we constructed the cascade $\Sigma(z)$ by using G_0 in (19). Alternatively, we could have also started with

$$G_0 \;=\; \begin{bmatrix} c_0 & 0 \\ c_1 & c_1 \\ c_2 & c_2 \\ \vdots & \vdots \\ c_n & c_n \end{bmatrix} . \tag{22}$$

This is because the resulting reflection coefficients, which fully characterize the cascade, will be the same as can be seen by using again the bilinear transformation (6).

Algorithm 2 (Solution of the Interpolation Problem) *Given $\{s_k, 0 \leq k \leq n\}$, we apply $(n+1)$ steps of Schur's recursion (20) to the matrix*

$$G_0 = \begin{bmatrix} 1 & s_0 \\ 0 & s_1 \\ 0 & s_2 \\ \vdots & \vdots \\ 0 & s_n \end{bmatrix},$$

and construct the scattering cascade that corresponds to the first $(n+1)$ sections. We then terminate the cascade with any strict Schur-type function. The transfer function at the input of the combined cascade solves the interpolation problem.

It is well-known that this procedure can be used to solve the original extension problem (which asks for the coefficients $\{c_k\}$ rather than the $\{s_k\}$) by means of a perfect reflection experiment, as shown in Figure 5 and as described in the following statement.

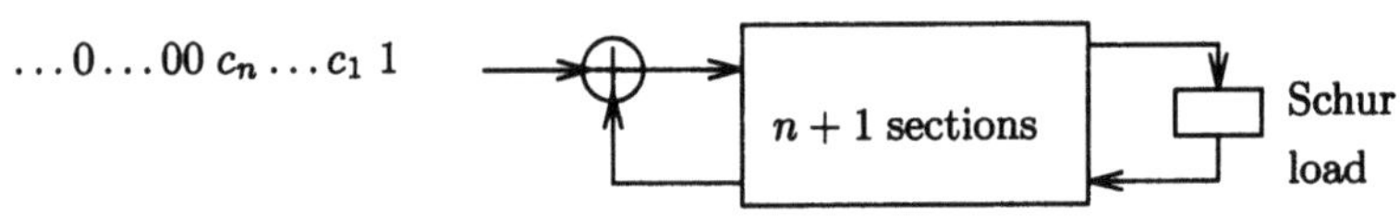

Figure 5 The perfect reflection experiment.

Algorithm 3 (Solution of the Extension Problem) *Given $\{c_k, c_0 = 1, |k| < n\}$, all valid extensions to a moment sequence can be obtained as follows:*

1. *Apply $(n+1)$ steps of Schur's recursion (15) to the matrix*

$$G_0 = \begin{bmatrix} 1 & 0 \\ c_1 & c_1 \\ c_2 & c_2 \\ \vdots & \vdots \\ c_n & c_n \end{bmatrix},$$

and construct the scattering cascade that corresponds to the first $(n+1)$ sections. We then terminate the cascade with any strict Schur-type function. The problem has a solution only if this step *can be performed and results in all γ_i strictly less than unity in magnitude.*

2. *Connect the left-most output through an adder to the left-most input (cf. Figure 5).*

3. *Excite the system with the input sequence $\{\ldots 0 \ldots 00\, c_n \ldots c_1 \boxed{1}\}$ and measure the values at the left-most output for time instants $k \geq n+1$. These values provide the successive extension values $\{c_j\}$.*

5. MAXIMUM ENTROPY EXTENSIONS

The discussion in the last section shows that the solution of the extension problem, and of its equivalent interpolation problem, is highly nonunique due to the freedom in choosing the Schur load. Different loads would lead to different solutions and we therefore have a multitude of solutions.

Among all possibilities, the one that is most natural (in the sense that it is the most random or imposes the least constraints on the solution) corresponds to the so-called *maximum entropy* choice. The maximum entropy principle has been often used in the literature to choose, among the many solutions of an underdetermined problem, the most natural (or the least-constrained) one — see [3, 11].

The maximum entropy criterion relies on the concept of differential entropy of a random variable. Let $f(y)$ denote the probability density function of a random variable y. Its differential entropy is defined by [11]

$$h(y) \triangleq -\int_{\text{SUP}} f(y)\ln(f(y))dy = -E\left[\ln f(y)\right], \tag{23}$$

where SUP is the support set of the random variable. When y has a normal distribution, say

$$f(y) = \frac{1}{\sqrt{2\pi\sigma^2}} \exp^{-\frac{y^2}{2\sigma^2}},$$

its differential entropy can be shown to be $h(y) = \frac{1}{2}\ln(2\pi e\sigma^2)$ (where $\ln e = 1$). If y is an n–dimensional vector-valued random variable that is still normally distributed with covariance matrix R_y, say

$$f(y) = \frac{1}{\left(\sqrt{2\pi}\right)^n \sqrt{\det R_y}} \exp^{-\frac{1}{2}y^T R_y^{-1} y},$$

then its differential entropy reduces to $h(y) = \frac{1}{2}\ln[(2\pi e)^n \det R_y]$. In the limit, when y becomes a Gaussian random process (*viz.*, a multivariable random vector with infinitely many components), its differential entropy reduces to

$$h(y) = \frac{1}{2\pi}\int_{-\pi}^{\pi} \ln[S_y(e^{j\omega})]d\omega \,, \tag{24}$$

where $S_y(e^{j\omega})$ is the power spectral density function of the process $\{y_i\}$.

This last expression motivates us to consider the following problem formulation. Given a finite-length sequence $\{c_k, c_0 = 1, 0 \leq k \leq n\}$, we know how to extend it to an infinitely-long valid moment sequence (that is strictly positive-real). Now since there are many solutions, one for each Schur-type load, we also obtain many possible z–spectra $S_y(z)$. Among all possible extensions of the moment sequence we seek that one whose z–spectrum is such that it has the most entropy.

Problem 5 (Maximum Entropy Problem) *Given a finite-length sequence*

$$\{c_k, c_0 = 1, 0 \leq k \leq n\} \,,$$

we wish to extend it, if possible, to a strictly positive-real moment sequence $\{c_k, |k| \geq 0\}$ that meets the following optimization criterion:

$$\max_{c_k, |k|>n} \left[\frac{1}{2\pi}\int_{-\pi}^{\pi} \ln[S_y(e^{j\omega})]d\omega\right] \,,$$

which is also equivalent to

$$\max_{\text{Schur load}} \left[\frac{1}{2\pi}\int_{-\pi}^{\pi} \ln[S_y(e^{j\omega})]d\omega\right] .$$

We can rewrite the cost function that appears in the above statement in a slightly different form that will be useful for our generalizations further ahead. For this purpose, we first remark that we can write

$$\frac{1}{2\pi}\int_{-\pi}^{\pi} \ln[S_y(e^{j\omega})]d\omega = 4\pi \ln\left[L(0)\right] , \tag{25}$$

where $L(z)$ is the so-called Szegö function of $S_y(z)$. It is defined as

$$L(z) \triangleq \exp\left(\frac{1}{4\pi}\int_{-\pi}^{\pi} \frac{e^{j\omega}+z}{e^{j\omega}-z} \ln\left[S_y(e^{j\omega})\right] d\omega\right) . \tag{26}$$

The Szegö function $L(z)$ is an outer function (*i.e.*, both $L(z)$ and its inverse are analytic in $|z| < 1$) that has the property that

$$|L(e^{j\omega})|^2 = S_y(e^{j\omega}),$$

almost everywhere. Consequently, it can be chosen as the spectral factor of $S_y(z)$.

Using the earlier relation (10) we also see that this spectral factor can be chosen as (recall that we are taking $c_0 = 1$)

$$L(z) = [c(z) + 1]L_s(z) \ , \tag{27}$$

where $L_s(z)$ is the spectral factor of the function $1 - s(z)\left[s\left(\frac{1}{z^*}\right)\right]^*$. Equation (27) therefore shows that $L(0) = 2L_s(0)$, and the maximum entropy problem becomes equivalent to the following:

$$\max_{\text{Schur load}} \ [4\pi \ln L(0)] \iff \max_{\text{Schur load}} \ [8\pi \ln L_s(0)] \tag{28}$$

In other words, we would like to choose the moments $\{c_k, |k| > n\}$ so as to maximize the value at zero of either of the spectral factors, $L(z)$ or $L_s(z)$.

It is shown in [3] that the solution to the above problem is obtained when the Schur load in the scattering cascade of Figure 4 is taken to be simply zero. This guarantees that there will be no direct path from the input to the output at the right-end of the scattering cascade; a condition that, as we shall explain further ahead, guarantees the maximum entropy property.

This construction that corresponds to the choice of the zero load is known as the *central solution*, and it arises in many contexts in systems, control, and estimation.

Algorithm 4 (Solution of Maximum Entropy Problem) *Simply use Algorithm 3 with the arbitrary Schur load in step 1 replaced by the zero load.*

6. GENERALIZED SCATTERING CASCADES

In this section we consider a generalization of the maximum entropy formulation of the previous section. The generalization is in two respects:

1. We study scattering cascades that involve multiple-input and multiple-output ports. In this case, the construction of maximum entropy solutions needs more care.

2. We replace the pure delay elements z that appear in the cascade of Figure 3 by more general Blaschke factors, *viz.*, first-order rational all-pass functions of the form $\frac{z-f_i}{1-f_i^* z}$, for some $|f_i| < 1$.

It turns out that a convenient framework to pursue this extension is within the displacement structure theory (see, *e.g.*, [1, 9] for recent surveys and examples of many applications). No major background in the theory is required to follow the discussions in the sequel, especially since we shall motivate most of the arguments. Some basic background in matrix theory is enough.

6.1 STRUCTURED MATRICES

The displacement structure concept provides a unifying framework for handling a variety of problems in signal processing, control, and mathematics, that involve some sort of matrix structure. For the moment and extension problems of the earlier sections, the matrix structure that we encountered was the Toeplitz structure in the matrices T_k in (3). It turns out that the Schur algorithm, when studied in the context of such structured matrices, can be extended and generalized rather significantly.

Consider again the Toeplitz matrix T_n of (3) — we continue to assume $c_0 = 1$ for simplicity. Let Z denote the lower triangular shift matrix (of appropriate dimensions). Then it can be verified that T_n satisfies the equation

$$T_n - ZT_nZ^* = G_0JG_0^*\,, \quad J = \begin{bmatrix} 1 & 0 \\ 0 & -1 \end{bmatrix}. \tag{29}$$

That is, the difference $T_n - ZT_nZ^*$ has low rank equal to 2 and independent of the dimension n, and its inertia is $\{1, -1, 0\}$. This is a consequence of the Toeplitz structure of T_n since it can be easily verified that

$$T_n - ZT_nZ^* = \begin{bmatrix} 1 & c_{-1} & c_{-2} & \cdots & c_{-n} \\ c_1 & & & & \\ \vdots & & \bigcirc & & \\ c_n & & & & \end{bmatrix}.$$

Moreover, the above difference can be factored as $G_0JG_0^*$, where G_0 is the same matrix that we encountered before in (22), *viz.*,

$$G_0 = \begin{bmatrix} 1 & 0 \\ c_1 & c_1 \\ c_2 & c_2 \\ \vdots & \vdots \\ c_n & c_n \end{bmatrix}. \tag{30}$$

We call G_0 a generator matrix for T_n since it contains all the information needed about T_n. Note further that while T_n has n^2 entries, G_0 has $2n$ entries; an order of magnitude less than T_n. Hence, algorithms that operate on G_0 in order to extract some information about T_n will tend to be an order of magnitude faster than algorithms that operate on the entries of T_n themselves. The Schur algorithm is one such procedure. It can be shown that it not only verifies the positive-definiteness of T_n (which is the context in which it was used earlier in this paper by checking whether $|\gamma_i| < 1$), but it also provides the Cholesky factorization of T_n using only a multiple of n^2 operations.

More specifically, it can be verified that the first columns of the successive $G_i\Theta_i$ in the array form (20) correspond to the columns of the Cholesky factor of T_n, say $T_n = \bar{L}\bar{L}^*$, where $\bar{L}$ is lower triangular with columns

$$\bar{l}_i = G_i\Theta_i \begin{bmatrix} 1 \\ 0 \end{bmatrix} . \tag{31}$$

Even more important, it turns out that the Schur algorithm (20) can be used to (Cholesky) factor more general structured matrices. For example, it can factor any positive-definite matrix R that satisfies an equation of the form

$$R - ZRZ^* = GJG^* ,$$

for some two-column matrix G and $J = (1 \oplus -1)$. That is, any matrix R for which the difference $R - ZRZ^*$ has rank 2 and inertia $\{1, -1, 0\}$. The factorization is achieved by simply applying (20) starting with $G_0 = G$ and using (31). Here, the only difference from the Toeplitz case is the initial general matrix G_0.

6.2 GENERALIZED SCHUR ALGORITHM

Generalizations of both structure and Schur's work have been studied extensively in the literature (see [1, 9] and the references therein). Here we focus on one such extension.

Let F be an $n \times n$ lower triangular stable matrix with diagonal entries $\{f_i,\ |f_i| < 1\}$, and let R be an $n \times n$ positive-definite Hermitian matrix. We shall say that R has *displacement structure* if it satisfies an equation of the form (also called a displacement equation):

$$R - FRF^* = G \underbrace{\begin{bmatrix} I_p & 0 \\ 0 & -I_q \end{bmatrix}}_{J} G^* , \tag{32}$$

where J is a signature matrix that specifies the displacement inertia of R, and G is an $n \times r$ generator matrix with $r \ll n$ and $r = (p+q)$. We say that R has structure to indicate that the difference $R - FRF^*$ is low rank; its rank is called the displacement rank of R. The condition $|f_i| < 1$ is sufficient to guarantee that the displacement equation (32) uniquely specifies R from knowledge of $\{F, G, J\}$. Note also that the

Toeplitz structure (29) is a special case of (32) by choosing $F = Z$, $G = G_0$, $p = q = 1$, and $J = (1 \oplus -1)$.

A major result concerning structured matrices R, is that the successive Schur complements of R, denoted by R_i, inherit a similar structure. That is, if R_i is the Schur complement of the leading $i \times i$ submatrix of R, then R_i also exhibits displacement structure of the form

$$R_i - F_i R_i F_i^* = G_i J G_i^* ,$$

where F_i is the submatrix obtained after deleting the first i rows and columns of F, and the generator G_i satisfies a recursive construction that we explain below.

Algorithm 5 (Generalized Schur Algorithm) *Generator matrices G_i for the successive Schur complements R_i of a positive-definite structured matrix R, as in (32), can be recursively constructed as follows. Start with $G_0 = G, F_0 = F$, and repeat for $i \geq 0$:*

1. *At step i we have F_i and G_i. Let g_i denote the top row of G_i.*
2. *Choose any J-unitary rotation Θ_i that reduces g_i to the form*

$$g_i \Theta_i = \begin{bmatrix} \delta_i & 0 & \ldots & 0 \end{bmatrix} \qquad (33)$$

   Such a rotation always exists in view of the positive-definiteness of R and it can be implemented in many different ways, e.g., *as a sequence of elementary unitary and hyperbolic rotations.*

3. *Apply Θ_i to G_i leading to the next generator as follows:*

$$\begin{bmatrix} \mathbf{0} \\ G_{i+1} \end{bmatrix} = G_i \Theta_i \begin{bmatrix} 0 & 0 \\ \mathbf{0} & I \end{bmatrix} + \Phi_i G_i \Theta_i \begin{bmatrix} 1 & 0 \\ \mathbf{0} & \mathbf{0} \end{bmatrix} , \qquad (34)$$

   where Φ_i denotes a so-called Blaschke matrix,

$$\Phi_i = (F_i - f_i I)(I - f_i^* F_i)^{-1} .$$

4. *The columns of the Cholesky factor of $R = \bar{L}\bar{L}^*$ are given by*

$$\bar{l}_i = \sqrt{1 - |f_i|^2}\, (I - f_i^* F_i)^{-1} G_i \Theta_i \begin{bmatrix} 1 \\ \mathbf{0} \end{bmatrix} . \qquad (35)$$

Recursion (34) is an extension of (20). Pictorially, we have the following (see Figure 6):

$$G_i = \begin{bmatrix} \times & \times & \times \\ \times & \times & \times \\ \times & \times & \times \\ \vdots & & \vdots \end{bmatrix} \overset{\Theta_i}{\rightarrow} \begin{bmatrix} \delta_i & 0 & 0 \\ \times' & \times' & \times' \\ \times' & \times' & \times' \\ \vdots & & \vdots \end{bmatrix} \overset{\Phi_i}{\rightarrow} \begin{bmatrix} 0 & 0 & 0 \\ \times'' & \times' & \times' \\ \times'' & \times' & \times' \\ \vdots & & \vdots \end{bmatrix} = \begin{bmatrix} \mathbf{0} \\ G_{i+1} \end{bmatrix}$$

In words:

- Choose an $r \times r$ J–unitary rotation Θ_i that reduces the top row of G_i as in (33). We say that G_i is reduced to proper form.
- Apply Θ_i to G_i.
- Multiply the first column of $G_i\Theta_i$ by Φ_i and keep all other columns unchanged.

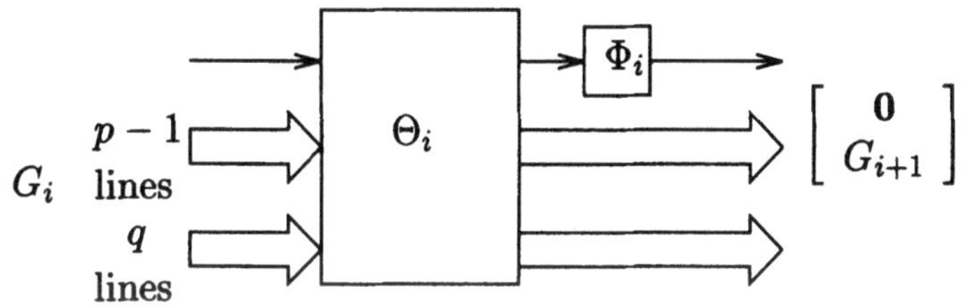

Figure 6 Pictorial representation of the generalized Schur algorithm.

A useful conclusion follows by combining the generator recursion(34) with the expression (35) for the i–th column of the Cholesky factor. Indeed, define

$$l_i \triangleq \sqrt{d_i}\,\bar{l}_i\ ,$$

where $1/\sqrt{d_i}$ is the top entry of $\bar{l}_i$ (and equal to $|\delta_i|^2/(1-|f_i|^2)$). That is, the top entry of l_i is normalized to unity. Then it can be verified that (34) and (35) lead to the following expression

$$\begin{bmatrix} l_i & \begin{matrix} 0 \\ G_{i+1} \end{matrix} \end{bmatrix} = \begin{bmatrix} F_i l_i & G_i \end{bmatrix} \begin{bmatrix} f_i^* & \frac{\delta_i}{d_i}\begin{bmatrix} 1 & 0 \end{bmatrix} \\ \Theta_i \begin{bmatrix} \delta_i \\ 0 \end{bmatrix} & \Theta_i \begin{bmatrix} -f_i & 0 \\ 0 & I_{r-1} \end{bmatrix} \end{bmatrix} \tag{36}$$

We can therefore regard the transformation

$$\begin{bmatrix} f_i^* & \frac{\delta_i}{d_i}\begin{bmatrix} 1 & 0 \end{bmatrix} \\ \Theta_i \begin{bmatrix} \delta_i \\ 0 \end{bmatrix} & \Theta_i \begin{bmatrix} -f_i & 0 \\ 0 & I_{r-1} \end{bmatrix} \end{bmatrix}$$

as the system matrix of a first-order state-space linear system; the rows of $\{G_i\}$ and $\{G_{i+1}\}$ can be regarded as inputs and outputs of this system, respectively, and the entries of $\{l_i, F_i l_i\}$ can be regarded as the corresponding current and future states. If we let $\Theta_i(z)$ denote the transfer function of the linear system (with inputs from the left), *viz.*,

$$\Theta_i(z) = \Theta_i \begin{bmatrix} -f_i & 0 \\ 0 & I_{r-1} \end{bmatrix} + \Theta_i \begin{bmatrix} \delta_i \\ 0 \end{bmatrix} (z^{-1} - f_i^*)^{-1} \frac{\delta_i}{d_i} \begin{bmatrix} 1 & 0 \end{bmatrix},$$

simple algebra will show that the above expression collapses to

$$\Theta_i(z) = \Theta_i \begin{bmatrix} B_i(z) & \mathbf{0} \\ \mathbf{0} & I_{r-1} \end{bmatrix}, \quad B_i(z) = \frac{z - f_i}{1 - f_i^* z}. \tag{37}$$

We therefore see that each step of the generalized Schur recursion gives us to a first-order section $\Theta_i(z)$ — see Figure 7. [Compare this with $\Theta_i(z)$ in (14), which is obtained as a special case by setting $f_i = 0$ and $r = 2$.]

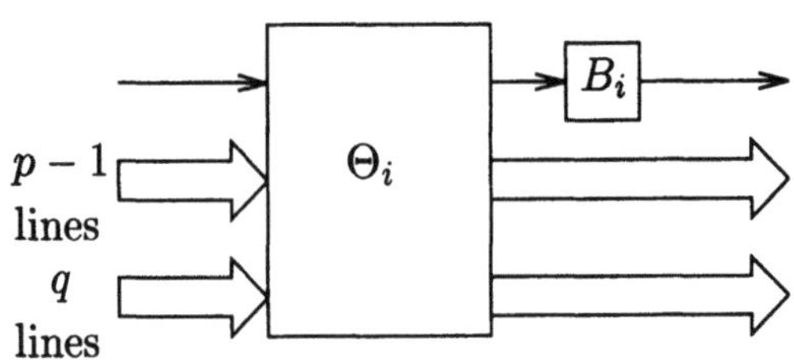

Figure 7 Elementary section $\Theta_i(z)$.

A succession of steps of the generalized Schur algorithm would therefore lead to a feedforward cascade of sections, say for $(n+1)$ steps,

$$\Theta(z) = \Theta_0(z)\Theta_1(z)\ldots\Theta_n(z) = \begin{bmatrix} \Theta_{11}(z) & \Theta_{12}(z) \\ \Theta_{21}(z) & \Theta_{22}(z) \end{bmatrix}, \tag{38}$$

which we partition accordingly with $J = (I_p \oplus -I_q)$. That is, $\Theta_{11}(z)$ is $p \times p$ $\Theta_{12}(z)$ is $p \times q$ $\Theta_{21}(z)$ is $q \times p$ $\Theta_{22}(z)$ is $q \times q$. The transformation implied by $\Theta(z)$ is depicted in Figure 8. It is also a J–lossless transformation.

The associated scattering cascade would then be (see Figure 9)

$$\Sigma(z) = \begin{bmatrix} \Theta_{11}(z) - \Theta_{12}(z)\Theta_{22}^{-1}(z)\Theta_{21}(z) & -\Theta_{12}(z)\Theta_{22}^{-1}(z) \\ \Theta_{22}^{-1}(z)\Theta_{21}(z) & \Theta_{22}^{-1}(z) \end{bmatrix}. \tag{39}$$

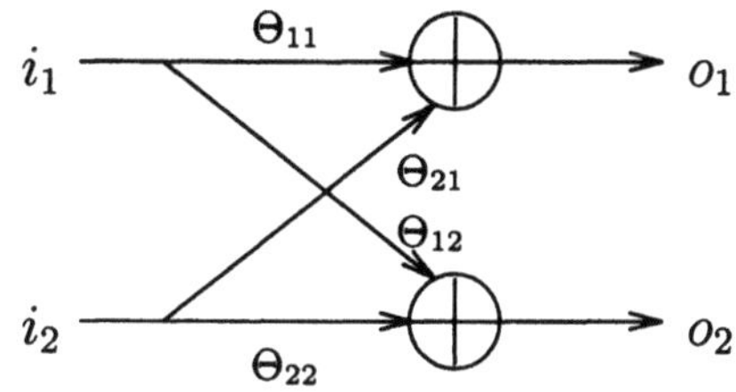

Figure 8 The feedforward cascade.

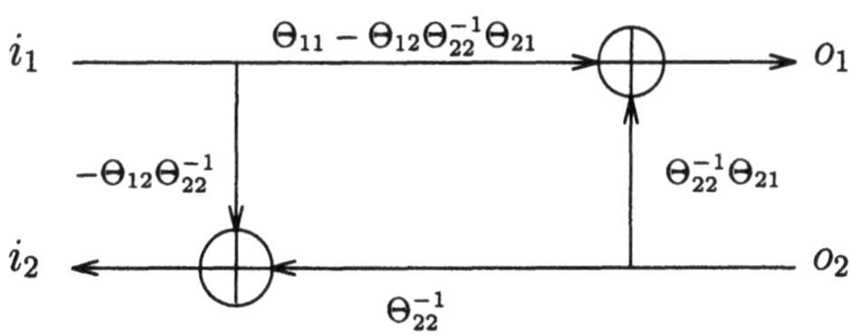

Figure 9 The scattering cascade.

The array form that we described for the the generalized Schur algorithm is in fact a special case of a more general description that also turns out to be relevant to our discussions.

Algorithm 6 (General Generator Recursion) *Generator matrices G_i for the successive Schur complements R_i of a positive-definite structured matrix R, as in (32), can be recursively constructed as follows. Start with $G_0 = G, F_0 = F$, and repeat for $i \geq 0$:*

$$\begin{bmatrix} \mathbf{0} \\ G_{i+1} \end{bmatrix} = \left\{ G_i + (\Phi_i - I_{n-i}) G_i \frac{J g_i^* g_i}{g_i J g_i^*} \right\} \Theta_i \,, \tag{40}$$

for any J-unitary Θ_i. Moreover,

$$\bar{l}_i = \frac{\sqrt{1-|f_i|^2}}{\sqrt{g_i J g_i^*}} (I - f_i^* F_i)^{-1} G_i J g_i^* \,, \tag{41}$$

where the positive-definiteness of R, and the fact that $|f_i| < 1$, guarantee $g_i J g_i^* > 0$. Also, the elementary sections $\Theta_i(z)$ that are generated by the algorithm are given by

$$\Theta_i(z) = \left\{ I_r + [B_i(z) - 1] \frac{J g_i^* g_i}{g_i J g_i^*} \right\} \Theta_i \,. \tag{42}$$

In the above description, Θ_i can be <u>any</u> J-unitary rotation matrix and we will again be led to a feedforward cascade $\Theta(z)$ and to a feedback cascade $\Sigma_i(z)$. The individual sections of $\Theta(z)$ will now be given by (42), while the individual sections of $\Sigma(z)$ will be obtained from (42) via an expression similar to (39). We may also note that (40) and (41) can be combined together as follows

$$\begin{bmatrix} l_i & 0 \\ & G_{i+1} \end{bmatrix} = \begin{bmatrix} F_i l_i & G_i \end{bmatrix} \begin{bmatrix} f_i^* & \frac{1}{d_i} g_i \Theta_i \\ J g_i^* & \left(I_r - (1+f_i) \frac{J g_i^* g_i}{g_i J g_i^*} \right) \Theta_i \end{bmatrix}, \tag{43}$$

and that $\Theta_i(z)$ in (42) is now the transfer matrix of the above state-space transformation, with

$$d_i = \frac{g_i J g_i^*}{1 - |f_i|^2} \,.$$

The array form (34) follows as a special case of (40) when we take Θ_i to be that rotation that reduces g_i to the proper form (33). In this case, the expression (42) for $\Theta_i(z)$ also collapses to (37). Why do we consider this more general Schur recursion? The point is, as we shall see, the freedom in choosing Θ_i turns out to be useful in solving maximum entropy problems. Rather than restricting ourselves to that particular choice of Θ_i that reduces the generator to proper form, we can consider other choices that force certain (maximum entropy) properties on the cascade, as we now explain.

7. MULTICHANNEL MAXIMUM ENTROPY CONSTRUCTION

The generalized scattering cascade $\Sigma(z)$ of Figure 9 now maps $p \times q$ Schur-type matrix-valued loads to $p \times q$ Schur-type transfer matrices $S(z)$ at the input of the cascade – this is the transfer matrix from input-to-output at the left-side of the cascade. Also, we now have a system with multiple inputs (p of them) and multiple outputs (q of them).

For any such matrix Schur function $S(z)$, we define its spectral factor as the maximal outer function $L_s(z)$ such that

$$I - S(z)\left[S\left(\frac{1}{z^*}\right)\right]^* \geq L_s(z)\left[L_s\left(\frac{1}{z^*}\right)\right]^* .$$

Using the analogy with the single channel case (28), we introduce the following maximum entropy problem.

Problem 6 (Multichannel Maximum Entropy Problem)
Consider a positive-definite structured matrix R satisfying (32), and let $\Theta(z)$ and $\Sigma(z)$ denote the feedforward and feedback cascades obtained by the generalized Schur algorithm. Every Schur-type load now leads to a Schur-type transfer function $S(z)$ at the input of the cascade. The maximum entropy problem is to choose the load that results in

$$\max_{\text{Schur load}} \left[\ln \det L_s(0)\right] \tag{44}$$

The solution to the earlier maximum entropy problem in Algorithm 4 (which corresponded to the special case $F = Z$ and $r = 2$), was obtained by terminating the corresponding cascade with the zero load. This solution does not work for general F. The following result is a special case of the main theorem we established in [12].

Algorithm 7 (Solution of Multichannel Version)
Given a positive-definite matrix R satisfying (32), let $\Theta(z)$ and $\Sigma(z)$ denote the feedforward and feedback cascades obtained by the generalized

Schur algorithm (either by applying (34) or (40)). The optimal load is obtained by terminating the scattering cascade with the following (constant and contractive) load

$$\text{Optimal Schur load} = [\Theta_{22}^{-1}\Theta_{21}]^*(0)\ . \qquad (45)$$

In other words, the result requires that we determine $\Theta_{22}^{-1}(z)\Theta_{21}(z)$ and evaluate its value at $z = 0$. Then the complex conjugate of this value should be taken as the load (see Figure 10).

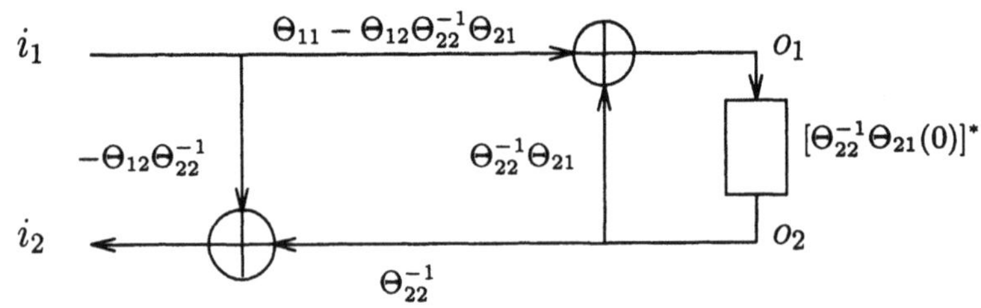

Figure 10 Maximum entropy load in the multichannel case.

This construction admits the following interpretation. Recall from expression (39) for the scattering cascade that the transfer matrix from the input to the output at the right side is equal to $\Theta_{22}^{-1}(z)\Theta_{21}(z)$. The value at zero gives us the value of the direct (*i.e.*, delayless) path that exists between these two ports. Therefore, in a sense, the load $[\Theta_{22}^{-1}\Theta_{21}]^*(0)$ compensates for this direct path. To see this more clearly, let us denote this load by A^* for simplicity of notation, $A^* = [\Theta_{22}^{-1}\Theta_{21}]^*(0)$. Then the gain of the direct path from o_2 to o_1 is obtained by combining the direct-path relations

$$o_2 = o_1 A^* \quad \text{and} \quad o_1 = o_2 A\ ,$$

which lead to $o_1 = o_1 A^* A$. But since A is strictly contractive, this equality holds only for $o_1 = 0$ and, consequently, $o_2 = 0$. That is, the response from o_2 to o_1 can only be strictly causal.

In the $F = Z$ case that we studied earlier (Algorithm 4), the elementary sections of the scattering cascade obtained via the array form (20) had the special form (14). Hence, note that each such section has the following special form at $z = 0$,

$$\Theta_i(0) = \begin{bmatrix} 0 & \times \\ 0 & \times \end{bmatrix}\ .$$

If we now form $\Theta(z)$ by combining several such $\Theta_i(z)$, say $\Theta(z) = \Theta_0(z)\ldots\Theta_n(z)$, then $\Theta(z)$ at $z = 0$ will have a similar form,

$$\Theta(0) = \begin{bmatrix} 0 & \times \\ 0 & \times \end{bmatrix}.$$

Consequently, the $(2,1)$ entry of $\Theta(z)$ will satisfy $\Theta_{21}(0) = 0$. This means that the resulting $\Theta_{21}(z)$ is strictly proper, which in turn implies that the resulting maximum-entropy load, according to the construction $[\Theta_{22}^{-1}\Theta_{21}]^*(0)$, will indeed be the zero load (as claimed earlier in the statement of Algorithm 4.)

A similar conclusion does not hold for the array form (34) for general lower triangular F, and even for $F = Z$ with higher displacement rank (higher than 2). That is, the cascade $\Theta(z)$ that results in these cases from applying the array form (34) (or even the general form (40) without special care) will not be such that $\Theta_{21}(z)$ is strictly proper. In such cases, a nonzero load will be necessary and, consequently, the central solution will not be the maximum entropy solution.

To demonstrate this, consider the simple example of a diagonal F and displacement rank 2, *i.e.*, G in (32) has two columns and $J = (1 \oplus -1)$. Using the array form (34) of the generalized Schur algorithm we obtain a cascade of sections of the form

$$\Theta_i(z) = \Theta_i \begin{bmatrix} \frac{z-f_i}{1-f_i^* z} & 0 \\ 0 & 1 \end{bmatrix},$$

which implies that $\Theta_i(0)$ will in general have the form

$$\Theta_i(0) = \begin{bmatrix} \times & \times \\ \times & \times \end{bmatrix}.$$

Therefore, by combining several such sections the resulting $\Theta_{21}(z)$ will not generally be strictly proper.

In any case, the solution (45) to the multichannel maximum entropy problem requires that we first determine the cascade $\Theta(z)$ and then evaluate the load $[\Theta_{22}^{-1}\Theta_{21}]^*(0)$. It does not matter which form of the generalized Schur algorithm we use (the array form (34) or the more general form (40)). In either case, we should determine the optimal load via (45).

There is an alternative solution of the maximum entropy problem that relies instead on the general form (40). It employs a different choice for the arbitrary J–unitary rotations Θ_i in order to guarantee that the cascade $\Theta(z)$ will always satisfy $\Theta_{21}(0) = 0$ and, therefore, that the zero load will be sufficient to solve the maximum entropy problem. This alternative solution is therefore such that the central solution will always be the maximum entropy solution. It avoids the additional burden of determining the right load via (45) since all the calculations are embedded into the recursive construction of the cascade.

The new choice for the arbitrary rotation Θ_i is one that enforces that the $(2,1)$ entry of the elementary section $\Theta_i(z)$ in (42) is itself strictly proper. In so doing, when these sections are combined together to yield $\Theta(z)$, the $(2,1)$ entry $\Theta_{21}(z)$ of the overall cascade will be strictly proper and the required load will then be the zero load.

The rotation Θ_i in (40) is chosen as follows. Consider the general expression (42) for $\Theta_i(z)$ and partition g_i accordingly with J, say

$$g_i \triangleq \begin{bmatrix} u_i & v_i \end{bmatrix} ,$$

where u_i is $1 \times p$ and v_i is $1 \times q$. Let also $|\delta_i|^2 \triangleq g_i J g_i^*$. Then some straightforward algebra shows that the value of $\Theta_i(z)$ of (42) at $z = 0$ is given by

$$\Theta_i(0) = \begin{bmatrix} I_p - \frac{1+f_i}{|\delta_i|^2} u_i^* u_i & -\frac{1+f_i}{|\delta_i|^2} u_i^* v_i \\ \frac{1+f_i}{|\delta_i|^2} v_i^* u_i & I_q + \frac{1+f_i}{|\delta_i|^2} v_i^* v_i \ . \end{bmatrix} \Theta_i \ .$$

This suggests that we could choose Θ_i in order to reduce the $r \times r$ matrix

$$\begin{bmatrix} I_p - \frac{1+f_i}{|\delta_i|^2} u_i^* u_i & -\frac{1+f_i}{|\delta_i|^2} u_i^* v_i \\ \frac{1+f_i}{|\delta_i|^2} v_i^* u_i & I_q + \frac{1+f_i}{|\delta_i|^2} v_i^* v_i \ . \end{bmatrix}$$

into the reversed upper triangular form

$$\begin{bmatrix} \times & \times \\ 0_{q \times p} & \times \end{bmatrix} .$$

That is, we should choose a J–unitary Θ_i that performs the triangularization

$$\begin{bmatrix} I_p - \frac{1+f_i}{|\delta_i|^2} u_i^* u_i & -\frac{1+f_i}{|\delta_i|^2} u_i^* v_i \\ \frac{1+f_i}{|\delta_i|^2} v_i^* u_i & I_q + \frac{1+f_i}{|\delta_i|^2} v_i^* v_i \ . \end{bmatrix} \Theta_i = \begin{bmatrix} \times & \times \\ 0_{q \times p} & \times \end{bmatrix} . \qquad (46)$$

Once this is done for each $\Theta_i(z)$, the resulting $\Theta(z)$ will be such that $\Theta_{21}(z)$ is strictly proper. We showed in [12] that such rotations Θ_i exist by considering a family of generalized reflection coefficients.

Algorithm 8 (Recursive Solution of Multichannel Version) *Given a positive definite structured matrix R satisfying (32), apply the generalized Schur algorithm (40) where the Θ_i are chosen so as to enforce the transformation (46). Now terminate the resulting scattering cascade with the zero load and the Schur function mapped at the input of the cascade will be such that it solves (45).*

Using the construction (46) for the Θ_i in (43), we see that each section $\Theta_i(z)$ now has a state-space description of the form

$$\begin{bmatrix} l_i & \begin{matrix} 0 \\ G_{i+1} \end{matrix} \end{bmatrix} = \begin{bmatrix} F_i l_i & G_i \end{bmatrix} \begin{bmatrix} f_i^* & \frac{1}{d_i} g_i \Theta_i \\ J g_i^* & \begin{bmatrix} \times & \times \\ 0 & \times \end{bmatrix} \end{bmatrix}. \tag{47}$$

8. CONCLUDING REMARKS

Several classical moment and extension problems, and the famous solutions of Toeplitz, Caratheodory, and Schur, were used to introduce the maximum entropy problem for scalar-valued covariance sequences. A recursive construction in terms of the generalized Schur algorithm was then presented that allows us to solve the maximum entropy problem for matrix-valued covariance sequences and some more general sequences.

References

[1] T. Kailath and A. H. Sayed, "Displacement Structure: Theory and Applications," *SIAM Review*, vol. 37, no. 3, pp. 297–386, September 1995.

[2] A. H. Sayed, *Displacement Structure in Signal Processing and Mathematics*. PhD thesis, Stanford University, Stanford, CA, August 1992.

[3] J. P. Burg, "Maximum Entropy Spectral Analysis," in *Proceedings 37th Meeting of Society of Exploration Geophysicists*, Oklahoma City, OK, October 1967.

[4] D. Z. Arov and M. G. Krein, "On the Evaluation of Entropy Functionals and their Minima in Generalized Extension Problems," *Acta Scientiarum Mathematicarum (Szeged)*, vol. 45, pp. 33–50, 1983.

[5] H. Dym, *J-Contractive Matrix Functions, Reproducing Kernel Hilbert Spaces and Interpolation*, vol. 71, Conference Board of the Mathematical Sciences, American Mathematical Society, Providence, RI, 1989.

[6] N. I. Akhiezer, *The Classical Moment Problem and Some Related Questions in Analysis*, Hafner Publishing Company, NY, 1965 (in Russian, 1961).

[7] L. Y. Geronimus, *Orthogonal Polynomials*, Transl. Consultants Bureau, NY, 1961 (in Russian, 1958).

[8] G. Szego, *Orthogonal Polynomials*, vol. 23, American Mathematical Society Colloquium Publishing, Providence, RI, 1939.

[9] T. Kailath and A. H. Sayed, editors, *Fast Reliable Algorithms for Matrices with Structure*, SIAM, PA, 1999.

[10] A. H. Sayed, T. Kailath, H. Lev-Ari, and T. Constantinescu, "Recursive Solutions of Rational Interpolation Problems Via Fast Matrix Factorization," *Integral Equations and Operator Theory*, vol. 20, pp. 84–118, September 1994.

[11] T. M. Cover and J. A. Thomas. *Elements of Information Theory.* John Wiley & Sons, 1991.

[12] T. Constantinescu, A. H. Sayed, and T. Kailath, "Displacement Structure and Maximum Entropy," *IEEE Transactions on Information Theory*, vol. 43, no. 3, pp. 1074–1080, May 1997.

Chapter 16

ON THE CAPACITY OF MULTIMODE OPTICAL FIBERS

Z. Zhao
University of Illinois at Urbana-Champaign
Coordinated Science Laboratory
1308 West Main Street
Urbana, Illinois 61801

R. E. Blahut
University of Illinois at Urbana-Champaign
Coordinated Science Laboratory
1308 West Main Street
Urbana, Illinois 61801

Abstract Multimode optical fiber links for local area networks have been challenged by the ever-increasing demand for higher bit rates. As a result, numerous research efforts have been devoted to such communication channels. The understanding of the fundamental limits of the multimode fiber channels thus becomes crucial. To this end, the capacity of an idealized multimode fiber channel is given here, and the relationships among the signal-to-noise ratio, the 'specific channel capacity', and the dimension of the channel are discussed. The space orthogonal signaling using a natural signal constellation can be used to utilize the channel capacity promised by information theory.

1. INTRODUCTION

The presence of hundreds or even thousands of modes in multimode optical fibers (MMF) has been thought to be the major limiting factor for the capacity of MMF communication channels. Numerous research efforts have been focused on how to improve the performance of such system by counteracting the effects of modal dispersion. However none of them has achieved substantial or promising improvements. But recently when people look at the modal dispersion from an information-theoretic point of view, this long-standing negative feature becomes a positive feature of MMF [1]. The existance of the large number of modes actually gives MMF more capacity than single-mode optical fiber (SMF). These modes are excited by the spatial distribution of light across the input face of the fiber and are observed in the spatial distribution of light across the output face of the fiber. However, there is no straightforward method to utilize this wonderful feature. Experimental study has shown that the large number of modes in MMF provide us with spatial freedom similar to that used in multiantenna radio systems. Space orthogonal signaling is introduced and studied elsewhere [2, 3].

The idea of utilizing the spatial freedom of MMF channels came from the newly developed multielement array (MEA) technology in the field of wireless communications. Motivated by the need for fundamental understanding of the ultimate limits of bandwidth efficient delivery for higher bit-rates in newer generation digital wireless communication systems, Foschini and Gans [4] studied the capacities of MEA systems from the information-theoretic point of view. They showed that for a MEA with a large number of elements, the capacity is proportional to the signal-to-noise ratio (SNR), while Shannon's classic formula implies that the capacity is proportional to the logarithm of SNR. The channel capacity of a MEA n_T-by-n_R additive white gaussian noise channel is:

$$C = \log_2 \det[I_{n_R} + (\rho/n_T) \cdot HH^{\dagger}] \text{ bps/Hz} \tag{1}$$

where n_T the number of transmitting antennas, n_R is the number of receiving antennas, H is the MEA transfer matrix, and ρ is the total signal to noise ratio. For an N by N channel, the capacity becomes:

$$C = \log_2 \det[I_N + (\rho/N) \cdot HH^{\dagger}] \text{ bps/Hz} \tag{2}$$

If the individual channels (corresponding to the rows of H) are orthogonal and identical, then the capacity becomes:

$$C = N \cdot \log_2(1 + \rho/N) \tag{3}$$

When N is large, the capacity goes to a limit regardless of the value of SNR:

$$\lim_{N \to \infty} C = \rho/\ln(2) \tag{4}$$

Figure 1 shows the SNR/capacity curves of N by N MEA channels for different numbers of antenna elements as given in Equation 3. The capacity of a MEA wireless system always increases when the number of elements increases. The capacity enhancement is due to two factors: the additional spatial freedom, and the increase of 'effective antenna aperture.' For enhancement due to more spatial freedom is easy to understand. It is the same as we use more telephone line channels. For the wireless case, the more antennas we use, the larger the total aperture and the more signal energy we can get at the receiver. This energy advantage is essentially for 'free.' So whenever possible, we would like to take this energy advantage by using MEA.

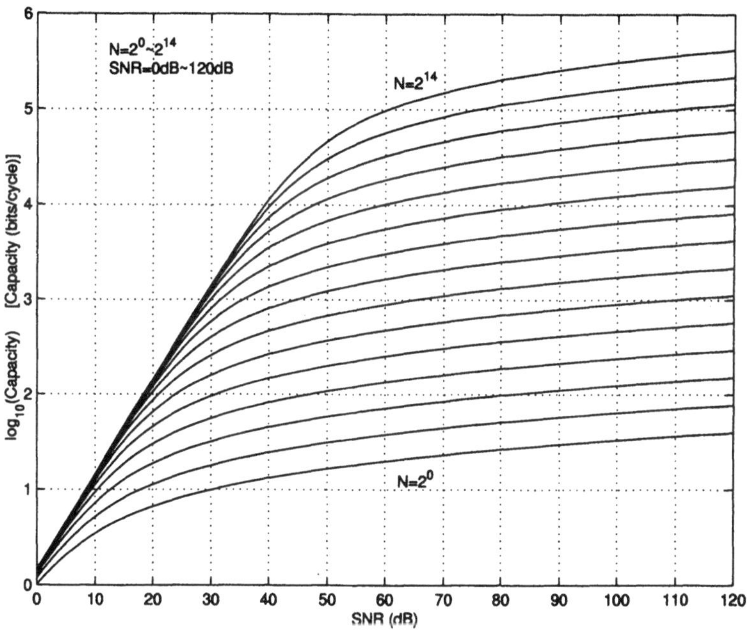

Figure 1 Capacity of a Multielement Array

We define space-time modulation as any modulation method utilizing both time and space freedoms. The channel capacity of a space-time modulated wireless system always increases as the number of transmitting/receiving elements increases.

2. THE CAPACITY OF MMF

The space-time modulated MMF channel is quite a different story. Although we can get capacity enhancement by utilizing the spatial freedom, we do not have the 'free' energy gain as in MEA, because the total optical aperture is fixed. The total signal power at the MMF receiving end is a fixed value. This energy is shared by all the sensors. Increasing the number of sensors cannot increase the total received signal power. This is the ultimate limit in channel capacity, eventually true even for MEA. When we increase the number of receiving antennas, we are still limited by the total power from the transmission antenna. But for the MEA case, we compare our capacity gain with the one-by-one case, so we can always do better by using MEA.

By carefully defining the SNR and noticing that the total amount of energy at the receiving end is independent of the number of elements in the multielement senser array (MESA), the channel capacity for a MMF channel is:

$$C = \log_2 \det[I_{n_R} + \frac{\rho}{n_R n_T} \cdot HH^{\dagger}] \text{ bps/Hz} \tag{5}$$

For an N by N space-time modulated MMF system, the capacity can be expressed as:

$$C = \log_2 \det[I_N + \frac{\rho}{N^2} \cdot HH^{\dagger}] \text{ bps/Hz} \tag{6}$$

If the individual channels (corresponding to the rows of H) are orthogonal and identical, then Equation 6 becomes:

$$C = N \cdot \log_2(1 + \rho/N^2) \text{ bps/Hz} \tag{7}$$

This equation behaves differently from the wireless case, given by Equation 3. If the the total power is fixed (ρ is fixed), the capacity goes to zero when the number of sensors goes to infinity:

$$\lim_{N\to\infty} C = \lim_{N\to\infty} \frac{1}{N} \cdot \log_2(1 + \frac{\rho}{N^2})^{N^2} = 0 \tag{8}$$

This result suggests that we cannot use arbitrarily large number of elements in MESA to increase the capacity for given SNR as in wireless case. However, this limit case does not stop us from achieving higher capacity. Now we examine Equation 6 carefully.

For a fixed number of array elements and high SNR, $\rho \gg N^2$, the capacity is:

$$C \approx N \log_2 \frac{\rho}{N^2} = N \log_2 \rho - 2N \log_2 N \tag{9}$$

We can see from Equation 9 that for every 3dB increase of SNR, the capacity increases by N bits/cycle, but for $N = 1$ case, the increase of capacity is only one bit/cycle.

For fixed number of MESA elements and low SNR, $\rho \ll N^2$, the capacity is:

$$C \approx \frac{\rho}{N \cdot \ln(2)} \tag{10}$$

It is clear that for low SNR region, the capacity is proportional to SNR and is inversely proportional to the number of MESA elements. So for low SNR, we increase SNR and use only the one-element array to achieve better performance.

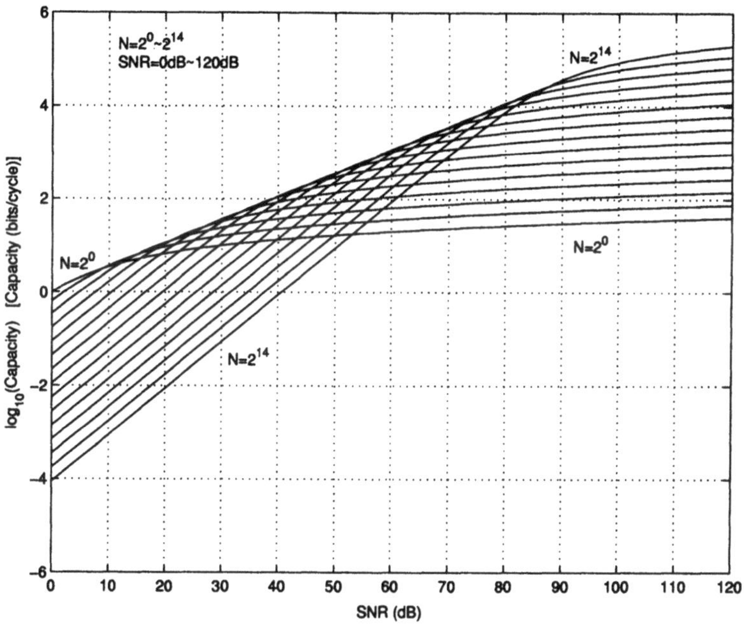

Figure 2 Capacity of MMF

Figure 2 shows the SNR/capacity curves with different number of elements. From this figure we can find out the capacity of the MMF channels. We can see that the capacity curves are upper-bounded by:

$$C \leq \kappa \cdot \rho^{1/2} \tag{11}$$

where

$$\kappa = \alpha^{-1/2} \cdot \ln(1 + \alpha) \tag{12}$$

and α is the root of the following equation:

$$(1 + \alpha)\ln(1 + \alpha) - 2\alpha = 0 \tag{13}$$

The numerical value of α is determined to be $\alpha \approx 3.9215535$, and hence $\kappa \approx 0.8047$. Therefore $C \leq 0.8047\rho^{1/2}$.

If we take the dimension N to be a positive integer, then the capacities of different array size N are upper-bounded by a curve composed of two parts from two different curves: curve 1 is Equation 11 and curve 2 is the specific capacity curve for $N = 1$:

$$C = \log_2(1 + \rho) \tag{14}$$

Curve 2 is always lower than curve 1, except at their junction point where they are tangent to each other. So the overall capacity of the MMF is the union of curve 1 to the right of the junction and curve 2 to the left of the junction:

$$C = \begin{cases} \log_2(1+\rho) & \text{if } \rho \leq \alpha \\ \kappa \cdot \rho^{1/2} & \text{if } \rho > \alpha \end{cases} \tag{15}$$

As shown in Equation 15, the SNR value at the junction is of critical importance, and it is equal to α. If SNR is lower than α, then we use one-dimensional channel, otherwise we should use higher dimensions to achieve better performance. The MMF capacity can be best approached if one picks the array size of the MESA channel to move the operating point as close as possible to the curve described by Equation 15. For high SNR, the capacity of the MMF is proportional to the square root of SNR instead of proportional to SNR as in the wireless MEA.

3. CONCLUSIONS

For a specific channel of N-dimensional MESA, the channel capacity can be increased N times higher than the corresponding MMF channel using conventional modulation methods by using space-orthoganl signaling. The capacity of the idealized MMF channels can be achieved by selecting the dimension of the channel for a fixed SNR. Unlike their wireless counterparts, the optimal dimension of a MMF channel is a finite number, and for very low SNR region, a single element array performs better than an array of a larger number of elements. In the range from 30dB to 40dB, an increase in capacity of almost an order of magnitude is possible by using $N > 1$.

4. DEDICATION

We dedicate this small paper to Dave Forney in the occasion of his sixtieth birthday in hope that we may one day see the evolution of space-time modems for the optical fiber channel.

References

[1] Howard R. Stuart. "Dispersive multiplexing in multimode optical fiber." *SCIENCE*, 289:281–283, July, 2000.

[2] Zhijun Zhao and Richard E. Blahut. "Space orthogonal signaling using natural signal constellations." *To Be Submitted to IEEE*, 2002.

[3] Zhijun Zhao and Richard E. Blahut. "The capacity of multimode optical fibers." *To Be Submitted to IEEE*, 2002.

[4] G. J. Foschini and M. J. Gans. "On limits of wireless communications in a fading environment when using multiple antennas." *Wireless Personal Communications*, 6:311–335, 1998.

Chapter 17

HUFFMAN SHAPING

G. Ungerboeck
Broadcom Corporation

Abstract Huffman shaping represents an alternative to trellis shaping and shell mapping. Shaping gains within 0.1 dB of the ultimate limit of 1.53 dB are achieved at low complexity. In the transmitter, the sequence of scrambled binary data bits is parsed into Huffman codewords, which are then mapped into modulation symbols. The Huffman code is designed to let the modulation symbols assume approximately a Gaussian distribution. The effectiveness of this technique is studied for uncoded PAM and QAM modulation. In a practical system, Huffman shaping would be combined with other coding techniques to achieve shaping gain and coding gain. Unlike trellis shaping or shell mapping, Huffman shaping is not a constant-rate-encoding scheme and decoding errors can lead to bit insertions and deletions. This can be acceptable for transmission of variable-length packets in burst mode. Transmission at constant rate requires some method of buffering and flow control. A simple framing scheme is presented and analyzed.

1. INTRODUCTION

At spectral efficiencies of three bits per dimension or higher, uncoded modulation with equiprobable PAM and QAM symbols exhibits an SNR gap of 9 dB from Shannon capacity, at a symbol error probability of 10^{-6} [1]. This gap comprises a 1.5 dB gap caused by the absence of constellation shaping (towards a Gaussian distribution). The remaining 7.5 dB gap stems from the lack of sequence coding to increase signal distances between permitted symbol sequences. Reductions of these gaps are referred to as *shaping gains* and *coding gains*, respectively. Prior to the invention of iterative soft decoding of serial- or parallel-concatenated codes (*turbo coding*), coding gains of up to 5 dB could be obtained with affordable complexities, e.g., by trellis coded modulation. Turbo coding schemes for high-spectral efficiencies will achieve coding gains about 1 dB from the theoretical maximum of 7.5 dB (the best combination of Turbo coding concepts with high-rate modulation is still debated). Once the ultimate coding gain is closely accomplished, it is natural to strive also for shaping gains near the *ultimate shaping gain* of 1.53 dB.

Two practical shaping techniques have been discussed in the literature: trellis shaping [2] and shell mapping [3, 4, 5, 6, 7]. With sixteen-dimensional shell mapping as employed in V.34 modems, a shaping gain of 0.8 dB is attained. Trellis shaping can provide a shaping gain of about 1 dB at affordable complexity. Trellis shaping is used in V.90 modems; however, for spectral shaping and not for obtaining shaping gain.

This paper describes a technique called Huffman shaping. In the transmitter, the sequence of scrambled binary data bits is parsed into Huffman codewords. The codewords are then mapped into modulation symbols. The Huffman code is designed to let the modulation symbols assume approximately a sampled Gaussian distribution. The effectiveness of this technique is studied for uncoded PAM and QAM modulation at high spectral efficiencies. For PAM constellations, the approximation of the Gaussian symbol distribution is relatively inaccurate. Nevertheless, shaping gains of up to $\approx$1.35 dB are achieved for some rates. For QAM constellations, shaping gains within 0.1 dB from the ultimate shaping gain are consistently obtained for rates of greater than three bit per dimension. Huffman shaping was first publicly disclosed in [8] in combination with trellis coding for the downstream channel in PCM modems. Only small shaping gains could be demonstrated because of the increasing symbol spacing at the boundaries of the PCM symbol constellations. The symbols of the constellations considered in this report are equally spaced.

Unlike trellis shaping or shell mapping, Huffman shaping is not a constant-rate-encoding scheme. Moreover, decoding errors can lead to bit insertion or deletion in the decoded binary data sequence. This can be acceptable for systems in which variable-length packets are transmitted in burst mode with an Ethernet-like medium access protocol. However, in many cases continuous transmission at constant rate is needed. The problem is dual to the transmission of variable-rate encoded voice and video streams over constant rate channels. A simple method is discussed for achieving constant rate and recovering from bit insertions and deletions. It is found that the framing overhead can be kept to a value equivalent to an SNR penalty of $\approx$ 0.1 dB.

2. SHAPING GAIN

Let $\mathbf{A}_M$ be a symmetric M-ary PAM constellation of equally spaced symbols. Adjacent symbols are spaced by 2, and M may be even or odd (usually M is even):

$$\mathbf{A}_M = \{a_i = -(M-1) + 2i, 0 \leq i \leq M-1\}.$$

$$: \mathbf{A}_8 = \{-7, -5, -3, -1, +1, +3, +5, +7\},$$
$$\mathbf{A}_5 = \{-4, -2, 0, +2, +4\} \tag{1}$$

If symbols are selected independently with probabilities $\mathbf{p} = \{p_i, 0 \leq i \leq M-1\}$, the *symbol entropy* $H(\mathbf{p})$ (rate) and the *average symbol energy* $E(\mathbf{p})$ become

$$H(\mathbf{p}) \quad = \quad -\sum_{i=0}^{M-1} p_i \log_2 p_i \quad \text{bit/symbol} \tag{2}$$

(If $p_i = 1/M$ for all i, then $i : H(\mathbf{p}) = H_M = \log_2 M$),

$$E(\mathbf{p}) \quad = \quad \sum_{i=0}^{M-1} p_i |a_i|^2. \tag{3}$$

(If for M-ary PAM, $p_i = 1/M$ for all $i : E(\mathbf{p}) = E_M = \frac{M^2-1}{3}$).

Shaping gain $G_s(\mathbf{p})$ expresses a savings in average symbol energy achieved by choosing symbols from $\mathbf{A}_M$ with probabilities $\mathbf{p}$ rather than selecting equiprobable symbols from a smaller constellation $\mathbf{A}_{M'}$, where $M' = 2^{H(\mathbf{p})}$ ($M' < M$, ignoring that M' may not be an integer):

$$G_s(\mathbf{p}) = \frac{E_{M'}}{E(\mathbf{p})} = \frac{2^{2H(\mathbf{p})} - 1}{3 \times E(\mathbf{p})}. \tag{4}$$

The maximum shaping gain is obtained by the probability distribution $\mathbf{p} = \tilde{\mathbf{p}}$, which minimizes $E(\mathbf{p})$ subject to the constraints $R = H(\mathbf{p})$ and $\sum_{i=0}^{M-1} p_i = 1$. Differentiation of

$$J(\mathbf{p}) = \sum_{i=0}^{M-1} p_i |a_i|^2 + \lambda_1 (-\sum_{i=0}^{M-1} p_i \log_2 p_i - R) + \lambda_2 (\sum_{i=0}^{M-1} p_i - 1), \tag{5}$$

with respect to the probabilities p_i yields the conditions

$$\frac{\partial J(\mathbf{p})}{\partial p_i}\Big|_{\mathbf{p}=\tilde{\mathbf{p}}} = |a_i|^2 - \frac{\lambda_1}{\ln 2}(\ln \tilde{p_i} + 1) + \lambda_2 = 0, \text{ for } 0 \leq i \leq M-1. \tag{6}$$

The parametric solution of (6), with the Lagrange multipliers λ_1, λ_2 transformed into the new variables α, s, becomes (see also [10])

$$\tilde{p_i} = \exp(-1 + \frac{\ln 2}{\lambda_1}(|a_i|^2 + \lambda_2)) = \alpha \exp(-s|a_i{}^2|), 0 \leq i \leq M-1. \quad (7)$$

The optimum distribution $\tilde{\mathbf{p}}$ is thus found to be a Gaussian distribution sampled at the symbol values of $\mathbf{A}_M$. This solution can also be obtained by maximizing the rate $R = H(\mathbf{p})$ subject to the constraints $E(\mathbf{p}) = S$ and $\sum_{i=0}^{M-1} p_i = 1$. The value of α follows from $\sum_{i=0}^{M-1} p_i = 1$. The value of s may be chosen to achieve a given rate $R \leq \log_2 M$ or a given average symbol energy $S \leq E_M$.

If M and R are increased, the optimum shaping gain tends towards the *ultimate shaping gain* $G_s{}^\infty = \pi e/6 = 1.423$ (1.53 dB) [1]. This gain can most easily be derived as the ratio of the variance of a uniform density over a finite interval and the variance of a Gaussian density, both with the same differential entropy [3].

One can easily see that (7) does not only hold for regular symmetric PAM constellations, but gives the optimum shaping probabilities for arbitrary one-dimensional and higher-dimensional signal constellations.

3. HUFFMAN CODES

The traditional application of Huffman codes (see, e.g., [10]) addresses the following problem: Given a sequence of M-ary source symbols, which occur independently with probability distribution $\mathbf{p}$, encode the source symbols into binary codewords of variable lengths such that (a) no codeword is a prefix of any other codeword (*prefix condition*), and (b) the expected length of the codewords is minimized.

An optimum set of codewords is obtained by Huffman's algorithm of 1952. Let a_i be a source symbol that occurs with probability p_i. The algorithm associates a_i with a binary codeword c_i of length ℓ_i such that $2^{-\ell_i} \approx p_i$. The algorithm guarantees that $\sum_{i=0}^{M-1} 2^{-\ell_i} = 1$ (Kraft's inequality is satisfied with equality), and that the expected value of the codeword length, $L = \sum_{i=0}^{M-1} p_i \ell_i$, approaches the entropy of the source symbols within one bit [10]:

$$H(\mathbf{p}) \leq L < H(\mathbf{p}) + 1. \quad (8)$$

In the limit for large H($\mathbf{p}$), the concatenated Huffman codewords yield a binary sequence of independent and equiprobable zeroes and ones with rate $R = L \cong H(\mathbf{p})$ bit per source symbol. However, for certain probability distributions L may be closer to $H(\mathbf{p}) + 1$ than $H(\mathbf{p})$ because of quantization effects inherent in the code construction. If $H(\mathbf{p})$ is small, the difference between L and $H(\mathbf{p})$ can be significant. The rate efficiency can be improved by constructing a Huffman code for blocks of $K > 1$ source symbols. Then (8) takes the form $H(\mathbf{p}) \leq L(K)/K = L \leq H(\mathbf{p}) + 1/K$, where $L(K)$ is the expected length of the Huffman codewords associated with K-symbol blocks. The code comprises M^K codewords and the rate expressed in bit per source symbol will be within $1/K$ bit from $H(\mathbf{p})$.

4. USING HUFFMAN CODES FOR SHAPING

With Huffman shaping the encoding problem is reversed. A Huffman code is generated for the optimum probability distribution $\tilde{\mathbf{p}}$ of the modulation symbols in a given M-ary constellation. In the transmitter, the sequence of data bits is suitably scrambled so that perfect randomness can be assumed. The scrambled sequence is buffered and segmented into Huffman codewords, as in traditional Huffman decoding. A codeword c_i is encountered with probability $2^{-\ell_i} \approx \tilde{p}_i$ and mapped into modulation symbol a_i. In the receiver, when a symbol a_i is detected codeword c_i is inserted into the binary output stream.

For the general case of K-dimensional modulation ($K = 1$: PAM, $K = 2$: QAM), it is appropriate to express rates and symbol energies per dimension, while $a_i, \tilde{p}_i$, and ℓ_i relate to K-dimensional symbols.

The mean value $\bar{R}^h$ and the standard deviation σ_R^h of the number of bits encoded per symbol dimension become

$$\bar{R}^h = \frac{1}{K}\sum_{i=0}^{M-1} 2^{-\ell_i}\ell_i \text{ bit/dimension } (\approx \frac{1}{K}H(\tilde{\mathbf{p}})), \tag{9}$$

$$\sigma_R^h = \sqrt{\frac{1}{K}\sum_{i=0}^{M-1} 2^{-\ell_i}(\ell_i - K\bar{R}^h)^2}. \tag{10}$$

The average symbol energy per dimension S^h and the shaping gain ${G_s}^h$ of the Huffman-shaped symbol sequence are given by

$$S^h = \frac{1}{K}\sum_{i=0}^{M-1} 2^{-\ell_i}|a_i|^2 \text{ energy per dimension } (\approx \frac{1}{K}E(\tilde{\mathbf{p}})), \tag{11}$$

$$G_s^h = \frac{2^{2\bar{R}^h} - 1}{3 \times E^h}. \tag{12}$$

The corresponding quantities obtained with optimum shaping probabilities $\tilde{\mathbf{p}}$ will be denoted, respectively, by $\tilde{R}$ and $\sigma_{\tilde{R}}$ (bit/dimension), $\tilde{S}$ (energy per dimension), and $\tilde{G}_s$ (optimum shaping gain).

5. NUMERICAL RESULTS

For numerical evaluations, uncoded modulation with M-PAM ($M = 2m$) and M-QAM ($M = 4m$) constellations have been considered. The M-QAM constellations are either *square* constellations $M - \text{QAM}_{sq} = \sqrt{M} - \text{PAM} \times \sqrt{M} - \text{PAM}$, or *lowest-energy* constellations M-QAM_{le} comprising the M points in the set $\{(1+2i_1, 1+2i_2), i_1, i_2, \in \mathbf{Z}\}$ nearest to the origin. The symmetries of the symbol constellations are enforced on the Huffman codes. In the PAM case, m codewords are constructed for positive symbols and then extended by a sign bit. Similarly, in the QAM case m codewords are constructed for symbols in the first quadrant and extended by two quadrant bits. The results of different numerical evaluations are depicted in Figures 1, 2, and 3.

Figure 1 shows shaping gain versus rate for PAM and QAM_{sq} constellations of different sizes. The solid curves indicate the shaping gains obtained with the optimum shaping probabilities $\tilde{\mathbf{p}}$. Every rate in the interval $1 \leq R \leq \log_2(M)/K$ can be accomplished (bits per dimension). The shaping gains vanish at $R = 1$ (constellations reduced to BPSK or QPSK) and $R = \log_2(M)/K$ (equiprobable M-QAM). The optimum shaping gains practically reach the ultimate shaping gain of 1.53 dB at R equal to four bits per dimension for signal constellations at least as large as 32-PAM or 1024-QAM_{sq}. Not every rate can be realized with Huffman shaping because of quantization effects in the construction of Huffman codes. For PAM, the approximation of the Gaussian distribution remains relatively coarse. Shaping gains of up to about 1.35 dB are achieved at some rates above three bits per dimension. The effects of quantization are significantly reduced in the QAM cases. With constellations at least as large as 256-QAM_{sq}, shaping gains within 0.1 dB from the ultimate shaping gain of 1.53 dB are consistently obtained at rates above three bits per dimension.

In Figure 2 shaping gains are plotted versus rate for square and lowest-energy 1024-QAM constellations. Minor differences occur in the region of diminishing shaping gains, at rates above 4.5 bits/dimension. The shaping gain of equiprobable 1024-QAM_{le} ($R = 5$ bits/dimension) is 0.2 dB [1].

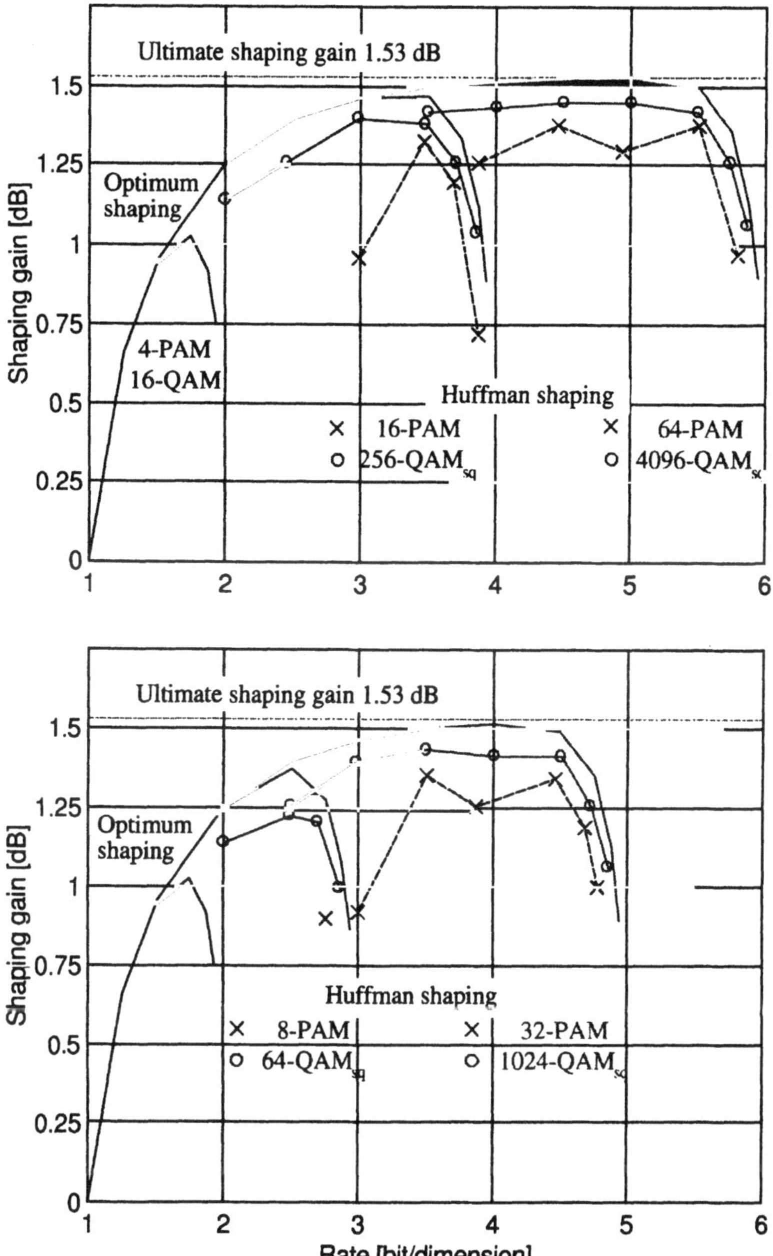

Figure 1 Optimum shaping gains and Huffman shaping gains versus rate: 4-PAM to 64-PAM, and 16-QAM$_{sq}$ to 4096-QAM$_{sq}$ (square).

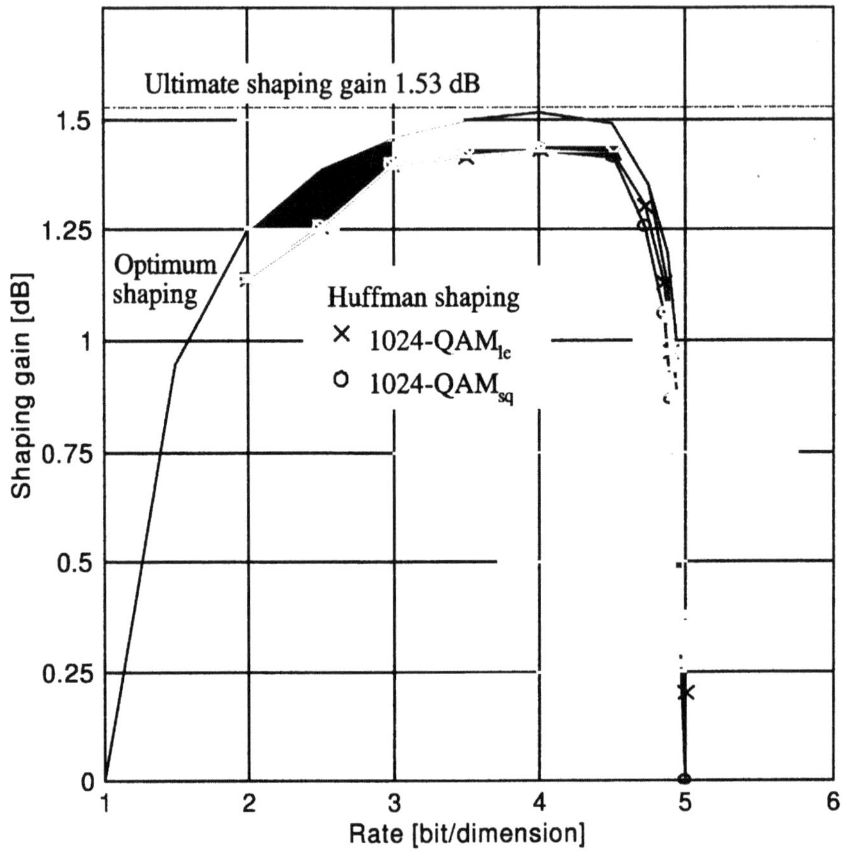

Figure 2 Shaping gain versus rate for 1024-QAM_{sq} (square) and 1024-QAM_{le} (lowest-energy).

Figure 3 depicts the mean and standard deviation of the rate in bits/dimension and the shaping gain accomplished for a nominal rate of $R = 4$ bits/dimension with QAM_{le} constellations of different sizes. The nominal rate is always closely achieved with Huffman shaping (with optimum shaping it is exactly achieved). The standard deviation increases with increasing constellation size to a final value of about 1 bit per dimension. The optimum shaping gain and the Huffman shaping gain increase rapidly when the initial 256-QAM constellation is enlarged. The respective final shaping gains of about 1.5 dB and about 1.4 dB are practically achieved with $M = 512$ (512-QAM_{le}: 1.495 dB and 1.412 dB, 1024-QAM_{le}: 1.516 dB and 1.432 dB).

A 128-QAM_{le} constellation with Huffman shaping for a nominal rate of three bits per dimension is illustrated in Figure 4. The codeword

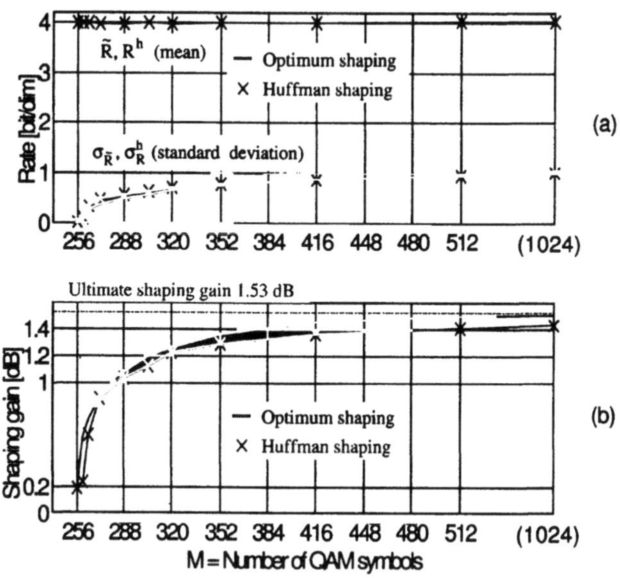

Figure 3 Optimum shaping and Huffman shaping for a nominal rate of $R = 4$ bits/dimension with M-QAM$_{le}$ (lowest-energy) constellations. (a) Mean and standard deviation of rate versus M. (b) Shaping gain versus M.

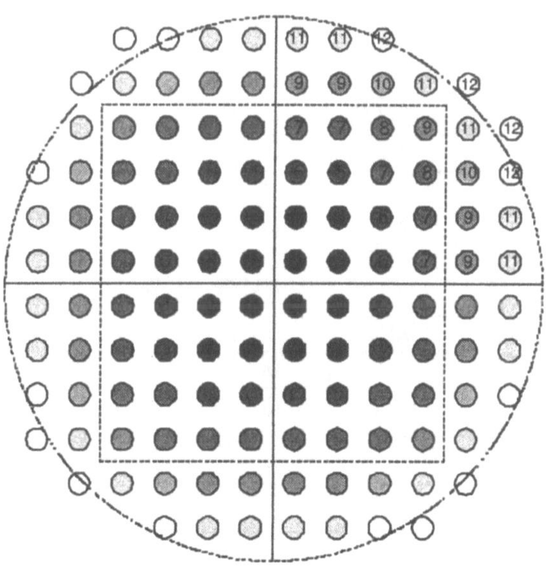

Figure 4 128-QAM$_{le}$ with Huffman shaping achieving $R^h = 2.975$ ($\sigma_R^h = 0.919$) bits/dimension and shaping gain $G_s^h = 1.378$ dB.

lengths ranging from five to twelve bits are indicated for the first-quadrant symbols. $\overline{\mathbf{R}}^h = 2.975(\sigma_R^h = 0.919)$ bits per dimension and $G_s^h = 1.378dB(\tilde{G}_s = 1.443dB)$ are achieved. The symbol energies, optimum shaping probabilities, codeword probabilities and lengths, and the codewords of the first quadrant symbols are listed below. The codewords for the first-quadrant symbols end with ...00.

i	$\|a_i\|^2$	$\tilde{p_i}$	$p_i^h = 2^{-\ell_i}$	ℓ_i	c_i
0	2	0.03872	0.03125	5	00000
1	10	0.02991	0.03125	5	10000
2	10	0.02991	0.03125	5	01100
3	18	0.02311	0.03125	5	11100
4	26	0.01785	0.01563	6	010000
5	26	0.01785	0.01563	6	001100
6	34	0.01379	0.01563	6	110000
7	34	0.01379	0.01563	6	101100
8	50	0.00823	0.00781	7	1010000
9	50	0.00823	0.00781	7	0101100
10	50	0.00823	0.00781	7	0101000
11	58	0.00636	0.00781	7	1101100
12	58	0.00636	0.00781	7	1101000
13	74	0.00379	0.00391	8	10101100
14	74	0.00379	0.00391	8	10101000
15	82	0.00293	0.00195	9	001001000
16	82	0.00293	0.00195	9	001000100
17	90	0.00226	0.00195	9	001010100
18	90	0.00226	0.00195	9	001010000
19	98	0.00175	0.00195	9	001011100
20	106	0.00135	0.00098	10	0010011100
21	106	0.00135	0.00098	10	0010011000
22	122	0.00081	0.00049	11	00100000100
23	122	0.00081	0.00049	11	00100000000
24	130	0.00062	0.00049	11	00101101000
25	130	0.00062	0.00049	11	00101100100
26	130	0.00062	0.00049	11	00101100000
27	130	0.00062	0.00049	11	00100001100
28	146	0.00037	0.00024	12	001000010100
29	146	0.00037	0.00024	12	001000010000
30	162	0.00022	0.00024	12	001011011000
31	170	0.00017	0.00024	12	001011011100

6. VARIABLE-RATE EFFECTS AND ACHIEVING CONSTANT RATE

The advantages of Huffman shaping are its simplicity and the achievement of shaping gains closer to the ultimate shaping gain than obtained by shell mapping or trellis shaping. Disadvantages stem from the variable-rate nature of this technique. In particular, in the receiver symbol-decoding errors can lead to insertion of codewords of wrong length into the output datastream. The insertion or deletion of bits causes subsequently received data to be displaced from their correct temporal positions. These characteristics can be acceptable for systems, in which variable-length packets are transmitted in burst mode with an Ethernet-like medium-access protocol. However, many systems require continuous transmission at constant data rate.

In the remaining part of this section, a generic method for achieving constant rate and recovering from bit insertions and deletions is discussed. The approach is essentially obvious and perhaps conceptually well known in other contexts, e.g., in the dual situation of transmitting variable-rate encoded voice and video streams over constant-rate channels.

The method is illustrated in Figure 5. Data frames of N_b bits are embedded into symbol frames of N_s modulation symbols. Every sequence of bits transmitted within a symbol frame begins with a S&P (synch & pointer) field of $n_{sp} = n_s + n_p$ bits, where n_s is the width of a synch subfield and n_p is the width of a pointer subfield. The synch subfield enables the receiver to acquire symbol-frame synchronization. In principle, sending a known pseudorandom binary sequence with one bit ($n_s = 1$) in every S&P field is sufficient (as in $T1$ systems). The pointer subfield of the nth symbol frame expresses the offset in bits of the nth data frame from the S&P field.

The method is now described by the example given in Figure 5. In the first symbol frame, the first data frame follows the S&P field with zero offset. The S&P field and first data frame are parsed into Huffman codewords, which are then mapped into modulation symbols indexed by $1, 2, 3, \ldots, N_s$. The end of the first data frame is reached before the N_sth modulation symbol has been determined. The data frame is padded with fill bits until the N_sth modulation symbol is obtained. The second data frame follows the S&P field of the second symbol frame again with zero offset. Now the last symbol of the second symbol frame is found before the second data frame is completely encoded. The S&P field of the third symbol frame is inserted and encoding of the remaining part of the second data frame is then continued, followed by encoding the third

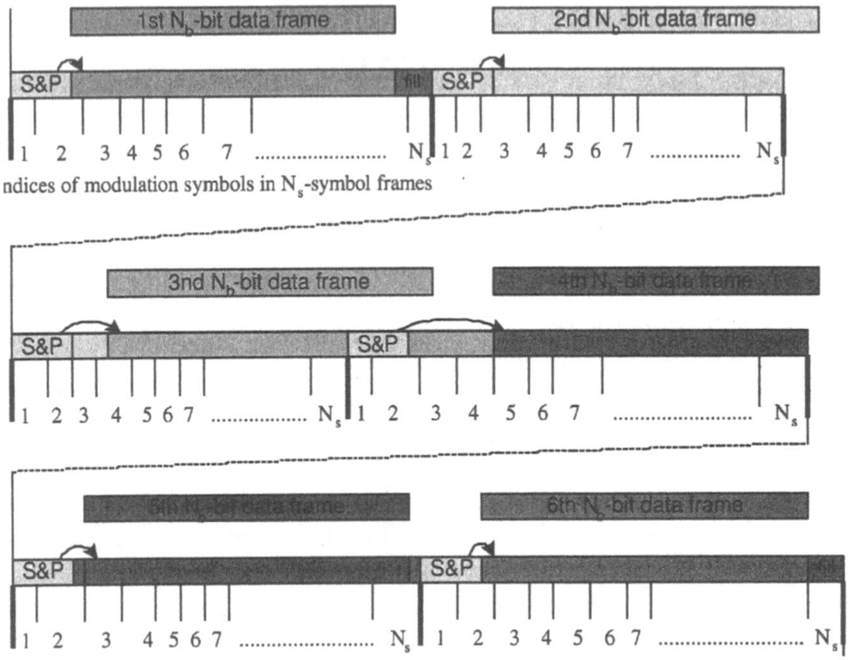

Figure 5 Framing scheme for achieving constant rate and recovering from bit insertions and deletions.

data frame. The pointer in the S&P field indicates the offset of the third data frame from the S&P field. The third data frame can again not completely be encoded in the third symbol frame. The fourth data frame becomes completely encoded in the fourth symbol frame and is padded with fill bits, and so on. The pointer information in the S&P fields enables a receiver to recover from bit insertion and deletion errors.

The dynamic behavior of the scheme will now be investigated in order to determine the overhead in framing bits per symbol needed to ensure correct operation.

Let B_n be the number of bits that are encoded into the N_s symbols of the nth symbol frame. Recall that the mean and standard deviation of the number of bits encoded per symbol dimension are R^h and σ_R^h, respectively, as given by (9) and (10). Then $B = N_s K R^h$ is the mean and the standard deviation of B_n. For large N_s, the probability distribution of B_n will accurately be approximated by the Gaussian distribution

$$Pr(B_n = x) \cong \frac{1}{\sqrt{2\pi}\sigma_B} \exp(-\frac{(x-B)^2}{2{\sigma_B}^2}), x = 0, 1, 2, 3, \ldots \qquad (13)$$

Let P_n be the pointer value in the S&P field of the nth symbol frame. The pointer values will remain bounded if $B > n_{sp} + N_b$. Equivalently, the average number of fill bits per frame, n_{fill}, must be nonzero:

$$n_{fill} = B - (n_{sp} + N_b) > 0. \tag{14}$$

Moreover, in a practical implementation the pointer values must remain limited to the values that can be represented in the n_p-bit pointer subfield, i.e., $0 \leq P_n \leq 2^{n_p} - 1$. Parameters must be chosen such that the probability of $P_n > 2^{n_p} - 1$ becomes negligible. From Figure 5, one can verify the recursive relation

$$P_n = \begin{cases} 0 & n_{sp} + P_{n-1} + N_b \leq B_{n-1} \\ n_{sp} + P_{n-1} + N_b - B_{n-1} & \text{otherwise.} \end{cases} \tag{15}$$

The temporal evolution of the pointer probabilities then becomes

$$Pr(P_n = 0) = \sum_{x \geq 0} Pr(P_{n-1} = x) Pr(B_{n-1} \geq n_{sp} + N_b + x), \tag{16}$$

$$Pr(P_n = y) = \sum_{\substack{x \geq 0 \text{ and} \\ x \geq y - n_{sp} - N_b}} Pr(P_{n-1} = x) Pr(B_{n-1} = n_{sp} + N_b + x - y) \quad y = 1, 2, 3, \ldots \tag{17}$$

The steady-state distribution $\Pr(P = x) = \Pr(P_{n \to \infty} = x)$ can be determined numerically (mathematically speaking $\Pr(P = x)$, is the eigensolution of (16) and (17) associated with eigenvalue one). $\Pr(P = x)$ and $\Pr(P \geq x)$ are plotted in Figure 6 for the following case.

- Lowest-energy 512-QAM, nominal rate $R = 4$ bit/dimension
- Huffman code design $\Rightarrow R^h = 4.015$, $\sigma_R^h = 0.927$ bit/dimension; shaping gain $G_s^h = 1.412$ dB
- Assume $N_s = 512$ QAM symbols/symbol frame, $N_b = 4094$ bit/data frame, $n_{sp} = 12$ $(n_s = 1,\ n_p = 11) \Rightarrow B = 4111.36$, $\sigma_B = 29.66$, $n_{fill} = 5.36$ bit/symbol frame.

The pointer field allows for a maximum pointer value of 2047. Figure 6 shows that $\Pr(P > 2047)$ is well below 10^{-10}. The pointer values exhibit a Pareto distribution [11], i.e., $\log(\Pr(P \geq x))$ decreases linearly for large x.

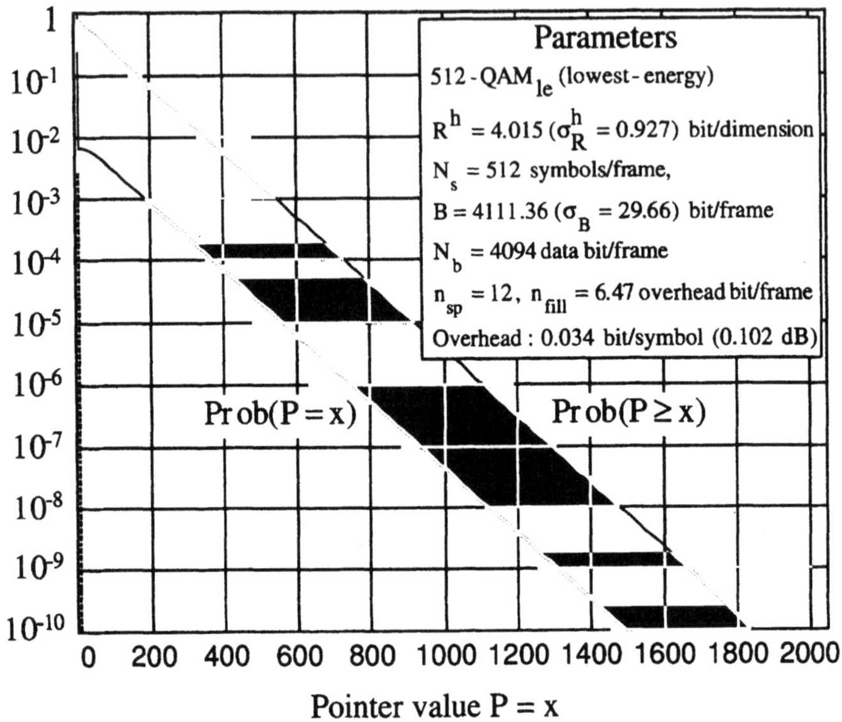

Figure 6 Probability distribution of pointer values.

A framing overhead of $(n_{sp} + n_{fill})/N_s = 0.034$ bit/QAM symbol is found, which is equivalent to an SNR penalty of 0.102 dB. The final net shaping gain becomes $1.412 - 0.102 = 1.310$ dB.

7. CONCLUSION

Huffman shaping has been found to provide shaping gains near the theoretical limit with low complexity. To approach channel capacity in the high-spectral efficiency regime, powerful coding techniques, like turbo coding, must be complemented with a powerful shaping technique. Huffman shaping is a promising candidate for this purpose. The problems arising from the variable-rate nature of Huffman shaping have been discussed. It has been shown that the framing overhead required to mitigate these problems can be small. However, it remains to be seen whether the proposed framing technique can pave a way for Huffman shaping into applications requiring constant rate transmission. One must also keep in mind that any form of shaping has systems implications that go beyond achieving good error performance in the presence of noise. For example, shaping increases the peak-to-average ratio of transmitted signals and essentially prevents the use of blind receiver training.

References

[1] G. D. Forney and G. Ungerboeck, "Modulation and Coding for Linear Gaussian Channels," *IEEE Transactions on Information Theory*, vol. 44, pp. 2384–2415, Oct. 1998.

[2] G. D. Forney, Jr., "Trellis shaping," *IEEE Transactions on Information Theory*, vol. 38, pp. 281–300, Mar. 1992.

[3] P. Fortier, A. Ruiz, and J. M. Cioffi, "Multidimensional Signal Sets Through the Shell Construction for Parallel Channels," *IEEE Transactions on Communications*, vol. 40, pp. 500–512, Mar. 1992.

[4] A. K. Khandani and P. Kabal, "Shaping Multidimensional Signal Spaces- Part I: Optimum Shaping, Shell Mapping," *IEEE Transactions on Information Theory*, vol. 39, pp. 1799–1808, Nov. 1993.

[5] F. R. Kschischang and S. Pasupathy, "Optimal Nonuniform Signaling for Gaussian Channels," *IEEE Transactions on Information Theory*, vol. 39, pp. 913–929, May 1993.

[6] G. R. Lang and F. M. Longstaff, "A Leech Lattice Modem," *IEEE Journal on Selected Areas in Communications*, vol. 7, pp. 968–973, Aug. 1989.

[7] R. Laroia, N. Farvardin and S. Tretter, "On Optimal Shaping of Multidimensional Constellations," *IEEE Transactions on Information Theory*, vol. 40, pp. 1044–1056, July 1994.

[8] G. Ungerboeck and Haixiang Liang, "Coding for V.92 Issue 2," Contribution TR-30.1/99-11-064R1, TIA TR30 meeting, Clearwater, Nov. 29, 1999.

[9] G. D. Forney and L.- F. Wei, "Multidimensional Constellations - Part I: Introduction, Figures of Merit, and Generalized Cross Constellations," *IEEE Journal on Selected Areas in Communications*, vol. 7, pp. 877–892, Aug. 1989.

[10] T. M. Cover and J. A. Thomas, *Elements of Information Theory*, John Wiley & Sons, 1991.

[11] J. M. Wozencraft and I. M. Jacobs, *Principles of Communication Engineering*, John Wiley & Sons, 1965.

V

LATTICES AND GEOMETRY

Chapter 18

LATTICES AND CRYPTOGRAPHY

I. F. Blake
Department of Electrical and Computer Engineering,
University of Toronto,
10 King's College Road,
Toronto, Ontario, Canada M5S 3A4

Abstract The use of the theory of lattices in communications for the construction of codes for the additive white Gaussian channel, is well established. Many elegant constructions, encoding and decoding algorithms, and structural properties of classes of lattices have been found. Lattices have also played an important role in cryptography where they are more often used for showing weaknesses of cryptographic algorithms. However, new public key systems were constructed using lattices based on ground breaking work of Ajtai. It is interesting to note that, in contrast to the situation in communications, the main lattice tool for cryptography is lattice reduction to determine short or close vectors. This note briefly reviews such applications of lattices to cryptography.

1. INTRODUCTION

The use of lattices in communications has been well established over the past three decades, both for coding for the additive white Gaussian noise channel and for the design of efficient quantizers [11]. In both cases, the goal is the design of lattices which lead to good sphere packings, as well as efficient encoding and decoding algorithms. These requirements have led to elegant constructions using results from many areas of mathematics.

In contrast, the use of lattices in cryptography, especially public key cryptography, has been largely in demonstrating weaknesses in cryptosystems, although recently they were also used to construct new and novel public key systems. The key problem of interest in this application is the shortest vector problem in large dimensional lattices. Determining such vectors is a known difficult problem and approximate solutions are often found via certain lattice reduction techniques to be discussed in Section 3.

The next section recalls certain topics of computational number theory and public key systems that will be useful for the subsequent discussion. The following section considers those aspects of lattices that will be required to consider their application to cryptography. This includes, in particular, the notion of lattice reduction.

Section 4 gives a brief overview of the celebrated Ajtai [3] public key system, as well as the related Goldreich et al system [19]. Although these systems seem at the moment too inefficient and therefore impractical for application, the notions on which they are based are ground breaking. A more sustained use of lattices in cryptosystems has been in the area of showing weaknesses or breaking cryptosystems. The technique is to formulate the problem in a matrix form in such a way that a short vector solution corresponds to demonstrating system weakness. A few of the numerous systems that this technique has been applied to successfully are noted in Section 5.

2. PUBLIC KEY CRYPTOSYSTEMS

A brief and informal discussion of a few central notions of public key cryptosystems, as needed for subsequent sections, is given. The notions of one-way and trapdoor one-way functions, as used in the public key systems, are noted. In particular the discrete logarithm problem, in a few of its many disguises, is considered as the original and still one of the most important one-way functions. As well, the integer factorization problem and its relation to the trapdoor one-way functions of modular square roots and the RSA function is discussed. Finally the subset sum or knapsack problem is introduced.

The original one-way function proposed was the discrete logarithm problem in the multiplicative group of a finite field [15]. The problem has received considerable attention over the past twenty years (e.g., [23, 29] and many others) and remains one of the interesting problems of computational number theory. Briefly stated, if α is a primitive element (or actually any element of large prime order) in the finite field F_q and $y = \alpha^x$ then x is called the discrete logarithm of y to the base α. Considering only prime order fields, it is important to choose the prime p so that $p-1$ has a large prime factor as otherwise there exists an efficient algorithm (the Pohlig-Hellman algorithm [30]) to determine discrete logarithms. In such a structure one can apply an index calculus method to determine discrete logarithms and the complexity of the process (in time and space) is approximately $L(c, 1/3)$ where

$$L(c,a) = \exp(c \log p^{a} \log\log p^{1-a})$$

for some constant c. The expression is subexponential (in $\log p$), since it is exponential when $a = 1$ and polynomial when $a = 0$. A sequence of algorithms to determine discrete logarithms has been derived over the past twenty years, beginning with the index calculus method and progressing through the number field sieve and the function field sieve algorithms.

The discrete logarithm can be posed in any finite group G which we assume to be cyclic with generator g. In an additive commutative group $G = \langle g \rangle = \{0, g, 2g, 3g, \ldots, (n-1)g\}$ the problem is to determine x given $y = x \cdot g$ and the generator g where 0 denotes the (additive) identity of the group. In the additive group of the finite field, the problem is trivial because the generator element (of order less than the characteristic of the field) itself has an inverse. However, the additive group of the points on an elliptic curve has proven (so far) to be very effective for the discrete logorithm problem and is in wide use in practice. We do not describe this setting further except to note that the complexity of that problem, using the best algorithms known to date, is on the order of the square root of the group order p i.e., $2^{\log p}$ and hence exponential in $\log p$ rather than the subexponential complexity noted above. It is this feature which makes such cryptography so attractive, allowing shorter key sizes than competing systems with the same level of security.

As a one-way function only, the discrete logarithm problem can be used for such cryptographic functions as key exchange and digital signatures. In most versions of the problem, there is no trapdoor that allows easy computation of the logarithm, while computing the exponential itself is very efficient using the standard square and multiply method—hence the term one-way function.

The two classical one-way trapdoor functions, where the inverse function is computationally difficult to compute unless additional (trapdoor) information is available, in common use for cryptographic protocols, involve modular arithmetic using the integers modulo a product of two odd primes. The two functions are square roots modulo n and the RSA function modulo n and are briefly noted here.

For the discussion let $n = pq$ be the product of two odd primes. Call x the square root of a mod n if

$$x^2 \equiv a \bmod n$$

and a given element a may have 0, 2, or 4 roots. It can be shown that the complexity of determining square roots is equivalent to factoring n.

On the other hand, if one knows the factorization of n then one is able to determine the square root of a modulo the primes and combine the solutions using the Chinese remainder theorem to find the square root modulo the product n. Determining the square roots modulo a prime p is easy if p is 3 modulo 4 or $p \equiv 1$ modulo 8—in each of these cases an efficient deterministic algorithm exists. However, an efficient random algorithm that works well in practice exists for any p. Thus without knowledge of the factorization of n the problem of square roots is computationally difficult while with the knowledge it is easy and hence the problem represents a simple example of a trapdoor one-way function. The problem of factoring integers has been well studied and by the currently most powerful method available, the general number field sieve, the complexity is $L(1.9229, 1/3)$. Interestingly, the general form of the expression is the same as that for the discrete logarithm problem.

The RSA one-way trapdoor function can be described as follows. Again let $n = pq$ be the product of two primes. By Fermat's little theorem it is known that

$$m^{\phi(n)} \equiv 1 \bmod n$$

where $\phi(pq) = (p-1)(q-1)$ is the Euler totient function for this case and we assume $(m, n) = 1$. If one chooses integers e, d so that $ed \equiv 1 \pmod{\phi(n)}$ then, if e and n are made public, one can encrypt a message m as $m^e \pmod{n} = c$. The receiver can then decrypt c by $c^d \equiv m \pmod{n}$. Breaking such an encryption is believed to be effectively equivalent to factoring n, although there is some evidence that this may not be so [8]. Again, it is clear the manner in which this RSA function is a trapdoor one-way function, depending (probably) on the integer factorization problem.

To describe the knapsack public key cryptosystem, define [25] the sequence of positive integers $b_1, b_2, \ldots, b_n$ to be superincreasing if it has the property that

$$b_i > \sum_{j=1}^{i-1} b_j\ , \quad 2 \leq i \leq n.$$

The property of superincreasing ensures that, given any sum $\sum_i x_i b_i = s$, $x_i \in \{0, 1\}$, the binary sequence $x_1, x_2, \ldots, x_n$ is easily recovered from s. To use such a 'subset sum' or 'knapsack' problem as a public key system, one has to hide the superincreasing character of the basis sequence. To do this, choose a modulus $m > \sum b_i$ and an integer w, $1 \leq w < m$, $\gcd(w, m) = 1$ and select a random permutation π of the integers $1, 2, \ldots, n$ and define

$$a_i = w b_{\pi(i)} \bmod m, \ i = 1, 2, \ldots, n.$$

The public key is then $a_1, a_2, \ldots, a_n$ while π, m, w and the original superincreasing sequence $b_1, b_2, \ldots, b_n$ remain private. To encrypt a message $m_1, m_2, \ldots, m_n$ of bits with A's public key, compute

$$c = m_1 a_1 + \ldots m_n a_n$$

which is sent to A. To decrypt this integer, A computes $d \equiv w^{-1} c \bmod m$ and applies the simple algorithm for superincreasing sequences to obtain the bits r_i where $d = r_1 b_1 + \ldots r_n b_n$ and finally

$$m_i = r_{\pi(i)}, \quad i = 1, 2, \ldots, n.$$

This knapsack scheme, the Merkle-Hellman scheme [26] was, for many years, the only credible alternative to the RSA public key scheme, until it was shown to be insecure, using lattice based techniques. Several variations of the scheme have been proposed and all, except one, the Chor-Rivest scheme, have been declared insecure. A few comments on the situation will be given in Section 5.

Thus the discrete logarithm problem (in different groups), square roots and the RSA function, remain the systems of choice to implement many cryptographic protocols, including digital signatures, key exchange, and authentication. A few of the attempts to compromise these schemes using lattices will be considered in Section 5. The next section introduces a few of the properties of lattices that will be needed for this purpose.

3. INTRODUCTION TO LATTICES

Let $\mathbb{R}^n$ denote the n-dimensional real Euclidean space with an inner product $\langle x, y\rangle = \sum_{i=1}^n x_i y_i$, $x, y \in \mathbb{R}^n$ and norm $|x|^2 = \langle x, x\rangle$. A lattice $\mathcal{L}$ will be a discrete subgroup of this space, consisting of the $\mathbb{Z}$-linear combinations of a set of basis vectors $B = \{w_1, \ldots, w_n\}$ i.e.,

$$\mathcal{L}(B) = \{\sum_{i=1}^n \alpha_i w_i, \ \alpha_i \in \mathbb{Z}\}.$$

We assume all lattices considered span $\mathbb{R}^n$. Let B be the matrix whose ith column is the basis vector w_i and by assumption B is nonsingular.

Define the parallelpiped

$$\mathcal{P} = \mathcal{P}(w_1, \ldots, w_n) \triangleq \left\{\sum_i \alpha_i w_i, \ \alpha_i \in [0,1), \ i = 1, \ldots, n\right\}.$$

The *width* of the parallelpiped $\mathcal{P}$ is the minimum over i of the distance between w_i and the hyperplane spanned by the other w_j's.

For an arbitrary vector $v \in \mathbb{R}^n$, denote by $v' \in \mathcal{P}$ the unique vector such that

$$v' = v + \sum_{i=1}^{n} \beta_i w_i$$

for some integers $\beta_i \in \mathbb{Z}$ and say that that $v' \equiv v \bmod \mathcal{P}$ i.e., there is precisely one lattice vector that translates the vector v to a point in $\mathcal{P}$. An invariant of the lattice is $|\det(B)|$ which is the volume of $\mathcal{P}$ i.e., if B' is another basis of $\mathcal{L}$ then $|\det(B)| = |\det(B')|$.

The notion of orthogonality defect [19] is defined by the expression

$$\frac{\prod_{i=1}^{n} |w_i|}{|\det(B)|}$$

and gives an indication of how far from orthogonal the basis vectors w_i are i.e., for a given set of basis vector lengths, the maximum volume of the parallelpiped is achieved when the vectors are orthogonal, in which case the orthogonal defect is unity. The orthogonality defect is otherwise greater than unity.

There are several problems on lattices that are purported to be difficult and perhaps the following two are of the most interest in applications. For a given lattice $\mathcal{L}$ and basis B, note the following problems:

The shortest vector problem (SVP):	determine the lattice vector with minimum nonzero length.
The closest vector problem (CVP):	given an arbitrary vector $v \in \mathbb{R}^n$, determine the lattice vector closest to v.

These lattice problems are believed to be difficult and no efficient algorithms exist for their solution. It has been known for some time that CVP is NP-complete [16]. More strongly, it was shown that even to approximate the CVP to within any constant factor is NP-hard [5], and, stronger yet that it is NP-hard to approximate within a factor $2^{\log^{1-\epsilon} n}$ (as noted in [27]).

Progress on the SVP problem has been a little more difficult. It was originally conjectured by van Emde Boas [16] that SVP was also NP-hard. Ajtai [4] has recently proved that SVP is NP-hard for randomized reductions. The problem continues to be investigated extensively and, for example, Micciancio [27] shows that SVP is NP-hard to approximate to within any constant less than $\sqrt{2}$.

With the above comments it is perhaps not surprising that use of these problems might be proposed for public key systems. Before showing how this was done, we first discuss the notion of a lattice reduction, which is central to the solution algorithms for the above two problems.

Lattice reduction algorithms [20] operate on the basis matrix using the simple column operations:

i) exchanging two columns of B
ii) adding to a given column an integer multiple of another one
iii) deleting zero columns

The general aim of basis reduction algorithms is to derive a new set of basis vectors that achieve some minimization. In the case of LLL [22] this is essentially to have the shortest vectors ahead and to make the columns as mutually orthogonal as possible. However, as noted, there is no efficient algorithm to find the shortest nonzero vector of a lattice.

If $\mathcal{L}$ is a lattice of full rank with associated basis matrix $B_{\mathcal{L}}$ then if B is the basis matrix after reduction by the LLL algorithm, with columns (basis vectors) b_i, and if λ_1 is the shortest nonzero vector of $\mathcal{L}$ then the LLL algorithm acting on $B_{\mathcal{L}}$ essentially achieves the following:

i) $b_1, \ldots, b_n$ is a basis of $\mathcal{L}$.
ii) $|b_1| \leq 2^{(n-1)/2}\lambda_1$
iii) $|b_1| \leq 2^{(n-1)/2}(\det(\mathcal{L}))^{1/n}$
iv) $\det(\mathcal{L}) \leq \prod_i |b_i| \leq 2^{n(n-1)/2}\det(\mathcal{L})$

Thus if a short vector in the lattice is needed, this algorithm will often perform satisfactorily. The algorithm often performs better in practice than the above constants might indicate and many variants of the algorithm are now known.

The LLL algorithm provides a partial solution to SVP. It runs in polynomial time and approximates the shortest vector in the lattice to within a factor of $2^{n/2}$. Schnorr [32] improved this constant to $(1+\epsilon)^n$.

4. THE AJTAI PUBLIC KEY CRYPTOSYSTEM

In this setting, Ajtai [2] showed that for lattices in $\mathbb{Z}^n$, if there is a probabilistic polynomial time algorithm that finds a short vector in a random lattice with probability at least 1/2 then there is a probabilistic polynomial time algorithm which solves the following three lattice problems for *every* lattice in $\mathbb{Z}^n$ with a probability exponentially close to 1:

(i) find the length of a shortest nonzero vector in an n-dimensional lattice, approximately, up to a polynomial factor
(ii) find the shortest nonzero vector in an n-dimensional lattice, where the shortest vector is unique in the sense that any other vector whose length is at most $n^c|v|$ is parallel to v where c is a sufficiently large constant
(iii) find a basis $w_1, \ldots, w_n$ such that $\max|w_i|$ is the smallest possible, up to polynomial factor

Thus to solve a random instance of one of these problems is essentially as difficult as a worst case instance of the problem. This is a remarkable result. It forms the basis of the Ajtai-Dwork public key system introduced below, where it translates into the notion that to break a random instance of the cryptosystem is as hard as breaking a worst-case instance, a desirable feature of any cryptosystem.

Based on the assumption that lattice-reduction is hard in the worst case, Ajtai constructed a one-way function and from this one-way function, a public key system was derived [3]. This scheme and a modification of it due to Goldreich et al [17] are sketched here.

The description of the Ajtai-Dwork public key system follows earlier publications [17, 28]. The system is in n-dimensional space $\mathbb{R}^n$. Let $m = n^3$ and $\rho_n = 2^{n \log n}$. Denote by B_n the n-dimensional cube with side ρ_nS in the positive quadrant, rooted at the origin. Denote by S_n the n-dimensional sphere of radius n^{-8}.

The cryptosystem chooses a private key u on the sphere of unit radius in $\mathbb{R}^n$.

For a given private key u generate a collection of vectors in B_n, and associated distribution $\mathcal{H}_u$ induced by the following construction:
(i) pick a point a uniformly and at random from the set $\{x \in B_n,\ \langle x, u\rangle \in \mathbb{Z}\}$
(ii) choose $\delta_i,\ i = 1, \ldots, n$ at random from S_n
(iii) output the vector $v = a + \sum_i \delta_i$

The public key corresponding to the private key u is obtained by choosing the points $w_1, \ldots, w_n, v_1, \ldots v_m$ independently and at random from the distribution $\mathcal{H}_u$ subject to the constraint that the width is at least $n^{-2}\rho_n$ (a condition that can be shown to be likely to be satisfied [2]).

Encryption with this public key system proceeds in a bit-by-bit fashion. To encrypt a zero, choose binary variables $b_i \in \{0, 1\},\ i = 1, \ldots, m$ at random and reduce the vector

$$v = \sum_{i=1}^{m} b_i v_i$$

mod $\mathcal{P} = \mathcal{P}(w_1, \ldots, w_n)$, say $x \equiv v \bmod \mathcal{P}$ and x is the cipherext of zero. To encipher one, simply choose a vector x in $\mathcal{P}$ at random.

It is possible to show that if x corresponds to an encrypted zero then the inner product of x with the private key u is always within $1/n$, in absolute value, of an integer. On the other hand if a one was encrypted, the inner product will be uniformly distributed over some interval of length unity. The decrypting rule then is to decrypt to a zero if the $\tau = \langle x, u \rangle$ is within $1/n$, in absolute value, to an integer, otherwise decrypt to a one. Clearly such a scheme will always decrypt a zero correctly and will have a probability of $2/n$ of decrypting a one as a zero, in error.

With a minor modification of this scheme, Goldreich et al [17] show how to avoid such decryption errors. With the same parameters, with high probability, one of the vectors v_i generated will have an odd integer-valued inner product with the private key u. Denote the subscript for which this happens i_1. The encryption of a zero is as before. The encryption of a one however is taken as the vector

$$x = \left(\frac{1}{2} v_{i_1} + \sum_{i=1}^{m} b_i v_i \right) \bmod \mathcal{P}.$$

The effect of this construction is to make the inner product of the private key with x to be within $1/n$ of an integer if x corresponds to an encrypted zero and to within $1/n$ of $\mathbb{Z} + 1/2$ if x corresponds to an encrypted one. Since there is no overlap between the possible intervals, there are no errors in the decryption process.

The Ajtai-Dwork public key cryptosystem, described above, is a remarkable result, relating average-case performance to worst-case performance. Recently, Nguyen and Stern [28] proved a converse to the cryptosystem by reducing the problem of distinguishing encryptions of zeros to that of approximating CVP or SVP. That is, they show that if one can approximate CVP within a factor of $cn^{1.33}$, then one can distinguish encryptions with a constant advantage d where c and d are related constants. They argue that this result rules out any form of NP-hardness for the Ajtai-Dwork cryptosystem. This shows that breaking the Ajtai-Dwork cryptosystem is essentially equivalent to approximating the shortest vector within a polynomial factor. Thus for dimensions where lattice reduction algorithms are efficient, there is cause for concern as to the security of the system. The dimensions required for reasonable security for the cryptosystem might be unreasonably large for effective use.

The one-way function constructed by Ajtai and used in the Ajtai-Dwork public key system uses the simple notion that adding a small error vector is simple but given the sum, it is difficult to retrieve the original lattice vector. It is provably a one-way function if the SVP is hard in the worst case. Although information can be recovered by using some private information, it is not a trapdoor one-way function. Goldreich et al [19] use similar notions to derive a true trapdoor one-way function (and hence a true public key system). In brief, their work depends on the observation that determining a close vector to a given vector is considerably easier for some bases than others. Thus a vector is chosen in a lattice with some fixed public basis and a small error vector is added. The CVP is supposed hard in this basis but much easier with a previously determined private basis. The details and justification of this approach are given in Goldreich, et al [19]. While clearly a novel and useful approach, such systems do not yet seem to have worked their way into practice.

5. LATTICE BASED ATTACKS ON PUBLIC KEY CRYPTOSYSTEMS

The lattice reduction techniques mentioned briefly earlier, and the resulting approximations to solutions to the SVP and CVP, have been used to attack many cryptosystems, often with remarkable success. In each case the problem at hand must be cast as a lattice problem and the solution to the cryptosystem must be made equivalent to a lattice problem such as SVP or CVP. How to achieve this is often not obvious and many of the solutions are ingenious. Unfortunately the complexity of equating the problem of compromising the cryptosystem with a lattice problem is often complex. We content ourselves here to giving the briefest of outlines of how this is achieved in two specific cases.

The first success at breaking a public key cryptosystem was that of Shamir [34] who used a linear programming technique to break the Merkle-Hellman knapsack system. However, shortly thereafter, Adleman [1] extended this work by casting the knapsack problem as a lattice problem, opening the way for a variety of other improvements. This demonstration of the insecurity of a system was a dramatic contribution to cryptography. We give a brief indication of how such an attack might be done.

Consider the simple subset sum (or knapsack) problem with given positive integer weights $a_1, a_2, \ldots, a_n$ and s. It is required to determine whether there is a subset of the a_i's that sum to s. This is equivalent to determining whether there are binary variables x_i such that

$$\sum_{i=1}^{n} x_i a_j = s, \quad x_i \in \{0, 1\} \text{ for all } i.$$

Consider the $(n+1) \times (n+1)$ matrix whose rows are taken to be the basis vectors of a lattice:

$$V = \begin{pmatrix} 1 & 0 & 0 & \dots & 0 & -a_1 \\ 0 & 1 & 0 & \dots & 0 & -a_2 \\ & & & \vdots & & \\ 0 & 0 & 0 & \dots & 1 & -a_n \\ 0 & 0 & 0 & \dots & 0 & s \end{pmatrix}.$$

If v_i is the ith row of this matrix and the $\{x_i\}$ give a $\{0,1\}$ solution of the subset sum, then

$$\sum_{i=1}^{n} x_i v_i + v_{n+1} = (x_1, x_2, \dots, x_n, 0)$$

and, since the x_i are binary, this is a vector of length at most $\sqrt{n}$ i.e., a short vector in the lattice. If the a_i are large then running the basis vectors through the LLL reduction might yield such a solution. Something can be said about the circumstances as to when this might happen. The density of the knapsack is defined as

$$d = \frac{n}{\log_2(\max_i a_i)}.$$

This parameter [20] determines the size of short vectors in the lattice, other than the $\{0,1\}$ solution. Originally to guarantee that the binary solution was obtained required that $d < .6463$, i.e., low density, but this was later improved to $d < .9408$ ([14] and [21]).

Another application of lattice reduction techniques concerns the relationship of finding solutions of modular polynomials to solving RSA systems [13]. Specifically, Coppersmith showed that if N is a large composite integer of unknown factorization and $p(x)$ is a monic integer polynomial of degree n, then if the polynomial has a small solution x_0, $|x_0| < N^{1/n}$, modulo N, an algorithm is given to find it. A similar result on bivariate polynomials is given. Finding such a solution in the integers is straightforward but the modular version is more difficult.

Let N be a large composite integer of unknown factorization and let

$$p(x) = \sum_{i=0}^{\delta} p_i x^i \in \mathbb{Z}[x], \; p_\delta = 1.$$

It will be indicated how to determine a solution x_0 to the polynomial modulo N, assuming that such a solution exists with $|x_0| < N^{1/\delta}$.

Consider a matrix

$$M = \begin{pmatrix} 1 & 0 & 0 & \dots & 0 & p_0 \\ 0 & x^{-1} & 0 & \dots & 0 & p_1 \\ 0 & 0 & x^{-2} & \dots & 0 & p_2 \\ & & & \vdots & & \\ 0 & 0 & 0 & \dots & x^{-\delta} & p_\delta \\ 0 & 0 & 0 & \dots & 0 & N \end{pmatrix}.$$

Suppose x_0, $y_0 \in \mathbb{Z}$ are such that $p(x_0) = y_0 N$ with $|x_0| < X$. Consider the row vector

$$\bar{s} = \bar{r}M = \left(1, \left(\frac{x_0}{X}\right), \left(\frac{x_0}{X}\right)^2, \dots, \left(\frac{x_o}{X}\right)^\delta, 0\right).$$

Clearly $\bar{s}$ is in the lattice generated by the rows of M and since each element is bounded by unity, the norm of $\bar{s}$ is bounded by $(\delta+1)^{1/2}$. If $\bar{s}$ is among the short vectors of the lattice, it might be found by lattice reduction.

The determinant $\det(M)$ is $N^{-\delta(\delta+1)/2}$ and for $\det > 1$

$$X < N^{2/\delta(\delta+1)}$$

which restricts the size of the possible root being sought, to ensure the short vector found by the lattice reduction yields the correct vector.

There are a few technical problems with this straightforward approach. These can be fixed by a slightly more involved set up (see [13]). However, this simplified approach gives the flavor of the technique. The immediate application of this result to RSA decryption by Coppersmith involved encryption exponents of 3. The technique is used to show how to recover certain messages in an RSA system using exponent 3 when the message m is composed of a known piece B and an unknown piece x, $m = B + x$ and x can be recovered as long as $|x| < N^{1/3}$ i.e., if about one third of the bits of x are known.

The result of Coppersmith on roots of modular polynomials has been put to other interesting uses. For example, Boneh [6, 7] shows, respectively, that

i) if the encryption exponent in RSA is a prime in the range $[N^{1/4}, N^{1/2}]$ then knowing half the bits of the private key suffice to reconstruct the entire key.

ii) if the private key of the RSA system is less than $N^{.292}$ then the system is insecure (this improves on an earlier result of Wiener of $N^{.25}$.

Lattice attacks have been successfully applied to many other crypto functions. For example the NTRU public key cryptosystem was recently described. The system is due to Hoffstein, Pipher, and Silverman and perhaps the most convenient reference to its description is the paper by Coppersmith and Shamir [12]. The security of the system is based on the difficulty of analyzing the result of polynomial arithmetic modulo two unrelated moduli. Its correctness is based on the clustering properties of sums of random variables. Coppersmith and Shamir use lattice reduction techniques to either discover the original key or another key which is also useful in decoding the ciphertexts. The original description of the cryptosystem did not seem to suggest such an attack and again the applicability of lattice based techniques is interesting.

Other applications of lattices includes linear congruential generators and Damgård's hash function. In addition, Schnorr [33] has shown how to pose both integer factorization and the discrete logarithm problem modulo a prime as a lattice problem, although these two formulations do not seem efficient enough to compete with the standard algorithms used.

6. COMMENTS

The formulation of new trapdoor one-way functions that would lead to viable alternatives to the existing public key cryptosystems, has proven to be a problem that is remarkably resistant to solution. The lattice-based approaches described in this work represent a significant and exciting development, yet have so far appeared to fall short of providing effective solutions to the problem. Other systems that have been proposed are based on ideal arithmetic in number fields [10], orders in imaginary quadratic fields [9], and Jacobians of hyperelliptic curves. As the level of mathematics required to understand and implement such systems increases, the number of researchers able to effectively attack them becomes more limited, posing a problem of confidence in their security.

References

[1] L. M. Adleman, "On Breaking Generalized Knapsack Public Key Cryptosystems," in *Proceedings of the 15th ACM Symposium on Theory of Computing*, pp. 402–412, 1983.

[2] M. Ajtai, "Generating Hard Instances of Lattice Problems," in *Proceedings of the 28th Annual ACM Symposium on Theory of Computing*, pp. 99–108, 1996.

[3] M. Ajtai and C. Dwork, "A Public-Key Cryptosystem with Worst-Case/Average-Case Equivalence," *Proceedings of the 29th Annual ACM Symposium on the Theory of Computing*, pp. 284–293, 1997.

[4] M. Ajtai, "The Shortest Vector Problem is NP-Hard for Randomized Reductions," in *Proceedings of the 30th ACM Symposium on Theory of Computing*, pp.10–19, 1998.

[5] S. Arora, L. Babai, J. Stern, and Z. Sweedyk, "The Hardness of Approximate Optima in Lattices, Codes and Systems of Linear Equations," *34th Foundations of Computer Science*, pp. 724–733, 1993.

[6] D. Boneh, "Exposing an RSA Private Key Given a Small Fraction of its Bits," *Asiacrypt '98, Lecture Notes in Computer Science*, vol. 1514, pp. 25–34, 1998.

[7] D. Boneh and G. Durfee, "Cryptanalysis of RSA with Private Key Less Than $N^{0.292}$," *Eurocrypt '99, Lecture Notes in Computer Science*, vol. 1592, pp. 1–11, 1999.

[8] D. Boneh and R. Venkatesan, "Breaking RSA May Not be Equivalent to Factoring," *Eurocrypt '98, Lecture Notes in Computer Science*, vol. 1403, pp. 59–71, 1998.

[9] J. Buchmann and H. C. Williams, "A Key-Exchange System Based on Imaginary Quadratic Fields," *Journal of Cryptology*, vol. 1, pp. 107–118, 1988.

[10] J. Buchmann and S. Paulus, "A One Way Function Based on Ideal Arithmetic in Number Fields," *Crypto '97, Lecture Notes in Computer Science*, vol. 1294, B. S. Kaliski ed., pp. 385–394, 1997.

[11] J. H. Conway and N. J. A. Sloane, *Sphere Packings, Lattices and Groups*, New York: Springer-Verlag, 1998.

[12] D. Coppersmith and A. Shamir, "Lattice Attacks on NTRU," *Eurocrypt '97, Lecture Notes in Computer Science*, W. Fumy ed., vol. 1123, pp. 52–61, 1997.

[13] D. Coppersmith, "Small Solutions to Polynomial Equations, and Low Exponent RSA Vulnerabilities," *Journal of Cryptology*, vol. 10, pp. 233–260, 1997.

[14] M. J. Coster, A. Joux, B. A. LaMacchia, A. Odlyzko, C. P. Schnorr, and J. Stern, "Improved Low-Density Subset-Sum Algorithms," *Computational Complexity*, vol. 2, pp. 11–28, 1992.

[15] W. Diffie and M. E. Hellman, "New Directions in Cryptography," *IEEE Transactions on Information Theory*, vol. IT-22, pp. 472–492, 1976.

[16] P. van Emde Boas, "Another NP-Complete Partition Problem and the Complexity of Computing Short Vectors in a Lattice," Technical Report 81-04, Department of Mathematics, University of Amsterdam, 1980.

[17] O. Goldreich, S. Goldwasser, and S. Halevi, "Eliminating Decryption Errors in the Ajtai-Dwork Cryptosystem," *Crypto '97, Lecture Notes in Computer Science*, vol. 1294, B. S. Kaliski ed., pp. 105-111, 1997.

[18] O. Goldreich, S. Goldwasser, and S. Halevi, "Collision-Free Hashing from Lattice Problems," ECCC Report TR96-042, available from `http://www.eccc.uni-trier.de/eccc/`

[19] O. Goldreich, S. Goldwasser, and S. Halevi, "Public-key Cryptosystems from Lattice Reduction Problems," *Crypto '97, Lecture Notes in Computer Science*, vol. 1294, B. S. Kaliski ed., pp. 112–131, 1997.

[20] A. Joux and J. Stern, "Lattice Reduction: A Toolbox for the Cryptanalyst," Dec. 1997. Preprint.

[21] J. C. Lagarias and A. M. Odlyzko, "Solving Low-Density Subset Sum Problems," *Journal of ACM*, vol. 32, pp. 229–246, 1985.

[22] A. K. Lenstra, H. W. Lenstra, and L. Lovász, "Factoring Polynomials with Rational Coefficients," *Mathematische Annalen*, vol. 261, pp. 515–534, 1982.

[23] K. S. McCurley, "The Discrete Logarithm Problem," *Proceedings of Symposia in Applied Mathematics*, vol. 42, pp. 49–74, 1990.

[24] A. J. Menezes, *Elliptic curve cryptosystems*, Kluwer Academic Publishers, 1993.

[25] A. J. Menezes, P. C. van Oorschot, and S. A. Vanstone, *Handbook of Applied Cryptography*, Boca Raton, FL: CRC press, 1997.

[26] R. Merkle and M. E. Hellman, "Hiding Information and Signatures in Trapdoor Knapsacks," *IEEE Transactions on Information Theory*, vol. IT 24, pp. 525–530, 1978.

[27] D. Micciancio, "The Shortest Vector in a Lattice is Hard to Approximate to Within Some Constant," *Elect. Coll. on Comp. Complexity*, Report no. 16, 1998.

[28] P. Nguyen and J. Stern, "A Converse to the Ajtai-Dwork Security Proof and its Cryptographic Implications," Nov. 1997, preprint.

[29] A. M. Odlyzko, "Discrete Logarithms in Finite Fields and Their Cryptographic Significance," *Eurocrypt '84, Lecture Notes in Computer Science*, vol. 209, pp. 224–314, 1984.

[30] S. C. Pohlig and M. E. Hellman, "An Improved Algorithm for Computing Logarithms over $GF(p)$ and its Cryptographic Significance," *IEEE Transactions on Information Theory*, vol. IT-24, pp. 106–110, 1978.

[31] C. P. Schnorr and H. H. Horner, "Attacking the Chor-Rivest Cryptosystem by Improved Lattice Reduction," in *Eurocrypt '95, Lecture Notes in Computer Science*, vol. 921, L. Guillou and J.-J. Quisquater eds., pp. 1–12, 1995.

[32] C. P. Schnorr, "A Hierarchy of Polynomial Time Lattice Basis Reduction Algorithm Algorithms," *Theoretical Computer Science*, vol. 53, pp. 201–224, 1987.

[33] C. P. Schnorr, "Factoring Integers and Computing Discrete Logarithms via Diophantine Approximation," *Eurocrypt '91, Lecture Notes in Computer Science* vol. 547, pp. 171–181, 1991.

[34] A. Shamir, "A Polynomial Time Algorithm for Breaking the Basic Merkle-Hellman Cryptosystem," in *Proceedings of the 23rd IEEE Symposium on Foundations of Computer Science*, pp. 145–152, 1982.

Chapter 19

A SIMPLE CONSTRUCTION FOR THE BARNES-WALL LATTICES

G. Nebe
Abteilung Reine Mathematik,
Universität Ulm, 89069 Ulm, Germany

E. M. Rains
Information Sciences Research, AT&T Shannon Labs,
180 Park Avenue, Florham Park, NJ 07932-0971, U.S.A.

N. J. A. Sloane
Information Sciences Research, AT&T Shannon Labs,
180 Park Avenue, Florham Park, NJ 07932-0971, U.S.A.

Abstract A certain family of orthogonal groups (called "Clifford groups" by G. E. Wall) has arisen in a variety of different contexts in recent years. These groups have a simple definition as the automorphism groups of certain generalized Barnes-Wall lattices. This leads to an especially simple construction for the usual Barnes-Wall lattices.

1. BACKGROUND

The Barnes-Wall lattices define an infinite sequence of sphere packings in dimensions 2^m, $m \geq 0$, which include the densest packings known in dimensions 1, 2, 4, 8, and 16 [1, 13]. In dimensions 32 and higher they are less dense than other known packings, but they are still interesting

*This is based on the third author's talk at the Forney-Fest, M.I.T., March 2000, which in turn is based on our paper "The Invariants of the Clifford Groups", Designs, Codes, Cryptography, **21** (2001), 99-121, to which the reader is referred for further details and proofs.

for other reasons — they form one of the few infinite sequences of lattices in which it is possible to do explicit calculations. For example, there is an explicit formula for their kissing numbers [13]. This talk will describe a beautifully simple construction for these lattices that we found in the summer of 1999. A more comprehensive account will appear elsewhere [20, 21]. Since Dave Forney is fond of the Barnes-Wall lattices (cf. [14, 15]) we hope he will like this construction as much as we do.

This work had its origin in 1995 when J. H. Conway, R. H. Hardin, and N. J. A. S. were studying packings in Grassmann manifolds — in other words, packings of Euclidean k-dimensional subspaces in n-dimensional space [12]. One of our nicest constructions was an optimal packing of 70 four-dimensional subspaces in $\mathbb{R}^8$. The symmetry group of this packing (the subgroup of the orthogonal group $O(8, \mathbb{R})$ that fixes the collection of subspaces) has order 5160960.

Shortly afterwards, the *same* eight-dimensional group arose in the work of P. W. Shor and others on quantum computers (cf. [2, 18]).

This astonishing coincidence — see [11] for the full story — drew attention to earlier work on the family to which this group belongs [5, 6, 7, 8, 32]. Following Wall, we call these Clifford groups, although these are not the groups usually referred to by this name [22]. Investigation of the representations of subgroups of these groups led to further constructions of optimal packings in Grassmann manifolds [9] and constructions of quantum error-correcting codes [10, 11].

Independently, and around the same time, these groups[1] also occurred in the work of V. M. Sidelnikov, in connection with the construction of spherical t-designs [17, 28, 29, 30, 31].

The complete account of our work [20, 21] describes the invariants of these Clifford groups and their connections with binary self-dual codes. Much of this work had been anticipated by Runge [24, 25, 26, 27]. In our two papers we clarify the connections with spherical t-designs and Sidelnikov's work, and also generalize these results to the complex Clifford groups and doubly-even binary self-dual codes. Again the main result was first given by Runge.

In recent years many other kinds of self-dual codes have been studied by a number of authors. Nine such families were named and surveyed in [23]. In [21] we give a general definition of the "type" of a self-dual code which includes all these families as well as other self-dual codes over rings and modules. For each "type" we investigate the structure of the associated "Clifford-Weil group" and its ring of invariants.

Some of the results in [20, 21] can be regarded as providing a general setting for Gleason's theorems [16, 19, 23] about the weight enumerator of a binary self-dual code, a doubly-even binary self-dual code and a self-dual code over $\mathbb{F}_p$. They are also a kind of discrete analogue of a long series of theorems going back to Eichler (see for example [4, 25, 26, 27]), stating that, under certain conditions, theta series of quadratic forms are bases for spaces of modular forms: here complete weight enumerators of generalized self-dual codes are bases for spaces of invariants of "Clifford-Weil groups".

2. A SIMPLE CONSTRUCTION FOR THE BARNES-WALL LATTICES

There is a pair of Barnes-Wall lattices L_m and L'_m in each dimension 2^m, $m \geq 0$. The two lattices are geometrically similar[2] and L_m is a sublattice of index 2^k, $k = 2^{m-1}$, in L'_m. In dimension 2 these lattices are shown in Figure 1, where L_1 consists of the points marked with solid circles and L'_1 consists of the points marked with either solid or hollow circles. Both are geometrically similar to the square lattice $\mathbb{Z}^2$.

Figure 1 The two Barnes-Wall lattices L_1 (solid circles) and L'_1 (solid or hollow circles) in two dimensions.

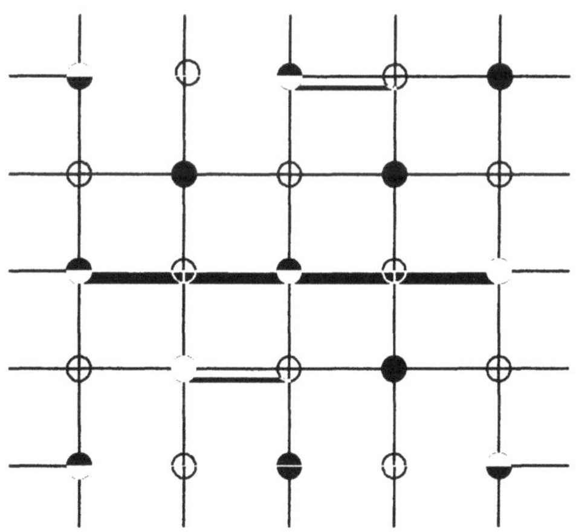

Suppose we multiply the points of L'_1 by $\sqrt{2}$. Then the eight minimal vectors of L_1 and $\sqrt{2}\, L'_1$ now have the same length and form the familiar configuration of points used in the 8-PSK signaling system (Figure 2).

We now define the generalized or "balanced" Barnes-Wall lattice M_1 to be the set of all $\mathbb{Z}[\sqrt{2}]$-integer combinations of the eight vectors in Figure 2. That is, we take integer combinations of these vectors where "integer" now means a number of the form $a + b\sqrt{2}$, $a, b \in \mathbb{Z}$. In more formal language, M_1 is a $\mathbb{Z}[\sqrt{2}]$-lattice (or $\mathbb{Z}[\sqrt{2}]$-module). Note that

Figure 2 The eight minimal vectors of L_1 and $\sqrt{2}\,L_1'$. The $\mathbb{Z}[\sqrt{2}]$ span of these points is the "balanced" Barnes-Wall lattice M_1.

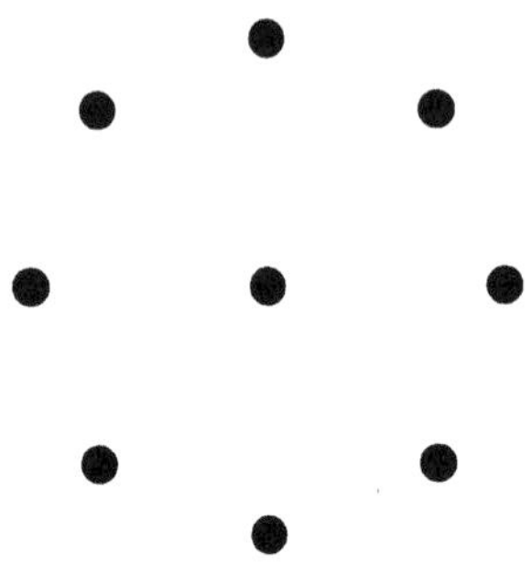

we can recover L_1 from M_1 by taking just those vectors in M_1 whose components are integers.

In general we define the rational part of a $\mathbb{Z}[\sqrt{2}]$-lattice Λ to consist of the vectors which have rational components, and the irrational part to consist of the vectors whose components are rational multiples of $\sqrt{2}$. We can now state the construction.

Theorem 1 *Define the balanced Barnes-Wall lattice M_m to be the m-fold tensor product $M_1^{\otimes m}$. Then the rational part of M_m is the Barnes-Wall lattice L_m, and the purely irrational part is $\sqrt{2}\,L_m'$.*

For the proof see [20].

To be quite explicit, note that we need only two of the vectors in Figure 2, and we can take

$$G_1 = \begin{pmatrix} \sqrt{2} & 0 \\ 1 & 1 \end{pmatrix}$$

as a generator matrix for M_1. Then the m-fold tensor power of this matrix,

$$G_m = G_1^{\otimes m} = G_1 \otimes G_1 \otimes \cdots \otimes G_1$$

is a generator matrix for M_m.

For example, $G_2 = G_1 \otimes G_1$ is

$$\begin{bmatrix} 2 & 0 & 0 & 0 \\ \sqrt{2} & \sqrt{2} & 0 & 0 \\ \sqrt{2} & 0 & \sqrt{2} & 0 \\ 1 & 1 & 1 & 1 \end{bmatrix}.$$

The rational part, L_2, is generated by

$$\begin{bmatrix} 2 & 0 & 0 & 0 \\ 2 & 2 & 0 & 0 \\ 2 & 0 & 2 & 0 \\ 1 & 1 & 1 & 1 \end{bmatrix}, \text{ or equivalently } \begin{bmatrix} 2 & 0 & 0 & 0 \\ 0 & 2 & 0 & 0 \\ 0 & 0 & 2 & 0 \\ 1 & 1 & 1 & 1 \end{bmatrix}.$$

This lattice is geometrically similar to D_4 ([13], Chapter 4, Equation (90)). The purely irrational part, $\sqrt{2}L_2'$, is generated by

$$\begin{bmatrix} 2\sqrt{2} & 0 & 0 & 0 \\ \sqrt{2} & \sqrt{2} & 0 & 0 \\ \sqrt{2} & 0 & \sqrt{2} & 0 \\ \sqrt{2} & \sqrt{2} & \sqrt{2} & \sqrt{2} \end{bmatrix}, \text{ or equivalently } \sqrt{2}\begin{bmatrix} 2 & 0 & 0 & 0 \\ 1 & 1 & 0 & 0 \\ 1 & 0 & 1 & 0 \\ 1 & 0 & 0 & 1 \end{bmatrix},$$

which is another version of D_4 ([13], Chapter 4, Equation (86)).

We may avoid the use of coordinates and work directly with Gram matrices or quadratic forms, provided we select an appropriate element ϕ of the Galois group. Let u_1 and u_2 be the generating vectors corresponding to the rows of the matrix G_1, and let ϕ negate u_1 and fix u_2.

Then M_1 is the $\mathbb{Z}[\sqrt{2}]$-lattice with Gram matrix

$$A_1 = G_1 G_1^{tr} = \begin{pmatrix} 2 & \sqrt{2} \\ \sqrt{2} & 2 \end{pmatrix},$$

L_1 is the sublattice of M_1 fixed by ϕ and $\sqrt{2}\,L_1'$ is the sublattice negated by ϕ.

Furthermore, ϕ has a natural extension to $M_m = M_1^{\otimes m}$, and L_m is the sublattice of M_n fixed by ϕ and $\sqrt{2}\,L_m'$ is the sublattice negated by ϕ.

3. THE CLIFFORD GROUPS AND THEIR INVARIANTS

The Clifford groups $\mathcal{C}_m$ mentioned at the beginning of the paper now have a very simple definition: for all $m \geq 1$, $\mathcal{C}_m$ is $Aut(M_n)$, i.e. the subgroup of $O(2^m, \mathbb{R})$ that preserves M_m.

For the proof that this definition is equivalent to the usual one given in [8, 9, 11], see Proposition 5.3 of [20].

An *invariant polynomial* of $\mathcal{C}_m$ is a polynomial in 2^m variables with real coefficients that is fixed by every element of the group [3]. The ring of invariant polynomials plays an important role in constructing spherical t-designs from the group (see for example [13], Chapter 3,

Section 4.2). The *Molien series* of the group is a generating function for the numbers of linearly independent homogeneous invariants of each degree [3, 19, 23].

In [8] it was asked "Is it possible to say something about the Molien series [of the groups $\mathcal{C}_m$], such as the minimal degree of an invariant?" Such questions also arise in the work of Sidelnikov [28, 29, 30, 31]. The answers are given by the following theorem of Runge [24, 25, 26, 27].

Theorem 2 (*Runge; [20]*). *Fix integers k and $m \geq 1$. The space of homogeneous invariants of $\mathcal{C}_m$ of degree $2k$ is spanned by the complete weight enumerators of the codes $C \otimes GF(2^m)$, where C ranges over all binary self-dual codes of length $2k$; this is a basis if $m \geq k-1$.*

We rediscovered this result in the summer of 1999. Our proof is somewhat simpler than Runge's as it avoids the use of Siegel modular forms [20].

Corollary 3 *Let $\Phi_m(t)$ be the Molien series of $\mathcal{C}_m$. As m tends to infinity, the series $\Phi_m(t)$ tend monotonically to*

$$\sum_{k=0}^{\infty} N_{2k} t^{2k} \ ,$$

where N_{2k} is the number of equivalence classes of binary self-dual codes of length $2k$.

Explicit calculations for $m = 1, 2$ show:

Corollary 4 *The initial terms of the Molien series of $\mathcal{C}_m$ are given by*

$$1 + t^2 + t^4 + t^6 + 2t^8 + 2t^{10} + O(t^{12}) \, ,$$

where the next term is $2t^{12}$ for $m = 1$, and $3t^{12}$ for $m > 1$.

Sidelnikov [29, 30] showed that the lowest degree of a harmonic invariant of $\mathcal{C}_m$ is 8. Inspection of the above Molien series gives the following stronger result.

Corollary 5 *The smallest degree of a harmonic invariant of $\mathcal{C}_m$ is 8, and there is a unique harmonic invariant of degree 8. There are no harmonic invariants of degree 10.*

There is a unique harmonic invariant of degree 8, which can be taken to be the complete weight enumerator of $H_8 \otimes GF(2^m)$, where H_8 is the

[8, 4, 4] binary Hamming code, minus a suitable multiple of the fourth power of the quadratic form.

From Corollary 5 we can easily show that appropriate orbits under $\mathcal{C}_m$ form spherical 7-designs, 11-designs, etc.

We conclude by mentioning one last result (Corollary 5.7 of [20]).

Theorem 6 *Let C be any binary self-dual code that is not generated by vectors of weight 2, and form the complete weight enumerator of $C \otimes GF(2^m)$. Then the subgroup of $O(2^m, \mathbb{R})$ that fixes this weight enumerator is precisely $\mathcal{C}_m$.*

As already mentioned, there are analogues of all these results for the complex versions of the Clifford group. Now "self-dual code" is replaced by "doubly-even self-dual code".

The case $m = 1$ of Theorem 6 and its complex analogue imply the following.

Let C be a binary self-dual code of length N and $W(x, y)$ its Hamming weight enumerator. Let G be the subgroup of $O(2, \mathbb{R})$ that fixes $W(x, y)$. Provided C is not generated by vectors of weight 2, $G \cong \mathcal{C}_1$, of order 16. If C is doubly-even, the subgroup of the unitary group $U(2, \mathbb{C})$ that fixes $W(x, y)$ is (apart from its center, which of course may contain complex N-th roots of unity) the familiar group of order 192 arising in Gleason's theorem.

It was of course known that $W(x, y)$ is fixed by these two groups, of order 16 and 192. But we were not aware before this of any proof that the group of $W(x, y)$ could never be bigger.

Notes

1. Although at that time they were not recognized as the Clifford groups.
2. In other words they differ only by a rotation and change of scale.

References

[1] E. S. Barnes, and G. E. Wall, "Some Extreme Forms Defined in Terms of Abelian Groups," *Journal of the Australian Mathematical Society*, vol. 1, pp. 47–63. (1959).

[2] C. H. Bennett, D. DiVincenzo, J. A. Smolin, and W. K. Wootters, "Mixed State Entanglement and Quantum Error Correction," *Physical Review A*, vol. 54, pp. 3824–3851, 1996.

[3] D. J. Benson, *Polynomial Invariants of Finite Groups*, Cambridge University Press, 1993.

[4] S. Böcherer, "Siegel Modular Forms and Theta Series," in *Theta functions (Bowdoin 1987), Proceedings of the Symposium on Pure Mathematics*, vol. 49, Part 2, Amererican Mathematical Society, Providence, RI, pp. 3–17, 1989.

[5] B. Bolt, "The Clifford Collineation, Transform and Similarity Groups III: Generators and Involutions," *Journal of the Australian Mathematical Society*, vol. 2, pp. 334–344, 1961.

[6] B. Bolt, T. G. Room, and G. E. Wall, "On Clifford Collineation, Transform and Similarity Groups I," *Journal of the Australian Mathematical Society*, vol. 2, pp. 60–79, 1961.

[7] B. Bolt, T. G. Room, and G. E. Wall, "On Clifford Collineation, Transform and Similarity Groups II," *Journal of the Australian Mathematical Society*, vol. 2, pp. 80–96, 1961.

[8] A. R. Calderbank, P. J. Cameron, W. M. Kantor, and J. J. Seidel, "$\mathbb{Z}_4$-Kerdock Codes, Orthogonal Spreads and Extremal Euclidean Line-Sets," *Proceedings of the London Mathematical Society*, vol. 75, pp. 436–480, 1997.

[9] A. R. Calderbank, R. H. Hardin, E. M. Rains, P. W. Shor, and N. J. A. Sloane, "A Group-Theoretic Framework for the Construction of Packings in Grassmannian Spaces," *Journal of Algebraic Combinatorics*, vol. 9, pp. 129–140, 1999.

[10] A. R. Calderbank, E. M. Rains, P. W. Shor, and N. J. A. Sloane, "Quantum Error Correction Orthogonal Geometry," *Physical Review Letters*, vol. 78, pp. 405–409, 1997.

[11] A. R. Calderbank, E. M. Rains, P. W. Shor, and N. J. A. Sloane, "Quantum Error Correction Via Codes Over $GF(4)$," *IEEE Transactions on Information Theory*, vol. 44, pp. 1369–1387, 1998.

[12] J. H. Conway, R. H. Hardin, and N. J. A. Sloane, "Packing Lines, Planes, etc.: Packings in Grassmannian Space," *Experimental Mathematics*, vol. 5, pp. 139–159, 1996.

[13] J. H. Conway and N. J. A. Sloane, *Sphere Packings, Lattices and Groups*, 3rd ed., Springer-Verlag, NY, 1998.

[14] G. D. Forney, Jr., "Coset codes - Part I: Introduction and Geometrical Classification," *IEEE Transactions on Information Theory*, vol. 34, pp. 1123–1151, 1988.

[15] G. D. Forney, Jr., "Coset codes - Part II: Binary Lattices and Related Codes," *IEEE Transactions on Information Theory*, vol. 34, pp. 1152–1186, 1988.

[16] A. M. Gleason, "Weight Polynomials of Self-Dual Codes and the MacWilliams Identities," in *Actes, Congrés International*

de Mathématiques (Nice, 1970), Gauthiers-Villars, Paris, vol. 3, pp. 211–215, 1971

[17] L. S. Kazarin, "On Groups Proposed by Sidelnikov (Russian)," *Mat. Sb.*, vol. 189, no. 7, pp. 131–144, 1998), English translation in *Sb. Math.*, vol. 189, pp. 1087–1100, 1998.

[18] A. Y. Kitaev, "Quantum Computations: Algorithms and Error Correction (in Russian), *Uspekhi Mat. Nauk.*, vol. 52, no. 6, pp. 53–112, 1997; English translation in *Russian Math. Surveys*, vol. 52, pp. 1191–1249, 1997.

[19] F. J. MacWilliams and N. J. A. Sloane, *The Theory of Error-Correcting Codes*, North-Holland, Amsterdam, 1977.

[20] G. Nebe, E. M. Rains, and N. J. A. Sloane, "The Invariants of the Clifford Groups," *Designs, Codes, Cryptography*, vol. 24, pp. 99–121, 2001.

[21] G. Nebe, E. M. Rains, and N. J. A. Sloane, "Generalized Self-Dual Codes and Clifford-Weil Groups," preprint.

[22] I. R. Porteous, *Clifford Algebras and the Classical Groups*, Cambridge, MA, 1995.

[23] E. M. Rains and N. J. A. Sloane, "Self-dual codes," in *Handbook of Coding Theory*, ed. V. Pless and W. C. Huffman, Elsevier, Amsterdam, pp, 177–294, 1998.

[24] B. Runge, The Schottky ideal, in *Abelian Varieties (Egloffstein, 1993)*, de Gruyter, Berlin, pp. 251–272, 1995.

[25] B. Runge, "On Siegel Modular Forms I," *Journal für die reine und angewandte Mathematik*, vol. 436, pp. 57–85, 1993.

[26] B. Runge, "On Siegel Modular Forms II, *Nagoya Mathematical Journal*, vol. 138, pp. 179–197, 1995.

[27] B. Runge, "Codes and Siegel Modular Forms," *Discrete Mathematics*, vol. 148, pp. 175–205, 1995.

[28] V. M. Sidelnikov, "On a Finite Group of Matrices and Codes on the Euclidean Sphere (in Russian)," *Problemy Peredachi Informatsii*, vol. 33, pp. 35–54, 1997; English translation in *Problems Inform. Transmission*, vol. 33, pp. 29–44, 1997.

[29] V. M. Sidelnikov, "On a Finite Group of Matrices Generating Orbit Codes on the Euclidean Sphere," in *Proceedings of the 1997 IEEE International Symposium on Information Theory*, IEEE Press, p. 436, 1997.

[30] V. M. Sidelnikov, "Spherical 7-designs in 2^n-Dimensional Euclidean Space," *Journal of Algebraic Combinatorics*, vol. 10, pp. 279–288, 1999.

[31] V. M. Sidelnikov, "Orbital Spherical 11-Designs Whose Initial Point is a Root of an Invariant Polynomial," (Russian), *Algebra i Analiz*, vol. 11, No. 4, pp. 183–203, 1999.

[32] G. E. Wall, "On Clifford Collineation, Transform and Similarity Groups IV: An Application to Quadratic Forms," *Nagoya Mathematical Journal*, vol. 21, pp. 199–222, 1962.

VI
BEHAVIORS AND CODES ON GRAPHS

Chapter 20

MINIMAL BASES OF RATIONAL VECTOR SPACES AND THEIR IMPORTANCE IN ALGEBRAIC SYSTEMS THEORY

J. Rosenthal*
Department of Mathematics
University of Notre Dame
Notre Dame, Indiana 46556, USA

Abstract Let $\mathcal{F}$ denote the field of rational functions over some base field. Every subspace V of $\mathcal{F}^n$ has a polynomial basis. A polynomial basis having minimal possible degrees is called a minimal basis of V. It was shown by G.D. Forney that minimal bases always exist and that these bases are of great importance in multivariable systems theory and convolutional coding theory.

Keywords: Vector spaces over the rationals, multivariable systems, vector bundles over the projective line.

1. INTRODUCTION

In 1975 Dave Forney wrote a paper [4] entitled *Minimal Bases of Rational Vector Spaces, with Applications to Multivariable Linear Systems*. This paper had an immense impact in the mathematical systems theory literature. According to the full citation database of the Institute for Scientific Information on DIALOG, [4] has been cited 227 times in the period 1976–1999 and it has been the most cited paper by Forney. In this way the paper is a citation classic (compare with Table 1 in the Appendix).

*Supported in part by NSF grant DMS-96-10389. This research has been carried out while the author was a guest professor at EPFL in Switzerland. The author would like to thank EPFL for its support and hospitality.

In this article we will survey the results of Forney's paper [4]. This paper contains several interesting contributions. The main theorem (see Theorem 2) on minimal polynomial bases of vector spaces over the rationals is a purely mathematical result. Its essence has been first derived by Forney in the context of coding theory [3]. The result connects to some classical work in algebra and algebraic geometry and in Section 2 we will review this result. In Section 3 we will relate Theorem 2 to a theorem of Grothendieck [8] and a famous classical paper by Dedekind and Weber [2].

A major contribution of paper [4] is a clear explanation on how to translate properties of a system from a polynomial matrix formulation into a state space formulation. In this way Forney was able to connect the controllability and the observability indices of a linear system to some minimal indices of a rational vector space. In Section 4, we will review these results.

One contribution of this paper was a concrete realization algorithm. In Section 5 we will explain how realization theory can be generalized to accommodate general behavioral systems [21] which have no predetermined input-output structure. The classical realization theory of a transfer function then becomes a special instance of this general realization theory.

The author would like to thank Thurston Miller from the University of Notre Dame for the citation search of Forney's paper [4].

2. MINIMAL POLYNOMIAL BASES OF RATIONAL VECTOR SPACES

Let $\mathbb{F}$ be an arbitrary field and consider the field $\mathcal{F} := \mathbb{F}(x)$ of rational functions in the indeterminant x. Let $\mathbb{F}[x]$ be the polynomial ring over $\mathbb{F}$.

Definition 1 One says $v = (v_1, \ldots, v_n) \in \mathcal{F}^n$ is a polynomial vector if $v \in \mathbb{F}^n[x]$. If v is a polynomial vector then one defines its degree as $\deg v := \max\{\deg v_i \mid i = 1, \ldots, n\}$.

Let $V \subset \mathcal{F}^n$ be a k-dimensional subspace. A basis $B := \{v^1, \ldots, v^k\}$ is called a *polynomial basis* if $B \subset \mathbb{F}^n[x]$. One says that B forms a *minimal polynomial basis* if the total degree (order) $\sum_{j=1}^k \deg v^j$ is minimal. It is clear that a polynomial basis exists. The following is the main theorem in [4] and it characterizes a minimal polynomial basis:

Theorem 2 ([4]) *Let $V \subset \mathcal{F}^n$ be a k-dimensional subspace and assume the rows of a $k \times n$ matrix G represent a polynomial basis. Assume the*

jth row degree is ν_j and the total degree (order) is $\nu = \sum_{j=1}^{k} \nu_j$. Then the following are equivalent.

1. *The rows of G form a minimal polynomial basis.*

2. *G is nonsingular modulo $p(x)$ for all irreducible polynomials $p(x) \in \mathbb{F}[x]$ and the high order coefficient matrix $[G]_h$ of G has full rank.*

3. *The greatest common divisor of the $k \times k$ full size minors is 1 and their greatest degree is ν.*

4. *If $y = xG$ is polynomial then x is polynomial and $\deg y = \max_{1 \le j \le k}\{\deg x_j + \nu_j\}$.*

5. *For $d \geq 0$ let $V_d \subset V$ be the subset of all polynomial vectors of degree at most d. Then V_d is a finite $\mathbb{F}$ vector space and the indices $\nu_1, \ldots, \nu_k$ have the property that $\dim_{\mathbb{F}} V_d = \sum_{\nu_j < d}(d - \nu_j)$.*

In coding theory [3, 10] a linear subspace $\mathcal{C} \subset \mathcal{F}^n$ defines a convolutional code. If the rows of a $k \times n$ matrix G represent a $\mathcal{F}$-basis for the code $\mathcal{C}$ then one says that G is an encoder. G defines the encoding map:

$$\varphi: \ \mathcal{F}^k \longrightarrow \mathcal{F}^n, \ m(x) \longmapsto c(x) = m(x)G(x). \tag{1}$$

Minimal polynomial encoders are of great significance because they describe feedforward encoders having a minimal number of delay elements. We will explain the significance of minimal polynomial bases in systems theory in Section 4.

It follows from the last conditions of Theorem 2 that the indices $\nu_1, \ldots, \nu_k$ are invariants of the subspace $V \subset \mathcal{F}^n$. These indices are of crucial importance both in convolutional coding theory [3, 10] and in systems theory and we follow McEliece [16] and call them the *Forney indices* of V. Without loss of generality one can assume that they are ordered $\nu_1 \geq \cdots \geq \nu_k$.

In [4, Remark 1] Forney asked if Theorem 2 has not already been derived earlier since the theorem spells out merely important properties of a subspace $V \subset \mathcal{F}^n$. Today it is clear that Theorem 2 was new at the time but that it is closely connected to some interesting results in mathematics. It has been recognized (compare with [5]), that minimal polynomial bases appeared already in 1908 by Plemelj [17] who did give a solution to Riemann's problem on functions with a given monodromic group. In the next section we show a connection to Grothendieck's theorem [8] and to a classical paper by Dedekind and Weber [2].

3. GROTHENDIECK'S THEOREM AND THE WORK OF DEDEKIND AND WEBER

In [8] Grothendieck provided a complete classification of vector bundles over the Riemann sphere. For this let $\mathbb{C}$ denote the complex numbers and let

$$\mathbb{P}^1_{\mathbb{C}} := \{\ell \subset \mathbb{C}^2 \mid \dim \ell = 1\} \hat{=} \{(x,1) \mid x \in \mathbb{C}\} \cup \{(1,0)\}$$

denote the projective line over $\mathbb{C}$, i.e. the Riemann sphere. Then one has:

Theorem 3 ([8]) *If ξ is a holomorphic vector bundle over $\mathbb{P}^1_{\mathbb{C}}$ then ξ decomposes as a sum of line bundles:*

$$\xi = O(\nu_1) \oplus \cdots \oplus O(\nu_k),$$

where $\nu_1, \ldots, \nu_k$ are the multiplicities of the line bundles. The nonnegative integers $\nu_1, \ldots, \nu_k$ depend up to order only on ξ.

The indices $\nu_1, \ldots, \nu_k$ are sometimes referred to as the Grothendieck indices of ξ. The integer $\nu = \sum_{j=1}^{k} \nu_j$ is called the *degree* of ξ. The relation to the Forney indices is now as follows:

Consider the Grassmann manifold

$$\text{Grass}(k, \mathbb{C}^n) := \{W \subset \mathbb{C}^n \mid \dim W = k\}$$

parameterizing all k-dimensional linear subspaces of $\mathbb{C}^n$. $\text{Grass}(k, \mathbb{C}^n)$ is a smooth, compact manifold of complex dimension $k(n-k)$. Consider a k-dimensional linear subspace $V \subset \mathcal{F}^n$ and assume that G is a $k \times n$ matrix over $\mathcal{F}$ such that $\text{rowspace}_{\mathcal{F}} G = V$. Consider the map:

$$h: \mathbb{P}^1_{\mathbb{C}} \longrightarrow \text{Grass}(k, \mathbb{C}^n), \; z \mapsto \text{rowspace}_{\mathbb{C}} G(z). \tag{2}$$

A priori h is not defined if $\text{rank}\, G(z_0) < k$ or if an entry of G has a pole at a number z_0. The map is also not defined at $z = \infty$. Each singularity is however isolated and it is a well known result of complex analysis that each singularity is removable since $\text{Grass}(k, \mathbb{C}^n)$ is compact. The map h extends therefore to a holomorphic map defined on all of $\mathbb{P}^1_{\mathbb{C}}$. In this way every subspace $V \subset \mathcal{F}^n$ describes a holomorphic map h and, vice versa, every holomorphic map h from $\mathbb{P}^1_{\mathbb{C}}$ to $\text{Grass}(k, \mathbb{C}^n)$ defines a linear subspace $V \subset \mathcal{F}^n$.

The Grassmann manifold is equipped with a natural vector bundle called the universal bundle U. Let U^* be its dual. The following theorem is due to Martin and Hermann [15]:

Theorem 4 *Let ξ be the pull back of the bundle U^* under the holomorphic map h. Then the Grothendieck indices $\nu_1, \ldots, \nu_k$ of ξ are up to order equal to the Forney indices of $V \subset \mathcal{F}^n$. Moreover the degree $\nu = \sum_{j=1}^k \nu_j$ of ξ is equal to the topological degree of the map h.*

The computation of a minimal polynomial basis works over an arbitrary base field $\mathbb{F}$. It is also known that Grothendieck's theorem and Theorem 4 are valid over an arbitrary base field. A simple proof of Grothendieck's theorem over a general field can be found in [9, 13].

In the situation of a general field, a morphism $h : \mathbb{P}^1_{\mathbb{F}} \longrightarrow \text{Grass}(k, \mathbb{F}^n)$ is given through a map

$$(s,t) \longmapsto \text{rowspace}_{\mathbb{F}} P(s,t),$$

where

$$P(s,t) = \begin{pmatrix} f_{11}(s,t) & f_{12}(s,t) & \ldots & f_{1,m+p}(s,t) \\ f_{21}(s,t) & f_{22}(s,t) & \ldots & f_{2,m+p}(s,t) \\ \vdots & \vdots & & \vdots \\ f_{p1}(s,t) & f_{p2}(s,t) & \ldots & f_{p,m+p}(s,t) \end{pmatrix} \tag{3}$$

is a $k \times n$ matrix whose entries $f_{ij}(s,t) \in \mathbb{F}[s,t]$ are homogeneous polynomials of degree ν_i, $i = 1, \ldots, k$.

For any base field $\mathbb{F}$ the Grassmannian $\text{Grass}(k, \mathbb{F}^n)$ is a smooth projective variety of dimension $k(n-k)$. Moreover the set of all morphisms of the form $h : \mathbb{P}^1_{\mathbb{F}} \longrightarrow \text{Grass}(k, \mathbb{F}^n)$ having total degree $\nu = \sum_{j=1}^k \nu_j$ forms itself a smooth projective variety of dimension $\nu n + k(n-k)$. In the algebraic geometry literature this variety is sometimes referred to as a *Quot scheme* and we refer to [18] for details.

Grothendieck did derive Theorem 3 using general results from the theory of holomorphic vector bundles like Serre duality and splitting theorems for subbundles. He was not aware that his result is also a straightforward consequence of some results by Dedekind and Weber [2]. The way to derive Grothendieck's theorem from [2] was shown by Geyer [7]. In the last part of this section we outline this connection.

Dedekind and Weber consider a field extension $\Omega \supset \mathbb{F}$ having transcendental degree 1. Let $x \in \Omega$ be an element which is not algebraic over $\mathbb{F}$ and consider the ring $R := \mathbb{F}[x]$ and its quotient field $\mathcal{F} = F(x)$. $\Omega \supset \mathcal{F}$ is then a finite field extension of degree $k := [\Omega : \mathcal{F}]$. Let R_∞ be the ring of proper rational functions of the form $r(x) = \frac{p(x)}{q(x)}$, $\deg q(x) \geq \deg p(x)$. We define a vector bundle as a pair (M, M_∞), where M is a R-submodule of Ω and M_∞ is a R_∞-submodule of Ω.

In [2, §22] Dedekind and Weber define a norm for each $x \in M$. They show how to iteratively construct a "Normalbasis" $\lambda_1, \ldots, \lambda_k$ of M having minimal norm $r_1, \ldots, r_k$. Moreover the elements $\lambda_1, \ldots, \lambda_k$ have the properties that

$$\frac{\lambda_1}{z^{r_1}}, \ldots, \frac{\lambda_k}{z^{r_k}}$$

form a normal basis of M_∞. Geyer [7] explains how Grothendieck's Theorem 3 is an immediate consequence once the existence of a normal basis of M has been shown. One finds similar arguments by Lomadze [13] who also shows how Theorem 2 relates to Grothendieck's theorem.

It is interesting to remark that Dedekind and Weber already used in their paper elements from valuation theory and maybe this paper started the development of valuation theory. The importance of valuation theory was stressed by Forney in [4] where he remarked that Theorem 2 might generalize to a broader context. For a historic account on the paper by Dedekind and Weber we refer to Geyer [7] and Strobl [20].

4. APPLICATIONS TO SYSTEMS THEORY

The theory of minimal polynomial bases is key for understanding the relation between a state space description of a linear system and the corresponding transfer function (respectively matrix polynomial) description. In the sequel we summarize some of these results. The reader who is not familiar with these results is advised to read Forney's article [4] or to have a look at the standard textbook [11].

Let $G(s)$ be a $p \times m$ proper transfer function. It is well known that the input-output map $\hat{y}(s) = G(s)\hat{u}(s)$ has a minimal state space realization of the form:

$$\dot{x} = Ax + Bu, \quad y = Cx + Du. \tag{4}$$

The size of the matrix A is called the McMillan degree of the transfer function $G(s) = C(sI - A)^{-1}B + D$ (respectively system (4)) and the left Kronecker indices of the pencil $\begin{bmatrix} sI - A \\ C \end{bmatrix}$ are called the observability indices of the transfer function $G(s)$ (respectively system (4)). See [11, page 413] for details. Similarly one defines the controllability indices of $G(s)$ as the right Kronecker indices of the pencil $[sI - A \mid B]$.

Consider the $\mathcal{F}$-vector space

$$V := \text{rowspace}_{\mathcal{F}} \, [I_p \mid G(s)] \subset \mathcal{F}^{m+p}.$$

Assume the rows of $[D(s) \mid N(s)]$ form a minimal polynomial basis of V having ordered Forney indices $\nu_1 \geq \cdots \geq \nu_p$ and total degree $\nu = \sum_{j=1}^{p} \nu_j$. Then one has:

Theorem 5 ([4]) *1. $D^{-1}(s)N(s) = G(s)$ forms a left coprime factorization.*

2. $\det D(s)$ *is equal to the characteristic polynomial of the transfer function $G(s)$, in particular $\nu = \deg \det D(s)$ is equal to the McMillan degree of $G(s)$.*

3. The indices ν_j are equal to the observability indices of $G(s)$.

The last theorem did relate the observability indices of $G(s)$ to the Forney indices of some subspace $V \subset \mathcal{F}^{m+p}$. Combining this result with Theorem 4 we obtain a corollary due to Martin and Hermann [15]:

Corollary 6 *Let ξ be the pullback of the bundle U^* under the Hermann-Martin map*

$$h: \mathbb{P}^1_{\mathbb{C}} \longrightarrow \text{Grass}(p, \mathbb{C}^{m+p}), \; z \mapsto \text{rowspace}_{\mathbb{C}} G(z). \qquad (5)$$

Then the Grothendieck indices of ξ are up to ordering equal to the observability indices of $G(s)$.

A minimal basis for the m-dimensional vector space

$$\tilde{V} := \text{rowspace}_{\mathcal{F}} \left[I_m \mid G(s)^t \right] \subset \mathcal{F}^{m+p}$$

results into the controllability indices of $G(s)$. In analogy to Theorem 5 one has:

Theorem 7 *Assume $Q(s)$ is a $m \times m$ polynomial matrix and the rows of $[Q(s) \mid R(s)]$ form a minimal polynomial basis of $\tilde{V}$ having ordered Forney indices $\kappa_1 \geq \cdots \geq \kappa_m$ and total degree $\kappa = \sum_{i=1}^m \kappa_i$. Then*

1. $R^t(s)(Q^t)^{-1}(s) = G(s)$ forms a right coprime factorization.

2. $\det Q(s)$ *is equal to the characteristic polynomial of the transfer function $G(s)$, in particular $\kappa = \deg \det Q(s)$ is equal to the McMillan degree of $G(s)$.*

3. The indices κ_i are equal to the controllability indices of $G(s)$.

5. BEHAVIORS AND GENERALIZED FIRST ORDER REPRESENTATIONS

The previous section showed how to transform results formulated in a state space formulation into results formulated in a polynomial formulation. Forney [4] also provided an algorithm on how to compute a minimal state space realization from a polynomial description $[D(s) \mid N(s)]$ of the system. At the time of writing of [4] there were several algorithms known for computing a state space realization of the form (4) from a transfer function respectively polynomial description and we refer to [1, 6, 22].

In this section we will take a slightly more modern (and general) point of view and we will show how it is possible to rewrite linear time invariant behaviors given through a polynomial formulation into a generalized first order form. We follow in the sequel mainly reference [19].

Let $P(s)$ be a $p \times (m+p)$ polynomial matrix with entries in the polynomial ring $\mathbb{R}[s]$. The smooth behavior associated with $P(s)$ is defined by

$$\mathcal{B}(P) = \{w \in C^{\infty}(\mathbb{R}; \mathbb{R}^{m+p}) \mid P(\tfrac{d}{dt})w = 0\}. \tag{6}$$

The following lemma is well known:

Lemma 8 *Assume $P(s)$ and $\tilde{P}(s)$ are both full rank polynomial matrices of the same size. Then $\mathcal{B}(P) = \mathcal{B}(\tilde{P})$ if and only if there is a unimodular matrix $U(s)$ such that $\tilde{P}(s) = U(s)P(s)$.*

Based on this lemma we can identify a smooth behavior with the row module of a polynomial matrix. There are classical canonical forms describing the row module of a polynomial matrix such as the Hermite's normal form and we refer to [11, 14]. In the sequel we will simply assume that the polynomial matrix $P(s)$ has been row reduced using unimodular operations only and that the row degrees of $P(s)$ are ordered in nonincreasing manner: $\nu_1 \geq \cdots \geq \nu_p$.

The indices $\nu_1, \ldots, \nu_p$ are in general different from the Forney indices of the rational row space of $P(s)$ and they are often referred to as the Kronecker indices of the row-module of $P(s)$. Clearly equivalent polynomial matrices $P(s)$ and $\tilde{P}(s)$ have the same rational row-space. The converse is in general not true. One has, however, the following:

Lemma 9 *Assume the rows of $P(s)$ and $\tilde{P}(s)$ form both a minimal polynomial basis of the same rational subspace $V \subset \mathcal{F}^n$. Then there is a unimodular matrix $U(s)$ such that $\tilde{P}(s) = U(s)P(s)$.*

In particular, this lemma allows one to derive canonical forms (such as the Hermite normal form) for minimal polynomial bases, a subject which is also treated in [4].

Consider now a triple of real matrices (F, G, H), where F and G have size $n \times (n+m)$ and H has size $(m+p) \times (n+m)$. This triple defines a smooth behavior through:

$$\mathcal{B}(F,G,H) = \{w \in C^{\infty}(\mathbb{R}; \mathbb{R}^{m+p}) \mid \exists z \in C^{\infty}(\mathbb{R}; \mathbb{R}^{n+m}) : G\dot{z} = Fz, w = Hz\}.$$

One says that (F, G, H) forms a generalized first order representation (or simply a realization) of $\mathcal{B}(P)$ if $\mathcal{B}(F, G, H) = \mathcal{B}(P)$. If (F, G, H) is a realization of $P(s)$ and S and T are nonsingular matrices of appropriate size then $(SFT^{-1}, SGT^{-1}, HT^{-1})$ forms a realization of $\mathcal{B}(P)$ as well.

The following theorem can be found in [12, Theorem 4.3] and it provides the conditions of minimality for a realization (F, G, H) of a behavior $\mathcal{B}(F, G, H) = \mathcal{B}(P)$.

Theorem 10 (F, G, H) *forms a minimal realization of the behavior* $\mathcal{B}(P)$ *if and only if:*

(i) G *has full row rank.*

(ii) $\begin{bmatrix} G \\ H \end{bmatrix}$ *has full column rank.*

(iii) $\begin{bmatrix} sG - F \\ H \end{bmatrix}$ *has full column rank for all* $s \in \mathbb{C}$.

The question now arises how to compute efficiently a minimal realization. The following lemma gives a way to compute a realization, not necessarily a minimal one.

Lemma 11 ([19]) *Let a polynomial matrix* $P(s) \in \mathbb{R}^{p\times(m+p)}[s]$ *and a triple of constant matrices* (F, G, H) *(*F *and* G *in* $\mathbb{R}^{n\times(n+m)}$, H *in* $\mathbb{R}^{(m+p)\times(n+m)}$*) be given. If there exists a polynomial matrix* $X(s) \in \mathbb{R}^{p\times n}[s]$ *such that the rows* $[X(s) \mid P(s)]$ *form a minimal polynomial basis and the equality*

$$\ker_{\mathbb{R}(s)} [X(s) \mid P(s)] \; = \; \operatorname{im}_{\mathbb{R}(s)} \begin{bmatrix} sG - F \\ H \end{bmatrix} \tag{7}$$

holds, then $\mathcal{B}(P) = \mathcal{B}(F, G, H)$*, so* (F, G, H) *is a realization of* $P(s)$.

In order to compute a minimal realization it is best to choose a particular polynomial matrix $X(s)$ such that (F, G, H) satisfying property (7) is most easily computed. For this assume that $P(s)$ is row reduced with Kronecker indices $\nu_1 \geq \cdots \geq \nu_p$. Define:

$$X_\nu(s) = \begin{bmatrix} 1 & s & \cdots & s^{\nu_1-1} & 0 & \cdots & \cdots & & \cdots & \cdots & 0 \\ 0 & \cdots & \cdots & 0 & 1 & \cdots & s^{\nu_2-1} & 0 & \cdots & \cdots & 0 \\ \vdots & & & & & \ddots & \ddots & & & & \vdots \\ 0 & \cdots & \cdots & \cdots & \cdots & \cdots & \cdots & 0 & 1 & \cdots & s^{\nu_p-1} \end{bmatrix}. \tag{8}$$

With this particular choice one has:

Theorem 12 **([19])** *Let $P(s)$, $X_\nu(s)$ be as above. Then:*

1. *A minimal polynomial basis of $\ker_{\mathbb{R}(s)}[X_\nu(s) \mid P(s)]$ is of the form $\mathrm{im}_{\mathbb{R}(s)} \begin{bmatrix} sG-F \\ H \end{bmatrix}$ and $\mathcal{B}(F,G,H)$ forms a minimal realization of $\mathcal{B}(P)$.*
2. *If the Kronecker indices of $P(s)$ have the property that $\nu_1 \geq \cdots \geq \nu_p \geq 1$ then $\ker_{\mathbb{R}(s)}[X_\nu(s) \mid P(s)]$ can be computed simply 'by inspection' this means by simply rearranging the entries of the polynomial matrix $P(s)$.*

We can now specialize Theorem 12 to the situation treated by Forney [4]. For this assume that $P(s)$ is partitioned into $P(s) = [D(s) \mid N(s)]$. Correspondingly, partition the vector w into y and u. We are now seeking a first order realization of the form

$$\begin{bmatrix} sG-F \\ H \end{bmatrix} = \begin{bmatrix} sI-A & B \\ C & D \\ 0 & I \end{bmatrix}$$

which then describes a classical behavior of the form $\dot{x} = Ax + Bu$, $y = Cx + Du$.

Assume the rows of $[D(s) \mid N(s)]$ form a minimal basis of $\mathrm{rowspace}_{\mathbb{R}(s)}[I_p \mid G(s)]$. For simplicity assume that $D^{-1}(s)N(s) = G(s)$ is strictly proper. Assume that the Forney indices satisfy $\nu_1 \geq \cdots \geq \nu_p \geq 1$ and assume that the high order coefficient matrix of $[D(s) \mid N(s)]$ is of the form $[I_p \mid 0]$. For $i,j = 1,\ldots,p$ let

$$d_{i,j}(s) = \sum_{k=0}^{\nu_i} d_{i,j}^k \, s^k$$

denote the polynomial entries of $D(s)$. Similarly let

$$n_i(s) = \sum_{k=0}^{\nu_i} n_i^k \, s^k$$

denote the ith row of $N(s)$. For $i = 1,\ldots,p$ define matrices of sizes $\nu_i \times \nu_i$, $\nu_i \times m$ and $1 \times \nu_i$ respectively:

$$A_{i,i} := \begin{bmatrix} 0 & \ldots & \ldots & \ldots & -d_{i,i}^0 \\ 1 & 0 & & & -d_{i,i}^1 \\ 0 & 1 & \ddots & & \vdots \\ \vdots & & \ddots & 0 & \vdots \\ 0 & \ldots & 0 & 1 & -d_{i,i}^{\nu_i-1} \end{bmatrix}, B_i := \begin{bmatrix} n_i^0 \\ n_i^1 \\ \vdots \\ n_i^{\nu_i-1} \end{bmatrix},$$

$$C_i := [0,\ldots,-1].$$

Finally for $i, j = 1, \ldots, p$, $i \neq j$ define matrices of size $\nu_i \times \nu_j$:

$$A_{i,j} := \begin{bmatrix} 0 & \ldots & 0 & -d_{i,j}^0 \\ \vdots & & \vdots & -d_{i,j}^1 \\ \vdots & & \vdots & \vdots \\ 0 & \ldots & 0 & -d_{i,j}^{\nu_i - 1} \end{bmatrix}.$$

Note that the matrices $A_{i,i}$, $A_{i,j}$, B_i and C_i were simply computed 'by inspection' from the data $[D(s) \mid N(s)]$.

Theorem 13 ([4, 19, 22]) *In the situation discussed above*

$$\begin{aligned} \dot{x}(t) &= \begin{bmatrix} A_{1,1} & \cdots & A_{1,p} \\ \vdots & \ddots & \vdots \\ A_{p,1} & \cdots & A_{p,p} \end{bmatrix} x(t) + \begin{bmatrix} B_1 \\ \vdots \\ B_p \end{bmatrix} u(t), \\ y(t) &= \begin{bmatrix} C_1 & & 0 \\ & \ddots & \\ 0 & & C_p \end{bmatrix} x(t) \end{aligned} \tag{9}$$

represents a (classical) minimal state space realization of the system

$$D(\tfrac{d}{dt})y(t) + N(\tfrac{d}{dt})u(t) = 0. \tag{10}$$

In particular one has $C(sI - A)^{-1}B = G(s)$.

6. CONCLUSION

In this paper we provided a survey about one of the most influential papers in systems theory. We showed how the result on minimal polynomial bases relates to some very classical results in mathematics. We also showed how the results on the relation between state space description of a system and polynomial description of a system can be treated in the broader context of behavioral theory.

Appendix

Table 1: Citations of [4] in the period 1976–1999:

Year	Cite's	Year	Cite's	Year	Cite's	Year	Cite's
1976	4	1982	3	1988	7	1994	14
1977	6	1983	11	1989	10	1995	7
1978	15	1984	12	1990	8	1996	8
1979	18	1985	7	1991	10	1997	10
1980	14	1986	10	1992	7	1998	7
1981	19	1987	6	1993	5	1999	9

References

[1] R. W. Brockett, *Finite Dimensional Linear Systems*, J. Wiley & Sons, 1970.

[2] R. Dedekind and H. Weber, "Theorie der algebraischen Functionen einer Veränderlichen," *Journal für die reine und angewandte Mathematik*, vol. 92, pp. 181–291, 1882.

[3] G. D. Forney, "Convolutional Codes I: Algebraic Structure," *IEEE Transactions on Information Theory*, vol. IT-16, no. 5, pp. 720–738, 1970.

[4] G. D. Forney, "Minimal Bases of Rational Vector Spaces, With Applications to Multivariable Linear Systems," *SIAM Journal on Control*, vol. 13, no. 3, pp. 493–520, 1975.

[5] G. D. Forney, "Algebraic Structure of Convolutional Codes, and Algebraic System Theory," in A.C. Antoulas, editor, *Mathematical System Theory: The Influence of R.E. Kalman*, pp. 527–557, Springer, Berlin, 1991.

[6] P. A. Fuhrmann, "Algebraic System Theory: An Analyst's Point of View," *Journal of The Franklin Institute*, vol. 301, pp. 521–540, 1976.

[7] W.-D. Geyer, "Die Theorie der Algebraischen Funktionen einer Veränderlichen nach Dedekind und Weber," in W. Scharlau, editor, *Richard Dedekind, 1831–1981*, pp. 109–133, Friedr. Vieweg & Sohn, Braunschweig, 1981. Eine Würdigung zu seinem 150. Geburtstag. [An appreciation on the occasion of his 150th birthday].

[8] A. Grothendieck, "Sur la classification des fibres holomorphes sur la sphere de Riemann," *American Journal of Mathematics*, vol. 79, pp. 121–138, 1957.

[9] M. Hazewinkel and C. Martin, "A Short Elementary Proof of Grothendieck's Theorem on Algebraic Vector Bundles Over the Projective Line," *Journal of Pure and Applied Algebra*, vol. 25, pp. 207–211, 1982.

[10] R. Johannesson and K. Sh. Zigangirov, *Fundamentals of Convolutional Coding*, IEEE Press, NY, 1999.

[11] T. Kailath, *Linear Systems*, Prentice-Hall, Englewood Cliffs, NJ, 1980.

[12] M. Kuijper, *First-Order Representations of Linear Systems*, Birkhäuser, Boston, MA, 1994.

[13] V. Lomadze, "Applications of Vector Bundles to Factorization of Rational Matrices. *Linear Algebra Applications*, vol. 288, no. 1-3, pp. 249–258, 1999.

[14] C. C. MacDuffee, *The Theory of Matrices*, Chelsea, NY, 1956. Reprint of original, 1933.

[15] C. F. Martin and R. Hermann, "Applications of Algebraic Geometry to System Theory: The McMillan Degree and Kronecker Indices as Topological and Holomorphic Invariants," *SIAM Journal on Control and Optimization*, vol. 16, pp. 743–755, 1978.

[16] R. J. McEliece, *The Algebraic Theory of Convolutional Codes*, in V. Pless and W. C. Huffman, editors, *Handbook of Coding Theory*, vol. 1, pp. 1065–1138, Elsevier Science Publishers, Amsterdam, The Netherlands, 1998.

[17] Plemelj. Riemannsche Funktionenscharen mit gegebener Monodromiegruppe. *Monath. Math.*, vol. 19, pp. 211–246, 1908.

[18] M. S. Ravi and J. Rosenthal, "A Smooth Compactification of the Space of Transfer Functions with Fixed McMillan Degree," *Acta Appl. Math.*, vol. 34, pp. 329 352, 1994.

[19] J. Rosenthal and J. M. Schumacher, "Realization by Inspection," *IEEE Transactions on Automatic Control*, vol. AC-42, no. 9, pp. 1257–1263, 1997.

[20] W. Strobl, "Über die Beziehungen zwischen der Dedekindschen Zahlentheorie und der Theorie der algebraischen Funktionen von Dedekind und Weber," *Abh. Braunschweig. Wiss. Ges.*, vol. 33, pp. 225–246, 1982.

[21] J. C. Willems, "Paradigms and Puzzles in the Theory of Dynamical Systems," *IEEE Transactions on Automatic Control*, vol. AC-36, no. 3, pp. 259–294, 1991.

[22] W. A. Wolovich, *Linear Multivariable Systems*, vol. 11 of *Appl. Math. Sc.* Springer-Verlag, NY, 1974.

Chapter 21

FINITE GEOMETRY LOW DENSITY PARITY-CHECK CODES: CONSTRUCTION, STRUCTURE, AND DECODING

S. Lin
Department of Electrical and Computer Engineering
University of California, Davis
Davis, CA 95616

Y. Kou
Department of Electrical and Computer Engineering
University of California, Davis
Davis, CA 95616

M. P.C. Fossorier
Department of Electrical Engineering
University of Hawaii at Manoa
Honolulu, HI 96822

(Dedicated to G. David Forney, Jr. for his outstanding contributions to coding theory and its practice.)

Abstract

This paper present an algebraic and systematic method for constructing low density parity-check (LDPC) codes from a geometric point of view. LDPC codes are constructed based on the lines and points of a finite geometry. Two classes of LDPC codes are constructed based on

*This research was supported by NSF under Grants CCR 0096191 and 0117891 and NASA under Grants NAG 5-9825 NAG 5-10480.

the well known Euclidean and projective geometries over finite fields. Codes of these two classes are cyclic and they have good minimum distances and structural properties. They can be decoded in various ways and perform very well with the iterative decoding based on belief propagation (IDBP). These finite geometry LDPC codes can be extended and shortened in various ways. Extension by splitting columns of the parity-check matrices of these codes results in good extended finite geometry LDPC codes. Long extended finite geometry LDPC codes have been constructed and they achieve an error performance only a few tenths of a dB away from the Shannon limit with IDBP. Finite geometry LDPC codes are strong competitors to turbo codes for error control in communication and digital data storage systems.

Five different decoding algorithms are described in this paper and they range from low to high decoding complexity and from reasonably good to very good decoding performance.

1. INTRODUCTION

LDPC codes were first discovered by Gallager [1,2] in the early 1960s and have recently been rediscovered and generalized [3-15]. It has been shown lately that these codes with iterative decoding based on belief propagation (IDBP) [10, 16-23] achieve a remarkable error performance that is very close to the Shannon limit [4, 9-15]. Consequently, these codes have become strong competitors to turbo codes [24-27] for error control in many communication and digital data storage systems where high reliability is required.

A LDPC code is defined as the *null space* of a parity-check matrix **H** with the following structural properties: (1) each row contains ρ ones and is otherwise zero; (2) each column contains γ ones; (3) the number of ones in common between any two columns, denoted λ , is not greater than one; and (4) both ρ and γ are small compared to the length of the code and the number of rows in **H** [1,2]. Because ρ and γ are small, **H** has a small density of ones. For this reason, the code specified by **H** is called a LDPC code. The LDPC code defined above is known as a *regular* LDPC code. If not all columns or all the rows of the parity-check matrix **H** have the same number of ones, a LDPC code is said to be *irregular*.

Although LDPC codes have been shown to achieve outstanding error performance, no analytic (algebraic or geometric) method has been found for constructing these codes. Gallager only provided a class of semirandom LDPC codes [1,2]. Good LDPC codes that have been found are largely computer generated, especially long codes. Encoding of these long computer generated LDPC codes is very complex due to the lack of understanding of their structure. Furthermore, their minimum distances are hard to determine.

In this paper, we investigate the construction of LDPC codes from a geometric approach. The construction is based on the points and lines of a finite geometry. For a given geometry with a finite number of points and a finite number of lines, we form a matrix whose columns correspond to the points of the geometry and whose rows correspond to the lines of the geometry. Under certain conditions, this matrix is a low density matrix and its null space is then a LDPC code. Well known finite geometries are the *Euclidean geometry* (*EG*) and *projective geometry* (*PG*) over finite fields. Based on these two finite geometries, two classes of LDPC codes are constructed. Codes of these two classes are cyclic and therefore, their encoding can be implemented with linear feedback shift registers with feedback connections based on their generator polynomials [28,29]. We call codes of these two classes finite geometry LDPC codes. The generator polynomial of a finite geometry LDPC code can be determined by finding its roots over a finite field. From these roots, a BCH bound [28-31] on its minimum distance can be derived. Finite geometry LDPC codes, especially high rate codes, perform very well with the IDBP decoding. These codes can also be decoded with *one-step majority decoding* [29-32], Gallager's *bit-flipping* (BF) decoding [2], and *weighted BF* decoding.

A finite geometry LDPC code can be extended by splitting each column of its parity-check matrix **H** into multiple columns. This column splitting results in a new low density parity-check matrix and hence a new LDPC code of longer length. If column splitting is done properly for a reasonably long finite geometry LDPC code, the extended code performs amazingly well using the IDBP decoding. An error performance only a few tenths of a dB away from its Shannon limit can be achieved. New LDPC codes can also be obtained by splitting each row of the parity-check matrix of a finite geometry LDPC code into multiple rows. Combining column and row splittings of parity-check matrices of finite geometry LDPC codes, we obtain a large class of LDPC codes with a wide ranges of code lengths and code rates. A finite geometry LDPC code can also be shortened by puncturing (or removing) the columns of its *parity-check matrix* that correspond to the points on a set of lines (or a subgeometry) of the geometry based on which the code is constructed. A shortened finite geometry LDPC code has at least the same minimum distance as the original code.

The paper is organized as follows. Section 2 presents a construction method of LDPC codes based on a finite geometry. Graph representations of these codes are discussed. Section 3 gives a brief presentation of

Euclidean and projective geometries over finite fields. Construction and characterization of EG-LDPC codes and PG-LDPC codes are given in Section 4. Various types of decoding of EG-LDPC codes and PG-LDPC codes are discussed in Section 5. Section 6 presents simulation results of error performance of some EG-LDPC codes and PG-LDPC codes using various decoding algorithms. Sections 7 and 8 present extension and shortening of EG-LDPC codes and PG-LDPC codes. Finally, Section 9 concludes this paper with some remarks and comparison with turbo codes.

2. A GEOMETRIC CONSTRUCTION OF LDPC CODES

This section presents a simple construction of LDPC codes based on the points and lines of finite geometries.

Let $\mathbf{G}$ be a finite geometry with n points and J lines which has the following fundamental structural properties: (1) every line consists of ρ points; (2) every point lies on γ lines (i.e., every point is intersected by γ lines); (3) two lines are either *parallel* (i.e., they have no point in common) or they intersect at one and only one point. Number these points from 1 to n. Let

$$\mathbf{v} = (v_1, v_2, \ldots, v_n)$$

be an n-tuple over GF(2) whose n components correspond to the n points of geometry $\mathbf{G}$, where the ith component v_i corresponds to the ith point in $\mathbf{G}$. Let $\mathcal{L}$ be a line in $\mathbf{G}$. Define a vector based on the points on $\mathcal{L}$ as follows:

$$\mathbf{v}_{\mathcal{L}} = (v_1, v_2, \ldots, v_n)$$

where

$$v_i = \begin{cases} 1, & \text{if } v_i \text{ corresponds to a point on } \mathcal{L}, \\ 0, & \text{otherwise.} \end{cases} \tag{1}$$

This vector $\mathbf{v}_{\mathcal{L}}$ is called the *incidence vector* of $\mathcal{L}$. It is clear that the weight of this incidence vector is ρ.

Form a $J \times n$ matrix $\mathbf{H}$ whose rows are the incidence vectors of the J lines of the finite geometry $\mathbf{G}$ and whose columns correspond to the n points of $\mathbf{G}$. Then it follows from the fundamental structural properties of the finite geometry $\mathbf{G}$ and the definition of the incidence vector of a line in $\mathbf{G}$ that the matrix $\mathbf{H}$ has the following properties: (1) each row has ρ ones; (2) each column has γ ones (this number is simply the number of lines in $\mathbf{G}$ that intersect at the point which corresponds to the column);

and (3) the number of common ones between any two columns is at most one, i.e., $\lambda = 0$ or 1 (this follows from the fundamental property that two lines are either parallel or intersect at one and only one point in **G**). Property 3 implies that no two rows of **H** have more than one one in common. The *density* of this matrix **H**, denoted r, is defined as the ratio of the total number of ones in **H** to the total number of entries of **H**. Then, we readily see that

$$r = \rho/n,$$

or

$$r = \gamma/J.$$

If ρ and γ are much smaller than n and J, then the density r of **H** is very small and hence **H** is a low density (or sparse) matrix. The null space of **H** is therefore a LDPC code of length n. Such a code is called the geometry-**G** LDPC code. The rows of **H** are not necessarily linearly independent over GF(2). Therefore, to determine the dimension of the code, it is necessary to determine the rank of **H**.

Let $\mathbf{h}_1, \mathbf{h}_2, \ldots, \mathbf{h}_J$ be the rows of **H** where

$$\mathbf{h}_j = (h_{j,1}, h_{j,2}, \ldots, h_{j,n}) \tag{2}$$

for $1 \leq j \leq J$. An n-tuple $\mathbf{v} = (v_1, v_2, \ldots, v_n)$ is a codeword of the geometry-**G** LDPC code specified by **H** if and only if

$$\mathbf{v} \cdot \mathbf{H}^T = (0, 0, \ldots, 0) \tag{3}$$

or the inner product

$$s_j = \mathbf{v} \cdot \mathbf{h}_j = \sum_{l=1}^{n} v_l h_{j,l} = 0 \tag{4}$$

for $1 \leq j \leq J$. The sum given by Equation (4) is called a *parity-check sum* (or simply check sum or parity-check equation), which is simply a linear sum of a subset of code bits. A code bit v_l is said to be checked by the sum $s_j = \mathbf{v} \cdot \mathbf{h}_j$ (or the row $\mathbf{h}_j$) if $h_{j,l} = 1$. The set of rows in **H** that check on the code bit v_l, denoted A_l, is said to be *orthogonal* on v_l [29,32]. Let S_l denote the set of check sums formed by the rows in A_l. Then the code bit v_l is contained in every check sum in S_l and any of the other $n-1$ code bits is contained in at most one check sum in S_l. The check sums in S_l are called the check sums orthogonal on the code bit v_i (or bit position i) [29,32]. These orthogonal check sums are used for majority decoding of finite geometry LDPC codes in Section 5.

For a linear block code of length n specified by a parity-check matrix $\mathbf{H}$ of J rows, a graph $\mathbf{Q}$ can be constructed to display the relationship between its code bits and the check sums that check them. This graph $\mathbf{Q}$ consists of two levels of vertices. The first level consists of n vertices that represent the n code bits of the code. These vertices, denoted $v_1, v_2, \ldots, v_n$, are called the code bit (or variable) vertices. The second level consists of J vertices which represented the J check sums (or equations), $s_1, s_2, \ldots, s_J$, that the code bits must satisfy. These vertices are call the check-sum vertices. A code bit vertex v_l is connected to a check sum vertex s_j by an edge, denoted (v_l, s_j), if and only if the code bit v_l is contained in the check sum s_j (or checked by s_j). No two code bit vertices are connected and no two check sum vertices are connected. Hence $\mathbf{Q}$ is a *bipartite graph* [33]. This graph was first proposed by Tanner [3] to study the structure of LDPC codes and hence it is called the Tanner graph. The number of edges that are connected to (or incident at) a code bit vertex v_l, called the *degree* of v_l, is simply the number of check sums that contain v_l. The number of edges that are incident at the check sum vertex s_j, called the degree of s_j, is simply the number of code bits that are checked by the check sum s_j. Then for a regular LDPC code, the degrees of all the code bit vertices are the same and the degrees of all the check sum vertices are the same. Such a Tanner graph is said to be regular. The geometry-$\mathbf{G}$ LDPC code constructed above is regular and hence its Tanner graph is regular.

A *cycle graph* of vertices and edges is defined as a sequence of connected edges which starts from a vertex and ends at the same vertex and satisfies the condition that no vertex (except the initial and the final vertex) appears more than once [33]. The number of edges on a cycle is called the *length* of the cycle. A graph that contains no cycles is said to be *acyclic*. The Tanner graph of a linear block code contains no cycle of length 2 and no cycle of odd length. It follows from the structural property (3) of the parity-check matrix of a LDPC code that the Tanner graph of a LDPC code contains no cycle of length 4.

Example 1 : *Consider a (7,3) linear block code which is the null space of the parity-check matrix shown in Figure 1. It is easy to check that this code is a LDPC code. Its Tanner graph is shown in Figure 2. The graph does not have any cycle of length 4 but it does have cycles of length 6, one such cycle is marked by the heavy line.*

Decoding a linear block code with IDBP, decoding performance very much depends on cycles of short lengths in its Tanner graph. These short cycles, especially cycles of length 4, make successive decoding iterations highly correlated and hence severely limit the decoding per-

$$\mathbf{H} = \begin{bmatrix} 1 & 1 & 0 & 1 & 0 & 0 & 0 \\ 0 & 1 & 1 & 0 & 1 & 0 & 0 \\ 0 & 0 & 1 & 1 & 0 & 1 & 0 \\ 0 & 0 & 0 & 1 & 1 & 0 & 1 \\ 1 & 0 & 0 & 0 & 1 & 1 & 0 \\ 0 & 1 & 0 & 0 & 0 & 1 & 1 \\ 1 & 0 & 1 & 0 & 0 & 0 & 1 \end{bmatrix}.$$

Figure 1 A Parity-check matrix.

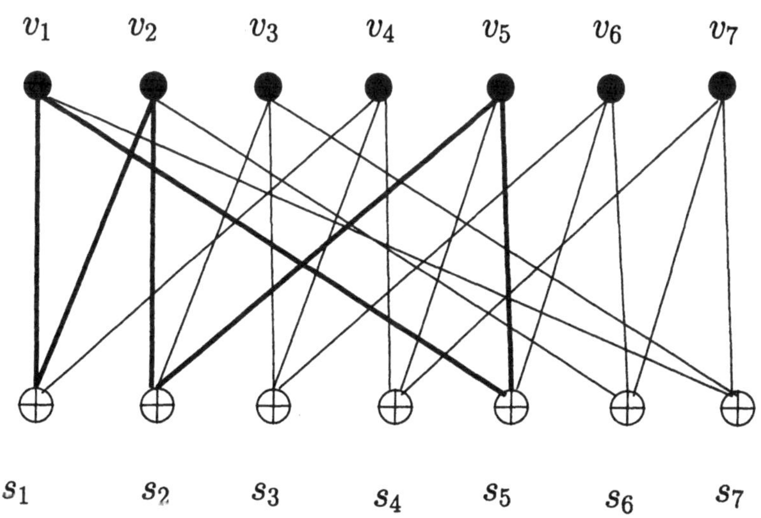

Figure 2 Tanner graph for the (7,3) code

formance [3,10,12,21,34-37]. Therefore, to use IDBP for decoding, it is important to design codes without short cycles. Long LDPC codes are generated by sparse parity-check matrices, and their Tanner graphs contain no cycles of length 4 and very likely they contain very few (or no) cycles of other short lengths, say lengths 6 and 8. As a result, these codes with IDBP perform very well and approach their Shannon limits [9,10,12-15]. In a later section, we will show how cycles of a Tanner graph can be broken.

3. EUCLIDEAN AND PROJECTIVE GEOMETRIES OVER FINITE FIELDS

This section gives a brief description of Euclidean and projective geometries over finite fields. The fundamental structural properties of points and lines for these two geometries are described without proofs. Any textbook for combinatorial mathematics contains detail exposition of this subject. Two such texts are references [38] and [39]. A simple discussion of this subject can also be found in [29].

3.1 EUCLIDEAN GEOMETRY

Let GF(2^s) be the Galois field with 2^s elements. Consider all the m-tuples

$$(a_0, a_1, \ldots, a_{m-1})$$

over GF(2^s). There are $(2^s)^m = 2^{ms}$ such m-tuples and they form a vector space over GF(2^s). The vector addition and scalar multiplication are defined as follows:

$$\begin{aligned}(a_0, a_1, \ldots, a_{m-1}) + (b_0, b_1, \ldots, b_{m-1}) &\triangleq (a_0 + b_0, a_1 + b_1, \ldots, \\ &\qquad a_{m-1} + b_{m-1}), \\ \beta \cdot (a_0, a_1, \ldots, a_{m-1}) &\triangleq (\beta \cdot a_0, \beta \cdot a_1, \ldots, \beta \cdot a_{m-1}),\end{aligned}$$

where additions $a_i + b_i$ and multiplication $\beta \cdot a_i$ are carried out in GF(2^s). In combinatorial mathematics, the 2^{ms} m-tuples over GF(2^s) are also known to form an m-dimensional *Euclidean geometry* over GF(2^s), denoted $EG(m, 2^s)$. Each m-tuple $\mathbf{a} = (a_0, a_1, \ldots, a_{m-1})$ is regarded as a *point*. The all-zero m-tuple, $\mathbf{0} = (0, 0, \ldots, 0)$, is considered as the *origin* of the geometry.

Let $\mathbf{a}$ be a nonorigin point in EG($m, 2^s$). Then the 2^s points,

$$\{\beta\mathbf{a}\} \triangleq \{\beta\mathbf{a} : \beta \in \mathrm{GF}(2^s)\}$$

form a *line* in EG($m, 2^s$). Since, for $\beta = 0$, $0 \cdot \mathbf{a} = \mathbf{0}$, every line contains the origin $\mathbf{0}$ as a point. We say that $\{\beta\mathbf{a}\}$ passes through the origin. Let $\mathbf{a}_0$ and $\mathbf{a}$ be two linearly independent points in EG($m, 2^s$) (i.e., $\beta_0\mathbf{a}_0 + \beta\mathbf{a} \neq \mathbf{0}$ unless $\beta_0 = \beta = 0$). Then the collection of the following 2^s points,

$$\{\mathbf{a}_0 + \beta\mathbf{a}\} \triangleq \{\mathbf{a}_0 + \beta\mathbf{a} : \beta \in \mathrm{GF}(2^s)\}$$

form a line in EG($m, 2^s$) that passes through the point $\mathbf{a}_0$. Lines $\{\beta\mathbf{a}\}$ and $\{\mathbf{a}_0 + \beta\mathbf{a}\}$ do not have any point in common and they are said to be *parallel.* Let $\mathbf{b}_0$ be a point not on line $\{\beta\mathbf{a}\}$ or on line $\{\mathbf{a}_0 + \beta\mathbf{a}\}$. The line $\{\mathbf{b}_0 + \beta\mathbf{a}\}$ passes through the point $\mathbf{b}_0$ and is parallel to both $\{\beta\mathbf{a}\}$ and $\{\mathbf{a}_0 + \beta\mathbf{a}\}$. In fact, any two lines that do not have any point in common are said to be parallel. Every line in $\text{EG}(m, 2^s)$ has

$$2^{(m-1)s} - 1 \tag{1}$$

lines parallel to it.

Let $\mathbf{a}_1$ and $\mathbf{a}_2$ be two linearly independent points in $\text{EG}(m, 2^s)$. The lines $\{\mathbf{a}_0 + \beta\mathbf{a}_1\}$ and $\{\mathbf{a}_0 + \beta\mathbf{a}_2\}$ have only one point, $\mathbf{a}_0$, in common. We say that they intersect at point $\mathbf{a}_0$. Given a point $\mathbf{a}_0$ in $\text{EG}(m, 2^s)$, there are

$$\frac{2^{ms} - 1}{2^s - 1} \tag{2}$$

lines in $\text{EG}(m, 2^s)$ that intersect at $\mathbf{a}_0$. The total number of lines in $\text{EG}(m, 2^s)$ is

$$\frac{2^{(m-1)s}(2^{ms} - 1)}{2^s - 1}. \tag{3}$$

Let $\text{GF}(2^{ms})$ be the extension field of $\text{GF}(2^s)$ (called the ground field). Each element in $\text{GF}(2^{ms})$ can be represented as an m-tuple over $\text{GF}(2^s)$. Therefore, the 2^{ms} elements in $\text{GF}(2^{ms})$ may be regarded as the 2^{ms} points in $\text{EG}(m, 2^s)$ and hence $\text{GF}(2^{ms})$ as the Euclidean geometry $\text{EG}(m, 2^s)$.

Let

$$\mathbf{v} = (v_0, v_1, v_2, \ldots, v_{2^{ms}-2})$$

be a $(2^{ms}-1)$-tuple over $\text{GF}(2)$. Let α be a primitive element of $\text{GF}(2^{ms})$. Number the components of $\mathbf{v}$ as follows: the component v_i is numbered α^i for $0 \le i \le 2^{ms} - 2$. Hence, α^i is the location number of v_i. If we regard $\text{GF}(2^{ms})$ as the Euclidean geometry $\text{EG}(m, 2^s)$, then $\alpha^0 = 1, \alpha^1, \alpha^2, \ldots, \alpha^{2^{ms}-2}$ are the $2^{ms} - 1$ nonorigin points of $\text{EG}(m, 2^s)$ and there is a *one-to-one correspondence* between a component in $\mathbf{v}$ and a nonorigin point in $\text{EG}(m, 2^s)$.

Let $\mathcal{L}$ be a line in $\text{EG}(m, 2^s)$ which does not pass the origin. Based on $\mathcal{L}$, form a $(2^{ms} - 1)$-tuple over $\text{GF}(2)$ as follows:

$$\mathbf{v}_{\mathcal{L}} = (v_0, v_1, \ldots, v_{2^{ms}-2}) \tag{4}$$

whose ith component

$$v_i = \begin{cases} 1, & \text{if } \alpha^i \text{ is a point on } \mathcal{L}, \\ 0, & \text{otherwise}. \end{cases} \tag{5}$$

The vector $\mathbf{v}_{\mathcal{L}}$ is the *incidence vector* of the line $\mathcal{L}$. The weight (or the number of ones) of this incidence vector is 2^s. We readily see that the incidence vectors of two lines in $\mathrm{EG}(m, 2^s)$ not passing through the origin either have no ones in common or have a single one in common.

Example 2 *Let $m = 2$ and $s = 2$. Consider the Galois field $GF(2^{2\times 2})$ generated by the primitive polynomial $p(X) = 1 + X + X^4$ which is given in Table 1. Regard this field as the two-dimensional Euclidean geometry $EG(2, 2^2)$ over $GF(2^2)$. Let α be a primitive element of $GF(2^{2\times 2})$ and $\beta = \alpha^5$. Then $\{0, 1, \beta, \beta^2\}$ form the subfield $GF(2^2)$. Every line in $EG(2, 2^2)$ consists of four points. The set of four points $\{\alpha^{14} + \eta\alpha\}$ with $\eta \in GF(2^2)$ is $\{\alpha^7, \alpha^8, \alpha^{10}, \alpha^{14}\}$ which forms a line in $EG(2, 2^2)$. The incidence vector for this line is (0 0 0 0 0 0 0 1 1 0 1 0 0 0 1). The line $\{\alpha^{14} + \eta\alpha^3\} = \{\alpha^0, \alpha^2, \alpha^6, \alpha^{14}\}$ intersects with line $\{\alpha^{14} + \eta\alpha\} = \{\alpha^7, \alpha^8, \alpha^{10}, \alpha^{14}\}$ at point α^{14}. The incidence vector of line $\{\alpha^{14} + \eta\alpha^3\}$ is (1 0 1 0 0 0 1 0 0 0 0 0 0 0 1). We see that the incidence vectors of line $\{\alpha^{14} + \eta\alpha\}$ and line $\{\alpha^{14} + \eta\alpha^3\}$ have a single one in common at location α^{14}. The line $\{\alpha^{14} + \eta\alpha^6\} = \{\alpha^0, \alpha^4, \alpha^{12}, \alpha^{13}\}$ is parallel to line $\{\alpha^{14} + \eta\alpha\} = \{\alpha^7, \alpha^8, \alpha^{10}, \alpha^{14}\}$. The incidence vector of line $\{\alpha^{14} + \eta\alpha^6\}$ is (1 0 0 0 1 0 0 0 0 0 0 0 1 1 0). We can easily see that the incidence vectors of line $\{\alpha^{14} + \eta\alpha^6\}$ and line $\{\alpha^{14} + \eta\alpha\}$ do not have a one in common.*

3.2 PROJECTIVE GEOMETRY

Consider the Galois field $\mathrm{GF}(2^{(m+1)s})$ which contains $\mathrm{GF}(2^s)$ as a subfield. Let α be a primitive element in $\mathrm{GF}(2^{(m+1)s})$. Then

$$\alpha^0, \alpha^1, \ldots, \alpha^{2^{(m+1)s}-2}$$

form all the nonzero elements of $\mathrm{GF}(2^{(m+1)s})$. Let

$$n = \frac{2^{(m+1)s} - 1}{2^s - 1} = 2^{ms} + 2^{(m-1)s} + \ldots + 2^s + 1 \tag{6}$$

and

$$\beta = \alpha^n.$$

Then the order of β is $2^s - 1$. The 2^s elements $0, 1, \beta, \beta^2, \ldots, \beta^{2^s-2}$ form the Galois field $\mathrm{GF}(2^s)$.

Consider the first n powers of α,

$$\Gamma = \{\alpha^0, \alpha^1, \alpha^2, \ldots, \alpha^{n-1}\}.$$

Table 1 The elements of $GF(2^4)$ generated by $p(X) = 1 + X + X^4$

Power representation	Polynomial representation	4-Tuple representation
0	0	$(0\ 0\ 0\ 0)$
1	1	$(1\ 0\ 0\ 0)$
α	α	$(0\ 1\ 0\ 0)$
α^2	α^2	$(0\ 0\ 1\ 0)$
α^3	α^3	$(0\ 0\ 0\ 1)$
α^4	$1 + \alpha$	$(1\ 1\ 0\ 0)$
α^5	$\alpha + \alpha^2$	$(0\ 1\ 1\ 0)$
α^6	$\alpha^2 + \alpha^3$	$(0\ 0\ 1\ 1)$
α^7	$1 + \alpha + \alpha^3$	$(1\ 1\ 0\ 1)$
α^8	$1 + \alpha^2$	$(1\ 0\ 1\ 0)$
α^9	$\alpha + \alpha^3$	$(0\ 1\ 0\ 1)$
α^{10}	$1 + \alpha + \alpha^2$	$(1\ 1\ 1\ 0)$
α^{11}	$\alpha + \alpha^2 + \alpha^3$	$(0\ 1\ 1\ 1)$
α^{12}	$1 + \alpha + \alpha^2 + \alpha^3$	$(1\ 1\ 1\ 1)$
α^{13}	$1 + \alpha^2 + \alpha^3$	$(1\ 0\ 1\ 1)$
α^{14}	$1 + \alpha^3$	$(1\ 0\ 0\ 1)$

No element $\alpha^i \in \Gamma$ can be a product of an element in $\mathrm{GF}(2^s)$ and another element $\alpha^j \in \Gamma$. Partition the nonzero elements of $\mathrm{GF}(2^{(m+1)s})$ into n disjoint subsets as follows:

$$\begin{array}{l} \{\alpha^0, \beta\alpha^0, \beta^2\alpha^0, \ldots, \beta^{2^s-2}\alpha^0\}, \\ \{\alpha^1, \beta\alpha^1, \beta^2\alpha^1, \ldots, \beta^{2^s-2}\alpha^1\}, \\ \{\alpha^2, \beta\alpha^2, \beta^2\alpha^2, \ldots, \beta^{2^s-2}\alpha^2\}, \\ \vdots \\ \{\alpha^{n-1}, \beta\alpha^{n-1}, \beta^2\alpha^{n-1}, \ldots, \beta^{2^s-2}\alpha^{n-1}\}, \end{array} \tag{7}$$

where $\beta = \alpha^n$, a primitive element in $\mathrm{GF}(2^s)$. Each set consists of $2^s - 1$ elements and each element is a multiple of the first element in the set. Represent each set by its first element as follows:

$$(\alpha^i) \triangleq \{\alpha^i, \beta\alpha^i, \ldots, \beta^{2^s-2}\alpha^i\}, \tag{8}$$

with $0 \leq i < n$. For any $\alpha^j \in \mathrm{GF}(2^{(m+1)s})$, if $\alpha^j = \beta^l \cdot \alpha^i$ with $0 \leq i < n$, then α^j is in (α^i) and represented by (α^i).

If we represent each element in $\mathrm{GF}(2^{(m+1)s})$ as an $(m+1)$-tuple over $\mathrm{GF}(2^s)$, then (α^i) consists of $2^s - 1$ $(m+1)$-tuples over $\mathrm{GF}(2^s)$. The

$(m+1)$-tuple for α^i represents the $2^s - 1$ $(m+1)$-tuples in (α^i). The $(m+1)$-tuple over GF(2^s) that represents (α^i) may be regarded as a *point* in a finite geometry over GF(2^s). Then the points,

$$(\alpha^0), (\alpha^1), (\alpha^2), \ldots, (\alpha^{n-1}), \tag{9}$$

form an m-dimensional projective geometry over GF(2^s), denoted PG$(m, 2^s)$ [29,38,39]. Note that the $2^s - 1$ elements in $\{\alpha^i, \beta\alpha^i, \ldots, \beta^{2^s-2}\alpha^i\}$ are considered to be the same point in PG$(m, 2^s)$.

Let (α^i) and (α^j) be any two *distinct* points in PG$(m, 2^s)$. Then the *line* (or called a 1-flat) passing through (α^i) and (α^j) consists of points of the following form:

$$(\eta_1\alpha^i + \eta_2\alpha^j), \tag{10}$$

where η_1 and η_2 are from GF(2^s) and are not both equal to zero. The points $(\eta_1\alpha^i + \eta_2\alpha^j)$ and $(\beta^l\eta_1\alpha^i + \beta^l\eta_2\alpha^j)$ are the same point. Therefore, each line in PG$(2^{(m+1)s})$ consists of

$$\frac{(2^s)^2 - 1}{2^s - 1} = 2^s + 1 \tag{11}$$

points.

Let (α^l) be a point not on the line $\{(\eta_1\alpha^i + \eta_2\alpha^j)\}$. Then the line $\{(\eta_1\alpha^i + \eta_2\alpha^j)\}$ and the line $\{(\eta_1\alpha^l + \eta_2\alpha^j)\}$ have (α^j) as a common point (the only common point). We say that they intersect at (α^j). The number of lines in PG$(m, 2^s)$ that intersect at a given point is

$$\frac{2^{ms} - 1}{2^s - 1} = 1 + 2^s + \ldots + 2^{(m-1)s}. \tag{12}$$

There are

$$\frac{(1 + 2^s + \ldots + 2^{ms})(1 + 2^s + \ldots + 2^{(m-1)s})}{1 + 2^s} \tag{13}$$

lines in PG$(m, 2^s)$. Two lines that do not intersect are said to be *parallel*.

Let $\mathbf{v} = (v_0, v_1, \ldots, v_{n-1})$ be an n-tuple over GF(2) where

$$n = \frac{2^{(m+1)s} - 1}{2^s - 1}.$$

We may number the components of $\mathbf{v}$ with the first n powers of α as follows: v_i is numbered α^i for $0 \leq i < n$. As usual, α^i is called the location number of v_i. Therefore, there is a one-to-one correspondence between the components of $\mathbf{v}$ and the points of the m-dimensional projective geometry PG$(m, 2^s)$. Let $\mathcal{L}$ be a line in PG$(m, 2^s)$. The incidence vector for $\mathcal{L}$ is an n-tuple over GF(2),

$$\mathbf{v}_{\mathcal{L}} = (v_0, v_1, \ldots, v_{n-1}) \tag{14}$$

whose ith component

$$v_i = \begin{cases} 1, & \text{if } (\alpha^i) \text{ is a point on } \mathcal{L} \\ 0, & \text{otherwise.} \end{cases} \tag{15}$$

The weight of $\mathbf{v}_{\mathcal{L}}$ is 2^s+1. The incidence vectors of two lines in PG$(m, 2^s)$ either have no ones in common or have a single one in common.

The incidence vectors of lines in the above two finite geometries will be used in the next section for constructing EG-LDPC codes an PG-LDPC codes.

4. EUCLIDEAN AND PROJECTIVE GEOMETRY LDPC CODES

This section presents two classes of LDPC codes whose constructions are based on Euclidean and projective geometries over finite fields, respectively. LDPC codes of these two classes are cyclic and hence their encoding can be implemented with linear feedback shift registers with connections based on their generator polynomials [28,29,31]. We call codes of these two classes the Euclidean geometry (EG) and projective geometry (PG) codes, respectively.

4.1 EG-LDPC CODES

Let $\mathbf{H}_{EG}$ be a matrix whose rows are the incidence vectors of all the lines in the m-dimensional Euclidean geometry EG$(m, 2^s)$ that do not pass through the origin. Then $\mathbf{H}_{EG}$ consists of $2^{ms} - 1$ columns and

$$(2^{(m-1)s} - 1)\frac{2^{ms} - 1}{2^s - 1} \tag{1}$$

rows. The columns of $\mathbf{H}_{EG}$ correspond to the $2^{ms} - 1$ nonorigin points of EG$(m, 2^s)$. It follows from the structural properties of lines and points in EG$(m, 2^s)$ that matrix $\mathbf{H}_{EG}$ has the following structures: (1) Each row has $\rho = 2^s$ ones; (2) Each column has $\gamma = (2^{ms} - 1)/(2^s - 1) - 1$ ones; (3) Any two columns have no more than a single one in common, i.e., $\lambda = 0$ or 1; and (4) Any two rows have no more than a single one in common. The density of $\mathbf{H}_{EG}$ is then

$$r = \frac{2^s}{2^{ms} - 1}. \tag{2}$$

For $m \geq 2, r$ is relatively small. For large m, r is approximately $1/2^{(m-1)s}$. Therefore, $\mathbf{H}_{EG}$ is a low-density matrix.

Let $\mathbf{C}$ be the null space of $\mathbf{H}_{EG}$. Then $\mathbf{C}$ is a regular LDPC code of length $n = 2^{ms} - 1$. We call this code an m-dimensional EG-LDPC code. It turns out that this EG-LDPC code is the $(0, s)$th order EG (Euclidean Geometry) code [29,40,41] and is the dual code of a polynomial code [41-44]. Therefore, it is cyclic and its generator polynomial is completely characterized by its roots in GF(2^{ms}). The $(0, s)$th order EG code is one-step majority decodable [29,40].

Let h be an nonnegative integer less than 2^{ms}. Then h can be expressed in radix-2^s form as follows:

$$h = \delta_0 + \delta_1 2^s + \ldots + \delta_{m-1} 2^{(m-1)s}$$

where $0 \leq \delta_i < 2^s$ for $0 \leq i < m$. The 2^s-*weight* of h, denoted $W_{2^s}(h)$, is defined as the following sum,

$$W_{2^s}(h) \triangleq \delta_0 + \delta_1 + \ldots + \delta_{m-1}. \tag{3}$$

For an nonnegative integer l, let $h^{(l)}$ be the remainder resulting from dividing $2^l h$ by $2^{ms} - 1$. Then $0 \leq h^{(l)} < 2^{ms} - 1$. Let $\mathbf{g}_{EG}(X)$ be the generator polynomial of the m-dimensional EG-LDPC code. Let α be a primitive element of GF(2^{ms}). Then α^h is a root of $\mathbf{g}_{EG}(X)$ if and only if [29,40,41]

$$0 < \max_{0 \leq l < s} W_{2^s}(h^{(l)}) \leq (m-1)(2^s - 1). \tag{4}$$

From the above characterization of the roots of $\mathbf{g}_{EG}(X)$, it has been shown [41] that $\mathbf{g}_{EG}(X)$ has the following sequence of consecutive powers of α,

$$\alpha, \alpha^2, \ldots, \alpha^{(2^{ms}-1)/(2^s-1)-1},$$

as roots. It follows from the BCH bound [28-31] that the minimum distance of the m-dimensional EG-LDPC code is lower bounded as follows:

$$d_{min} \geq \frac{2^{ms} - 1}{2^s - 1}. \tag{5}$$

This bound can also be derived from the majority-decoding point of view (see Section 5).

The number of parity-check symbols of the m-dimensional EG-LDPC code is of course equal to the degree of its generator polynomial $\mathbf{g}_{EG}(X)$. However, a combinatorial expression for this number has been derived by Lin [43]. It can be shown that the rate R of the m-dimensional EG-LDPC code is greater than 1/2 for $m < s$ and smaller than 1/2 for $m \geq s$.

A special subclass of EG-LDPC codes is the class of two-dimensional EG-LDPC codes ($m = 2$). for any positive integer $s \geq 2$, the two-dimensional EG-LDPC code has the following parameters [29,41]:

$$\begin{array}{ll} \text{Length} & n = 2^{2s} - 1, \\ \text{Number of parity bits} & n - k = 3^s - 1, \\ \text{Dimension} & k = 2^{2s} - 3^s, \\ \text{Minimum distance} & d_{min} = 2^s + 1, \\ \text{Density} & r = \frac{2^s}{2^{2s}-1}. \end{array} \tag{6}$$

For this special case, the geometry EG$(2, 2^s)$ contains $2^{2s} - 1$ lines that do not pass through the origin. Therefore, the parity-check matrix $\mathbf{H}_{EG}$ of the two-dimensional EG-LDPC code is a $(2^{2s} - 1) \times (2^{2s} - 1)$ square matrix. Actually, $\mathbf{H}_{EG}$ can be constructed easily by taking the incidence vector $\mathbf{v}_{\mathcal{L}}$ of a line $\mathcal{L}$ in EG$(2, 2^s)$ that does not pass through the origin and then cyclically shifting this incidence vector $\mathbf{v}_{\mathcal{L}}$ $2^{2s} - 2$ times. This results in $2^{2s}-1$ incidence vectors for the $2^{2s}-1$ distinct lines in EG$(2, 2^s)$ that do not pass through the origin. The incidence vector $\mathbf{v}_{\mathcal{L}}$ and its $2^{2s} - 2$ cyclic shifts form the rows of the parity-check matrix $\mathbf{H}_{EG}$.

A list of two-dimensional EG-LDPC codes is given in Table 2.

Table 2 A list of two-dimensional EG-LDPC codes

n	k	d_{min}	ρ	γ	r
15	7	5	4	4	0.267
63	37	9	8	8	0.127
255	175	17	16	16	0.0627
1023	781	33	32	32	0.0313
4095	3367	65	64	64	0.01563
16383	14197	129	128	128	0.007813

Example 3 *Consider the two-dimensional Euclidean geometry EG$(2, 2^2)$ given in Example 2. Let α be a primitive element of GF$(2^{2\times 2})$. The incidence vector for the line $\mathcal{L} = \{\alpha^7, \alpha^8, \alpha^{10}, \alpha^{14}\}$ is* (0 0 0 0 0 0 0 1 1 0 1 0 0 0 1). *This vector and its 14 cyclic shifts form the parity-check matrix $\mathbf{H}_{EG}$ shown in Figure 3. The null space of this matrix is the (15,7) two-dimensional EG-LDPC code, the first code given in Table 2.*

$$\mathbf{H} = \begin{bmatrix} 0&0&0&0&0&0&0&1&1&0&1&0&0&0&1 \\ 1&0&0&0&0&0&0&0&1&1&0&1&0&0&0 \\ 0&1&0&0&0&0&0&0&0&1&1&0&1&0&0 \\ 0&0&1&0&0&0&0&0&0&0&1&1&0&1&0 \\ 0&0&0&1&0&0&0&0&0&0&0&1&1&0&1 \\ 1&0&0&0&1&0&0&0&0&0&0&0&1&1&0 \\ 0&1&0&0&0&1&0&0&0&0&0&0&0&1&1 \\ 1&0&1&0&0&0&1&0&0&0&0&0&0&0&1 \\ 1&1&0&1&0&0&0&1&0&0&0&0&0&0&0 \\ 0&1&1&0&1&0&0&0&1&0&0&0&0&0&0 \\ 0&0&1&1&0&1&0&0&0&1&0&0&0&0&0 \\ 0&0&0&1&1&0&1&0&0&0&1&0&0&0&0 \\ 0&0&0&0&1&1&0&1&0&0&0&1&0&0&0 \\ 0&0&0&0&0&1&1&0&1&0&0&0&1&0&0 \\ 0&0&0&0&0&0&1&1&0&1&0&0&0&1&0 \end{bmatrix}$$

Figure 3 The parity-check matrix of the (15, 7) two-dimensional EG-LDPC code constructed based on $\mathrm{EG}(2, 2^2)$.

4.2 PG-LDPC CODES

The construction of LDPC codes based on a projective geometry is similar to that based on an Euclidean geometry. For any two positive integers, m and s, with $m \geq 2$, there exists an m-dimensional PG-LDPC code whose construction is based on the lines and points of $\mathrm{PG}(m, 2^s)$.

Form a matrix $\mathbf{H}_{PG}$ whose rows are the incidence vectors of the lines of $\mathrm{PG}(m, 2^s)$ and whose columns correspond to the points of $\mathrm{PG}(m, 2^s)$. Then $\mathbf{H}_{PG}$ has

$$\frac{(1 + 2^s + \ldots + 2^{(m-1)s})(1 + 2^s + \ldots + 2^{ms})}{1 + 2^s}$$

rows and $(2^{(m+1)s} - 1)/(2^s - 1)$ columns. It follows from the structural properties of lines and points of $\mathrm{PG}(m, 2^s)$ developed in Section 3 that matrix $\mathbf{H}_{PG}$ has the following structural properties: (1) Each row has $\rho = 2^s + 1$ ones; (2) Each column has $\gamma = (2^{ms} - 1)/(2^s - 1)$ ones; (3) Any two columns have at most one a single one in common, i.e., $\lambda = 0$ or 1; and (4) Any two rows have at most a single one in common. We can easily find that the density of matrix $\mathbf{H}_{PG}$ is

$$r = \frac{(2^s - 1)(2^s + 1)}{2^{(m+1)s} - 1}. \tag{7}$$

For $m \geq 2$, r is relatively small. For large m, the density r is approximately $1/2^{(m-1)s}$ which is very small. Therefore, $\mathbf{H}_{PG}$ is a very sparse matrix.

Let $\mathbf{C}$ be the null space of $\mathbf{H}_{PG}$. Then $\mathbf{C}$ is an m-dimensional PG-LDPC code of length $n = (2^{(m+1)s} - 1)/(2^s - 1)$. This regular LDPC code turns out to be the (1,s)th order PG code [29,45,46] and is the dual of a nonprimitive polynomial code [41-44]. It is cyclic and one-step majority decodable.

Let h be an nonnegative integer less than $2^{(m+1)s} - 1$. For an nonnegative integer l, let $h^{(l)}$ be the remainder resulting from dividing $2^l h$ by $2^{(m+1)s} - 1$. The 2^s-weight of h, $W_{2^s}(h)$, is defined by Equation (3). Let $\mathbf{g}_{PG}(X)$ be the generator polynomial of the m-dimensional PG-LDPC code constructed based on PG$(m, 2^s)$. Let α be a primitive element of GF$(2^{(m+1)s})$. Then $\mathbf{g}_{PG}(X)$ has α^h as a root if and only if h is divisible by $2^s - 1$ and

$$0 \leq \max_{0 \leq l < s} W_{2^s}(h^{(l)}) = j(2^s - 1) \tag{8}$$

with $0 \leq j \leq m - 1$ [29,41,45]. Let $\xi = \alpha^{2^s - 1}$. The order of ξ is then $n = (2^{(m+1)s} - 1)/(2^s - 1)$. From the characterization of the roots of $\mathbf{g}_{PG}(X)$ given by Equation (8), it can be shown [41] that $\mathbf{g}_{PG}(X)$ has the following consecutive powers of ξ,

$$\xi, \xi^2, \ldots, \xi^{(2^{ms} - 1)/(2^s - 1)} \tag{9}$$

as roots. Therefore, it follows from the BCH bound that the minimum distance of the m-dimensional PG-LDPC code is lower bounded as follows:

$$d_{min} \geq \frac{2^{ms} - 1}{2^s - 1} + 1. \tag{10}$$

This bound can also be derived from majority decoding point of view (see Section 5).

The number of parity-check symbols of the m-dimensional PG-LDPC for a given s can be enumerated by determining the roots of its generator polynomial. A complicated combinatorial expression for this number can be found in [43].

A special subclass of PG-LDPC codes is the class of two-dimensional PG-LDPC codes constructed based on PG$(2, 2^s)$ for various s. For any positive integer $s \geq 2$, the two-dimensional PG-LDPC code has the following parameters [29,43,47,48]:

$$\begin{array}{ll} \text{Length} & n = 2^{2s} + 2^s + 1, \\ \text{Number of parity bits} & n - k = 3^s + 1, \\ \text{Dimension} & k = 2^{2s} + 2^s - 3^s, \\ \text{Minimum distance} & d_{min} = 2^s + 2, \\ \text{Density} & r = \frac{2^s+1}{2^{2s}+2^s+1}. \end{array} \tag{11}$$

It is a difference-set code [29,47]. The parity-check matrix $\mathbf{H}$ of this code is a $(2^{2s}+2^s+1) \times (2^{2s}+2^s+1)$ matrix, which can be formed by taking the incidence vector of a line in PG$(2, 2^s)$ and its $2^{2s} + 2^s$ cyclic shifts as rows.

A list of two-dimensional PG-LDPC codes is given in Table 3.

Table 3 A list of PG-LDPC codes

n	k	d_{min}	ρ	γ	r
21	11	6	5	5	0.2381
73	45	10	9	9	0.1233
273	191	18	17	17	0.0623
1057	813	34	33	33	0.0312
4161	3431	66	65	65	0.0156
16513	14326	130	129	129	0.0078

Both classes of EG-LDPC codes and PG-LDPC codes are regular LDPC codes. Their Tanner graphs do not contain cycles of length 4. From Equations (6) and (11), we see that for the same s, the two-dimensional EG-LDPC code and the two-dimensional PG-LDPC code have about the same length, the same dimension and the same density. Their important parameters, d_{min}, ρ, and γ are different by 1. In Section 5, we show that they have almost identical error performance.

5. DECODING OF FINITE GEOMETRY CODES

A finite geometry LDPC code can be decoded in various ways, namely: majority decoding, bit flipping (BF), weighted BF decoding, a posteriori probability (APP) decoding, and iterative decoding based on belief propagation (IDBP). The first two types are hard-decision decoding, last two are soft-decision decoding and the third one is in-between. Majority decoding is the simplest in decoding complexity. BF decoding requires a little more decoding complexity but gives better error performance than the majority decoding. APP decoding and IDBP provide much better error performance but require much more decoding complexity than the majority and BF decodings. The weighted BF decoding allows a good trade between error performance and decoding complexity. IDBP gives the best error performance among the five types of decoding of finite geometry LDPC codes.

Suppose a finite geometry LDPC code $\mathbf{C}$ is used for error control over an AWGN channel with zero mean and power spectral density $N_0/2$. Assume BPSK signaling with unit energy. A codeword $\mathbf{v} = (v_0, v_1, \ldots, v_{n-1})$ is mapped into a bipolar sequence $\mathbf{x} = (x_0, x_1, \ldots, x_{n-1})$ before its transmission where $x_l = (2v_l - 1) = +1$ for $v_l = 1$ and $x_l = -1$ for $v_l = 0$ with $0 \leq l \leq n-1$. Let $\mathbf{y} = (y_0, y_1, \ldots, y_{n-1})$ be the soft-decision received sequence at the output of the receiver matched filter. For $0 \leq l \leq n-1, y_l = \pm 1 + n_l$ where n_l is a Gaussian random variable with zero mean and variance $N_0/2$. Let $\mathbf{z} = (z_0, z_1, \ldots, z_{n-1})$ be the binary hard-decision received sequence obtained from $\mathbf{y}$ as follows:

$$z_l = \begin{cases} 1, & \text{for } y_l > 0, \\ 0, & \text{for } y_l \leq 0. \end{cases}$$

Let $\mathbf{H}$ be the parity-check matrix of the finite geometry LDPC code $\mathbf{C}$ with J rows and n columns. Let $\mathbf{h}_1, \mathbf{h}_2, \ldots, \mathbf{h}_J$, denote the rows of $\mathbf{H}$, where

$$\mathbf{h}_j = (h_{j,0}, h_{j,1}, \ldots, h_{j,n-1})$$

for $1 \leq j \leq J$. Then

$$\mathbf{s} = (s_1, s_2, \ldots, s_J) = \mathbf{z} \cdot \mathbf{H}^T. \tag{1}$$

gives the syndrome of the received sequence $\mathbf{z}$, where the jth syndrome component s_j is given by the check-sum,

$$s_j = \mathbf{z} \cdot \mathbf{h}_j = \sum_{l=0}^{n-1} z_l h_{j,l} \tag{2}$$

The received vector $\mathbf{z}$ is a codeword if and only if $\mathbf{s} = \mathbf{0}$. If $\mathbf{s} \neq \mathbf{0}$, errors in $\mathbf{z}$ are detected. A *nonzero* syndrome component s_j indicates a *parity failure*. The total number of parity failures is equal to the number of nonzero syndrome components in $\mathbf{s}$. Let

$$\begin{aligned} \mathbf{e} &= (e_0, e_1, \ldots, e_{n-1}) \\ &= (v_0, v_1, \ldots, v_{n-1}) + (z_0, z_1, \ldots, z_{n-1}) \end{aligned} \tag{3}$$

Then $\mathbf{e}$ is the error pattern in $\mathbf{z}$. This error pattern $\mathbf{e}$ and the syndrome $\mathbf{s}$ satisfy the condition,

$$\mathbf{s} = (s_1, s_2, \ldots, s_J) = \mathbf{e} \cdot \mathbf{H}^T \tag{4}$$

where

$$s_j = \mathbf{e} \cdot \mathbf{h}_j = \sum_{l=0}^{n-1} e_l h_{j,l} \tag{5}$$

for $1 \leq j \leq J$.

5.1 MAJORITY DECODING

The first majority decoding algorithm was devised by Reed [49] for decoding Reed-Muller codes [28]. Later Reed's algorithm was reformulated and generalized by Massey for decoding both block and convolutional codes [32,29]. Finite geometry LDPC codes presented in the previous section are one-step majority decodable and decoding is based on the syndrome equations of Equation (5) and the structural properties of the parity-check matrix $\mathbf{H}$.

It follows from the structural properties of the parity-check matrix $\mathbf{H}$ of a finite geometry LDPC code that for every bit position l, there is a set

$$A_l = \{\mathbf{h}_1^{(l)}, \mathbf{h}_2^{(l)}, \ldots, \mathbf{h}_\gamma^{(l)}\} \tag{6}$$

of γ rows in $\mathbf{H}$ which are orthogonal on this bit position, i.e., the lth component of each row in A_l is a one and no two rows in A_l have a common one in any other position. Form the following set of syndrome equations based on the rows in A_l:

$$S_l = \{s_j^{(l)} = \mathbf{e} \cdot \mathbf{h}_j^{(l)} : \mathbf{h}_j^{(l)} \text{ for } 1 \leq j \leq \gamma\} \tag{7}$$

where

$$s_j^{(l)} = \mathbf{e} \cdot \mathbf{h}_j^{(l)} = \sum_{i=0}^{n-1} e_i h_{j,i}^{(l)}. \tag{8}$$

Due to the structural properties of A_l, every syndrome equation in S_l checks (or contains) the error digit e_l and no other error digit in $\mathbf{e}$ is checked by more than one syndrome equation in S_l. Suppose there are $\lfloor \gamma/2 \rfloor$ or fewer errors in $\mathbf{e}$. There are two cases to be considered. The first case is that the lth digit of $\mathbf{e}$ is an erroneous bit, i.e., $e_l = 1$. In this case, the other errors can distribute among at most $\lfloor \gamma/2 \rfloor - 1$ syndrome equations in S_l and hence they affect at most $\lfloor \gamma/2 \rfloor - 1$ syndrome digits and cancel the effect of e_l. Consequently, at least $\gamma - \lfloor \gamma/2 \rfloor + 1$ syndrome digits, a *majority*, are equal to $e_l = 1$. The second case is that $e_l = 0$. In this case, the errors in **e** can distribute among at most $\lfloor \gamma/2 \rfloor$ syndrome equations in S_l. Hence, at least $\gamma - \lfloor \gamma/2 \rfloor$ of the syndrome digits are zero, the value of e_l. Thus, the value of e_l is equal to the value assumed by a majority of the syndrome digits in S_l; if no value is assumed by a majority of the syndrome digits (i.e., there is a tie), the value of e_l is zero.

The above decoding algorithm is known as one-step majority decoding [29,32]. Correct decoding of e_l for $0 \leq l \leq n-1$ is guaranteed if there are $\lfloor \gamma/2 \rfloor$ or fewer errors in **e**. It can be seen that if **e** contains more than $\lfloor \gamma/2 \rfloor$ errors, correct decoding is not guaranteed by the above simple majority decoding based on the orthogonal set A_l. Since up to $\lfloor \gamma/2 \rfloor$ errors can be corrected, the minimum distance of the **C** is at least $\gamma+1$. For the m-dimensional EG-LDPC code, $\gamma = (2^{ms}-1)/(2^s-1)-1$. Therefore, its minimum distance is lower bounded by

$$d_{min} \geq \gamma + 1 = (2^{ms} - 1)/(2^s - 1),$$

which is exactly the same bound given by Equation (5). For the m-dimensional PG-LDPC code, $\gamma = (2^{ms} - 1)/(2^s - 1)$ and its minimum distance is lower bounded by

$$d_{min} \geq \gamma + 1 = (2^{ms} - 1)/(2^s - 1) + 1$$

which is the same bound given by Equation (10).

5.2 BIT FLIPPING DECODING ALGORITHM

Bit flipping (BF) decoding of LDPC codes was devised by Gallager in early 1960's [1,2]. When detectable errors occur during the transmission, there will be parity-check failures in the syndrome $\mathbf{s} = (s_1, s_2, \ldots, s_J)$ and some of the syndrome bits are equal to 1. The BF decoding is based on the change of the number of parity failures in $\{\mathbf{z} \cdot \mathbf{h}_j : 1 \leq j \leq J\}$ when a bit in the received sequence **z** is changed.

Again consider the decoding of a finite geometry LDPC code **C** with parity-check matrix **H**. Every code bit position l is checked by γ parity-check equations. Consider the set of γ parity-check equations, $\{\mathbf{z} \cdot \mathbf{h}_j^{(l)} : \mathbf{h}_j^{(l)} \in A_l \text{ for } 1 \leq j \leq \gamma\}$, that check the bit position l. If there is a single error in the received sequence **z** at position l, then γ parity-check equations that check the bit position l will fail. If there are two errors in the received vector $\mathbf{z}$, because of $\lambda \ll \gamma$, the number of parity-check failures will increase, and the number of failed parity-check equations that contain either of the two erroneous bits will be greater than $\lfloor \gamma/2 \rfloor$, provided that the error correcting capability of the code $t = \lfloor (d_{min}(C) - 1)/2 \rfloor \geq 2$. The number of parity-check failures increases as the number of errors in $\mathbf{z}$ increases until it reaches the error correcting capability t of the code. When the number of errors exceeds t, there will be error patterns that result in a decrease in the number of parity-check failures. In this case, an erroneous bit position may be checked by less than $\lfloor \gamma/2 \rfloor$ failed parity-check equations or by none of the failed parity-check equations. We conclude that if the number of errors in the received vector is t or less, changing an erroneous bit results in a decrease in the number of parity-check equation failures, and if the number of errors exceeds the error correcting capability t, changing an erroneous bit may result in an increase in the number of parity-check failures.

Accordingly, Gallager devised the following decoding scheme for LDPC codes. The decoder computes all the parity-check sums based on Equations (1) and (2) and then changes any bit in the received vector $\mathbf{z}$ that is contained in more than some fixed number δ of unsatisfied parity-check equations. Using these new values, the parity-check sums are recomputed, and the process is repeated until the parity-check equations are all satisfied.

Bit flipping is an iterative decoding algorithm. The parameter δ, called *threshold*, is a design parameter that should be chosen to optimize the error performance while minimizing the number of computations of parity-check sums. The value of δ depends on the code parameters $\rho, \gamma, \lambda, d_{min}(C)$ and the signal-to-noise ratio (SNR).

If decoding fails for a given value of δ, then the value of δ is reduced to allow further decoding iterations. For error patterns with number of errors less than or equal to the error correcting capability of the code, the decoding will be completed in one or a few iterations. Otherwise, more decoding iterations are needed. Therefore, the number of decoding iterations is a random variable and is a function of the channel SNR. A limit maybe set on the number of iterations. When this limit is reached, the decoding process is terminated.

Due to the nature of low-density parity-checks, the above decoding algorithm corrects many error patterns with number of errors exceeding the error correcting capability of the code.

A very simple BF decoding algorithm is given below:

Step 1 Compute the parity-check sums (syndrome bits) based on Equations (1) and (2). If all the parity-check equations are satisfied (i.e., all the syndrome bits are zero), stop the decoding.

Step 2 Find the number of unsatisfied parity-check equations for each bit, denoted f_i, $i = 0, 1, \ldots, n-1$.

Step 3 Identify the set $\mathcal{S}$ of bits for which f_i is the largest.

Step 4 Flip the bits in set $\mathcal{S}$.

Step 5 Repeat steps 1 to 4 until all the parity-check equations are satisfied (for this case, we stop the iteration in step 1) or a predefined maximum iteration number is reached.

The above simple BF decoding algorithm can be improved by using adaptive thresholds δ's. Of course, this improvement is achieved at the expense of more computations.

5.3 WEIGHTED DECODING

The simple hard-decision one-step majority decoding and the BF decoding can be improved to achieve better error performance by including some kind of reliability information (or measure) of the received symbols in their decoding decisions. Of course, additional decoding complexity is required for such performance improvement.

Consider the soft-decision received sequence $\mathbf{y} = (y_0, y_1, \ldots, y_{n-1})$ at the output of the receiver matched filter. For the AWGN channel, a simple measure of the reliability of a received symbol y_l is its magnitude, $|y_l|$. The larger the magnitude $|y_l|$ is, the more reliable is the hard decision z_{lo}.

Again consider a finite geometry LDPC code specified by a parity-check matrix $\mathbf{H}$ with J rows, $\mathbf{h}_1, \mathbf{h}_2, \ldots, \mathbf{h}_J$. For $0 \leq l \leq n-1$ and $1 \leq j \leq J$, define

$$|y_j|_{min}^{(l)} \triangleq \{\min\{|y_i|\} : 0 \leq i \leq n-1, h_{j,i} = 1\} \tag{9}$$

and

$$E_l \triangleq \sum_{s_j^{(l)} \in S_l} (2s_j^{(l)} - 1)|y_j|_{min}^{(l)}. \tag{10}$$

E_l is simply a weighted check sum that is orthogonal on the code bit position l. Let $\mathbf{e} = (e_0, e_1, \ldots, e_{n-1})$ be the error pattern to be estimated. Then the one-step majority decoding can be modified based on the weighted check sum E_l as follows:

$$e_l = \begin{cases} 1, & \text{for } E_l > 0, \\ 0, & \text{for } E_l \leq 0. \end{cases} \tag{11}$$

for $0 \leq l \leq n-1$. The above decoding algorithm is called weighted majority decoding [50].

The decision rule given by Equation (11) can be used in BF decoding. In this case the decoding is carried out as follows:

Step 1. Compute the parity-check sums (syndrome bits) based on Equations (1) and (2). If all the parity-check equations are satisfied (i.e., all the syndrome bits are zero), stop the decoding.

Step 2. Compute E_l based on Equation (10), for $0 \leq l \leq n-1$.

Step 3. Find the bit position l for which E_l is the largest.

Step 4. Flip the bit z_l.

Step 5. Repeat Step 1 to 4. This process of bit flipping continues until all the parity-check equations are satisfied or a preset maximum number of iterations is reached.

This modified BF algorithm is called weighted BF decoding algorithm.

The above weighted decoding algorithms requires some real number computations.

5.4 ITERATIVE DECODING BASED ON BELIEF PROPAGATION

Iterative decoding based on belief propagation (IDBP) is extremely efficient for decoding LDPC codes [10,12,21-23]. Like the MAP decoding algorithm [51], it is a symbol-by-symbol soft-in/soft-out decoding algorithm. It processes the received symbols iteratively to improve the reliability of each decoded code symbol based on the parity-check sums computed from the hard decisions of the received symbols and the parity-check matrix **H**. Reliability of a decoded symbol can be measured by its marginal posterior probability, its log-likelihood ratio or the value of its corresponding received symbol. The computed reliability measures of code symbols at the end of each decoding iteration are used as inputs for the next iteration. The decoding iteration process continues until a certain stopping condition is satisfied. Then based on the computed reliability measures of code symbols, hard decision are made.

Again we consider a finite geometry LDPC code $\mathbf{C}$ of length n specified by a parity-check matrix $\mathbf{H}$ with J rows, $\mathbf{h}_1, \mathbf{h}_2, \ldots, \mathbf{h}_J$, where

$$\mathbf{h}_j = (h_{j,0}, h_{j,1}, \ldots, h_{j,n-1}).$$

For $1 \leq j \leq J$, define the following index set for $\mathbf{h}_j$:

$$B(\mathbf{h}_j) = \{l : h_{j,l} = 1, 0 \leq l < n\}, \tag{12}$$

which is called the *support* of $\mathbf{h}_j$.

In the following we describe two versions of IDBP, one in terms of extrinsic information and the other in terms of probability.

Consider a code bit position l. Let $\mathbf{h}$ be a row in the set A_l that is orthogonal on bit position l. Then the extrinsic information provided to the transmitted code bit v_l by other code bits checked by $\mathbf{h}$ (or the check sum $\mathbf{z} \cdot \mathbf{h}$) is [1,2,23,52]

$$\varepsilon_l(\mathbf{h}) = \log \frac{1 + \prod_{t \in B(\mathbf{h}) \setminus l} \tanh(\frac{4}{N_0} y_t)}{1 - \prod_{t \in B(\mathbf{h}) \setminus l} \tanh(\frac{4}{N_0} y_t)}. \tag{13}$$

The total extrinsic information provided to code bit v_l by the rows in A_l is then

$$\begin{aligned} \varepsilon_l &= \sum_{\mathbf{h} \in A_l} \varepsilon_l(\mathbf{h}) \\ &= \sum_{\mathbf{h} \in A_l} \log \frac{1 + \prod_{t \in B(\mathbf{h}) \setminus l} \tanh(\frac{4}{N_0} y_t)}{1 - \prod_{t \in B(\mathbf{h}) \setminus l} \tanh(\frac{4}{N_0} y_t)}. \end{aligned} \tag{14}$$

Define a $J \times n$ matrix

$$\mathbf{E} = [E_{j,l}]_{1 \leq j \leq J}^{0 \leq l < n} \tag{15}$$

where

$$E_{j,l} = \begin{cases} \varepsilon_l - \varepsilon_l(\mathbf{h}_j), & \text{for } l \in B(\mathbf{h}_j) \\ 0, & \text{otherwise.} \end{cases} \tag{16}$$

$\mathbf{E}$ is called the *extrinsic matrix*. Define another $J \times n$ matrix

$$\mathbf{Y} = [Y_{j,l}]_{1 \leq j \leq J}^{0 \leq l < n} \tag{17}$$

where

$$Y_{j,l} = \begin{cases} \frac{4}{N_0} y_l, & \text{for } l \in B(\mathbf{h}_j) \\ 0, & \text{otherwise.} \end{cases} \tag{18}$$

This matrix is called the *channel information matrix.* In IDBP, each decoding iteration is to update (or improve) the extrinsic values $\varepsilon_l(\mathbf{h}_j)$, ε_l and $E_{j,l}$ for $0 \leq l < n$ and $1 \leq j \leq J$. These values computed during the ith decoding iteration are denoted $\varepsilon_l^{(i)}(\mathbf{h}_j)$, $\varepsilon_l^{(i)}$, and $E_{j,l}^{(i)}$. For $0 \leq l < n$ and $1 \leq j \leq J$, the extrinsic value $E_{j,l}^{(i)}$ is used to modify the channel information $Y_{j,l}$ by setting

$$Z_{j,l}^{(i+1)} = Y_{j,l} + E_{j,l}^{(i)}. \tag{19}$$

Form the channel value matrix at the end of the ith iteration as follows:

$$\mathbf{Z}^{(i)} = [Z_{j,l}^{(i)}]_{1 \leq j \leq J}^{0 \leq l < n} \tag{20}$$

where

$$Z_{j,l}^{(i)} = Y_{j,l} + E_{j,l}^{(i-1)}. \tag{21}$$

For $i = 0$, $\mathbf{Z}^{(0)} = \mathbf{Y}$. The entries of $\mathbf{Z}^{(i)}$ will be used to compute the extrinsic values for the $(i+1)$th decoding iteration. For $0 \leq l < n$, $\varepsilon_l^{(i)}$ will be used to modify the received vector $\mathbf{y} = (y_0, y_1, \ldots, y_{n-1})$. Let

$$\begin{aligned} \mathbf{r}^{(0)} &= (r_0^{(0)}, r_1^{(0)}, \ldots, r_{n-1}^{(0)}) \\ &= (\tfrac{4}{N_0} y_0, \tfrac{4}{N_0} y_1, \ldots, \tfrac{4}{N_0} y_{n-1}). \end{aligned}$$

Let

$$\mathbf{r}^{(i)} = (r_0^{(i)}, r_1^{(i)}, \ldots, r_{n-1}^{(i)})$$

be the modified received vector at the ith decoding iteration. Then the modified received vector at the next iteration is

$$\mathbf{r}^{(i+1)} = (r_0^{(i+1)}, r_1^{(i+1)}, \ldots, r_{n-1}^{(i+1)})$$

where

$$r_l^{(i+1)} = r_l^{(i)} + \varepsilon_l^{(i)}$$

and $r_l^{(0)} = \frac{4}{N_0} y_l$ for $0 \leq l < n$.

Let

$$\mathbf{z}^{(i)} = (z_0^{(i)}, z_1^{(i)}, \ldots, z_{n-1}^{(i)})$$

be the hard-decision vector obtained from $\mathbf{r}^{(i)} = (r_0^{(i)}, r_1^{(i)}, \ldots, r_{n-1}^{(i)})$ with

$$z_l^{(i)} = \begin{cases} 1 & \text{for } r_l^{(i)} > 0, \\ 0 & \text{for } r_l^{(i)} \leq 0. \end{cases}$$

This hard-decision vector can be tested for code membership by computing $\mathbf{z} \cdot \mathbf{H}^T$. If $\mathbf{z}^{(i)} \cdot \mathbf{H}^T = \mathbf{0}$, $\mathbf{z}^{(i)}$ is a codeword and the decoding iterations can be stopped. Otherwise, iterations continue until either the above condition is satisfied or a preset maximum number of iterations is reached.

The IDBP, in terms of extrinsic information, is specified as follows [23]:

Initialization: Set $i = 0$, maximum number of iterations to I_{max}, $\mathbf{r}^{(0)} = \frac{4}{N_0}\mathbf{y}$, and $\mathbf{Z}^{(0)} = \mathbf{Y}$.

Step 1: For $0 \leq l < n, 1 \leq j \leq J$ and each $\mathbf{h}_j \in A_l$, compute

$$\varepsilon_l^{(i)}(\mathbf{h}_j) = \log \frac{1 + \prod_{t \in B(\mathbf{h}_j) \backslash l} \tanh(Z_{j,l}^{(i)})}{1 - \prod_{t \in B(\mathbf{h}_j) \backslash l} \tanh(Z_{j,l}^{(i)})}, \quad (22)$$

$$E_{j,l}^{(i)} = \sum_{\mathbf{h}_t \in A_l \backslash \mathbf{h}_j} \varepsilon_l^{(i)}(\mathbf{h}_t), \quad (23)$$

$$\varepsilon_l^{(i)} = E_{j,l}^{(i)} + \varepsilon_l^{(i)}(\mathbf{h}_j), \quad (24)$$

Go to Step 2.

Step 2: Form $\mathbf{Z}^{(i+1)} = \mathbf{Y} + \mathbf{E}^{(i)}, \mathbf{r}^{(i+1)} = \mathbf{r}^{(i)} + \mathbf{e}^{(i)}$ and $\mathbf{z}^{(i+1)}$ where $\mathbf{e}^{(i)} = (e_0^{(i)}, e_1^{(i)}, \ldots, e_{n-1}^{(i)})$. Test $\mathbf{z}^{(i+1)} \cdot \mathbf{H}^T$. If $\mathbf{z}^{(i+1)} \cdot \mathbf{H}^T = \mathbf{0}$ or the maximum iteration number I_{max} is reached, stop the decoding iteration and go to Step 3. Otherwise set $i := i + 1$ and go to Step 1.

Step 3: Output $\mathbf{z}^{(i+1)}$ as the decoded codeword.

Another implementation of IDBP [10] is based on the computation of marginal posterior probabilities,

$$P(v_l|\mathbf{y})$$

for $0 \leq l < n$. Then the log-likelihood ratio (LLR) for each code bit is given by

$$L(v_l) = \log \frac{P(v_l = 1|\mathbf{y})}{P(v_l = 0|\mathbf{y})}. \quad (25)$$

Let $p_l^0 = P(v_l = 0)$ and $p_l^1 = P(v_l = 1)$ be the prior probabilities of $v_l = 0$ and $v_l = 1$, respectively.

For $0 \leq l < n, 1 \leq j \leq J$, and each $\mathbf{h}_j \in A_l$, let $q_{j,l}^{x,(i)}$ be the conditional probability that the transmitted code bit v_l has value x, given the check sums computed from the check vectors in $A_l \backslash \mathbf{h}_j$ at the ith decoding iteration. For $0 \leq l < n, 1 \leq j \leq J$, and $\mathbf{h}_j \in A_l$, let $\sigma_{j,l}^{x,(i)}$ be the conditional probability that the check sum s_j is satisfied, given $v_l = x$ and the other code bits in $B(\mathbf{h}_j)$ have a separable distribution $\{q_{j,l'}^{x_{l'},(i)} : l' \in B(\mathbf{h}_j) \backslash l\}$, i.e.,

$$\sigma_{j,l}^{x,(i)} = \sum_{\{v_{l'}:l' \in B(\mathbf{h}_j)\backslash l\}} P(s_j | v_l = x, \{v_{l'} : l' \in B(\mathbf{h}_j) \backslash l\}) \cdot \prod_{l' \in B(\mathbf{h}_j)\backslash l} q_{j,l'}^{x_{l'},(i)}. \tag{26}$$

The computed values of $\sigma_{j,l}^{x,(i)}$ are then used to update the values of $q_{j,l}^{x,(i+1)}$ as follows:

$$q_{j,l}^{x,(i+1)} = \alpha_{j,l}^{(i+1)} p_l^x \prod_{\mathbf{h}_{j'} \in A_l \backslash \mathbf{h}_j} \sigma_{j',l}^{x,(i)} \tag{27}$$

where $\alpha_{j,l}^{(i+1)}$ is chosen such that

$$q_{j,l}^{0,(i+1)} + q_{j,l}^{1,(i+1)} = 1.$$

At the ith iteration step, the pseudo-posterior probabilities are given by

$$P^{(i)}(v_l = x | \mathbf{s}) = \alpha_l^{(i)} p_l^x \prod_{\mathbf{h}_j \in A_l} \sigma_{j,l}^{x,(i-1)}$$

where α_l^i is chosen such that $P^{(i)}(v_l = 0|\mathbf{y}) + P^{(i)}(v_l = 1|\mathbf{y}) = 1$. Based on these probabilities, we can form the following vector as the decoded candidate,

$$\mathbf{z}^{(i)} = (z_0^{(i)}, z_1^{(i)}, \ldots, z_{n-1}^{(i)})$$

with

$$z_l^{(i)} = \begin{cases} 1, & \text{for } P^{(i)}(v_l = 1|\mathbf{y}) > 0.5 \\ 0, & \text{otherwise.} \end{cases}$$

Then compute $\mathbf{z}^{(i)} \cdot \mathbf{H}^T$. If $\mathbf{z}^{(i)} \cdot \mathbf{H}^T = \mathbf{0}$, stop the iterations and output $\mathbf{z}^{(i)}$ as the decoded codeword.

IDBP in terms of probability consists of the following steps [10]:

Initialization: Set $i = 0$, maximum number of iteration to I_{max}. For every (j, l) such that $h_{j,l} = 1$ with $1 \leq j \leq J$ and $0 \leq l < n$, set $q_{i,j}^{0,(0)} = p_l^0$ and $q_{i,j}^{1,(0)} = p_l^1$.

Step 1: For $0 \leq l < n, 1 \leq j \leq J$ and each $\mathbf{h}_j \in A_l$, compute the probabilities, $\sigma_{j,l}^{0,(i)}$ and $\sigma_{j,l}^{1,(i)}$. Go to Step 2.

Step 2: For $0 \leq l < n, 1 \leq j \leq J$ and each $\mathbf{h}_j \in A_l$, compute the values of $q_{j,l}^{0,(i+1)}$ and $q_{j,l}^{1,(i+1)}$ and the values of $P^{(i+1)}(v_l = 0|\mathbf{y})$ and $P^{(i+1)}(v_l = 1|\mathbf{y})$. Form $\mathbf{z}^{(i+1)}$ and test $\mathbf{z}^{(i+1)} \cdot \mathbf{H}^T$. If $\mathbf{z}^{(i+1)} \cdot \mathbf{H}^T = \mathbf{0}$ or the maximum iteration number I_{max} is reached, go to Step 3. Otherwise, set $i := i + 1$ and go to Step 1.

Step 3: Output $\mathbf{z}^{(i+1)}$ as the decoded codeword and stop.

IDBP algorithm is also known as sum-product decoding [18,19,21,34]. The APP decoding is not as effective as IDBP and is not discussed here.

6. PERFORMANCE OF FINITE GEOMETRY LDPC CODES

To demonstrate the error performance of finite geometry LDPC codes, we select several EG-LPDC code and PG-LDPC codes of various lengths and decode them with various decoding algorithms. Figures 4 to 9 show the bit and block error probabilities of these codes.

Figure 4 gives the error performance of the two-dimensional (255,175) EG-LDPC code and the two-dimensional (273,191) PG-LDPC code given in Tables 2 and 3, respectively. The PG-LDPC code is decoded with the simple one-step majority decoding, BF decoding, weighted BF decoding and IDBP algorithm. The EG-LDPC code is only decoded with the IDBP algorithm. From Tables 2 and 3, we find that the two codes have about the same parameters and minimum distance. From Figure 4, we see that these two codes have almost the same bit error performance with the IDBP algorithm. We also see that the IDBP algorithm gives the best error performance among the four decoding algorithms at the expense of extensive computational complexity. The hard-decision BF decoding achieves good error performance with much less computational complexity. It outperforms the simple majority decoding. With some additional computational complexity, the weighted BP decoding gives 0.7 dB coding gain over the hard-decision BF decoding. Therefore, both BF and weighted BF algorithms provide a trade between the error performance of the IDBP algorithm and the complexity of the simple majority decoding algorithm. Figure 5 gives a comparison of the error performance of the two finite geometry codes to that of two best computer-generated (273,191) Gallager LDPC codes with γ equal to 3 and 4. All are decoded with the IDBP algorithm. For the two finite geometry codes, the maximum number of iterations is set to 50, however for the Gallager codes, the maximum number of iterations is set to 200. We see that both finite geometry LDPC codes outperform their corresponding computer-generated Gallager LDPC codes. The Gallager code with γ equal to 3 also shows an error floor indicating that the code has poor minimum distance.

Figure 6 shows the bit error performance of the two-dimensional (1023,781) EG-LDPC code and the two-dimensional (1057,813) PG-LDPC code, given in Tables 2 and 3, respectively. These two codes have about the same code rate, 0.764 and 0.769. Again the PG-LDPC code is decoded with all four decoding algorithms, but the EG-LDPC code is only decoded with the IDBP algorithm. The two codes perform almost the same with the IDBP algorithm. At a bit error rate of 10^{-4}, we see the performance of both codes is 1.4 dB away from the Shannon limit for the same code rate. We also see that both BF and weighted BF decoding algorithms perform well. The block error performance of both codes with IDBP decoding is also shown in Figure 6. Figure 7 gives a comparison of the error performance of the two finite geometry LDPC code to that of two computer generated (1057,813) Gallager LDPC codes with γ equal to 3 and 4. All are decoded with the IDBP algorithm. We see that the two finite geometry codes slightly outperform their corresponding Gallager codes.

Next two codes being evaluated are the two-dimensional (4095,3367) EG-LDPC code and the two-dimensional (4161,3431) PG-LDPC code, the fifth codes given in Tables 2 and 3, respectively. They both have rates about 0.83 and low densities of 0.01563. Their error performances with various types of decoding are shown in Figure 8. With IDBP, they achieve a bit error performance which is 1.35 dB away from the Shannon limit.

Figure 9 gives the error performance of the two-dimensional (16383,14197) EG-LDPC code. It displays the sharp waterfall bit and block error performances of the code.

In decoding of finite geometry codes with IDBP algorithm, we set the maximum number of iterations to 50. Many finite geometry LDPC codes have been simulated. Simulation results of all these codes show that IDBP converges very fast. For example, consider the two-dimensional

(4161,3367) PG-LDPC code, the fifth code given in Table 3. Figure 10 shows the convergence of the IDBP algorithm for this code. We see that at the bit error rate of 10^{-4}, the performance gap between five iterations and 99 iteration is less than 0.2 dB, and the performance gap between ten iterations and 99 iterations is less than 0.05 dB.

All the finite geometry LDPC codes being evaluated have very good block error performance, better than their corresponding computer generated Gallager LDPC codes.

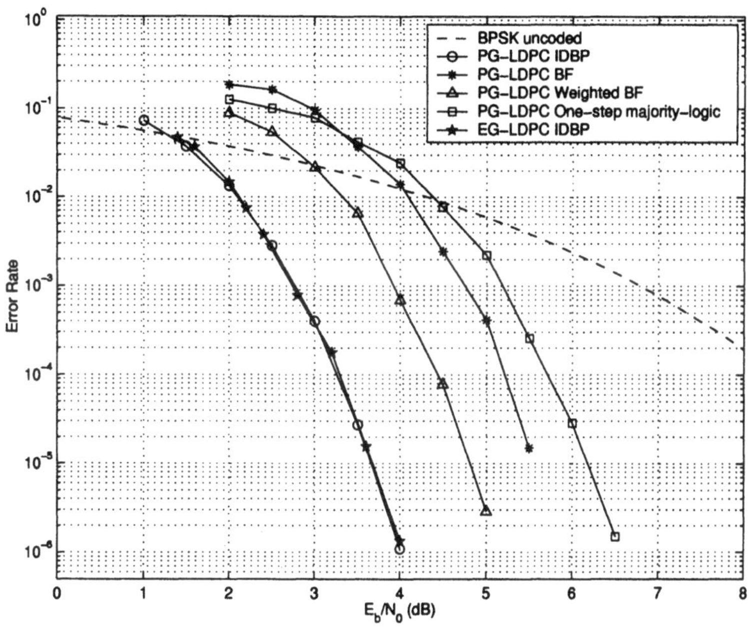

Figure 4 Bit-error probabilities of the (255, 175) EG-LDPC code and (273,191) PG-LDPC code based on different decoding algorithms.

7. EXTENDED FINITE GEOMETRY LDPC CODES

A finite geometry LDPC code **C** of length n can be extended by splitting each column of its parity-check matrix **H** into multiple columns. This results in a new parity-check matrix with smaller density and hence a new LDPC code. If column splitting is done properly, very good extended finite geometry LDPC codes can be obtained. Some of the extended finite geometry LDPC codes constructed perform amazingly well with IDBP. They achieve an error performance only a few tenths of

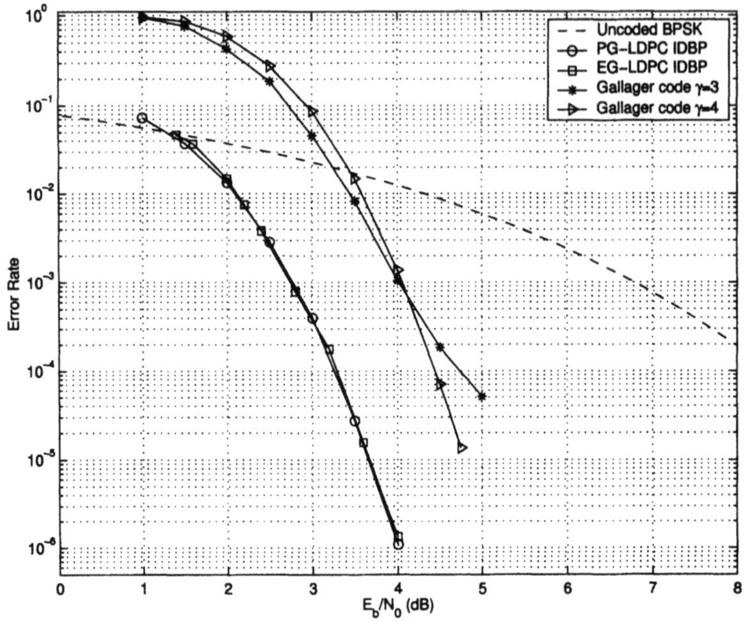

Figure 5 Bit-error probabilities of the (255, 175) EG-LDPC code, (273,191) PG-LDPC code and two computed searched (273,191) Gallager codes.

a dB away from the Shannon limit. They are first known algebraically constructed codes approaching the Shannon limit.

Let $\mathbf{g}_0, \mathbf{g}_1, \ldots, \mathbf{g}_{n-1}$ denote the columns of the parity-check matrix $\mathbf{H}$. First we consider splitting each column of $\mathbf{H}$ into the same number of columns. All the new columns have the same length as the original column. The weight of the original column is distributed among the new columns. A regular column weight distribution can be done as follows. Let q be a positive integer such that $2 \leq q \leq \gamma$. Dividing γ by q, we have

$$\gamma = q \times \gamma_{ext} + b$$

where $0 \leq b < q$. Split each column $\mathbf{g}_i$ of $\mathbf{H}$ into q columns $\mathbf{g}_{i,1}, \mathbf{g}_{i,2}, \ldots, \mathbf{g}_{i,q}$ such that the first b columns, $\mathbf{g}_{i,1}, \mathbf{g}_{i,2}, \ldots, \mathbf{g}_{i,b}$, have weight $\gamma_{ext} + 1$ and the next $q - b$ columns, $\mathbf{g}_{i,b+1}, \mathbf{g}_{i,b+2}, \ldots, \mathbf{g}_{i,q}$, have weight γ_{ext}. The distribution of γ ones of $\mathbf{g}_i$ into $\mathbf{g}_{i,1}, \mathbf{g}_{i,2}, \ldots, \mathbf{g}_{i,q}$ is carried out in a rotating manner. In the first rotation, the first one of $\mathbf{g}_i$ is put in $\mathbf{g}_{i,1}$, the second one of $\mathbf{g}_i$ is put in $\mathbf{g}_{i,2}$, and so on. In the second rotation, the $(q+1)$th one of $\mathbf{g}_i$ is put in $\mathbf{g}_{i,1}$, the $(q+2)$th one of $\mathbf{g}_i$ is put in $\mathbf{g}_{i,2}$ and so on. This rotating distribution of the ones of

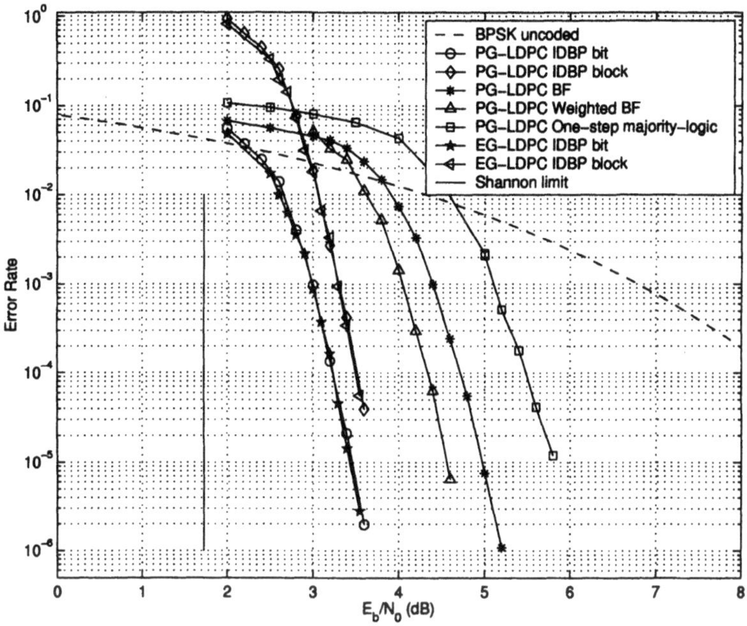

Figure 6 Bit- and block-error probabilities of the (1023, 781) EG-LDPC code and (1057, 813) PG-LDPC code based on different decoding algorithms.

$\mathbf{g}_i$ continues until all the ones of $\mathbf{g}_i$ have been distributed into the q new columns.

The above column splitting results in a new parity-check matrix $\mathbf{H}_{ext}$ with qn columns that has the following structural properties: (1) Each row has weight ρ ; (2) Each column either has weight γ_{ext} or has weight $\gamma_{ext}+1$; (3) Any two columns have at most a single one in common. If the density of $\mathbf{H}$ is r, the density of $\mathbf{H}_{ext}$ is then r/q. Therefore, the above column splitting results in a new parity-check matrix with smaller density. The null space of $\mathbf{H}_{ext}$ gives an extended finite geometry LDPC code $\mathbf{C}_{ext}$. We call $\mathbf{C}_{ext}$ the qth extension of $\mathbf{C}$. If γ is not divisible by q, then the columns of $\mathbf{H}_{ext}$ have two different weights, γ_{ext} and $\gamma_{ext}+1$. Therefore, a code bit of the extended code $\mathbf{C}_{ext}$ is either checked by γ_{ext} check sums or by $\gamma_{ext}+1$ check sums. In this case, the extended finite geometry LDPC code $\mathbf{C}_{ext}$ is an irregular LDPC code. An extended finite geometry LDPC code $\mathbf{C}_{ext}$ with regular column weight distribution is not cyclic but can be put into quasi-cyclic form, i.e., cyclically shifting a codeword in $\mathbf{C}_{ext}$ q bits at a time results in another codeword in $\mathbf{C}_{ext}$.

Example 4 *For $m = 2$ and $s = 6$, the two-dimensional EG-LDPC code is a (4095,3367) code, the fifth code given in Table 2. This code*

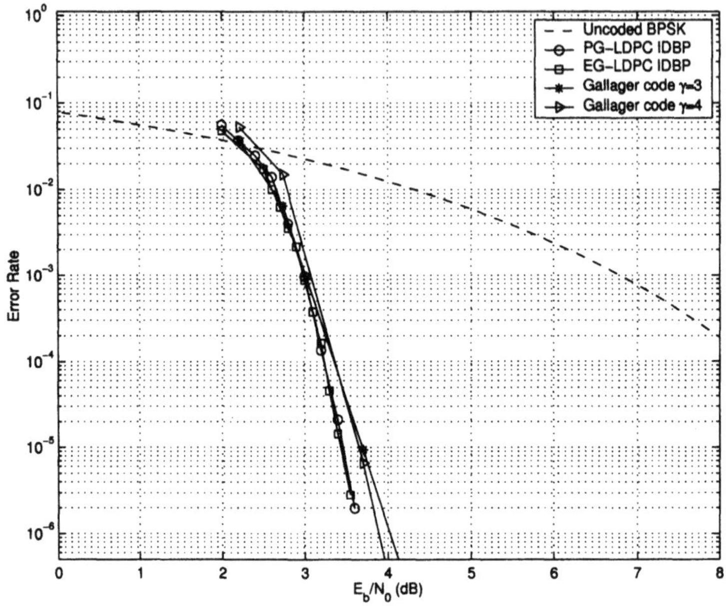

Figure 7 Bit-error probabilities of the (1023, 781) EG-LDPC code, (1057, 813) PG-LDPC code and two computed searched (1057, 813) Gallager codes.

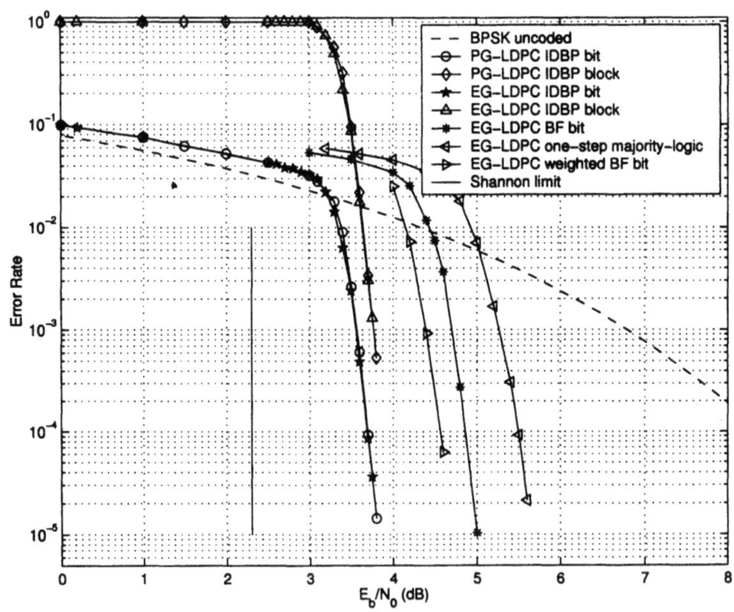

Figure 8 Bit- and block-error probabilities of the (4095, 3367) EG-LDPC code and (4161, 3431) PG-LDPC code based on different decoding algorithms.

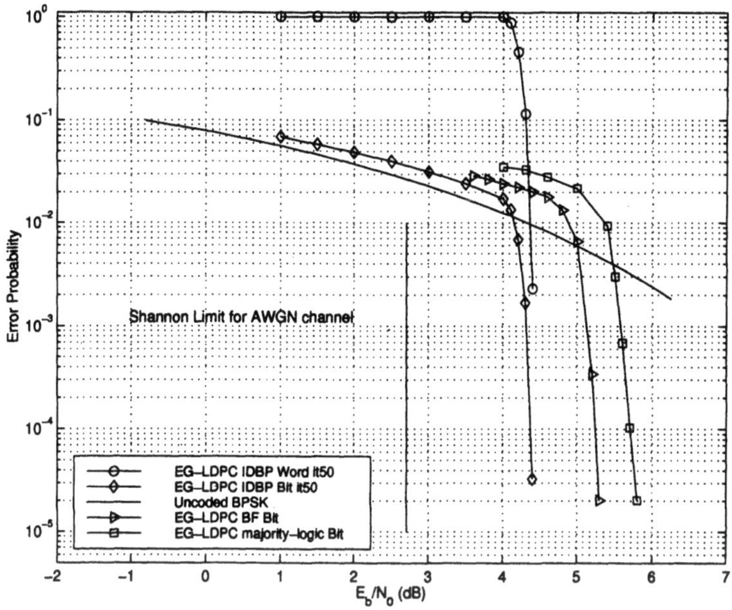

Figure 9 Bit and block-error probabilities of the (16383,14179) EG-LDPC code.

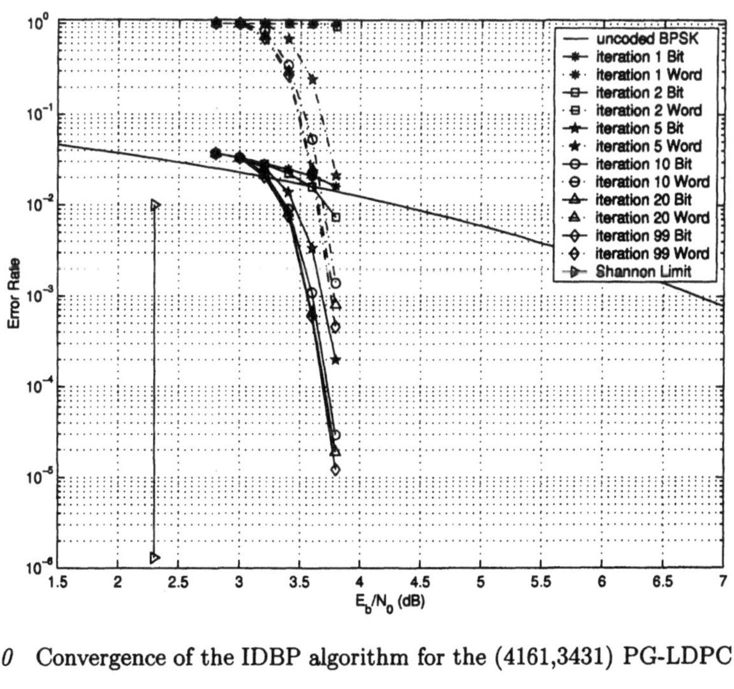

Figure 10 Convergence of the IDBP algorithm for the (4161,3431) PG-LDPC code.

has the following parameters: $\rho = 64, \gamma = 64, d_{min} = 65$ *and density* $r = 0.01563$. *Its error performance has been discussed in the previous section and is shown in Figure 8. At a bit error rate of* 10^{-4}, *the required SNR is 1.4 dB away from the Shannon limit. Suppose we split each column of the parity-check matrix of this code into sixteen columns with rotating column weight distribution. This column splitting results in a (65520,61425) extended EG-LDPC code with* $\rho = 64, \gamma_{ext} = 4, \lambda = 0$ *or* 1 *and density* $r = 0.00098$. *The rate of this code is 0.9375. We see that the parity-check matrix of this code is very sparse. This code decoded with IDBP algorithm achieves an error performance which is only 0.42 dB away from the Shannon limit at a bit error rate of* 10^{-4} *as shown in Figure 11. We see that it has a sharp waterfall error performance. In decoding, the maximum number of iterations is set to 50, but the decoding converges very fast. The performance gap between ten iterations and fifty iterations is less than 0.1 dB. The two-dimensional PG-LDPC code equivalent to the two-dimensional (4095,3367) EG-LDPC code is a (4161,3431) code. If we split each column of the parity-check matrix of this PG-LDPC code into sixteen columns, we obtain a (66576,62415) whose parity-check matrix has two column weights, 4 and 5. Its error performance is almost identical to that of the (65520,61425) extended EG-LDPC code.*

Given a base finite geometry LDPC code **C**, it can be extended into codes of many different lengths. All these extended codes have different rates and behave differently. Consider the two-dimensional (4095,3367) EG-LDPC code discussed in Example 4. Suppose we split each column of its parity-check matrix into various numbers of columns from 2 to 23. Figure 12 and Table 4 show the performances of all the extended codes in terms of SNR's required to achieve a bit error rate of 10^{-4} and the gaps between the required SNR's and their corresponding Shannon limits. We see that splitting each column of the parity-check matrix of the base code into sixteen or seventeen columns gives the best performance in terms of the Shannon limit gap.

Example 5 *For* $m = 2$ *and* $s = 7$, *the two-dimensional EG-LDPC code is a (16383,14197) code with* $\rho = 128, \gamma = 128, d_{min} = 129$ *and density* $r = 0.007813$. *Suppose we split each column of the parity-check matrix of this code into 32 columns. We obtain an extended LDPC code of length 524,256 and rate approximately 31/32. The density of this extended code is 0.000244. The error performance of this extended code is shown in Figure 13. At a bit error rate of* 10^{-4}, *the performance of the extended code is 0.3 dB away from the Shannon limit.*

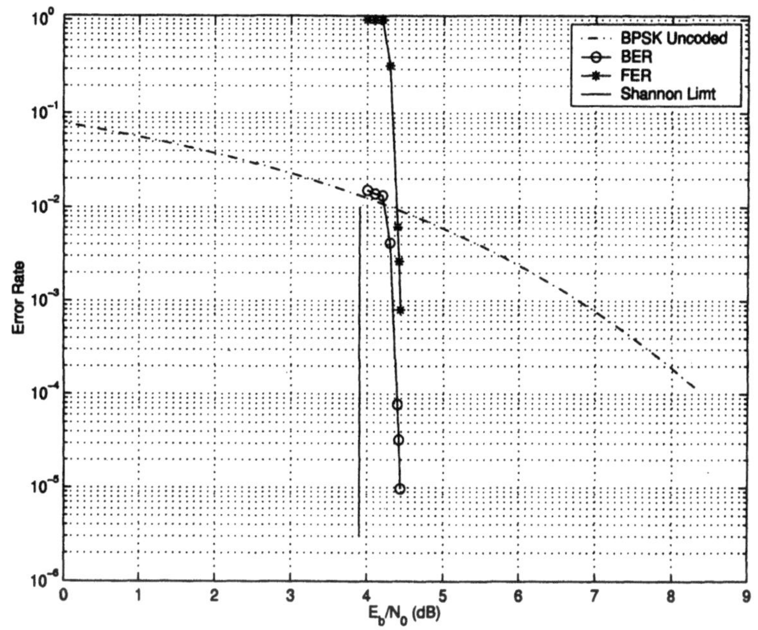

Figure 11 Bit- and block-error probabilities of the (65520,61425) EG-LDPC code based on IDBP.

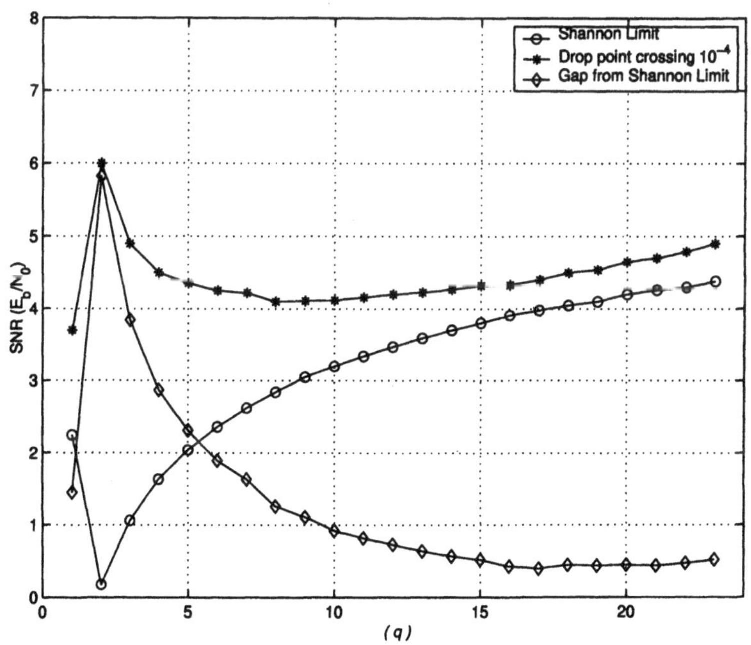

Figure 12 Error performance and Shannon limits of extended codes obtained from the (4095,3367) EG-LDPC Code.

Table 4 Comparison of extended (4095,3367) EG-LDPC codes

q	n	k	ρ	γ	rate	Shannon Limit (dB)	Drop Point Crossing Bit Error Rate of 10^{-4} (dB)	The Gap Between Drop Point and Shannon Limit
1	4095	3367	64	64	0.822	2.25	3.7	1.45
2	8190	4095	64	32	0.5	0.18	6	5.82
3	12285	8190	64	21 or 22	0.666	1.06	4.9	3.84
4	16380	12285	64	16	0.75	1.63	4.5	2.87
5	20475	16380	64	12 or 13	0.8	2.04	4.35	2.31
6	24570	20475	64	10 or 11	0.833	2.36	4.25	1.89
7	28665	24570	64	9 or 10	0.857	2.62	4.22	1.63
8	32760	28665	64	8	0.875	2.84	4.1	1.26
9	36855	32760	64	7 or 8	0.888	3.05	4.11	1.11
10	40950	36855	64	6 or 7	0.9	3.2	4.12	0.92
11	45045	40950	64	5 or 6	0.909	3.34	4.16	0.82
12	49140	45045	64	5 or 6	0.916	3.47	4.2	0.73
13	53235	49140	64	4 or 5	0.923	3.59	4.23	0.64
14	57330	53235	64	4 or 5	0.928	3.7	4.27	0.57
15	61425	57330	64	4 or 5	0.933	3.8	4.32	0.52
16	65520	61425	64	4	0.937	3.91	4.33	0.42
17	69615	65520	64	3 or 4	0.941	3.98	4.4	0.4
18	73710	69615	64	3 or 4	0.944	4.05	4.5	0.45
19	77805	73710	64	3 or 4	0.947	4.1	4.54	0.44
20	81900	77805	64	3 or 4	0.95	4.2	4.65	0.45
21	85995	81900	64	3 or 4	0.952	4.26	4.7	0.44
22	90090	85995	64	2 or 3	0.954	4.3	4.79	0.49
23	94185	90090	64	2 or 3	0.956	4.38	4.9	0.52

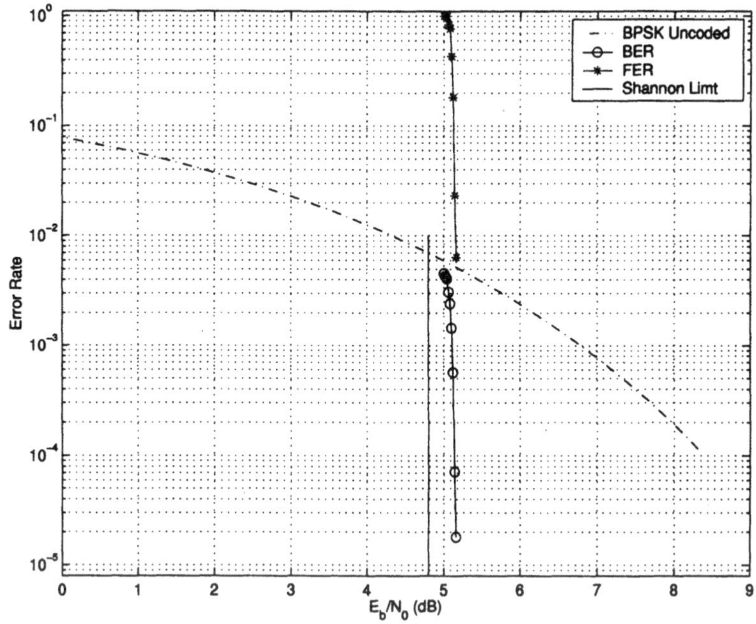

Figure 13 Error Performance of the (524256,507873) LDPC code.

Example 6 *Let $m = s = 3$. The three-dimensional EG-LDPC code* **C** *constructed based on lines and points of EG(3, 2^3) is a (511,139) code with $d_{min} = 73$ and rate $R = 0.272$. It is a very low rate code. Its parity-check matrix is a 4599×511 matrix with $\rho = 8, \gamma = 72, \lambda = 0$ or 1 and density $r = 0.01565$. Suppose we extended this code by splitting each column of its parity-check matrix* **H** *into 24 columns. Then the extended code* $\mathbf{C}_{ext}$ *is a (12264,7665) LDPC code with rate $R_{ext} = 0.625$. The extension results in a high rate code. The density of the parity-check matrix* $\mathbf{H}_{ext}$ *is $r_{ext} = 0.000652$ and column weight of* $\mathbf{H}_{ext}$ *is $\gamma_{ext} = 3$. The bit error performances of this extended LDPC code and its base code are shown in Figure 14. The performance of the extended code is 1 dB away from its corresponding Shannon limit at a bit error rate of 10^{-4}, however its base code is 3.7dB away from its corresponding Shannon limit. From Figure 14, we also see that the overall error performance of the extended code is much better than its base code.*

The last three examples show that splitting each column of the parity-check matrix **H** of a finite geometry LDPC code **C** into multiple columns properly results in an extended LDPC code $\mathbf{C}_{ext}$ which performs very close to the Shannon limit with the IDBP decoding. A reason for this

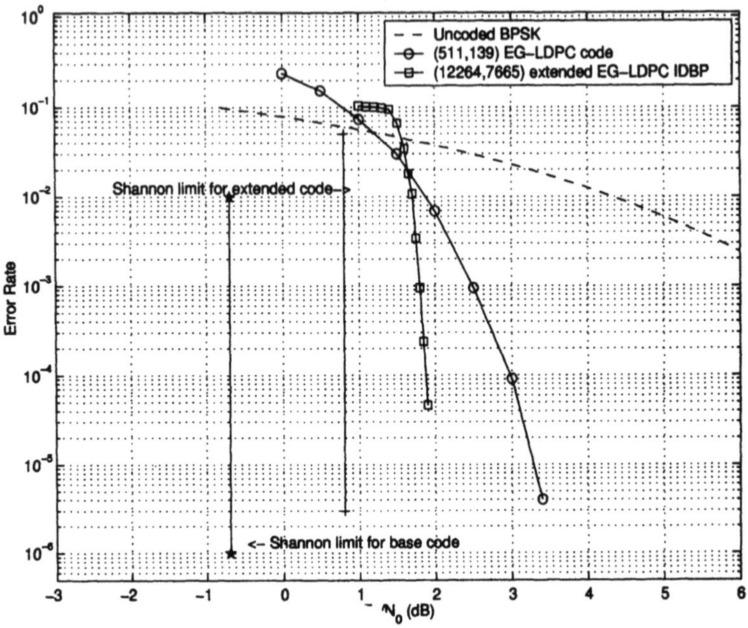

Figure 14 Error Performance of the (511,139) EG-LDPC code and the (12264,7665) extended EG-LDPC code.

is that column splitting reduces the degree of each code bit vertex in the Tanner graph **G** of the base code and hence reduces the number of cycles in the graph. Splitting a column of **H** into q columns results in splitting a code bit vertex of the Tanner graph **G** of the base code into q code bit vertices in the Tanner graph $\mathbf{G}_{ext}$ of the extended code $\mathbf{C}_{ext}$. Each code bit vertex in $\mathbf{G}_{ext}$ is connected to a smaller number of check sum vertices than in **G**. Figure 15(a) shows splitting a column in **H** into two columns results in splitting a code bit vertex in the Tanner graph **G** into two code bit vertices in the Tanner graph $\mathbf{G}_{ext}$. The original code bit vertex has a degree of 4 but each code bit after splitting has a degree of 2. This code bit splitting breaks some cycles that exist in the Tanner graph **G** of the base code **C**. Figures 16(a) and 17 show the breaking of cycles of lengths four and six. Therefore, column splitting of a base finite geometry LDPC code breaks many cycles of its Tanner graph and results in an extended LDPC code whose Tanner graph has many fewer cycles. This reduction in cycles in the Tanner graph improves the performance of the code with IDBP. In fact, breaking cycles with column splitting of the parity-check matrix can be applied to any linear block code.

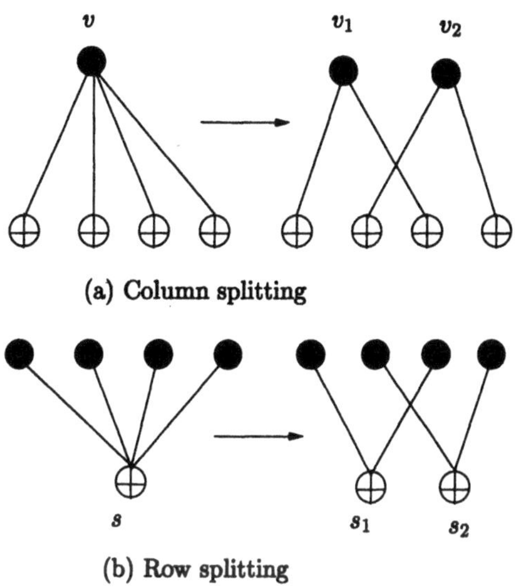

Figure 15 Graph decomposition by column/row splittings.

Example 7 *Consider the (7,4) cyclic Hamming code generated by polynomial* $g(X) = 1+X+X^3$. *Its parity-check polynomial is* $h(X) = 1+X+X^2+X^4$, *and its dual code is generated by* $X^4h(X^{-1}) = 1+X^2+X^3+X^4$. *A parity-check matrix for the (7,4) cyclic Hamming code can be formed by cyclic shifting the vector* (1 0 1 1 1 0 0) *six times as shown in Figure 18.*

The Tanner graph for this code is shown in Figure 19. It contains 21 cycles of length four and each code bit vertex is on six of these cycles. If we split each column of $\mathbf{H}$ *into two columns, we obtain a new parity-check matrix* $\mathbf{H}_{ext}$ *as shown in Figure 20. The null space of* $\mathbf{H}_{ext}$ *gives a (14,8) code. From* $\mathbf{H}_{ext}$, *we can easily check that no four ones appear at the four corners of a rectangle in* $\mathbf{H}_{ext}$. *This implies that the Tanner graph of the extended code does not have any cycle of length four. Figure 21 shows the error performance of this extended Hamming code using IDBP decoding which is almost identical to the performance of maximum likelihood decoding (MLD).*

Given a finite geometry LDPC code specified by a parity-check matrix $\mathbf{H}$, each column of $\mathbf{H}$ can be split in different manner and into different number of columns. Consequently, many extended finite geometry

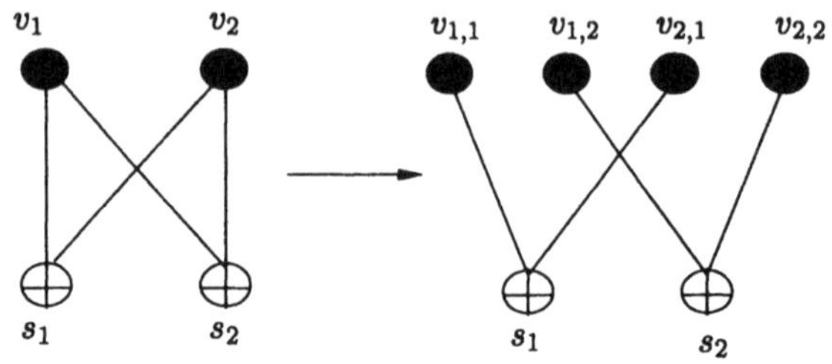

(a) Breaking a cycle of length four by column splitting operation.

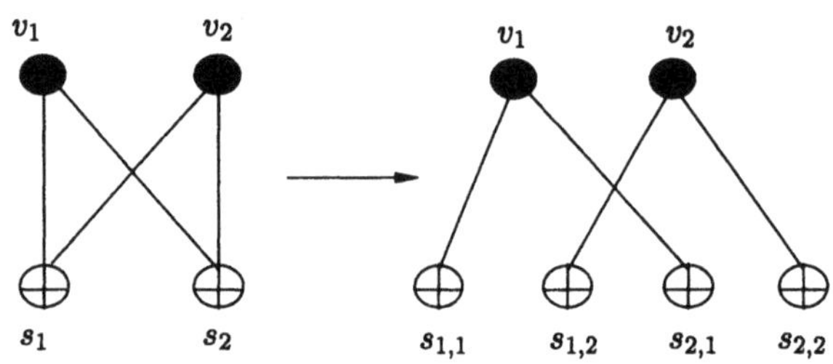

(b) Breaking a cycle of length four by row splitting operation.

Figure 16 Cycle decomposition.

LDPC codes can be obtained by split columns of the parity-check matrix **H**. Of course, if columns of **H** are split differently, the resultant extended code is an irregular LDPC code.

LDPC codes can also be obtained by splitting each row of the parity-check matrix **H** of a base finite geometry LDPC code into multiple rows. The resultant code has the same length as the base code but has a lower code rate. If each row is splitted regularly, the resultant LDPC code is also cyclic. Clearly, LDPC codes can be obtained by splitting both columns and rows of the parity-check matrix of a base finite geometry code.

Splitting a row in the **H** matrix is equivalent to splitting a check sum vertex in the Tanner graph of the code and hence reduces the degree of the vertex as shown in Figure 15(b). Therefore, row splitting of the parity-check matrix of a base code can also break many cycles in the Tanner graph of the base code. An example of cycle breaking by check sum vertex splitting is shown in Figure 16(b). Clearly a combination of column and row splittings will break many cycles in the Tanner graph of the base code. This may result in a very good LDPC code.

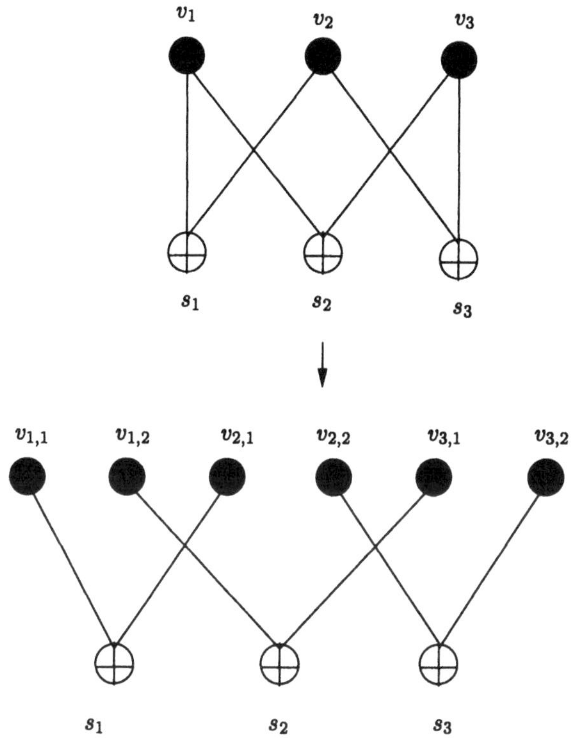

Figure 17 Decomposition of a cycle of length 6 by column splitting.

$$\mathbf{H} = \begin{bmatrix} 1\,0\,1\,1\,1\,0\,0 \\ 0\,1\,0\,1\,1\,1\,0 \\ 0\,0\,1\,0\,1\,1\,1 \\ 1\,0\,0\,1\,0\,1\,1 \\ 1\,1\,0\,0\,1\,0\,1 \\ 1\,1\,1\,0\,0\,1\,0 \\ 0\,1\,1\,1\,0\,0\,1 \end{bmatrix}$$

Figure 18 A Parity-check Matrix for (7,4) Hamming Code Generated by $g(X) = 1 + X + X^3$.

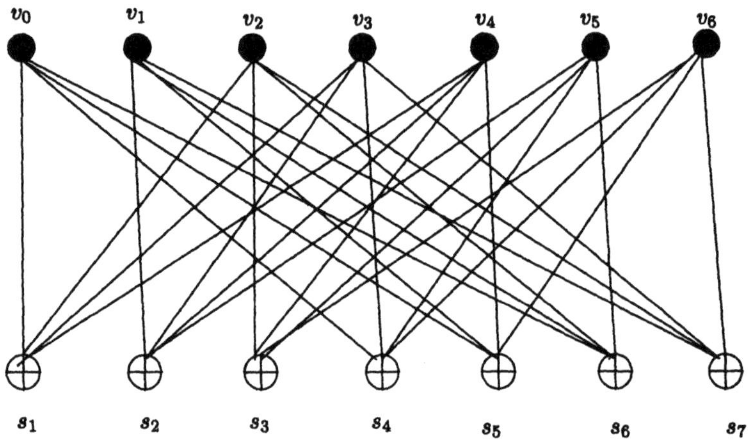

Figure 19 Tanner for the (7,4) Hamming Code

$$\mathbf{H}_{ext} = \begin{bmatrix} 1\,0\,0\,0\,1\,0\,1\,0\,1\,0\,0\,0\,0\,0 \\ 0\,0\,1\,0\,0\,0\,0\,1\,0\,1\,1\,0\,0\,0 \\ 0\,0\,0\,0\,0\,1\,0\,0\,1\,0\,0\,1\,1\,0 \\ 0\,1\,0\,0\,0\,0\,1\,0\,0\,0\,1\,0\,0\,1 \\ 1\,0\,0\,1\,0\,0\,0\,0\,0\,1\,0\,0\,1\,0 \\ 0\,1\,1\,0\,1\,0\,0\,0\,0\,0\,0\,1\,0\,0 \\ 0\,0\,0\,1\,0\,1\,0\,1\,0\,0\,0\,0\,0\,1 \end{bmatrix}$$

Figure 20 Parity-check matrix for (14,8) code

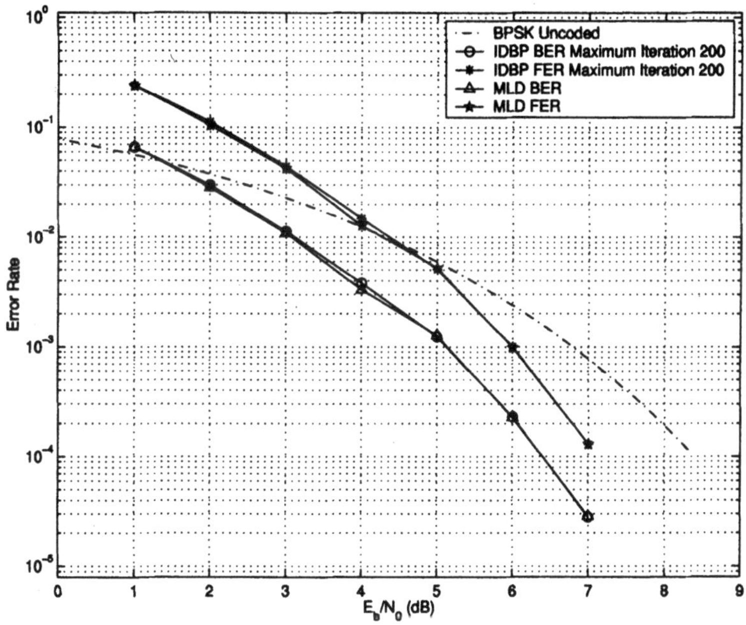

Figure 21 Error Performance of (14,8)Extended LDPC Code.

Example 8 *For $s = 4$. The two-dimensional EG-LDPC code is a (255,175) code, the third code in Table 2. It has the following parameters: $\rho = 16, \gamma = 16, d_{min} = 17$, and density $r = 0.0627$. Its error performance in shown Figure 4. If each column of its parity-check matrix* **H** *is splitted into five columns and each row is splitted into two rows, we obtain a parity-check matrix* **H**$'$ *whose columns have two different weights, three and four and whose rows have the same weight eight. The null space of this new parity-check matrix is a (1275,765) LDPC code with density $r' = 0.00627$. The performance of this new LDPC code is shown in Figure 22. The performance of the base code is also included for comparison.*

Column and row splitting operations can be used to construct LDPC codes from many known codes.

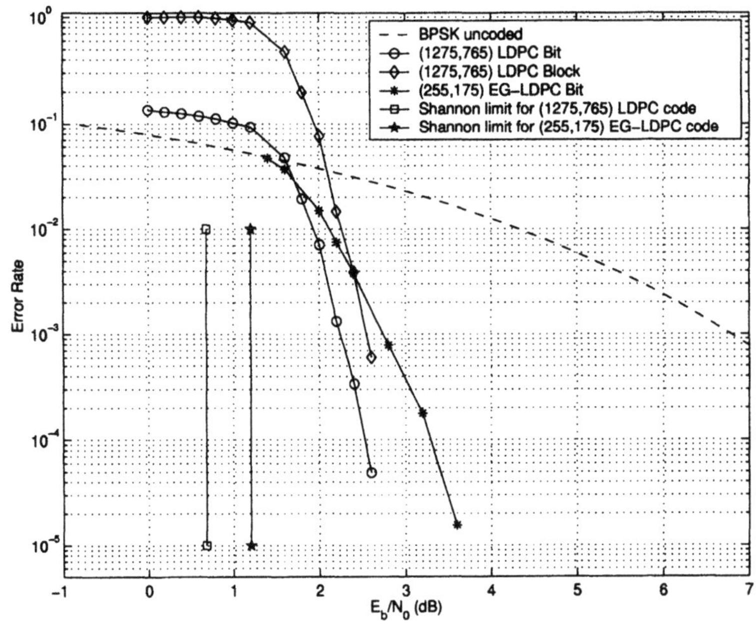

Figure 22 Bit- and block-error probabilities of the expanded (1275,765) LDPC code and its base (275,175) EG-LDPC code based on IDBP.

8. SHORTENED FINITE GEOMETRY LDPC CODES

Finite geometry LDPC codes can be punctured in various ways to obtain good shortened LDPC codes. One approach is to delete the columns of the parity-check matrix **H** of a finite geometry LDPC code that correspond to a properly chosen set of points in the finite geometry based on which the code is constructed. If the deletion of the chosen columns results in all-zero rows (rows of zeros), these rows are removed. The resultant parity-check matrix $\mathbf{H}_s$ also generates a LDPC code. As an example, consider the EG-LDPC code constructed based on the m-dimensional Euclidean geometry EG$(m, 2^s)$. Let EG$(m-1, 2^s)$ be an $(m-1)$-dimensional subspace (also called an $(m-1)$-flat) of EG$(m, 2^s)$ [29,38,39]. If the points in EG$(m-1, 2^s)$ are removed from EG$(m, 2^s)$, we obtain a system **S**, denoted EG$(m, 2^s)\backslash$EG$(m-1, 2^s)$, that contains

$$2^{ms} - 2^{(m-1)s} \tag{1}$$

points. Every line (or 1-flat) contained in EG$(m-1, 2^s)$ is deleted from EG$(m, 2^s)$. Every line that is completely outside of EG$(m-1, 2^s)$ remains in **S** and still contains 2^s points. Every line not completely contained in **S** contains only $2^s - 1$ points, since by deleting an EG$(m-1, 2^s)$ from EG$(m, 2^s)$ we also delete a point in EG$(m-1, 2^s)$ from each such line. The columns of **H** that correspond to the points in the chosen $(m-1)$-flat EG$(m-1, 2^s)$ are deleted, the rows in **H** that correspond to the lines contained in EG$(m-1, 2^s)$ become rows of zeros in the punctured matrix, the rows of **H** that correspond to the lines contained in **S** become rows in the punctured matrix with weight 2^s, and the row of **H** that correspond to lines not completely contained in **S** become rows in the punctured matrix with weight $2^s - 1$. Removing the rows of zeros from the punctured matrix, we obtain a new matrix $\mathbf{H}_s$ that has

$$\frac{2^{(m-1)s}(2^{ms}-1) - 2^{(m-2)s}(2^{(m-1)s}-1)}{2^s - 1} \tag{2}$$

rows and $2^{ms} - 2^{(m-1)s}$ columns. Every column of $\mathbf{H}_s$ still has weight 2^s, but the rows of $\mathbf{H}_s$ have two different weights, $2^s - 1$ and 2^s. $\mathbf{H}_s$ still has low density of ones and the null space of $\mathbf{H}_s$ gives a shortened EG-LDPC code whose minimum distance is at least the same as the original EG-LDPC code.

Consider the EG-LDPC code constructed based on the two-dimensional Euclidean geometry EG$(2, 2^s)$. Its low-density parity-check matrix **H** is a $(2^{2s}-1) \times (2^{2s}-1)$ matrix whose rows are the incidence vectors of the lines in EG$(2, 2^s)$ that do not pass through the origin. The weight of each column of **H** is $\gamma = 2^s$ and the weight of each row of **H** is $\rho = 2^s$. Let $\mathcal{L}$ be a line in EG$(2, 2^s)$ that does not pass through the origin. Delete the columns in **H** that correspond to the 2^s points on $\mathcal{L}$.

This results in a matrix $\mathbf{H}'$ with $2^{2s} - 2^s - 1$ columns. The row in $\mathbf{H}$ that corresponds to the line $\mathcal{L}$ becomes a row of zeros in $\mathbf{H}'$. Removing this zero row from $\mathbf{H}'$, we obtain a $(2^{2s} - 2) \times (2^{2s} - 2^s - 1)$ matrix $\mathbf{H}_s$. Each column of $\mathbf{H}_s$ still has weight $\gamma = 2^s$. Removing a column of $\mathbf{H}$ that correspond a point $\mathbf{p}$ on $\mathcal{L}$ will delete a one from $2^s - 1$ rows in $\mathbf{H}$ which are the incidence vectors of the lines that intersect with line $\mathcal{L}$ at the point $\mathbf{p}$. Therefore, there are $2^s(2^s - 1)$ rows in $\mathbf{H}_s$ with weight $\rho_1 = 2^s - 1$. There are $2^s - 2$ lines in EG$(2, 2^s)$ not passing through the origin of EG$(2, 2^s)$ that are parallel to $\mathcal{L}$. Deleting the columns of the $\mathbf{H}$ that correspond to the points on $\mathcal{L}$ does not change the weights of the rows that are the incidence vectors of the $2^s - 2$ lines parallel to $\mathcal{L}$. Therefore, there are $2^s - 2$ rows in $\mathbf{H}_s$ with weight $\rho_2 = 2^s$. Any two columns in $\mathbf{H}_s$ still have at most a single one in common, i.e., $\lambda = 0$ or 1. The density of $\mathbf{H}_s$ is $2^s/(2^{2s} - 2)$. Therefore, $\mathbf{H}_s$ is still a low density matrix. The null space of $\mathbf{H}_s$ is a shortened EG-LDPC code with minimum distance at least $2^s + 1$.

Example 9 *Let* $m = 2$ *and* $s = 4$. *The two-dimensional EG-LDPC code with* $s = 4$ *is a (255, 175) code (the third code in Table 2) with rate* $R = 0.6863, d_{min} = 17, \rho = \gamma = 16$ *and density* $r = 0.0627$. *A line in EG(2,*2^4*) has sixteen points. Puncturing this EG-LDPC code based on a line in EG(2,*2^4*) not passing through the origin results in a (239,160) LDPC code with rate* $R = 0.667$, *minimum distance at least 17 and density* $r = 0.0630$. *Note that the punctuation removes fifteen information bits and one parity-check bit from the (255,175) EG-LDPC code. Figure 23 shows that the error performance of this punctured code is slightly better than that of the original code.*

An EG-LDPC code can be punctured based on the points of a set of q parallel lines with $1 \leq q \leq 2^{(m-1)s}$. This results in a shortened LDPC code of length $n = 2^{ms} - q2^s - 1$. its parity-check matrix $\mathbf{H}_s$ has

$$\frac{(2^{(m-1)s} - 1)(2^{ms} - 1)}{2^s - 1} \tag{3}$$

rows and $2^{ms} - q2^s - 1$ columns.

Example 10 *If we puncture the (255,175) EG-LDPC code (given in Example 9) based on the points of two parallel lines in EG(2,*2^4*), we obtained a (223,145) LDPC code with rate* $R = 0.650$ *and* $d_{min} \geq 17$. *The parity-check matrix* $\mathbf{H}_s$ *of this code is a* 253×223 *matrix. Thirteen rows of* $\mathbf{H}_s$ *have weight 16 and all the other rows have weight 14. All the columns of* $\mathbf{H}_s$ *have weight 16. The density of* $\mathbf{H}_s$ *is* r*=0.0632.*

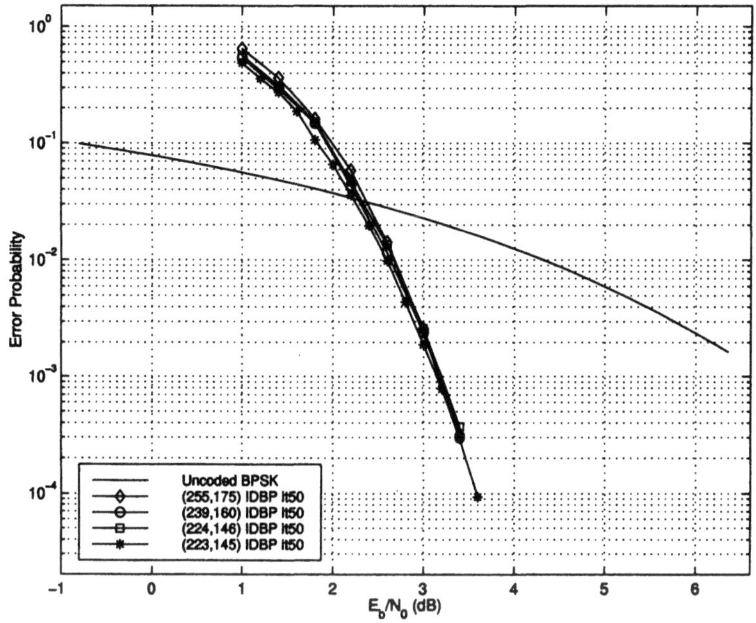

Figure 23 Bit-error probabilities of the (255,175) EG-LDPC code and its (239,160), (223,145) and (224,146) punctured codes.

The puncturing removes thirty information bits and two parity-check bits from the original code. The error performance of the shortened code is shown in Figure 23.

Puncturing can also be achieved by removing columns and rows of the low density parity-check matrix **H**. For example, let **Q** be a set of l lines in EG(m,2^s) not passing through the origin that intersect at a common point α^i, where $1 \leq l \leq 2^s$. Let **P** be the set of lines in EG(m,2^s) that are parallel to the lines in **Q**. Suppose we puncture **H** as follows: (1) remove all the rows in **H** that are the incidence vectors of the lines in **Q** and **P**; and (2) remove the columns that correspond to the points on the lines in **Q**. The total number of distinct points on the lines in **Q** is $l \cdot (2^s - 1) + 1$. The total number of lines in **Q** and **P** is

$$l \cdot (2^{(m-1)s} - 1).$$

Therefore, puncturing results in matrix $\mathbf{H}_s$ with

$$(2^{(m-1)s} - 1)(\frac{2^{ms} - 1}{2^s - 1} - l)$$

rows and

$$2^{ms} - l(2^s - 1) - 2$$

columns. Both columns and rows of $\mathbf{H}_s$ do not have constant weights any more. The null space of $\mathbf{H}_s$ gives an irregular LDPC code.

Example 11 *Consider puncturing the (255,175) EG-LDPC code. Let $\mathcal{L}_1$ and $\mathcal{L}_2$ be two lines in EG(2,2^4) not passing through the origin that intersect at the point α^i. There are 28 lines not passing through the origin parallel to either $\mathcal{L}_1$ or $\mathcal{L}_2$. Puncturing the parity-check matrix* $\mathbf{H}$ *of the (255,175) EG-LDPC code based on $\mathcal{L}_1$, $\mathcal{L}_2$ and their parallel lines results in a* 225×224 *matrix* $\mathbf{H}_s$. *Fourteen rows in* $\mathbf{H}_s$ *have weight 15 and all the other rows have weight 14. Twenty-nine columns of* $\mathbf{H}_s$ *have weight 15 and all the other columns have weight 14. The density of* $\mathbf{H}_s$ *is* $r = 0.06278$. *The LDPC code generated by* $\mathbf{H}_s$ *is a (224,146) code with minimum distance at least 15. Its error performance is shown in Figure 23.*

All the above example shortened EG-LDPC codes have about the same error performance as shown in Figure 23.

PG-LDPC codes can be shortened in a similar manner. In fact, the m-dimensional EG-LDPC code constructed from EG$(m, 2^s)$ can be obtained by puncturing the m-dimensional PG-LDPC code constructed from PG$(m, 2^s)$. Let PG$(m-1, 2^s)$ be an $(m-1)$-dimensional subspace (or $(m-1)$- flat) of PG$(m, 2^s)$. If the points in PG$(m-1, 2^s)$ are removed from PG$(m, 2^s)$, we obtain the m-dimensional Euclidean geometry EG$(m, 2^s)$ [38,39]. Let $\mathbf{H}_{PG-LDPC}$ be the parity-check matrix of the m-dimensional PG-LDPC code. If we delete the columns that correspond to the points in PG$(m-1, 2^s)$ from $\mathbf{H}_{PG-LDPC}$, we obtain a parity-check matrix $\mathbf{H}_0$ with 2^{ms} columns. Each column of $\mathbf{H}_0$ corresponds to a point in EG$(m, 2^s)$, including the origin. The first column of $\mathbf{H}_0$ corresponds to the origin of EG$(m, 2^s)$. The rows of $\mathbf{H}_0$ are the incidence vectors of all the lines in EG$(m, 2^s)$. If we remove from $\mathbf{H}_0$ the first column and the rows that are the incidence vectors of the lines in EG$(m, 2^s)$ which intersect at the origin, we obtain the parity-check matrix $\mathbf{H}_{EG-LDPC}$ of the m-dimensional EG-LDPC codes. Therefore, the EG-LDPC codes are *descendants* of the PG-LDPC codes.

Clearly, the null space of $\mathbf{H}_0$ is also an EG-LDPC code but not cyclic.

9. CONCLUSION

In this paper, a geometric approach to the construction of LDPC codes has been presented. This approach allows the construction of LDPC codes based on the lines and points of a finite geometry, algebraically and systematically. Two classes of finite geometry LDPC codes have been constructed based on the well known Euclidean and projective geometries over finite fields. These finite geometry codes are cyclic and can be generated easily based on their generator polynomials. They have relatively good minimum distances and large majority error correction capabilities. Their Tanner graphs do not ontain cycles of length four. They can be decoded with various decoding algorithms, ranging from low to high decoding complexity and from reasonably good to very good error performance. They can be extended or shortened in various ways to form many other good LDPC codes of various lengths and rates. Extension by column splitting the parity-check matrix of a finite geometry LDPC code is a powerful method to construct long, high performance LDPC codes. Some long extended finite geometry LDPC codes have been constructed and they achieve a performance that is only a few tenths of a dB away from the Shannon limit.

Due to their outstanding error performance, finite geometry LDPC codes are strong competitors to turbo codes for error control in communication systems and digital data storage systems. Finite geometry LDPC codes can be constructed analytically and systematically for various minimum distances and various rates, high and low. However, this is not the case for turbo codes. Well designed turbo codes do provide better bit error performance than LDPC codes for bit error rates above 10^{-4} but they exhibit an error floor generally around bit error rate of 10^{-5} due to their poor minimum distances which finite geometry LDPC codes do not have because of their good minimum distances. Turbo codes in general have very poor block error probability, but finite geometry LDPC codes exhibit good block error performance. In turbo decoding, only information bits are decoded and they can not be used for error detection. Consequently, some stopping criteria must be devised to terminate the decoding iteration process. These stopping criteria in general requires extensive real number (such as the cross entropy stopping criteria) or logic computations. However, in decoding of LDPC codes, all the code bits are decoded and they can be used for error detection by computing the syndrome of the decoded word at the end of each decoding iteration. This error detection can be used as a stopping condition to terminates the decoding iteration process. Computation of syndromes for a EG-LDPC code or a PG-LDPC code requires only a division circuit which is a linear shift register with feedback connections based on the generator polynomial of the code. Finally, a reasonably long interleaver is required for a turbo code to achieve good bit error performance, but it is not needed in LDPC codes.

It is possible to combine the advantage of good error performance of the turbo codes in the range of bit error rate above 10^{-4} and the advantages of finite geometry LDPC codes in a concatenated coding system.

This marriage of turbo and LDPC codes may provide extremely good error performance. Research in this direction is now being conducted.

References

[1] R. G. Gallager, "Low Density Parity-Check Codes," *IRE Transactions on Information Theory*, IT-8, pp. 21–28, January 1962.

[2] R. G. Gallager, *Low Density Parity-Check Codes*, MIT Press, Cambridge, MA, 1963.

[3] R. M. Tanner, "A Recursive Approach to Low Complexity Codes," *IEEE Transactions on Information Theory*, IT-27, pp. 533–547, Sep. 1981.

[4] D. J. C. MacKay and R. M. Neal, "Near Shannon Limit Performance of Low Density Parity-Check Codes," *Electronics Letters*, vol. 32 no. 18, pp. 1645–1646, 1996.

[5] M. Sipser and D. Spielman, "Expander Codes," *IEEE Transactions on Information Theory*, vol. 42, no. 6, pp. 1710–1722, Nov. 1996.

[6] D. Spielman, "Linear-time Encodable Error-correcting Codes," *IEEE Transactions on Information Theory*, vol. 42, no. 6, pp. 1723–1731, Nov. 1996.

[7] M. C. Davey and D. J. C. MacKay, "Low Density Parity Check Codes over GF(q)," *IEEE Communications Letters*, June 1998.

[8] M. G. Luby, M. Mitzenmacher, M. A. Shokrollahi, and D. A. Spielman, "Improved Low-Density Parity-Check Codes Using Irregular Graphs and Belief Propagation," *Proceedings of 1998 IEEE International Symposium on Information Theory*, p. 171, Cambridge, MA, Aug 1998.

[9] D. J. C. Mackay, "Gallager Codes that are Better Than Turbo Codes," *Proceedings of the 36th Allerton Conference on Communication, Control, and Computing*, Monticello, IL, Sep. 1998.

[10] D. J. C. MacKay,"Good Error-Correcting Codes Based on Very Sparse Matrices," *IEEE Transactions on Information Theory*, vol. IT-45, pp. 399–432, Mar. 1999.

[11] D. J. C. Mackay, "Sparce Graph Codes," *Proceedings of the 5th International Symposium on Communication Theory and Applications*, Ambleside, UK, pp. 2–4, July 1999.

[12] T. Richardson, A. Shokrollahi, and R. Urbanke, "Design of Capacity-Approaching Irregular Low-Density Parity-Check Codes," *IEEE Transactions on Information Theory*, vol. IT-47, pp. 619–638, Feb. 2001.

[13] T. Richardson and R. Urbanke, "The Capacity of Low-Density Parity-Check Codes under Message-Passing Decoding," *IEEE Transactions on Information Theory*, vol. IT-47, pp. 599–618, Feb. 2001.

[14] Y. Kou, S. Lin and M. Fossorier, "Low Density Parity-Check Codes based on Finite Geometries: A Rediscovery," *Proceedings of the IEEE International Symposium on Information Theory*, Sorrento, Italy, June 2000.

[15] S. Lin and Y. Kou, "A Geometric Approach to the Construction of Low Density Parity-Check Codes," presented at *the IEEE 29th Communication Theory Workshop*, Haines City, FL, May 2000.

[16] J. Pearl, *Probabilistic Reasoning in Intelligent Systems: Networks of Plausible Inference*, Morgan Kaufmann, San Mateo, 1988.

[17] S. L. Lauritzen and D. J. Spiegelhalter, "Local Computations with Probabilities on Graphical Structures and Their Application to Expert Systems," *Journal of the Royal Statistical Society B*, vol. 50, pp. 157–224, 1988.

[18] N. Wiberg, H.-A. Loeliger, and R. Kötter, "Codes and Iterative Decoding on General Graphs," *European Transactions on Telecommunications*, vol. 6, pp. 513–526, 1955.

[19] R. J. McEliece, D. J. C. MacKay, and J.-F. Cheng,"Turbo Decoding as an Instance of Pearl's Belief Propagation Algorithm," *IEEE Journal on Selected Areas*, vol. 16, pp. 140–152, Feb. 1998.

[20] F. R. Kschischang and B. J. Frey, "Iterative Decoding of Compound Codes by Probability Propagation in General Models," *IEEE Journal on Selected Areas in Communications*, vol. 16, no. 12, pp. 219–230, Feb. 1998.

[21] F. R. Kschischang, B. J. Frey, and H.-A. Loeliger, "Factor Graphs and the Sum-Product Algorithm," *IEEE Transactions on Information Theory*, vol. IT-47, pp. 498–519, Feb. 2001.

[22] M. Fossorier, M. Mihaljevic, and H. Imai,"Reduced Complexity Iterative Decoding of Low Density Parity-Check Codes," *IEEE Transactions on Communications*, vol. 47, pp. 673–680, May 1999.

[23] R. Lucas, M. Fossorier, Y. Kou, and S. Lin, "Iterative Decoding of One-Step Majority Logic Decodable Codes Based on Belief Propagation," accepted by *IEEE Transactions on Communications*, 1999.

[24] C. Berrou, A. Glavieux, and P. Thitimajshima, "Near Shannon Limit Error-Correcting Coding and Decoding: Turbo Codes," *Proc. 1993 IEEE International Conference on Communications*, Geneva, Switzerland, pp. 1064–1070, May 1993.

[25] C. Berrou and A. Glavieux, "Near Optimum Error Correcting Coding and Decoding: Turbo-Codes," *IEEE Transactions on Communications*, vol. 44, pp. 1261–1271, Oct. 1996.

[26] S. Benedetto and G. Montorsi, "Unveiling Turbo Codes: Some Results on Parallel Concatenated Coding Schemes," *IEEE Transactions on Information Theory*, vol. 42, no. 2, pp. 409–428, Mar. 1996.

[27] J. Hagenauer, E. Offer, and L. Papke, "Iterative Decoding of Binary Block and Convolutional Codes," *IEEE Transactions on Information Theory*, vol. 42, pp. 429–445, Mar. 1996.

[28] W. W. Peterson and E. J. Weldon, Jr., *Error-Correcting Codes*, 2nd ed., MIT Press, Cambridge, MA, 1972.

[29] S. Lin and D. J. Costello, Jr., *Error Control Coding: Fundamentals and Applications*, Prentice Hall, Englewood Cliffs, NJ, 1983.

[30] R. C. Bose and D. J. Ray-Chaudhuri, "On a Class of Error Correcting Binary Group Codes,". *Information and Control*, no. 3, pp 68–79, Mar. 1960.

[31] E. R. Berlekamp, *Algebraic Coding Theory*, McGraw-Hill, NY, 1968.

[32] J. L. Massey, *Threshold Decoding*, MIT Press, Cambridge, MA 1963.

[33] N. Deo, *Graph Theory with Applications to Engineering and Computer Science*, Prentice Hall, Englewood Cliffs, NJ, 1974.

[34] N. Wiberg, "Codes and Decoding on General Graphs," *Ph.D Dissertation*, Department of Electrical Engineering, University of Linköping, Linköping, Sweden, April 1996

[35] Tuvi Etzion, Ari Trachtenberg, and A. Vardy, "Which Codes Have Cycle-Free Tanner Graphs," *IEEE Transactions on Information Theory*, vol. 45, no. 6, pp. 2173–2181, Sep. 1999.

[36] G. D. Forney, Jr., "Codes on Graphs: A Survey for Algebraists," *Proc. 13-th International Symposium on Applied Algebra, Algebraic Algorithms and Error-Correcting Codes*, Honolulu, HI, Nov. 1999.

[37] G. D. Forney, Jr., "Codes on Graphs: Normal Realizations," *submitted to IEEE Transactions on Information Theory*, 1998.

[38] R. D. Carmichael, *Introduction to the Theory of Groups of Finite Order*, Dover Publications, Inc., New York, NY, 1956.

[39] A. P. Street and D. J. Street, *Combinatorics of Experimental Design*, Oxford Science Publications, Clarendon Press, Oxford, 1987.

[40] E. J. Weldon, Jr., "Euclidean Geometry Cyclic Codes," *Proceedings of the Symp. Combinatorial Math.*, University of North Carolina, Chapel Hill, NC, 1967.

[41] T. Kasami, S. Lin, and W. W. Peterson, "Polynomial Codes," *IEEE Transactions on Information Theory*, vol. 14, pp. 807–814, 1968.

[42] S. Lin, "On a Class of Cyclic Codes," *Error Correcting Codes*, (Edited by H. B. Mann), John Wiley & Sons, Inc., New York, 1968.

[43] S. Lin, "On the Number of Information Symbols in Polynomial Codes," *IEEE Transactions on Information Theory*, vol. IT-18, pp. 785–794, Nov. 1972.

[44] T. Kasami and S. Lin, "On Majority-Logic Decoding for Duals of Primitive Polynomial Codes," *IEEE Transactions on Information Theory*, vol. IT-17, no. 3, pp. 322–331, May 1971.

[45] E. J. Weldon, Jr., "New Generations of the Reed-Muller Codes, Part II: Non-primitive Codes," *IEEE Transactions on Information Theory*, vol. IT-14, pp. 199–205, May 1968.

[46] L. D. Rudolph, "A Class of Majority Logic Decodable Codes," *IEEE Transactions on Information Theory*, vol. IT-13, pp. 305-307, April 1967.

[47] E. J. Weldon, Jr.,"Difference-Set Cyclic Codes," *Bell System Technical Journal*, vol. 45, pp. 1045-1055, Sep. 1966.

[48] F. L. Graham and J. MacWilliams, "On the Number of Parity-Checks in Difference-Set Cyclic Codes," *Bell System Technical Journal*, vol. 45, pp. 1056–1070, Sep. 1966.

[49] I. S. Rees, "A Class of Multiple-Error-Correcting Codes and the Decoding Scheme," *IRE Transactions*, vol. IT-4, pp. 38–49, Sep. 1954.

[50] V. D. Kolesnik, "Probability Decoding of Majority Codes," *Prob. Peredachi Inform.*, vol. 7, pp. 3–12, July 1971.

[51] L. R. Bahl, J. Cocke, F. Jelinek, and J. Raviv, "Optimal Decoding of Linear Codes for Minimizing Symbol Error Rate," *IEEE Transactions on Information Theory*, vol. 20, pp. 284–287, Mar. 1974.

[52] R. Lucas, M. Bossert, and M. Breitbach, "On Iterative Soft-Decision Decoding of Linear Block Codes and Product Codes," *IEEE Journal on Selected Areas in Communications*, vol. 16, pp. 276-296, Feb. 1998.

Chapter 22

THE TETRAHEDRAL GOLAY CODE

R. M. Tanner
Baskin Engineering 117
University of California, Santa Cruz
Santa Cruz, CA 95064

Abstract Forney conjectured the existence of a "normal realization" (NR) for the [24,12,8] Golay code in which the code appears as a tetrahedron, with six output channel bits at each of the four vertices and two state bits carried by each of the six edges. Although the conjecture was dashed by Vardy, we here give a quasi-cyclic normal realization exhibiting tetrahedral symmetry, confirming the common wisdom that Forney is never (far) wrong.

1. INTRODUCTION

Research on codes on graphs and iterative decoding algorithms begun in [6, 9, 10] has stimulated interest in expanding and generalizing the theoretical framework for graphical models [3, 4]. Forney [2] has presented a model called "normal realization" (NR) in which state information is passed by ordinary edges representing states, leaf edges represent symbols that taken together form the codeword, and vertices represent local constraints on the incident edges. For any global solution to the set of constraints, the symbols incident on a constraint vertex are determined unambiguously by values for the state variables for edges incident on the vertex. The requirement that the symbols depend only on this local information is essential to the implementation of graph message-passing decoding. The union of all the local constraints defines the code.

The graph representation of a code can make visually explicit underlying symmetries of the code, the automorphism group of the constraint graph being a subgroup of the automorphism group of the code itself. Calderbank, Forney, and Vardy [1] present the [24,12,8] binary Golay code as a tail-biting trellis with minimal state complexity. In view of

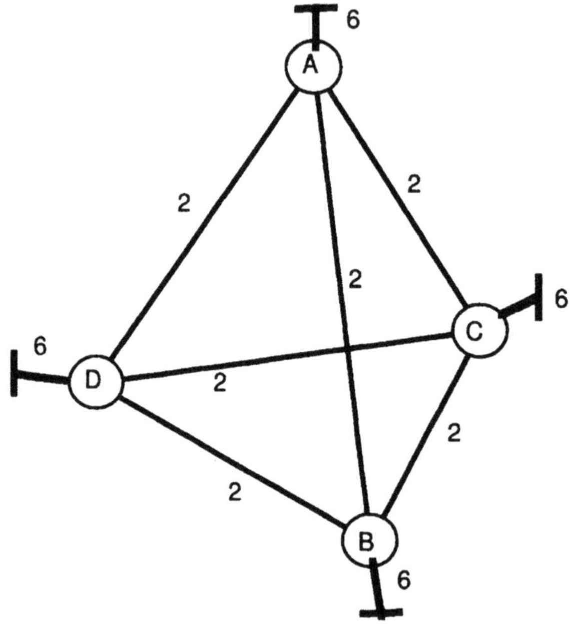

Fig. 1 Conjectured [24,12,8] Golay Normal Realization as a Tetrahedron

the powerful Mathieu group of automorphisms of the [24,12,8] Golay code, one might expect that it could be given a number of attractive realizations. Forney conjectured that it could be put into a tetrahedral form as shown in Figure 1. The states associated with each of the six edges might carry two bits of information, a total of twelve bits, so that each constraint vertex carries out a mapping from six information bits to six output code bits. Finding a linear realization of this form is tantamount to finding a generator matrix for the code that can be partitioned into four blocks of six, with two generator rows being nonzero in only two of the four blocks. Unfortunately, Vardy [8] showed that such a realization cannot exist by mapping the conjectured basis vectors to a canonical form and comparing the result with the Miracle Octad Generators for the Golay code. The form that would be required for the generator is shown in Figure 2. Each edge of the tetrahedron would have to carry two bits of information, having four possible states determined by the information bits. No information from one edge could be conveyed along another edge, since each edge must be determined by the independent information bits. Vardy showed that the Golay code cannot have a generator matrix that is nonzero in blocks of six bits as shown in Figure 2. One concludes that information flows are too restricted by the conjectured form.

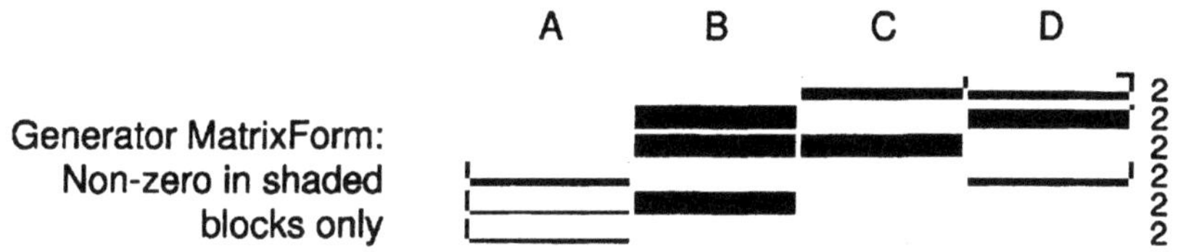

Fig. 2 Generator form for a tetrahedron

In this brief paper we will show that the Golay code can be presented with a tetrahedral symmetry as given in Figure 3. To each of the four vertices of the tetrahedron are attached five channel bits, and each of the six state edges of the tetrahedron carries two bits of state information. But at the center of the tetrahedron is an additional vertex with four attached channel bits that is connected by an edge carrying a single bit to each of the four corners. For encoding one can envision each of the four tetrahedron vertices as accepting two message bits and the center vertex as accepting four.

This normal realization was found by recasting the Golay code so as to exploit a [20,8,8] quasi-cyclic code consisting of four cycles of five bits. For those wishing to leap immediately to the conclusion, the generator matrix that admits this normal realization is given in Figure 7. The theory behind it is sketched in the following sections.

2. DOUBLE CIRCULANT FORM REDUX

The [24,12,8] Golay code can be considered to be a quadratic residue code with an overall parity check. Using the code automorphism that maps $c(x)$ to $c(x^2)$, it is known that the code can be put in the form of

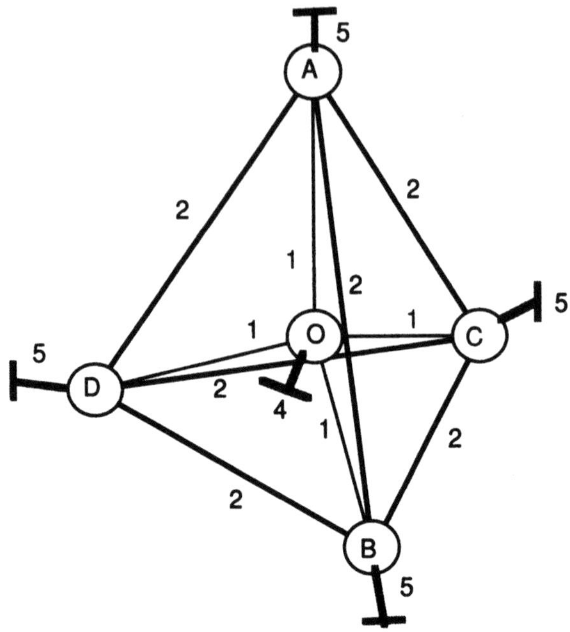

Fig. 3 [24,12,8] Golay Normal Realization with Tetrahedral Symmetry

a bordered double circulant with two 11×11 circulants [5](p.65):

$$\left[\begin{array}{c|ccc}
1 & 10000000000 & 11011100010 & 0 \\
1 & 01000000000 & 01101110001 & 0 \\
1 & 00100000000 & 10110111000 & 0 \\
1 & 00010000000 & 01011011100 & 0 \\
1 & 00001000000 & 10101101110 & 0 \\
1 & 00000100000 & 01010100111 & 0 \\
1 & 00000010000 & 10101010011 & 0 \\
1 & 00000001000 & 11010101001 & 0 \\
1 & 00000000100 & 11101010100 & 0 \\
1 & 00000000010 & 01110101010 & 0 \\
1 & 00000000001 & 00111010101 & 0 \\
0 & 00000000000 & 11111111111 & 1
\end{array}\right] \tag{1}$$

The bit on the far left will be labeled $*$, the overall parity-check bit on the far right will be labeled &. Suppressing the twelfth row and bits $*$ and & from consideration for the moment, we examine the 11×22 double circulant matrix, $\mathbf{G}_{11}$. The code spanned by $\mathbf{G}_{11}$ can be defined as pairs of code polynomials $\mathbf{c}(x) = [c_1(x), c_2(x)] = u(x)[1, g_2(x)]$, with $g_2(x) = 1 + x + x^3 + x^4 + x^5 + x^9$ the polynomial defined by the first row of the second circulant, and multiplication is in the ring of polynomials mod $x^{11} = 1$. Observe that $g_2(x)$ has the important property that $g_2(x^4) = 1 + x^4 + x^1 + x^5 + x^9 + x^3 = g_2(x)$. Consequently, for any codeword $\mathbf{c}(x)$, $\mathbf{c}(x^4) = u(x^4)[1, g_2(x^4)] = u(x^4)[1, g_2(x)]$ is also a codeword. The mapping $\mathbf{c}(x) = [c_1(x), c_2(x)] \to \mathbf{c}(x^4) = [c_1(x^4), c_2(x^4)]$ is therefore in the automorphism group of the length 22 code of $\mathbf{G}_{11}$. This automorphism simultaneously permutes the positions in both blocks of eleven bits. If the bit positions are labeled l in the left block and k in the right, $0 \leq l, k \leq 10$, then the automorphism maps $l \to 4l, k \to 4k$, mod 11. This automorphism has order five and can be used to decompose the two 11×11 circulants into four 5×5 circulants, plus two extra bits. Each circulant is turned into two 5×5 circulants with one fixed bit, the zero position. This mirrors on a smaller scale the original conversion of the cyclic length 23 code into two circulants.

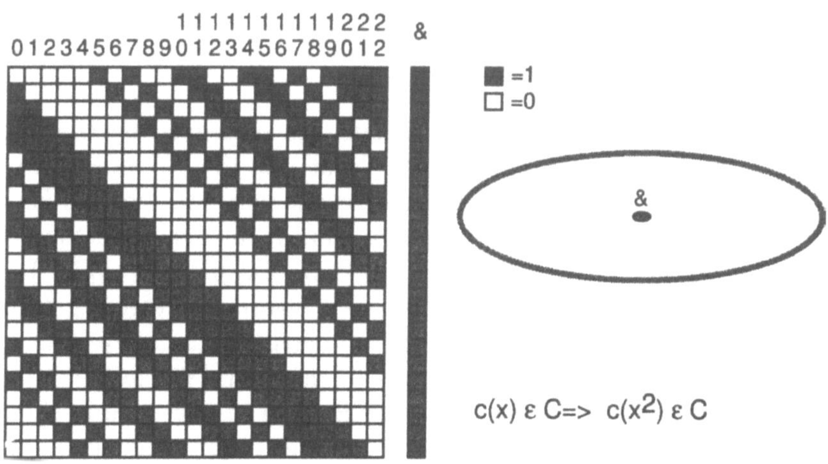

Fig. 4 23-Circulant Generator Matrix

With this as motivation, the code realization will be successively transformed from the standard cyclic code of length 23 with an overall parity bit into a quasi-cyclic code with four cycles of five bits with four extra bits. In Figure 4 the standard 23-cyclic form is shown using a circulant generator matrix based on an idempotent polynomial. In addition to the 23-cyclic permutation, the code is invariant under the permutation that takes position i to position $2i$. Using this permutation the Golay code of Figure 4 can be mapped to a double circulant as shown in Figure 5. The required reordering of the columns is indicated above the 11×11 circulants.

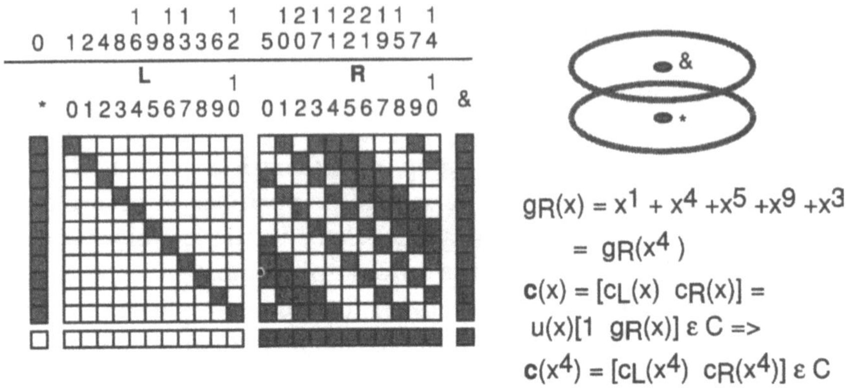

Fig. 5 Double 11-Circulant Generator Matrix

The columns of this double circulant are reindexed by the integers mod 11, with left and right halves. What was the "0" position in the 23-cycle form is now fixed under the 11-cyclic permutation, and is shown as "*". Now observe that the polynomial defined by the top row of the right circulant, $g_R(x)$ has the property that $g_R(x) = g_R(x^4)$. This implies that the double circulant is itself invariant under the mapping $i \to 4i$ operating simultaneously on both halves. This creates four cycles of length five, with the left and right "0" positions left fixed by this new permutation. The resulting quasi-cyclic form of the code is shown in Figure 6, with the reordering of the columns required indicated at the top of the columns. To complete the 5-circulant matrices, two redundant rows have been added.

Setting aside the four extra bits for the moment, the last ten rows of Figure 6 span a [20,8,8] quasi-cyclic code. In the normal realization, the four blocks of five bits will be located at the four vertices of the tetrahedron, the A bits at the vertex labeled A, the B bits at the B vertex, and so forth, as in Figure 3.

For the curious, we comment that the 2×4 block matrix of 5×5 circulants, $\mathbf{G}_{QC}$, can be analyzed as a quasi-cyclic code using the methods of [7]. Let $\mathbf{F}_5$ be the Fourier transform based on a primitive fifth root of unity $\beta, \beta^5 = 1$ on $GF(16)$. Specifically let $\mathbf{F}_5 = [f_{i,j}] = \beta^{ij}$ for $0 \le i, j \le 4$. $\mathbf{F}_5$ is invertible, and multiplying the circulant blocks from the left by $\mathbf{F}_5$ transforms the row space of $\mathbf{G}_{QC}$ into the span of a set of vectors over $GF(16)$. Viewed as a code over $GF(16)$, the code

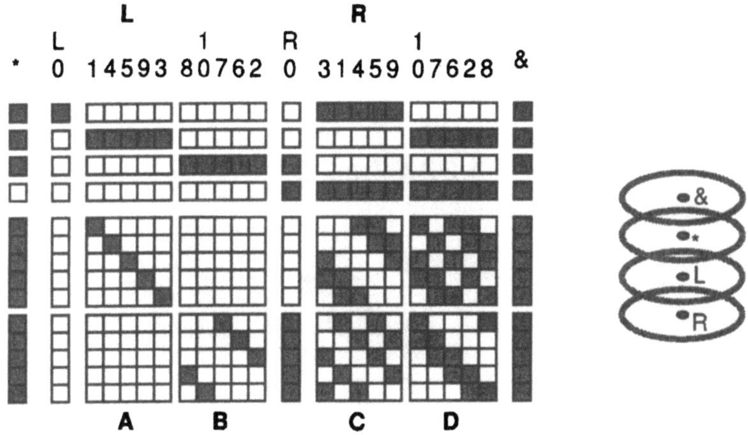

Fig. 6 Quadruple 5-Circulant Generator Matrix

space is spanned by the vectors

$$\begin{bmatrix} 1 & 0 & \omega^1 & \omega^2 \\ 0 & 1 & \omega^2 & \omega^1 \end{bmatrix} \otimes \begin{bmatrix} \beta^0 & \beta^1 & \beta^2 & \beta^3 & \beta^4 \end{bmatrix}, \tag{2}$$

and all the vectors obtained by componentwise squaring these vectors, where ω is a third root of unity on the subfield $GF(4)$. This representation was used to solve for the exact basis vectors needed for the normal realization.

3. TETRAHEDRALLY SYMMETRIC REALIZATION

Figure 7 displays the generator matrix for the code needed for the normal realization. It can be easily confirmed that the bottom eight rows of the matrix also span the row space of $\mathbf{G}_{QC}$. The basis vectors are purposefully chosen to map to a normal realization as given in Figure 3. What remains to define the normal realization is to give the definitions for the state variables of the edges of Figure 3 and how they act and are constrained in each of the constraint vertices A, B, C, D, and O, the central vertex.

The state variables emanating from vertex O are the simplest. The rows of Figure 7 labeled Oa, Ob, Oc and Od are multiplied by the state variable of the edges from O to each of the other four vertices. For example, if the binary state variable edge from O to A has value one, then $[1,1,1,1,1]$ is added to the five-bit codeword emitted by A and $[0,1,1,1]$ is added to the four-bit codeword emitted by O. Clearly four information bits are encoded in this way, each radiating state variable edge carrying a single bit.

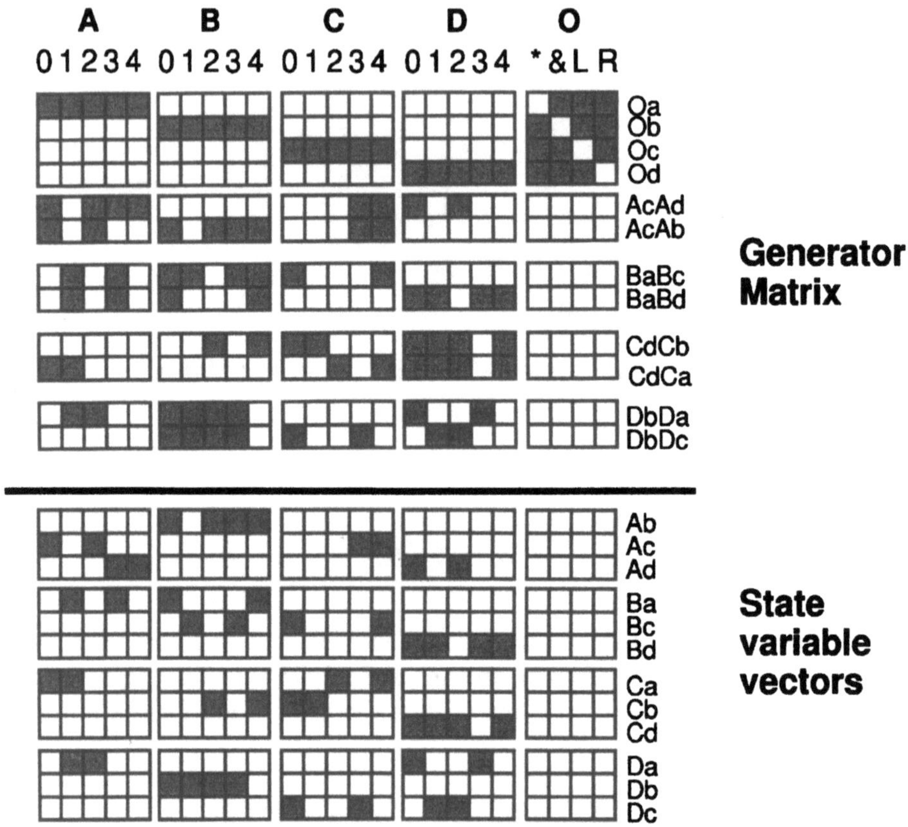

Fig. 7 Generator Matrix for Tetrahedrally Symmetric Form and Normal Realization State Variable Vectors

The other twelve state variables are slightly more subtle. Each of the four vertices dictate three state variables, one of the two from each of the three edges incident on the vertex. Let Ab and Ba be the two state variables of the edge $[A, B]$, with Ab dictated by A and Ba dictated by B; that is, state variable Ab is determined by constraints at A, whereas Ba is determined by constraints at B, and similarly for the other vertices. Each of constraint vertices A, B, C, D imposes the constraint that the parity of the three state variables it dictates must be even. For example, vertex A imposes the condition $Ab + Ac + Ad = 0 \bmod 2$. The rows below the line specify the channel vectors that are added to the channel word when each of the twelve tetrahedral edge state variables has value one. A row labeled Xy is nonzero in only the X vertex components and the y vertex components. This means that a value of one for Xy will cause the X five-bit vector component of row Xy to be added to the channel output by X and the Y component to be added to the Y output. For example, if Ac is one, then vertex A will add $[1,0,1,0,0]$ to its channel output and vertex C will add $[0,0,0,1,1]$ to its channel output. If Ab is one, then vertex A will add $[0,0,0,0,0]$ to its channel output and vertex B will add $[1,0,1,1,1]$ to its channel output. When both Ac and Ab are one, these are both added, which matches the row of the generator matrix labeled $AcAb$ above the line. With the requirement that $Ab+Ac+Ad=0 \mod 2$ imposed by constraint vertex A, the span of the three rows Ab, Ac, Ad collapses to the two-dimensional subspace spanned by the rows labeled $AcAd$ and $AcAb$ in the generator matrix.

The minor subtlety is the specific choice of the Figure 7 generator matrix to allow this localized definition of constraints. For example, the choice of the rows $AcAb$ and $AcAd$ so that they have the same C component vector allows the implicit creation of a row $AbAd$ that is zero in its C component and therefore can be implemented by a normal realization definition that has state variable Ac equal to zero. There is some latitude in the choice of the vectors in the A components of the normal realization variables Ab, Ac, Ad. Since they must satisfy $Ab + Ac + Ad = 0 \mod 2$, there are multiple ways that the A components could be assigned to achieve the same result. This latitude could have some influence on the functioning of a graph-message passing decoder, depending on the way the parity-check constraints are implemented.

For purposes of encoding, one can envision two information bits entering into each of the four vertices A, B, C, D, and a [3,2,2] parity-check code operative at each vertex determines the state variable values of the three state variables constrained at the vertex.

4. CONCLUSION

Forney's conjectured simple tetrahedral form of the normal realization for the [24,12,8] Golay code was shown by Vardy to be too restrictive. A normal realization that is a tetrahedron cannot exist. But by communicating one extra bit of state information to each of the four corners from a central vertex, a generator matrix can be created that endows the resulting normal realization with a tetrahedral symmetry.

References

[1] A. R. Calderbank, G. D. Forney, and A. Vardy, "Minimal Tail-Biting Trellises: The Golay Code and More," *IEEE Transactions on Information Theory*, vol. 45, no. 5, pp. 1435–1455, July 1999

[2] G. D. Forney, "Codes on Graphs: Normal Realizations," *IEEE Transactions on Information Theory*, vol. 47, no. 2, pp. 520-548, Feb. 2001.

[3] F. Kschischang and B. Frey, "Iterative Decoding of Compound Codes by Probability Propagation in Graphical Models," *IEEE Journal Selected Areas in Communications*, vol. 16, no. 2, pp. 219–230, Feb. 1998.

[4] F. Kschischang, B. Frey, and H-A. Loeliger, "Factor Graphs and the Sum-Product Algorithm," *IEEE Transactions on Information Theory*, vol. 47, no. 2, pp. 498-519, Feb. 2001.

[5] F. J. MacWilliams and N. J. A. Sloane, *The Theory of Error-Correcting Codes*, North-Holland, Second Printing, 1978.

[6] R. M. Tanner, "A Recursive Approach to Low Complexity Codes," *IEEE Transactions on Information Theory*, vol. 27, pp. 533–547, Sep. 1981.

[7] R. M. Tanner, "A Transform Theory for a Class of Group-Invariant Codes," *IEEE Transactions on Information Theory*, vol. 34, no. 4, pp. 752–775, July 1988.

[8] A. Vardy, private correspondence, Sept. 1998.

[9] N. Wiberg, *Codes and Decoding on General Graphs*, Linkoeping Studies in Science and Technology, no. 440, 1996.

[10] N. Wiberg, H.-A. Loeliger, and R. Koetter, "Codes and Iterative Decoding on General Graphs," *European Transactions on Telecommunications*, vol. 6, no. 5, pp. 513–526, Sep 1995.

Chapter 23

ON THE REPRESENTATION OF CODES IN FORNEY GRAPHS

R. Koetter
Coordinated Science Lab.
University of Illinois at Urbana/Champaign

Abstract We investigate the representation of codes in graphical models. In particular, we use the notion of a trellis formation on a Forney graph to visualize the structure of a code on a given graph. We focus on the question of whether a given trellis formation contains mergeable vertices and whether the description of a code in terms of local behaviors on the Forney graph can be made smaller. Necessary and sufficient conditions for mergeability are given leading to a polynomial time algorithm that decides if a given trellis formation contains mergeable vertices. One of our main tools is a duality theorem by Forney, for which we give a short proof in the context of binary codes.

1. INTRODUCTION

Codes on graphs are a field of coding theory that is emerging in the wake of the enormous success of iterative decoding algorithms. It is by now well understood that message passing algorithms, like the sum-product algorithm (belief propagation) or min-sum algorithm, have a natural description in terms of a graph that represents the code in consideration. The theme of codes on graphs is recurrent in the coding theory literature. Gallager's low density parity-check (LDPC) codes [1] were the earliest appearance of iteratively decodable codes and much of modern coding theory could have been derived in the early 1960s. Other notable contributions to the area of codes on graphs are the work of Tanner [2], Wiberg et al. [3, 4], and a number of papers exploring

*Work supported under NSF grant CAREER: CCR-9984515.

the interplay between iterative decoding and other graph-based algorithms [5, 6, 7, 8, 9].

Trellis representations of codes were introduced by Forney [10] in order to explain the workings of the Viterbi algorithm. However it soon became clear that the representation of codes in trellis-like structures is a discipline that provides much insight into the structure of the underlying code itself. Questions about minimal realizations and canonical trellis representations are by now well understood in the context of conventional trellises (we shall define this term shortly). An excellent overview of this topic is given by Vardy [11]. Connections between system theory and trellises for linear systems, or equivalently linear codes, are explored in [12].

The combination of the areas of codes on graphs and trellis representation of codes (or linear systems) is a natural unification between the two fields, undertaken for the first time by Wiberg [4]. The traditional problems concerning minimality and other system theoretic notions are considerably more difficult in this context. It is by now understood [13, 14] that state space realizations that are generalized to loopy time-axis (a time-ordering that is not linearly evolving but obeys more general relationships) usually do not exhibit a *minimal* realization. Indeed, it is nontrivial even to define the meaning of minimal in this context (cf. [13, 14]). In general, the different realizations of the same linear system form a partially ordered set.

Nevertheless, most construction techniques for conventional trellises, like the product construction [15] can be applied to the construction of state space realizations on loopy time axes. In particular, for the important case of tail-biting trellises, Koetter and Vardy could exhibit a polynomial time algorithm that characterizes the minimal (for practically all interesting definitions of minimality) linear state space realizations of a given linear system [16]. However, progress for more complex graph structures has remained sparse. In particular, not even the simplest questions about generalized state space realizations, like the mergeability of vertices, have a satisfactory answer.

In 1999, Forney [17] derived the notion of "normal graphs" and "normal factor graphs", which introduces an important normalization to the problem of codes on graphs by restricting the degree of the state space nodes. Acknowledging the importance of this normalization we will refer to these normalized graphs as Forney graphs. It is known [17] that any factor graph representing a linear system has an equivalent Forney graph. Moreover, Forney derives a theorem that relates the state space size of a Forney graph describing a linear system and the state space size of its dual.

In this paper we review a slightly specialized form of Forney's basic results [17] and address the simplest form of minimality of a given system. In particular, we give a polynomial time algorithm to decide if a trellis formation (defined in Section 2.1) on a given Forney graph contains a pair of mergeable vertices. This surprisingly difficult question in the context of general state space realizations is crucial for any further investigation of minimality of Forney graphs.

In Section 2 of the paper we review the basic notions of factor graphs and Forney graphs. The connection between trellis representations of codes and Forney graphs is described in Section 2.1. Section 3 gives a simplified form of Forney's duality theorem. Section 4 gives the main results concerning the mergeability of vertices in trellis formation. The algorithm for finding a trellis formation that does not contain mergeable vertices is given in Section 5.

2. FORNEY GRAPHS

A *factor graph* $\mathcal{G} = (V, E)$ with vertex set V and edge set E is a bipartite graph consisting of a set of state/symbol nodes V_S, a set of function nodes V_f , (i.e., $V = V_S \cup V_f$,) and a set of edges $E \subseteq V_S \times V_f$. We consider edges as sets of pairs of vertices and hence $\{v, u\} \in E$ implies $\{u, v\} \in E$. We will use the terms "node" and "vertex" interchangeably.

To each vertex $u \in V_S$ we associate a set of symbols $\mathscr{A}_u$ called an alphabet. In the context of this paper we will assume that all alphabets are finite.

The *configuration space* for the factor graph $\mathcal{G}$ is defined as the Cartesian product $\mathscr{A} = \bigotimes_{u \in V_S} \mathscr{A}_u$. The restriction or projection of the configuration space to a subset $V' \subset V_S$ of vertices is denoted by $\mathscr{A}_{|V'}$. Similarly, the restriction of any element $\underline{a} \in \mathscr{A}$ to V' is denoted $\underline{a}_{|V'}$.

For any $v \in V$ let $\Gamma(v)$ denote the neighborhood of v, i.e.,

$$\Gamma(v) = \Big\{v' : \{v, v'\} \in E\Big\}.$$

We assume a fixed ordering of the elements of $\Gamma(v)$ for all $v \in V_f$, and associate to each $v \in V_f$ a *local behavior* $\mathscr{C}_v \subseteq \bigotimes_{u \in \Gamma(v)} \mathscr{A}_u$.

For a given set of local behaviors, a *global behavior* $\mathscr{C}$ is obtained as a subset of $\mathscr{A}$ by requiring that the restrictions of any element of $\mathscr{C}$ to $\Gamma(v)$ is an element in the local behavior $\mathscr{C}_v$ for all $v \in V_f$, i.e.,

$$\mathscr{C} = \{\underline{a} \in \mathscr{A} : \underline{a}_{|\Gamma(v)} \in \mathscr{C}_v, \forall v \in V_f\}.$$

Let $I_{\mathscr{C}}$ and $I_{\mathscr{C}_v}$ be set indicator functions for sets $\mathscr{C}$ and $\mathscr{C}_v$, respectively, that is $I_{\mathscr{C}}(\underline{a})$ evaluates to one if $\underline{a} \in \mathscr{C}$ and evaluates to zero otherwise. The notion of a factor graph reflects the factorization of the function $I_{\mathscr{C}}$ as $I_{\mathscr{C}}(\underline{a}) = \prod_{v \in V_f} I_{\mathscr{C}_v}(\underline{a}_{|\Gamma(v)})$. We note that this factorization is not restricted to indicator functions [8] in general factor graphs.

Up to this point all the definitions and notions are sufficiently general in order to describe a variety of general factor graphs. A specialization was introduced in [13, 14] where the degree of vertices in V_f is restricted to equal two. Forney [17] introduced a different specialization. Similar to Wiberg's graphical model [4], Forney distinguishes nodes in V_S as symbol nodes and state nodes, i.e., $V_S = V_O \cup V_H$. Following Forney's normalization we define a Forney graph as follows:

Definition 1 *A Forney graph is a factor graph with vertex set* $V = V_S \cup V_f$, $V_S = V_O \cup V_H$ *such that all vertices in* V_H *have degree exactly two and all vertices in* V_O *have degree exactly one.*

The underlying understanding is that symbol nodes V_O are *observable* while state nodes V_H are *hidden* and unobservable. We are particularly interested in the restriction of $\mathscr{C}$ to the vertices V_O.

Definition 2 *Let* $\mathcal{G} = (V_f \cup \{V_O \cup V_H\}, E)$ *be a Forney graph with global behavior* $\mathscr{C}$. *We say that the Forney graph* $\mathcal{G}$ *represents a code* C *if* $C = \mathscr{C}_{|V_O}$.

Definition 1 is suitable for the purpose of this paper. In [17], Forney introduced the notion of normal graph, which is equivalent to the notion of Forney graphs. In normal graphs the vertices $u \in V_H$ are absorbed in edges between vertices in V_f and symbol nodes are represented as "half-edges" that are incident with vertices in V_f. However, here it is more convenient to explicitly keep the vertices in V_H because we will be concerned with the minimization of the alphabets $\mathscr{A}_u$, $v \in V_H$. Also in order to achieve a streamlined notation we will assume that exactly one symbol node $w \in V_O$ is adjacent to every node $v \in V_f$. We note that the associated alphabet $\mathscr{A}_w$ may be the empty set.

It is known [17] that there exist straightforward mappings between the mentioned different graphical frameworks.

Let $\mathbb{F}_2$ be the binary field. We will restrict the remainder of this paper to the case that for any $u \in V_S$ the alphabet $\mathscr{A}_u$ is defined as a subset of the binary vectors of a given length $m(u)$, i.e., $\mathscr{A}_u = \mathbb{F}_2^{m(u)}$. This will allow us to succinctly derive our results and only relatively minor modifications to the arguments are necessary to extend our findings to more general scenarios where the configuration space $\mathscr{A}$ is equipped with only a group structure. The more involved technical concepts that are necessary to treat the group case are given by Forney in his landmark paper [17].

In addition to the specialization to the binary case we will restrict our attention to the situation where all local behaviors are vector spaces over the binary field $\mathbb{F}_2$ in which case we say that a Forney graph is *linear* over $\mathbb{F}_2$. If all local behaviors are vector spaces then the global behavior $\mathscr{C}$ is also clearly a $\mathbb{F}_2$-linear space. Moreover, the code C represented by $\mathcal{G}$ is linear. Unless stated otherwise, we will always assume that the Forney graphs in this paper are linear. The total length of the code $\mathscr{C}$ is given as $n = \sum_{u \in V_S} m(u)$.

Let $v \in V_f$ be adjacent to δ vertices $\{u_1, u_2, \ldots, u_\delta\}$ in V_H and to one vertex $w \in V_O$. Let the local behavior $\mathscr{C}_v$ be a k_v dimensional code of length $n_v = \sum_{u \in \Gamma(v)} m(u)$ and let H_v be a parity-check matrix for $\mathscr{C}_v$, i.e., H_v is a $(n_v - k_v) \times n_v$ matrix. We assume a fixed ordering of positions in H_v such that H_v can be written as $H_v = (H_{v,u_1} | H_{v,u_2} | \ldots, | H_{v,u_\delta} | H_{v,w})$, where $H_{v,u}$ is a $(n_v - k_v) \times m(u)$ matrix associated to the alphabet $\mathscr{A}_u = \mathbb{F}_2^{m(u)}$. Given a collection of local behaviors $\mathfrak{C} = \{\mathscr{C}_v : v \in V_f\}$, we define a $\sum_{v \in V_f} (n_v - k_v) \times \sum_{u \in V_S} m(u)$ matrix $H(\mathcal{G}, \mathfrak{C})$ as a block matrix with rows indexed by vertices $v \in V_f$ and columns indexed by vertices $u \in V_S$ such that

$$H(\mathcal{G}, \mathfrak{C})_{v,u} = \begin{cases} H_{v,u} & \{v,u\} \in E \\ \mathbf{0} & \text{otherwise.} \end{cases}$$

holds.

Lemma 1 *Let a Forney graph $\mathcal{G} = (V_f \cup V_S, E)$ and a collection of linear, local behaviors $\mathfrak{C} = \{\mathscr{C}_v : v \in V_f\}$ be given. The matrix $H(\mathcal{G}, \mathfrak{C})$ is a parity-check matrix for the global behavior $\mathscr{C}$.*

Proof. The proof follows immediately by checking that any word in $\mathscr{C}$ is characterized by satisfying all local behaviors. Exactly this condition is checked by $H(\mathcal{G}, \mathfrak{C})$.

We illustrate the introduced concepts with an example.

Example 1 *Let a Forney graph $\mathcal{G} = (\{V_O \cup V_H\} \cup V_f, E)$ be given as in Figure 1. The alphabets associated with vertices in $V_H = \{u_1, u_2, u_3, u_4, u_5, u_6\}$ are chosen as $\mathscr{A}_{u_1} = \mathscr{A}_{u_2} = \mathscr{A}_{u_3} = \mathscr{A}_{u_4} = \mathscr{A}_{u_6} = \mathbb{F}_2$ and $\mathscr{A}_{u_5} = \mathbb{F}_2^2$. The alphabets associated with vertices in $V_O = \{w_1, w_2, w_3, w_4\}$ are all chosen as $\mathbb{F}_2^2$.*

We choose the local behaviors $\mathfrak{C} = \{\mathscr{C}_{v_j} : j = 1, 2, \ldots, 4\}$ as $\mathbb{F}_2$-linear codes with generator matrices (the top row in each generator matrix

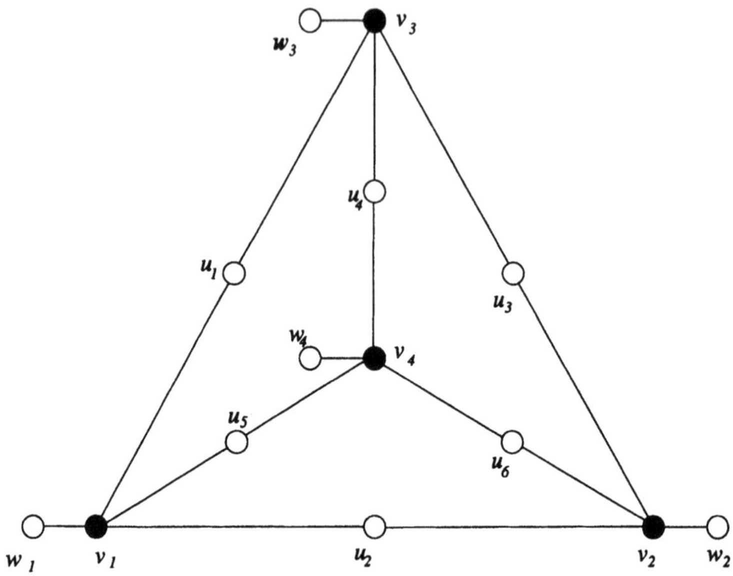

Figure 1 A Forney graph

is included to indicate the ordering of symbols and is not part of the matrix):

$$G_{v_1} = \begin{pmatrix} u_1 & w_1 & u_5 & u_2 \\ \hline 1 & 11 & 00 & 0 \\ 0 & 11 & 10 & 0 \\ 0 & 01 & 01 & 1 \end{pmatrix} \qquad G_{v_4} = \begin{pmatrix} u_4 & w_4 & u_5 & u_6 \\ \hline 1 & 01 & 01 & 0 \\ 0 & 11 & 10 & 0 \\ 0 & 11 & 00 & 1 \end{pmatrix}$$

$$G_{v_2} = \begin{pmatrix} u_2 & w_2 & u_6 & u_3 \\ \hline 0 & 11 & 1 & 0 \\ 1 & 01 & 0 & 1 \end{pmatrix} \qquad G_{v_3} = \begin{pmatrix} u_3 & w_3 & u_4 & u_1 \\ \hline 1 & 01 & 1 & 0 \\ 0 & 11 & 0 & 1 \end{pmatrix}$$

The parity-check matrices corresponding to matrices G_{v_i} are then given by

$$H_{v_1} = \begin{pmatrix} u_1 & w_1 & u_5 & u_2 \\ \hline 0 & 11 & 00 & 1 \\ 0 & 00 & 01 & 1 \\ 1 & 01 & 11 & 0 \end{pmatrix} \qquad H_{v_4} = \begin{pmatrix} u_4 & w_4 & u_5 & u_6 \\ \hline 1 & 00 & 01 & 0 \\ 0 & 11 & 01 & 0 \\ 0 & 01 & 11 & 1 \end{pmatrix}$$

$$H_{v_2} = \left(\begin{array}{cccc} u_2 & w_2 & u_6 & u_3 \\ \hline 0 & 01 & 1 & 1 \\ 1 & 11 & 0 & 0 \\ 1 & 00 & 0 & 1 \end{array}\right) \qquad H_{v_3} = \left(\begin{array}{cccc} u_3 & w_3 & u_4 & u_1 \\ \hline 1 & 11 & 0 & 0 \\ 0 & 01 & 1 & 1 \\ 1 & 00 & 0 & 1 \end{array}\right)$$

The overall parity-check matrix $H(\mathcal{G}, \mathfrak{C})$ for the binary linear code $\mathscr{C}$ is given by (blank entries in $H(\mathcal{G}, \mathfrak{C})$ represent zero entries):

$$H(\mathcal{G}, \mathfrak{C}) = \left(\begin{array}{c|cccccccccc} & u_1 & u_2 & u_3 & u_4 & u_5 & u_6 & w_1 & w_2 & w_3 & w_4 \\ \hline v_1 & 0 & 1 & & & 00 & & 11 & & & \\ v_1 & 0 & 1 & & & 01 & & 00 & & & \\ v_1 & 1 & 0 & & & 11 & & 01 & & & \\ v_2 & & 0 & 1 & & & 1 & & 01 & & \\ v_2 & & 1 & 0 & & & 0 & & 11 & & \\ v_2 & & 1 & 1 & & & 0 & & 00 & & \\ v_3 & 0 & & 1 & 1 & & & & & 11 & \\ v_3 & 1 & & 0 & 1 & & & & & 01 & \\ v_3 & 0 & & 1 & 1 & & & & & 00 & \\ v_4 & & & & 1 & 01 & 0 & & & & 00 \\ v_4 & & & & 0 & 01 & 0 & & & & 11 \\ v_4 & & & & 0 & 11 & 1 & & & & 01 \end{array}\right)$$

A generator matrix for $\mathscr{C}$ is found from $H(\mathcal{G}, \mathfrak{C})$ as

$$G = \left(\begin{array}{cccccccccc} u_1 & u_2 & u_3 & u_4 & u_5 & u_6 & w_1 & w_2 & w_3 & w_4 \\ \hline 1 & 1 & 1 & 1 & 01 & 1 & 10 & 11 & 00 & 10 \\ 1 & 1 & 1 & 1 & 11 & 1 & 01 & 11 & 00 & 01 \\ 1 & 0 & 0 & 0 & 00 & 0 & 11 & 01 & 01 & 00 \\ 0 & 0 & 0 & 0 & 10 & 1 & 11 & 10 & 10 & 00 \end{array}\right)$$

From the generator matrix G it is straightforward to obtain a generator matrix for the restriction of $\mathscr{C}$ to V_O. A generator matrix for $\mathscr{C}_{|V_O}$ is given as

$$G_{|V_f} = \left(\begin{array}{cccc} w_1 & w_2 & w_3 & w_4 \\ \hline 10 & 11 & 00 & 10 \\ 01 & 11 & 00 & 01 \\ 11 & 01 & 01 & 00 \\ 11 & 10 & 10 & 00 \end{array}\right)$$

It is easily verified that $G_{|V_f}$ is a generator matrix for a binary [8, 4, 4] *extended Hamming code, i.e., the Forney graph $\mathcal{G}$ represents the* [8, 4, 4] *extended Hamming code.* □

In addition to the projection of the linear space $\mathscr{C}$ to a subset V' of V_S we are interested in the subspaces of $\mathscr{C}$ that are obtained by requiring that the linear conditions $\underline{c}_{|v} = 0$ are satisfied for all $v \in V'$. We define $\mathscr{C}_{V'}$ as

$$\mathscr{C}_{V'} = \{\underline{c}_{|V'} | \underline{c} \in \mathscr{C} \textbf{ and } \underline{c}_{|\{V_S \setminus V'\}} = \mathbf{0}\}$$

Lemma 1 and Example 1 provide a strategy to obtain a generator matrix of the code $C_{|V_O}$ from a given Forney graph and a description of the local behaviors. The approach is comprised of the three steps of first finding a parity-check matrix $H(\mathcal{G}, \mathfrak{C})$, second, finding a generator matrix G from $H(\mathcal{G}, \mathfrak{C})$ and, finally, restricting the generator matrix G to positions that correspond to nodes in V_O. In the remainder of this section we give a second approach to the problem of constructing a generator matrix from a given list of local behaviors. To this end we construct a matrix $G(\mathcal{G}, \mathfrak{C})$ similarly to the construction of $H(\mathcal{G}, \mathfrak{C})$. Let G_v be a *generator matrix* of the local behavior $\mathscr{C}_v$. We can write G_v as $G_v = (G_{v,u_1}|G_{v,u_2}|\dots,|G_{v,u_\delta}|G_{v,w})$ where $G_{v,u}$ is a $k_v \times m(u)$ matrix associated to the alphabet $\mathscr{A}_u = \mathbb{F}_2^{m(u)}$. We define the block matrix $G(\mathcal{G}, \mathfrak{C})$ as

$$G(\mathcal{G}, \mathfrak{C})_{v,u} = \begin{cases} G_{v,u} & \{v,u\} \in E \\ \mathbf{0} & \text{otherwise.} \end{cases}$$

We emphasize that $G(\mathcal{G}, \mathfrak{C})$ is not a generator matrix of the code $\mathscr{C}$. However, we have the following theorem:

Theorem 2 *Let a Forney graph $\mathcal{G} = (\{V_O \cup V_H\} \cup V_f, E)$ and a collection of local linear behaviors $\mathfrak{C} = \{\mathscr{C}_v : v \in V\}$ be given such that $\mathcal{G}$ represents the linear code C. The matrix $G(\mathcal{G}, \mathfrak{C})$ generates a code $\bar{\mathscr{C}}$ such that $\bar{\mathscr{C}}_{V_O}$ equals $C = \mathscr{C}_{|V_O}$.*

Proof. *Let $\mathfrak{X}$ be a collection of $|V_f|$ codewords such that we have exactly one codeword from every local code in $\mathfrak{C}$. Let $v \in V_f$ and $v' \in V_f$ be two nodes such that a vertex $u \in V_H$ is a neighbor of both v and v' and let $\underline{x} \in C_v$ and $\underline{x}' \in C_v'$ be the two codewords in $\mathfrak{X}$ that correspond to v and v'. Denote by $\underline{x}_u$ and $\underline{x}'_u$ the part of $\underline{x}$ and $\underline{x}'$ that is indexed by the vertex u. We call $\underline{x}$ and $\underline{x}'$ mutually consistent if $\underline{x}_u$ equals $\underline{x}'_u$. The collection $\mathfrak{X}$ is called globally consistent if all codewords corresponding to two v and v', that share a mutually adjacent vertex u, are mutually consistent. A globally consistent collection $\mathfrak{X}$ clearly corresponds to a codeword in $\mathscr{C}$ and, vice versa, we can find a globally consistent collection $\mathfrak{X}$ for any codeword in $\mathscr{C}_{|V_O}$. In particular, given a globally consistent $\mathfrak{X}$, we can find the values of x_w for all $w \in V_O$ to find a codeword in $\mathscr{C}_{|V_f}$. Let $\underline{i}_v$ be the information vector such that $\underline{x} = \underline{i}_v G_v$. Moreover, let $\underline{i}$ be the information vector given by the concatenation of all $\underline{i}_v$, $v \in V_f$.*

Interpreting $\underline{i}$ as an information vector for $G(\mathcal{G},\mathfrak{C})$ yields a word $\underline{\bar{c}} = \underline{i}G(\mathcal{G},\mathfrak{C})$. If and only if the collection $\mathfrak{X}$ is globally consistent we find that $\underline{\bar{c}}_{|V_H}$ equals zero and also $\underline{\bar{c}}_{|V_O} \in \mathscr{C}_{|V_O}$. Hence every word in $\bar{\mathscr{C}}_{V_O}$ gives rise to a codeword in $\mathscr{C}_{|V_O}$ and every word in $\mathscr{C}_{|V_O}$ corresponds to a word in $\bar{\mathscr{C}}_{V_O}$.

Theorem 2 is the cornerstone of Forney's duality theorem that we will present in Section 3. In the following section we introduce a generalization of the trellis representation of a code.

2.1 FORNEY GRAPHS AND TRELLIS REPRESENTATIONS

Figure 1 gives the Forney graph for Example 1. However, the graph is not explicit in characterizing the code in consideration, i.e., a Forney graph describes a code only jointly with a description of the local behaviors. The notion of a trellis formation [13] combines the behaviors and the underlying graphical structure into one graphical model at the expense of an enlarged graph. Given a Forney graph $\mathcal{G} = (V_S \cup V_f, E)$ we construct a trellis formation on $\mathcal{G}$ as follows: To each vertex $u \in V_S$ we associate a vertex set $\mathscr{V}_u$ such that $|\mathscr{V}_u| = |\mathscr{A}_u|$ and each vertex in $\mathscr{V}_u$ represents one element of $\mathscr{A}_u$. Similarly, to each vertex $v \in V_f$ we associate a vertex set $\mathscr{V}_v$ such that $|\mathscr{V}_v| = |\mathscr{C}_v|$ and we associate to each codeword in the local code $\mathscr{C}_v$ exactly one vertex in $\mathscr{V}_v$.

Definition 3 *A trellis formation $\mathscr{G} = (\mathscr{V},\mathscr{E},\mathcal{G})$ on a Forney graph $\mathcal{G} = (\{V_O \cup V_H\} \cup V_f, E)$ is a $|V_H|+|V_f|$-partite vertex-labeled graph with vertex classes $\mathscr{V}_u$, $u \in V_H$ and $\mathscr{V}_v$ $v \in V_f$, i.e., $\mathscr{V} = \bigcup_{u \in V_H} \mathscr{V}_u \cup \bigcup_{v \in V_f} \mathscr{V}_v$. The edge set $\mathscr{E}$ contains an edge $\{s,x\}$ between a vertex $s \in \mathscr{V}_u$ and $x \in \mathscr{V}_v$ if and only if $u \in \Gamma(v)$ and the codeword x associated to $x \in \mathscr{V}_v$ satisfies $\underline{x}_{|u} = s$. The labels of vertices $s \in \mathscr{V}_u$ equal the corresponding element of $\mathscr{A}_u$. The label of $x \in \mathscr{V}_v$ equals $\underline{x}_{|w}$ where w is the symbol vertex adjacent to v.*

The notion of a trellis formation on a Forney graph is illustrated in Figure 2 for the special case that the underlying Forney graph is a simple chain. In this case the trellis formation is equivalent to the well-known trellis representation of a block code [11]. The labeled edges of a conventional trellis representation are replaced by labeled vertices in $\mathscr{V}_v$ $v \in V_f$. The graphical representation in a trellis formation allows an identification of codewords in $\mathscr{C}$ with subgraphs of the trellis formation $\mathscr{G}$. Let $\mathcal{R} \subseteq \mathscr{V}$ be a set of vertices such that $|\mathcal{R} \cap \mathscr{V}_v| = 1$ for all v in $V_f \cup V_H$. We call $\mathcal{R}$ a *reduction* of $\mathscr{G}$ to $\mathcal{G}$ if the subgraph of $\mathscr{G}$ induced by the vertex set $\mathcal{R}$ is isomorphic to the subgraph of the original Forney graph $\mathcal{G}$ that is induced by vertices $V_H \cup V_f$. The isomorphism is given by the identification of a vertex in class $\mathscr{V}_v$ with the vertex v in $\mathcal{G}$. It is clear that the set of reductions in a trellis formation is in a bijective correspondence with the codewords of $\mathscr{C}$. The codewords in $\mathscr{C}$ can be obtained from the reduction as the sequence of vertex labels. The code $\mathscr{C}_{|O}$ is obtained by reading the sequence of vertex labels of vertices in $\mathscr{V}_v$ $v \in V_f$. In general the set of reductions is not in one-to-one correspondence with the represented code $C = \mathscr{C}_{|V_O}$. We call a trellis formation *one-to-one* if there exists a bijective map from the set $\mathscr{R}$ of reductions of $\mathscr{G}$ to the code $C = \mathscr{C}_{|V_O}$. Also, we will encounter trellis formations that contain vertices that are not contained in any reduction. We will call a trellis formation $\mathscr{G}$ *reduced* if any vertex in $\mathscr{G}$ is contained in at least one reduction of $\mathscr{G}$ to $\mathcal{G}$.

Example 2 (Example 1 continued) *We consider the Forney graph $\mathcal{G}$ of Example 1 together with the local behaviors given there. A trellis formation $\mathscr{G}$ on $\mathcal{G}$ is depicted in Figure 3. Figure 4 gives a reduction of $\mathscr{G}$ to $\mathcal{G}$. The reduction corresponds to the all-one codeword. It is straightforward (though tedious) to verify that the trellis formation is one-to-one.*

2.2 STATE SPACE SIZES IN TRELLIS FORMATIONS

Given a trellis-formation $\mathscr{G}$ on a Forney graph $\mathcal{G}$, we can associate a state space size vector $\Theta(\mathscr{G}) = (|\mathscr{V}_{u_1}|, |\mathscr{V}_{u_2}|, \ldots, |\mathscr{V}_{u_{|V_H|}}|)$ with $\mathscr{G}$. Given two trellis formations $\mathscr{G}$ and $\mathscr{G}'$ that represent the same code we say that $\mathscr{G}$ is greater than $\mathscr{G}'$ ($\Theta(\mathscr{G}) \geq \Theta(\mathscr{G}')$), if $\Theta(\mathscr{G})$ is component-wise greater than or equal to $\Theta(\mathscr{G}')$. Moreover, we say that $\Theta(\mathscr{G})$ is greater than $\Theta(\mathscr{G}')$ ($\Theta(\mathscr{G}) > \Theta(\mathscr{G}')$) if $\Theta(\mathscr{G}) \geq \Theta(\mathscr{G}')$ holds *and* strict inequality holds in at least one position. For a variety of reasons [11], it is desirable to find a trellis formation that gives rise to a small state space size vector. Indeed the size of the state spaces has a natural interpretation relating to the amount of information that has to be communicated from one node in V_f to another node in V_f in order to use, e.g., iterative decoding algorithms. The size of the vertex classes $\mathscr{V}_v$ for $v \in V_f$ is neglected in this set-up (cf. [14, 13]).

It is remarkable that for linear codes and a fixed ordering of positions we have a unique (up to graph isomorphism) *minimal* trellis representing the code if the underlying Forney graph is a tree.

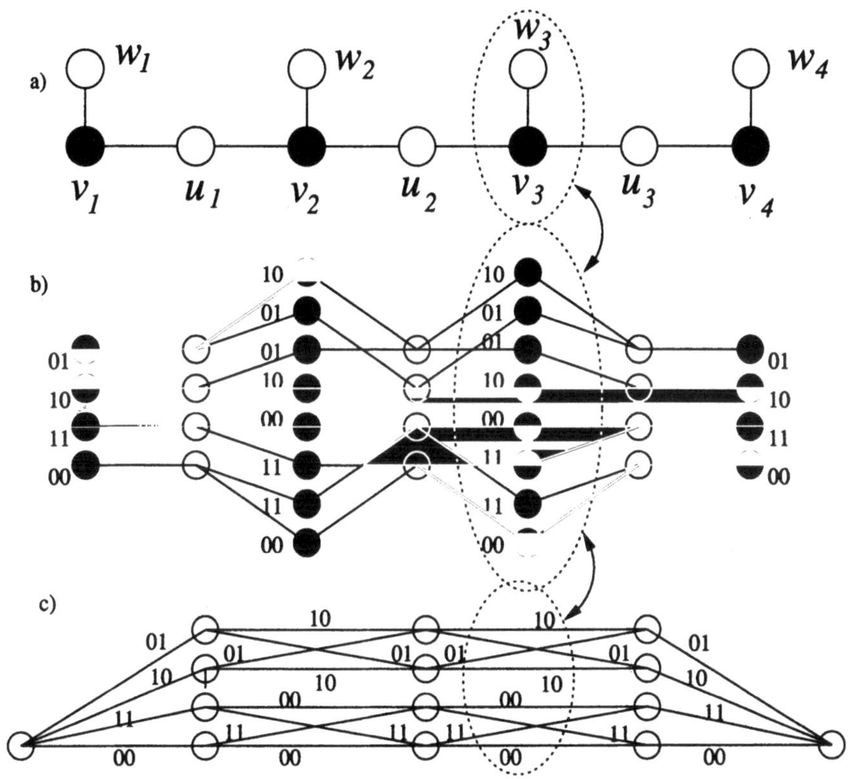

Figure 2 Graphical models for the [8,4,4] extended binary Hamming code: 2a) A Forney graph that establishes the adjacency relationships between the code symbols; 2b) A trellis formation on the Forney graph of figure 2a; 2c) A well known conventional trellis for the [8,4,4] extended binary Hamming code. The eight marked labeled branches of the conventional trellis correspond to the eight labeled vertices of the trellis formation, which in turn are a visualization of the code words in $\mathcal{C}_{v_3}$.

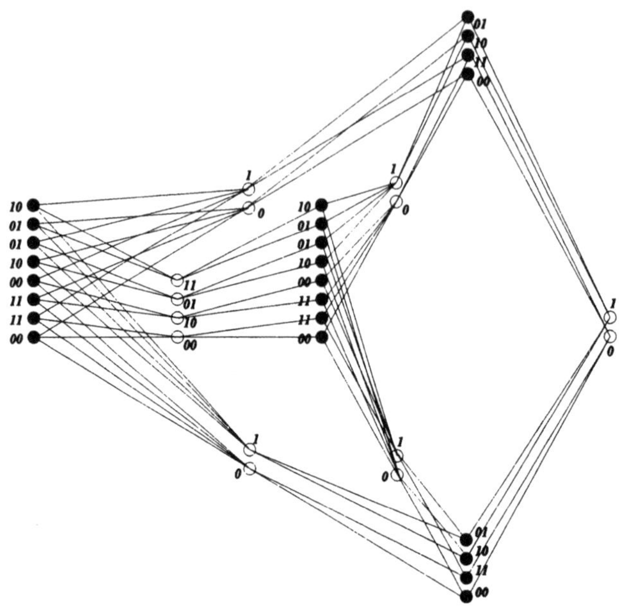

Figure 3 A trellis formation for the [8,4,4] extended binary Hamming codes represented by the Forney graph of Example 1.

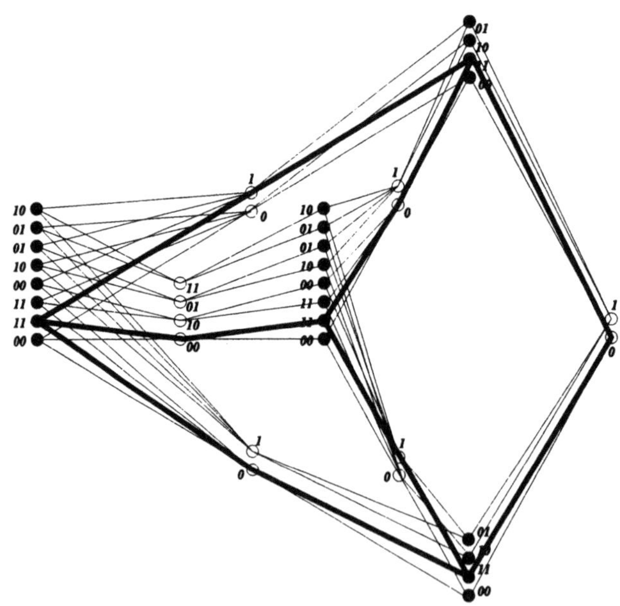

Figure 4 A reduction in the trellis formation. The highlighted reduction is isomorphic to the subgraph of Forney graph of Figure 1 that is induced by vertices in $V_H \cup V_f$ and represents the all one word.

For trellis formations on general Forney graphs, all that is known is that there usually does *not* exist a unique minimal trellis formation (cf. [14, 13]). In fact, not even the much simpler issue of vertex mergeability (a term we will define shortly) of a trellis formation is solved. Deciding the mergeability of a conventional trellis for a linear code is relatively easy and related to the notion of *properness* of the trellis [15]. However, no algorithm has been known that efficiently decides this issue for general trellis formations.

Definition 4 *Let $\mathscr{G} = (\mathscr{V}, \mathscr{E}, \mathcal{G})$ be a trellis formation on a Forney graph $\mathcal{G} = (V_f \cup \{V_O \cup V_H\}, E)$. We say that two vertices $v_1, v_2 \in \mathscr{V}_u$, $u \in V_H$ are mergeable if merging v_1 and v_2 into a new node v', that inherits all edges incident to either v_1 or v_2, does not change the code that is represented by $\mathscr{G}$.*

It is the primary goal of the present paper to derive an algorithm that decides if a given vertex class in a trellis formation contains mergeable vertices. The complexity of the problem becomes apparent by trying to answer the question for the trellis formation depicted in Figure 3. The main tool in deriving an efficient algorithm towards this end is Forney's duality theorem, which we will present in the following section.

3. CANONICAL DUAL GRAPHS

Given a Forney graph $\mathcal{G}$ and a collection of local behaviors $\mathfrak{C} = \{\mathscr{C}_v : v \in V_f\}$ it is a natural question to consider the collection of dual codes $\mathfrak{C}^\perp = \{\mathscr{C}_v^\perp : v \in V_f\}$ and to ask which code is represented by this new set of local behaviors on the graph $\mathcal{G}$. Clearly, this new code is intimately related to the originally represented code and one might expect that the new set of local behaviors represents the code that is dual to the original code. That this is indeed the case is one of the most pleasing results in the theory of codes on graphs [17]. We formulate the important relations in the form of a commutative diagram (Figure 5). The relations are proven in Theorem 3.

Theorem 3 (Forney) *Let a Forney graph $\mathcal{G}$ be given together with a collection of local behaviors $\mathfrak{C} = \{\mathscr{C}_v : v \in V_f\}$ representing the code C. The Forney graph $\mathcal{G}$ together with the collection of local behaviors $\mathfrak{C}^\perp = \{\mathscr{C}_v^\perp : v \in V_f\}$ represents the dual code $C^\perp$.*

Proof. *The theorem follows if we can prove all the relations of the commutative diagram in Figure 5 and if we can prove that $C = \mathscr{C}_{|V_O}$ and that $C^\perp = (\bar{\mathscr{C}}^\perp)_{|V_O}$. We redraw the diagram in Figure 5 in an abbreviated form and assign identifiers to the corresponding relations.*

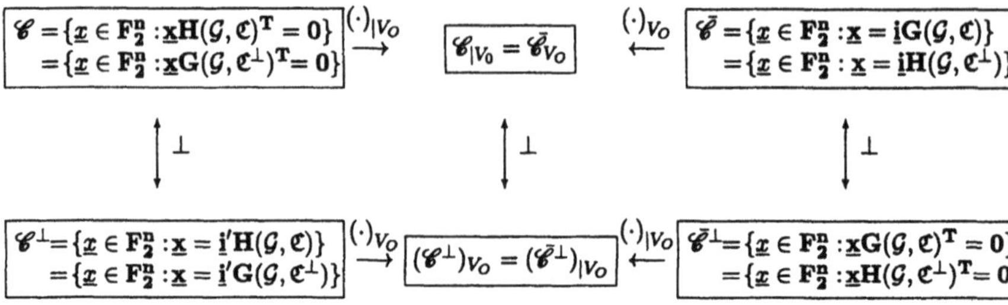

Figure 5

$$
\begin{array}{ccccc}
\mathscr{C} & \overset{(\cdot)_{|V_O}}{\underset{\textbf{(a)}}{\longrightarrow}} & \mathscr{C}_{|V_O} = \bar{\mathscr{C}}_{V_O} & \overset{(\cdot)_{V_O}}{\underset{\textbf{(b)}}{\longleftarrow}} & \bar{\mathscr{C}} \\
\textbf{(c)}\updownarrow\perp & & \textbf{(d)}\updownarrow\perp & & \textbf{(e)}\updownarrow\perp \\
\mathscr{C}^{\perp} & \overset{(\cdot)_{V_O}}{\underset{\textbf{(f)}}{\longrightarrow}} & (\mathscr{C}^{\perp})_{V_O} = (\bar{\mathscr{C}}^{\perp})_{|V_O} & \overset{(\cdot)_{|V_O}}{\underset{\textbf{(g)}}{\longleftarrow}} & \bar{\mathscr{C}}^{\perp}
\end{array}
\qquad (1)
$$

Relations **(a)** *and* **(g)** *follow by definition. Relations* **(b)** *and* **(f)** *are a consequence of Theorem 2. The duality relations* **(c)** *and* **(e)** *are a consequence of the definition of duality. Finally the well known duality of the operations puncturing* $(\cdot)_{|V_O}$ *and shortening* $(\cdot)_{V_O}$ *implies the relation* **(d)**. *Hence, we have* $C = \mathscr{C}_{|V_O}$ *and* $C^{\perp} = (\bar{\mathscr{C}}^{\perp})_{|V_O}$, *which proves the theorem.*

Given a Forney graph $\mathcal{G}$ and a set of local behaviors $\mathfrak{C}$ that represents a code C, there are many different ways to find a set of local behaviors that represents the dual code $C^{\perp}$. Theorem 3 suggests that choosing the collection of local behaviors as $\mathfrak{C}^{\perp}$ is a natural choice. We translate this property into a property of a trellis formation on a Forney graph:

Definition 5 *Let* $\mathscr{G}$ *be a trellis formation on a Forney graph* $\mathcal{G}$ *derived from a collection of local behaviors* $\mathfrak{C}$. *The canonical dual trellis formation* $\mathscr{G}^{\perp}$ *on* $\mathcal{G}$ *is obtained from the collection of local behaviors* $\mathfrak{C}^{\perp}$.

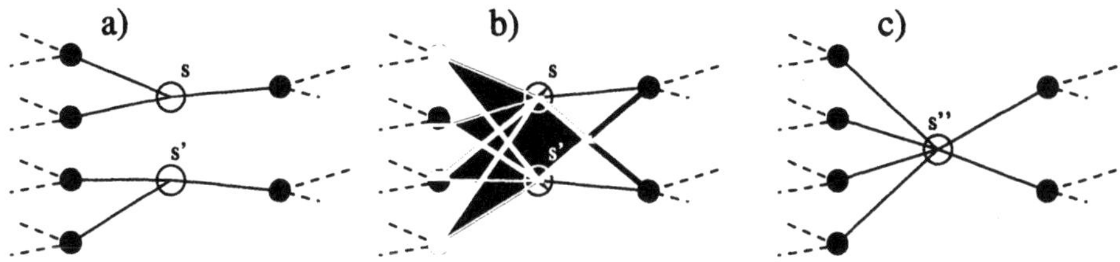

Figure 6 Merging two vertices s and s'. a) The local description of the trellis formation before merging; b) Newly introduced edges (bold) to give s and s' identical neighborhoods; c) The vertices s and s' in b) are indistinguishable and can be merged into a new vertex s''.

4. VERTEX MERGING IN TRELLIS FORMATIONS

Definition 4 gives a condition for the mergeablility of two vertices in a trellis formation. However it is not clear how this condition can be checked efficiently for any two given vertices. The brute force approach to the problem is to merge the vertices and to investigate the set of reductions and corresponding codewords that are described by the new trellis formation. However, this approach requires a complexity that in general grows exponentially with the codelength. This is clearly infeasible for larger trellis formations. In order to develop an efficient procedure to test for mergeability we need to follow a different approach. Also we focus only on the merging of state vertices. However, we note that the merging of vertices corresponding to local behaviors is an intriguing and important question.

Let s and s' be two vertices that are mergeable in a given vertex class $\mathscr{V}_u$, $u \in V_H$. Merging two vertices is equivalent to introducing edges so that the vertices s and s' have an identical set of neighbors. Figure 6 illustrates the operation.

Lemma 4 *Let $\mathscr{G}$ be a linear trellis formation with a vertex class $\mathscr{V}_u$ of size $2^{m(u)}$. Assume that $\mathscr{V}_u$ contains two mergeable vertices s and s' corresponding to elements $\underline{s}$ and $\underline{s}'$ in $\mathscr{A}_u = \mathbb{F}_2^{m(u)}$. We can merge any of the $2^{m(u)-1}$ pairs of vertices t and t' in $\mathscr{V}_u$ provided that $\underline{t} - \underline{t}' = \underline{s} - \underline{s}'$ holds. Merging all of these vertex pairs again yields a linear trellis formation.*

Proof. Consider a reduction $\mathcal{R}_s$ passing vertex s. Assume that s' is mergeable with s and, as suggested in Figure 6, introduce all edges necessary to give s and s' identical neighborhoods, yielding a new trellis formation $\mathscr{G}'$. This new trellis formation contains two reductions that *only differ in the vertices s and s'. $\mathcal{G}'$ represents a code $\mathcal{C}'$ whose restriction to the positions indicated by V_O yields a linear code, (namely C) by the definition of mergeability (while $\mathcal{C}'$ is typically not linear). Taking a projection is a linear map and it is easy to see that $C = (\mathcal{C}')_{|V_O}$ implies $C = span(C) = (span(\mathcal{C}'))_{|V_O}$. Let $\mathcal{C}''$ be the linear code obtained by taking the linear span of codewords in $\mathcal{C}'$. The restriction of $\mathcal{C}''$ to the positions indicated by V_O is still the same linear code, i.e., $\mathcal{C}_{|V_O} = \mathcal{C}'_{|V_O} = \mathcal{C}''_{|V_O}$. The linear code $\mathcal{C}''$ contains a word that is zero everywhere except in the position that corresponds to u, where it equals $\underline{s} - \underline{s}'$. Let $\mathcal{G}''$ be the trellis formation that represents the code $\mathcal{C}''$. Clearly, for $\mathcal{G}''$ all nodes in $\mathcal{V}_u$ that differ by the same difference $s - s'$ now have identical neighborhoods and can be merged, which proves the lemma.*

One of the consequences of Lemma 4 is that, provided a vertex class $\mathcal{V}_u$ contains a pair of mergeable vertices s and s', we can choose one of the vertices, (i.e., s'), as the vertex corresponding to $\mathbf{0} \in \mathbb{F}^{m(u)}$. We will assume this situation in the sequel and focus on the case of merging a state node s corresponding to $\underline{s} \in \mathbb{F}^{m(u)}$ to the state node s' corresponding to $\mathbf{0}$. Also, once we have merged the $2^{m(u)-1}$ pairs of vertices there is no need for labels of length $m(u)$ any more. We will always assume that a vertex class $\mathcal{V}_u$ with $2^{m'}$ vertices is associated with sequences from $\mathbb{F}_2^{m'}$. In particular if we say that we merge vertices in a vertex class we tacitly assume that we also perform the relabeling of vertices with sequences of smaller length.

In Figure 6 we sketched the graphical process involved in merging two vertices s and s' in $\mathcal{V}_u$, $u \in V_H$. We note that, strictly speaking, the situation depicted in Figure 6b does not correspond to a trellis formation on a Forney graph because Definition 3 requires exactly one vertex in $\mathcal{V}_f$ for each element of the local behavior $\mathcal{C}_v$. This condition is satisfied in Figures 6a and 6b. In order to get a succinct description of the process we map the graphical procedure to algebraic conditions. Let v and v' be the two vertices in V_f that are adjacent to u. Any edge in $\mathcal{V}_u \times \mathcal{V}_v$ corresponds to a local behavior $\underline{x} \in \mathcal{C}_v$. We have the following central lemma which is connected to the notion of properness of a conventional trellis [11].

Lemma 5 *Let $\mathcal{G} = (V, E)$ be a Forney graph with a collection of local behaviors $\mathfrak{C} = \{\mathcal{C}_v : v \in V_f\}$. Assume a vertex u is adjacent to both v and v' in $\mathcal{G}$ and assume $\mathcal{C}_v$ contains a codeword $\underline{x}$ such that $\underline{x}_{|\{\Gamma(v)\setminus u\}} = \mathbf{0}$ and $\underline{x}_{|u} \neq \mathbf{0}$. We can add a codeword $\underline{x}'$ to $\mathcal{C}_{v'}$ such that $\underline{x}'_{|\{\Gamma(v')\setminus u\}} = \mathbf{0}$ and $\underline{x}'_{|u} = \underline{x}_u$ without changing the code represented by $\mathcal{G}$.*

Proof. The most elegant proof of this statement uses Forney's duality theorem. Observe that the code $\bar{\mathscr{C}}$ remains unchanged by adding the codeword to $\mathscr{C}_{v'}$. Hence, the code $\bar{\mathscr{C}}^{\perp}$ remains unchanged and consequently the code $\bar{\mathscr{C}}^{\perp}{}_{|V_O}$ equals $C^{\perp}$. However, if the canonical dual Forney graph still represents the same code after adding the codeword to $\mathscr{C}_{v'}$ then, by Forney's duality result, the code represented in the original Forney graph also remains unchanged.

The following lemma gives the simplest condition for mergeability of two vertices:

Lemma 6 *Let $\mathscr{G} = (\mathscr{V}, \mathscr{E}, \mathcal{G})$ be a trellis formation on a Forney graph $\mathcal{G} = (V, E)$. The following two conditions are equivalent:*

A: *The code $\bar{\mathscr{C}}$ contains a word $\underline{x}$ such that $\underline{x}_{|u} \neq \mathbf{0}$ and $\underline{x}_{|\{V_S \setminus \{u\}\}} = \mathbf{0}$ holds for a vertex $u \in V_H$.*

B: *The canonical dual trellis $\mathscr{G}^{\perp}$ is not reduced and the vertex class $\mathscr{V}_u$ in $\mathscr{G}^{\perp}$ contains a vertex that is not contained in any reduction.*

Moreover, if any of the conditions A or B are satisfied, we can merge the vertices in $\mathscr{V}_u$ corresponding to the symbols $\mathbf{0} \in \mathbb{F}_2^{m(u)}$ and $\underline{x}_{|u} \in \mathbb{F}_2^{m(u)}$.

Proof.

$A \Rightarrow B$ Assume that $\bar{\mathscr{C}}$ contains a word $\underline{x}$ such that $\underline{x}_{|u} \neq \mathbf{0}$ and $\underline{x}_{|\{V_S \setminus \{u\}\}} = \mathbf{0}$ holds for a vertex $u \in V_S$. This implies that any word $\underline{y}$ in $\bar{\mathscr{C}}^{\perp}$ satisfies $\underline{x} \perp \underline{y}$. But the codewords in $\bar{\mathscr{C}}^{\perp}$ correspond exactly to the reductions in the canonical dual trellis. Hence, not all vertices in $\mathscr{G}^{\perp}$ are contained in a reduction and $\mathscr{G}^{\perp}$ is not reduced.

$B \Rightarrow A$ Assume $\mathscr{G}^{\perp}$ is not reduced and that $\mathscr{V}_u$ in $\mathscr{G}^{\perp}$ does contain a vertex that is not contained in any reduction. Due to linearity we know that the vertices in $\mathscr{V}_u$ that are part of a reduction form a subspace of $\mathbb{F}_2^{m(u)}$. Hence, there exists a vector $\underline{z} \in \mathbb{F}_2^{m(u)}$ that is orthogonal to the labels of all vertices in this subspace. Clearly, a vector $\underline{x} \in \mathbb{F}_2^{n}$ with $\underline{x}_{|u} \neq \mathbf{0}$ and $\underline{x}_{\{|V \setminus \{u\}\}} = \underline{z}$ is in the code $\bar{\mathscr{C}}$.

In order to see that the vertices o and z in $\mathscr{V}_u$ corresponding to the symbols $\mathbf{0} \in \mathbb{F}_2^{m(u)}$ and $\underline{x}_{|u} \in \mathbb{F}_2^{m(u)}$ are mergeable, we again employ Forney's duality theorem. Let v and v' be the two vertices in $\mathcal{G}$ that are adjacent to u. Assume that condition A of the Lemma is satisfied. Adding a local behavior $\underline{c}$ to $\mathscr{C}_v$ such that $\underline{c}_{|\{\Gamma(v) \setminus \{u\}\}} = \mathbf{0}$ and $\underline{c}_{|u} = \underline{x}_{|u}$

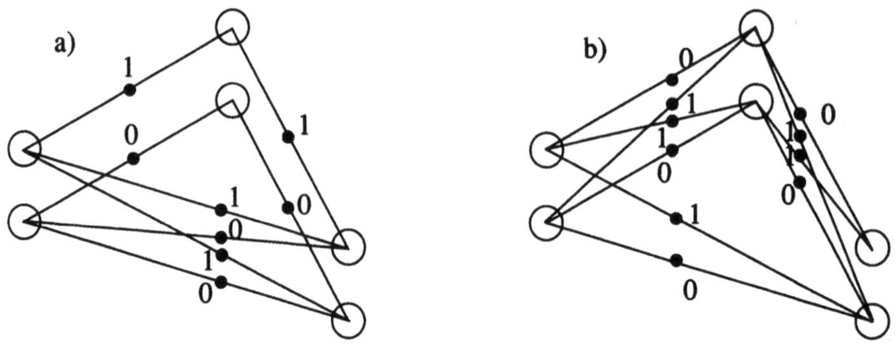

Figure 7 A simple trellis formation $\mathcal{G}$ for the repetition code of length 3 (figure (a)) and its canonical dual $\mathcal{G}^{\perp}$ — the parity-check code of length 3 (figure (b)). The trellis formation $\mathcal{G}$ satisfies the conditions of Lemma 6 and, hence, the canonical dual $\mathcal{G}^{\perp}$ is not reduced.

(similar for $\mathcal{C}_{v'}$) does not change the code $\bar{\mathcal{C}}$ (although, it renders a disconnected canonical dual trellis formation $\mathcal{G}^{\perp}$.) But now the vertices corresponding to the symbols $\mathbf{0} \in \mathbb{F}_2^{m(u)}$ and $\underline{x}_{|u} \in \mathbb{F}_2^{m(u)}$ have identical neighbors in the trellis formation and can, hence, be merged, which completes the proof of the lemma.

We illustrate a few of the key notions relating to Lemma 6 in Figure 7. It can be shown that in the case of linear codes and trellis formations on conventional trellisses the mergeability of vertices implies that the condition described in Lemma 6 is met for some vertex class $\mathcal{V}_u$. Indeed, the condition is closely related to the notion of *properness* [11] and we can successively identify and eliminate nonproper situations in the case of conventional trellises. It is a well known result that a proper conventional trellis is the unique minimal trellis for a linear system. However, this classical construction breaks down for two reasons in the context of general Forney graphs. First, condition A of Lemma 6 does not imply a similar consequence for any local behavior. A second complication is that Lemma 6 does not capture all scenarios in which a trellis formation contains mergeable vertices. We also note that for general topologies the notion of properness becomes ill-defined because the related concepts of the *future* and *past* of a code are not applicable.

Figure 7 gives an example of trellis formations such that the conditions of Lemma 6 apply. Figure 8 depicts a simple trellis formation that obviously contains mergeable vertices while the conditions of Lemma 6

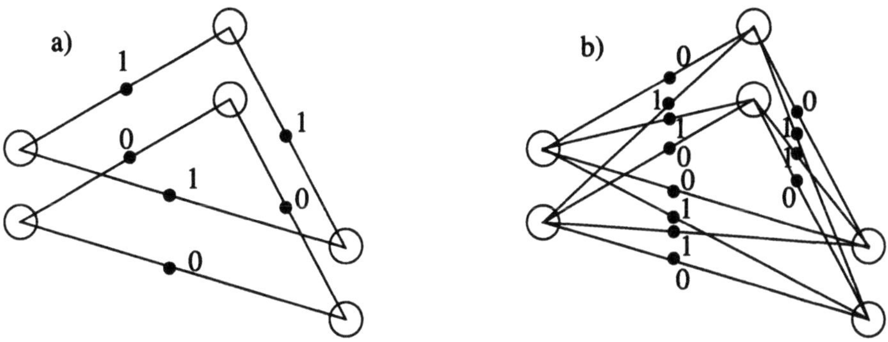

Figure 8 A simple trellis formation $\mathscr{G}$ for the repetition code of length 3 (figure (a)) and its canonical dual $\mathscr{G}^{\perp}$ for the parity-check code of length 3 (figure (b)).

are not met. A more substantial example of this phenomenon is the trellis formation of Figure 2 which will be treated in Section 6.

For the case of general topologies we have the following lemma which states a necessary and sufficient condition for mergebility of vertices:

Lemma 7 *Let $\mathscr{G}$ be a trellis formation on a Forney graph with local behaviors $\mathfrak{C}$ such that $\mathscr{G}$ represents a code C. Moreover, let $\underline{x} \in \mathbb{F}_2^n$ be a word in $\mathbb{F}_2^n$ such that $\underline{x}_{|u} \neq \mathbf{0}$ and $\underline{x}_{|\{V_S \setminus \{u\}\}} = \mathbf{0}$ holds. Let $\hat{\mathscr{C}}$ be the linear code obtained as the span of $\bar{\mathscr{C}}$ and $\underline{x}$. The vertices s and t, with $\underline{s} - \underline{t} = \underline{x}_{|u}$, are mergeable if and only if the condition $\hat{\mathscr{C}}_{V_O} = C$ holds.*

Proof. *The lemma is an algebraic reformulation of the merging procedure sketched in Figure 6. Merging s and t is equivalent to the vector $\underline{x}$ being added to $\bar{\mathscr{C}}$. By definition the vertices s and t are mergeable if and only if the represented code C remains unchanged. The represented code before the merging is $\bar{\mathscr{C}}_{V_O}$ and after the merging it is $\hat{\mathscr{C}}_{V_O}$. Hence, $\bar{\mathscr{C}}_{V_O} = \hat{\mathscr{C}}_{V_O}$ is a necessary and sufficient condition for mergeability.* ∎

Lemma 7 states a simple general condition for mergeability of vertices. There are two ways in which we can guarantee that the condition $\hat{\mathscr{C}}_{V_O} = C$ is met. The simplest criterion is that $\hat{\mathscr{C}}$ equals $\bar{\mathscr{C}}$ which is the condition treated in Lemma 6. In the following lemma we investigate the other possible case that $\hat{\mathscr{C}}$ is not equal to $\bar{\mathscr{C}}$ but we still have $\hat{\mathscr{C}}_{V_O} = C$.

Lemma 8 *Let $\mathscr{G} = (\mathscr{V}, \mathscr{E}, \mathcal{G})$ be a trellis formation on a Forney graph $\mathcal{G} = (V, E)$. Let s be a vertex in $\mathscr{V}_u$ labeled with $\underline{s} \neq \mathbf{0}$ and assume that $\bar{\mathscr{C}}$ does not contain a word $\underline{x}$ such that $\underline{x}_{|u} = \underline{s}$ and $\underline{x}_{|\{V_S \setminus \{u\}\}} = \mathbf{0}$. The following two conditions are equivalent:*

A: *We can merge vertices s and t in $\mathscr{V}_u$ where t corresponds to the zero element in $\mathbb{F}_2^{m(u)}$.*

B: *The canonical dual trellis $\mathscr{G}^\perp$ is not one-to-one and the code $\bar{\mathscr{C}}^\perp$ contains a word $\underline{y}$ such that $\underline{y}_{|u} \not\perp \underline{s}$ and $\underline{y}_{|V_O} = \mathbf{0}$*

Proof.

$A \Rightarrow B$ From Lemma 7 we know that merging two vertices is equivalent to introducing a word $\underline{x}$ to $\bar{\mathscr{C}}$ such that $\underline{x}_{|u} = \underline{s}$ and $\underline{x}_{|\{V\setminus\{u\}\}} = \mathbf{0}$ holds. By assumption this word was not already contained in $\bar{\mathscr{C}}$ and hence by considering the span $\underline{x}$ and $\bar{\mathscr{C}}$ the dimension of $\bar{\mathscr{C}}$ increases by one. Consequently, the dimension of $\bar{\mathscr{C}}^\perp$ is decreased by one. However, the words in $\bar{\mathscr{C}}^\perp$ are in one-to-one correspondence with the reductions of $\mathscr{G}^\perp$. By the mergeability of the vertices, it follows that $(\bar{\mathscr{C}}^\perp)_{|V_O}$ is unchanged and hence we must have decreased the dimension of the kernel of the map $(\cdot)_{|V_O}$. This implies that $\mathscr{G}^\perp$ must contain a non-trivial reduction which is not orthogonal to $\underline{x}$ and which maps to the zero word under $(\cdot)_{|V_O}$.

$B \Rightarrow A$ Assume a reduction corresponding to a word $\underline{y}$ as in the Lemma exists. $\underline{y}$ is in the kernel of the map $(\cdot)_{|V_O}$ and by introducing a word $\underline{x}$ to $\bar{\mathscr{C}}$ we do not change the code represented by the trellis formation $\mathscr{G}^\perp$. But introducing $\underline{x}$ to $\bar{\mathscr{C}}$ implies mergeability of the corresponding vertices by Lemma 7.

Before we use our findings in this section, in order to give an algorithm that efficiently identifies all pairs of mergeable vertices in a given vertex class $\mathscr{V}_u$, we summarize our result in the following theorem:

Theorem 9 *Let $\mathscr{G}$ be a trellis formation on a Forney graph with local behaviors $\mathfrak{C}$ such that $\mathscr{G}$ represents a code C. We can merge any two vertices s and t in $\mathscr{V}_u$ associated with $\underline{s}$ and $\underline{t}$ in $\mathbb{F}_2^{m(u)}$ if and only if either one of the following conditions hold:*

Condition A: $\underline{s} - \underline{t} \in \bar{\mathscr{C}}_u$

Condition B: $\underline{s} - \underline{t} \not\perp (\bar{\mathscr{C}}^\perp_{V_H})_{|u}$

Proof. The theorem is a concise formulation of the results of Lemma 6, 7, and 8. In particular Condition A corresponds to the case treated in Lemma 6 and Condition B is proved in Lemma 8. Also, it should be observed that the two conditions are mutually exclusive.

It is an interesting observation that the conditions of Theorem 9 impose conditions on the code $\bar{\mathscr{C}}$ and its dual. Indeed, similar conditions can be applied to the code $\mathscr{C}$ and its dual in order to identify opportunities to obtain smaller state spaces in a trellis formation. Also, applying the conditions of Theorem 9 to the canonical dual trellis $\mathscr{G}^{\perp}$ rather than to $\mathscr{G}$ may reveal mergable vertices in $\mathscr{G}^{\perp}$ while $\mathscr{G}$ may not contain mergable vertices. However, we note that applying the conditions of Theorem 9 to $\mathscr{C}$ and $\mathscr{C}^{\perp}$ does not yield any further decreased state space size as long as the original trellis formation is reduced.

5. CONSTRUCTION OF NONMERGEABLE TRELLIS FORMATIONS

Before we state an algorithm to construct nonmergeable trellis formations, we would like to emphasize that this algorithm in general does not achieve a trellis formation in which the state space sizes are minimal. Indeed, even the notion of minimality is difficult to apply (cf. [13, 14, 16]). Here, we present a greedy algorithm that successively merges randomly chosen mergeable vertices in randomly chosen vertex classes. We continue this greedy algorithm until no further vertices can be merged. Also, we only prove that the running time of the algorithm is bounded by a polynomial. Efficient implementations and speed-ups of the procedure will be discussed elsewhere.

Algorithm 1

Input: *A Forney graph $\mathcal{G} = (V, E)$; Generator matrices for a collection of local behaviors $\mathfrak{C}$; All vertices in V_H are unmarked.*

Iterations: *1. Choose an arbitrary vertex u in V_H that is not marked as nonmergeable. If no such vertex exists go to the Output step.*

2. Compute matrices H_A and G_B such that H_A is a parity-check matrix for $\bar{\mathscr{C}}_u$ and G_B generates $(\bar{\mathscr{C}}^{\perp}_{\bar{V}_H})_{|u}$.

3. Provided at least one of the codes $\bar{\mathscr{C}}_u$ and $(\bar{\mathscr{C}}^{\perp}_{\bar{V}_H})_{|u}$ has dimension greater than zero, choose an arbitrary element $\underline{t}$ in $\mathbb{F}_2^{m(u)}$ such that either $H_A\underline{t}^T = \mathbf{0}$ or $G_B\underline{t}^T \neq \mathbf{0}$, otherwise go to step (1) and mark vertex u as nonmergeable.

4. Merge all vertices in $\mathscr{V}_u$ such that the difference of their associated elements in $\mathbb{F}_2^{m(u)}$ equals $\underline{t}$. Go to step (1).

Output: *Output a collection of generator matrices for a new collection of local behaviors $\mathfrak{C}'$.*

Theorem 10 *Given a trellis formation $\mathscr{G}$ and associated code $\mathscr{C}$, Algorithm 1 yields a set of local behaviors describing a trellis formation $\mathscr{G}'$ representing $\mathscr{C}$ that does not contain mergeable vertices. The algorithm runs in time polynomial in the length n of the code $\mathscr{C}$.*

Proof. The main part of the algorithm is proved in Section 4. In particular if we cannot find a vertex u in V_H such that we can perform step (4) of Algorithm 1 then neither Condition A nor Condition B of Theorem 9 can be satisfied, and, hence, the trellis formation does not contain mergeable vertices. It should also be clear from Lemma 7 that once a vertex class is identified as not containing mergeable vertices this vertex class will not contain mergeable vertices at a later stage of the algorithm. We have to show that the algorithm runs in polynomial time. This follows from the fact that all operations are matrix operations on matrices of dimensions less than n. Moreover, every time we perform step (4) of Algorithm 1 we reduce the dimension of the alphabet $\mathbb{F}_2^{m(u)}$ or equivalently the length of the code $\mathscr{C}$ by one. It follows that we perform at most n iterations of Algorithm 1.

6. AN EXAMPLE

In this section we investigate the trellis formation of Examples 1 and 2. We follow Algorithm 1 in order to find mergeable vertex pairs in the trellis formation. The input required by Algorithm 1 is given in Example 1. We start with the first iteration step and choose the vertex u_5 in order to determine if the vertex class $\mathscr{V}_{u_5}$ in the trellis formation of Figure 2 contains mergeable vertices. A generator matrix for code $\bar{\mathscr{C}}$ is given as

$$G = \left(\begin{array}{c|cccccccccc}
 & u_1 & u_2 & u_3 & u_4 & u_5 & u_6 & w_1 & w_2 & w_3 & w_4 \\
\hline
v_1 & 1 & 0 & & & 00 & & 11 & & & \\
v_1 & 0 & 0 & & & 10 & & 11 & & & \\
v_1 & 0 & 1 & & & 01 & & 01 & & & \\
v_2 & & 0 & 0 & & & 1 & & 11 & & \\
v_2 & & 1 & 1 & & & 0 & & 01 & & \\
v_3 & 0 & & 1 & 1 & & & & & 01 & \\
v_3 & 1 & & 0 & 0 & & & & & 11 & \\
v_4 & & & & 1 & 01 & 0 & & & & 01 \\
v_4 & & & & 0 & 10 & 0 & & & & 11 \\
v_4 & & & & 0 & 00 & 1 & & & & 11
\end{array}\right)$$

We bring G into row reduced echelon form to obtain the generator matrix G'

$$G' = \left(\begin{array}{cccccccccc}
u_1 & u_2 & u_3 & u_4 & u_5 & u_6 & w_1 & w_2 & w_3 & w_4 \\
\hline
1 & 0 & 0 & 0 & 00 & 0 & 00 & 00 & 00 & 11 \\
0 & 1 & 0 & 0 & 01 & 0 & 00 & 01 & 01 & 01 \\
0 & 0 & 1 & 0 & 01 & 0 & 00 & 00 & 01 & 01 \\
0 & 0 & 0 & 1 & 01 & 0 & 00 & 00 & 00 & 01 \\
0 & 0 & 0 & 0 & 10 & 0 & 00 & 00 & 00 & 11 \\
0 & 0 & 0 & 0 & 00 & 1 & 00 & 00 & 00 & 11 \\
0 & 0 & 0 & 0 & 00 & 0 & 10 & 01 & 01 & 10 \\
0 & 0 & 0 & 0 & 00 & 0 & 01 & 01 & 01 & 01 \\
0 & 0 & 0 & 0 & 00 & 0 & 00 & 11 & 00 & 11 \\
0 & 0 & 0 & 0 & 00 & 0 & 00 & 00 & 11 & 11
\end{array}\right)$$

From G' it is easy to see that the space $\bar{\mathscr{C}}_{u_5}$ has dimension zero and hence Condition A of Theorem 9 is not satisfied.

From G' we can find a parity-check matrix for $\bar{\mathscr{C}}$ using standard methods of linear algebra:

$$H = \left(\begin{array}{cccccccccc}
u_1 & u_2 & u_3 & u_4 & u_5 & u_6 & w_1 & w_2 & w_3 & w_4 \\
\hline
1 & 0 & 0 & 0 & 10 & 1 & 10 & 10 & 10 & 10 \\
0 & 1 & 0 & 0 & 00 & 0 & 11 & 11 & 00 & 00 \\
0 & 0 & 1 & 0 & 00 & 0 & 00 & 11 & 11 & 00 \\
0 & 0 & 0 & 1 & 00 & 0 & 00 & 00 & 11 & 11 \\
0 & 0 & 0 & 0 & 01 & 0 & 11 & 00 & 00 & 11
\end{array}\right)$$

It is easy to verify that the word $(0,1,1,1,01,0,00,00,00,00)$ is in $\bar{\mathscr{C}}^{\perp}$ and hence the canonical dual trellis formation $\mathscr{G}^{\perp}$ is not one-to-one. From this condition we can immediately deduce that $\mathscr{G}$ contains mergeable vertices. In order to check Condition B of Theorem 9 we find $\bar{\mathscr{C}}^{\perp}_{V_H}$ which equals the one-dimensional space spanned by $(0,1,1,1,01,0)$. Hence, a generator matrix for $(\bar{\mathscr{C}}^{\perp}_{V_H})_{|u_5}$ equals $G_B = (01)$. It follows from Condition B of Theorem 9 that we can merge all vertices in $\mathscr{V}_{u_5}$ such that their difference is *not* orthogonal to (01). We choose as this difference 01 (equally well we could have taken (11)). After merging these vertices, we obtain a modified generator matrix G'' for the modified code $\bar{\mathscr{C}}''$ as

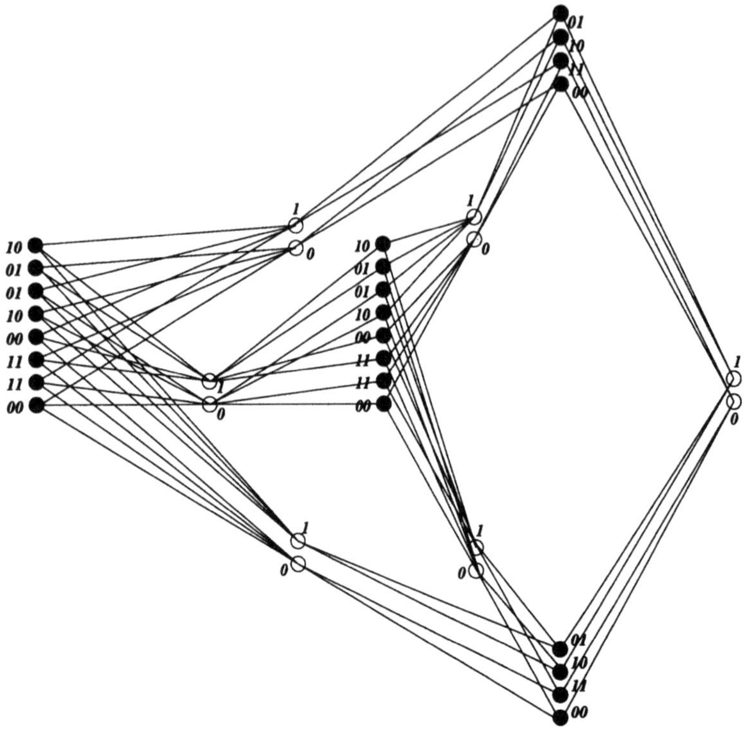

Figure 9 A trellis formation for the binary code of type [8,4,4]. The trellis formation does not contain mergeable vertices.

$$G'' = \left(\begin{array}{c|cccccccccc} & u_1 & u_2 & u_3 & u_4 & u_5 & u_6 & w_1 & w_2 & w_3 & w_4 \\ \hline v_1 & 1 & 0 & & & 0 & & 11 & & & \\ v_1 & 0 & 0 & & & 1 & & 11 & & & \\ v_1 & 0 & 1 & & & 0 & & 01 & & & \\ v_2 & & 0 & 0 & & & 1 & & 11 & & \\ v_2 & & 1 & 1 & & & 0 & & 01 & & \\ v_3 & 0 & & 1 & 1 & & & & & 01 & \\ v_3 & 1 & & 0 & 0 & & & & & 11 & \\ v_4 & & & & 1 & 0 & 0 & & & & 01 \\ v_4 & & & & 0 & 1 & 0 & & & & 11 \\ v_4 & & & & 0 & 0 & 1 & & & & 11 \end{array}\right)$$

The trellis formation $\mathcal{G}''$ corresponding to G'' is depicted in Figure 9. $\mathcal{G}''$ does not contain further mergeable pairs of vertices.

References

[1] R. G. Gallager, *Low-Density Parity-Check Codes.* Cambridge, MA: MIT Press, 1963.

[2] R. M. Tanner, "A Recursive Approach to Low Complexity Codes," *IEEE Transactions on Information Theory*, vol. IT-27, pp. 533–547, Sept. 1981.

[3] N. Wiberg, H.-A. Loeliger, and R. Koetter, "Codes and Iterative Decoding on General Graphs," *European Transactions on Telecommunications*, vol. 6, pp. 513–525, Sept./Oct. 1995.

[4] N. Wiberg, *Codes and Decoding on General Graphs*, Ph.D. Thesis, Linköping University, Sweden, 1996.

[5] S. M. Aji and R. J. McEliece, "The Generalized Distributive Law," *IEEE Transactions on Information Theory*, vol. 46, pp. 325-343, March 2000.

[6] B. J. Frey, R. Koetter and A. Vardy, "Skewness and Pseudocodewords in Iterative Decoding," *Proceedings of the 1998 IEEE International Symposium on Information Theory*, Cambridge, MA, p. 148, Aug. 1998.

[7] F. R. Kschischang and B. J. Frey, "Iterative Decoding of Compound Codes by Probability Propagation in Graphical Models," *IEEE Journal on Selected Areas in Communications*, vol. 16, pp. 219–230, Feb. 1998.

[8] F. R. Kschischang, B. J. Frey, and H.-A. Loeliger, "Factor Graphs and the Sum-Product Algorithm," *IEEE Transactions on Information Theory*, vol. IT-47, pp. 498–519, Feb. 2001.

[9] R. J. McEliece, D. J. C. MacKay, and J.-F. Cheng, "Turbo Decoding as an Instance of Pearl's 'Belief Propagation' Algorithm," *IEEE Journal on Selected Areas in Communications*, vol. 16, pp. 140–152, Feb. 1998.

[10] G. D. Forney, "Final Report on a Coding System Design for Advanced Solar Missions," Contract NAS2-3637, NASA Ames Research Center, CA, Dec. 1967.

[11] A. Vardy, "Trellis Structure of Codes", chapter in "Handbook of Coding Theory", edited by V. S. Pless, W. C. Huffman, and R. A. Brualdi, Elsevier Science Publications, Amsterdam.

[12] G. D. Forney and M. D. Trott, "The Dynamics of Group Codes: State Spaces, Trellises and Canonical Encoders", *IEEE Transactions on Information Theory*, vol. 39, pp. 1491–1513, 1993.

[13] R. Koetter and A. Vardy, "Factor Graphs: Construction, Classification, and Bounds," in *Proceedings of the 1998 IEEE International Symposium on Information Theory*, Cambridge, MA, 1998.

[14] R. Koetter and A. Vardy, "Minimality of Factor Graphs," in *Proceedings of the International Symposium on Mathematical Theory of Networks and Systems 1998*, Padua, Italy, 1998.

[15] F. R. Kschischang and V. Sorokine, "On the Trellis Structure of Block Codes," *IEEE Transactions on Information Theory*, vol. 41, pp. 1924–1937, 1995.

[16] R. Koetter and A. Vardy, "Construction of Minimal Tail-Biting Trellises," in *Proceedings of the 1998 Information Theory Workshop*, (Killarney), Ireland, pp. 72–74, June 1998.

[17] G. D. Forney, "Codes on Graphs: Normal Realization," *IEEE Transactions on Information Theory*, vol. IT-47, pp.520–548, Feb. 2001.

Index

MIX
Papier aus verantwortungsvollen Quellen
Paper from responsible sources
FSC® C105338
www.fsc.org

If you have any concerns about our products,
you can contact us on
ProductSafety@springernature.com

In case Publisher is established outside the EU,
the EU authorized representative is:
Springer Nature Customer Service Center GmbH
Europaplatz 3, 69115 Heidelberg, Germany

Printed by Libri Plureos GmbH
in Hamburg, Germany